Measurement Uncertainty in Chemical Analysis

Springer
Berlin
Heidelberg
New York
Hong Kong
London
Milan
Paris
Tokyo

Paul De Bièvre · Helmut Günzler (Eds.)

Measurement Uncertainty in Chemical Analysis

Springer

Prof. Dr. Paul De Bièvre
Duineneind 9
2460 Kasterlee
Belgium

Prof. Dr. Helmut Günzler
Bismarckstr. 4
69469 Weinheim

ISBN 978-3-642-07884-2

Cataloging-in-Publication Data applied for
A catalog record for this book is available from the Library of Congress.
Bibliographic information published by Die Deutsche Bibliothek
Die Deutsche Bibliothek lists this publication in the Deutsche Nationalbibliografie;
detailed bibliographic data is available in the Internet at http://dnb.ddb.de

Springer-Verlag is a part of Springer Science+Business Media
Springer-Verlag Berlin Heidelberg 2010
Printed in Germany
springeronline.com

Coverdesign: Design & Production, Heidelberg

Preface

For over six years, the journal "Accreditation and Quality Assurance" (ACQUAL) has been publishing contributions from the chemical measurement community on various aspects of *reliability in chemical measurement*, the key mission of ACQUAL.

One of these aspects is *uncertainty*. Although even its very concept is still controversial, ACQUAL authors are quite proficient in writing about it. Their papers show that uncertainty is interpreted – and used – in many divergent ways.

It seemed a good idea to the publisher, Springer-Verlag, to present some of the most prominent contributions on the topic that have appeared in ACQUAL in the course of the years. The result lies before you.

It should be clear to the reader that this is a collection of papers and not an "integrated" book. We are still far from a homogeneous, internationally accepted common perception of what uncertainty means in chemical measurement. The result is that we use it in different ways. But dramatic *changes* in the perception and interpretation of "uncertainty" amongst chemists are becoming visible. They are already reflected in the papers selected. Clearly more time is necessary for the implementation of end-of-20th century uncertainty concepts, and be accepted by beginning-of-21-st century minds.

The spectrum of what we read and hear in matters of uncertainty in chemical measurement is very broad: it goes from interpreting uncertainty as a mere repeatability of measurement results obtained from replicate measurements of the *quantity subject to measurement**, all the way to the full uncertainty of the result of a measurement procedure applied to a *quantity intended for measurement*.

* quantity (German: "Grösse", French: "grandeur", Dutch: "grootheid") is not used here in the meaning 'amount', but as the generic term for the quantities we measure: concentration, volume, mass, temperature, time, etc.

The latter interpretation has the consequence that the uncertainty of any quantity influencing the result, including the chemical sample preparation prior to the measurement, must be included in the final uncertainty evaluation, thus yielding a *combined uncertainty*. That, however, entails almost invariably an increase in the size of the uncertainty bar of the measurement result, previously called the "error bar". Most of the chemists do not yet agree on this. Thus, uncertainty is increased as the result of more work because the *whole* measurement process must be evaluated for possible uncertainty contributions. All of this makes uncertainty evaluation more elaborate but more realistic, and therefore more responsible. This marks a truly dramatic change.

It would be very helpful if the ongoing revision of the "International Vocabulary of Basic and General Terms in Metrology" (VIM), would define "measurement result" unequivocally in order to promote one meaning of the term in our work as well as in international discussions. A common language is absolutely essential in this matter.

It gives us great pleasure to deliver what we consider a useful compendium from ACQUAL authors and editors to ACQUAL readers and to other colleagues in the art and science of chemical measurement. For those among the readers who consider themselves as newcomer in the field, I propose to read the articles by Dube and / or Kadis as an introduction to the current state and to the problems involved.

Prof. Dr. P. De Bièvre
Editor-in-Chief
Accreditation and Quality Assurance
Kasterlee
2002-09-20

Contents

Contributors

Anton Alink
Nederlands Meetinstituut,
P.O. Box 654, 2600 AR Delft, The Netherlands

Hans Andersson
SP Swedish National Testing and Research Institute,
P.O. Box 857, 501 15 Borås, Sweden

Thomas Anglov
Department of Metrology, Novo Nordisk A/S,
Krogshøjvej 51, 2880 Bagsværd, Denmark

Sabrina Barbizzi
Agenzia Nazionale per la Protezione dell'Ambiente – Unità Interdipartimentale di Metrologia Ambientale,
Via Vitaliano Brancati 48, 00144 Rome, Italy

Vicky J. Barwick
Laboratory of the Government Chemist,
Queens Road, Teddington, Middlesex, TW11 0LY, UK

Maria Belli
Agenzia Nazionale per la Protezione dell'Ambiente – Unità Interdipartimentale di Metrologia Ambientale,
Via Vitaliano Brancati 48, 00144 Rome, Italy

Ricard Boqué
Department of Analytical and Organic Chemistry,
Institute of Advanced Studies, Rovira i Virgili
University of Tarragona, Pl. Imperial Tàrraco, 1,
43005 Tarragona, Catalonia, Spain

David Bradley
ASTM Headquarters, West Conshohocken, Pa., USA

A.J.M. Broos
Nederlands Meetinstituut, P.O. Box 654, 2600 AR Delft,
The Netherlands

Lutz Brüggemann
UFZ Centre for Environmental Research Leipzig-Halle,
Department of Analytical Chemistry, Permoserstrasse
15, 04318 Leipzig, Germany

Mirella Buzoianu
National Institute of Metrology, Reference Materials
Group, Sos. Vitan-Bârzesti No.11, 75669 Bucharest,
Romania

M. Filomena G. F. C. Camões
CECUL, Faculdade de Ciências da Universidade de
Lisboa, P-1700 Lisbon, Portugal

Ivo Cancheri
Agrarian Institute, I-38010 San Michele all'Adige,
Trento, Italy

John R. Cowles
LGC (Teddington) Ltd., Teddington, England TW11
0LY, UK

Larry F. Crawford
Bethlehem Steel Corporation, Bethlehem, Pa., USA

Ricardo J. N. Bettencourt da Silva
CECUL, Faculdade de Ciências da Universidade de
Lisboa, P-1700 Lisbon, Portugal

Simon Daily
LGC (Teddington) Ltd., Teddington, England TW11
0LY, UK

Paolo de Zorzi
Agenzia Nazionale per la Protezione dell'Ambiente
(ANPA) – Unità Interdipartimentale di Metrologia
Ambientale, Via Vitaliano Brancati 48,
00144 Rome, Italy

Andrea Delusa
Ente Regionale per lo Sviluppo Agricolo del Friuli
Venezia-Giulia (ERSA), Via Sabbatini 5,
33050 Pozzuolo del Friuli (UD), Italy

Elias Diaz
Beca de Ampliación de Estudios del F.I.S., Instituto de
Salud Carlos III, Madrid, Spain

Gunther Dube
Physikalisch-Technische Bundesanstalt, Bundesallee 100, 38116 Braunschweig, Germany

René Dybkaer
Copenhagen Hospital Corporation, Department of Standardization in Laboratory Medicine, H:S Kommunehospitalet, Øster Farimagsgade 5, DK-1399 Copenhagen K, Denmark

João Seabra e Barros
Instituto Nacional de Engenharia e Tecnologia Industrial, Estrada do Paço do Lumiar, P-1699 Lisbon Codex, Portugal

Stephen L.R. Ellison
Laboratory of the Government Chemist, Queens Road, Teddington, Middlesex, TW11 0LY, UK

Osvaldo Failla
Faculty of Agriculture, University of Milan, Milan, Italy

John Fleming
LGC, Queens Road, Teddington, Middlesex TW 11 0LY, UK

Dean A. Flinchbaugh
Flinchbaugh Consulting, Bethlehem, Pa., USA

Blandine Fourest
Institut de Physique Nucléaire, 91406 Orsay Cedex, France

Michel Gerboles
European Reference Laboratory of Air Pollution (ERLAP), Commission of the European Communities, Joint Research Centre, I-21020 Ispra, Italy

Rattanjit S. Gill
Laboratory of the Government Chemist, Queens Road, Teddington, Middlesex, TW11 0LY, UK

Manfred Golze
Federal Institute for Materials Research and Testing (BAM), Unter den Eichen 87, 12205 Berlin, Germany

William A. Hardcastle
Laboratory of the Government Chemist, Queens Road, Teddington, Middlesex, TW11 0LY, UK

Werner Hässelbarth
Federal Institute for Materials Research and Testing (BAM), 12200 Berlin, Germany

André Henrion
Physikalisch-Technische Bundesanstalt, Bundesallee 100, D-38116 Braunschweig, Germany

Kaj Heydorn
Department of Chemistry, Technical University of Denmark, 2800 Lyngby, Denmark

Zhengzhi Hu
Chinese National Center for Food Quality Supervision & Testing, 32 Xiaoyun Road, Chaoyang District, Beijing 100027, China

Rouvim Kadis
D. I. Mendeleyev Institute for Metrology, 19 Moskovsky pr., 198005 St. Petersburg, Russia

Elena Kardash-Strochkova
The National Physical Laboratory of Israel, Danciger A Building, Givat Ram, Jerusalem 91904, Israel

Jesper Kristiansen
The National Institute of Occupational, Health, Lersø Parkallé 105, DK-2100 Copenhagen, Denmark

Daniela Kruh
Rafael Calibration Laboratories, P.O. Box 2250, 31021 Haifa, Israel

Stephan Küppers
Schering AG, In-Process-Control, Müllerstrasse 170–178, 13342 Berlin, Germany

Ilya Kuselman
The National Physical Laboratory of Israel, Danciger A Building, Givat Ram, Jerusalem 91904, Israel

Andrée Lamberty
European Commission, Joint Research Centre, Institute for Reference Materials and Measurements (IRMM), B-2440 Geel, Belgium

Ján Lauda
Department of Analytical Chemistry, CHTF STU, Radlinského 9, 812 37 Bratislava, Slovak Republic

Yunqiao Li
National Research Center for Certified Reference Material, No.18 Bei San Huan Dong Lu, Chaoyang Qu, Beijing, 100013, P. R. China

Thomas Linsinger
European Commission, Joint Research Centre, Institute for Reference Materials and Measurements (IRMM), B-2440 Geel, Belgium

Li Liu
Chinese National Center for Food Quality Supervision & Testing, 32 Xiaoyun Road, Chaoyang District, Beijing 100027, China

John L. Love
Institute of Environmental Sciences and Research, P.O. Box 29 181, Christchurch, New Zealand

Xiaohua Lu
National Research Center for Certified Reference Material, No.18 Bei San Huan Dong Lu, Chaoyang Qu, Beijing, 100013, P. R. China

Hans Malissa
University of Salzburg, Institute of Chemistry and Biochemistry, Hellbrunnerstrasse 34, 5020 Salzburg, Austria

Alicia Maroto
Department of Analytical and Organic Chemistry, Institute of Advanced Studies, Rovira i Virgili University of Tarragona, Pl. Imperial Tàrraco, 1, 43005 Tarragona, Catalonia, Spain

Sandro Menegon
Ente Regionale per lo Sviluppo Agricolo del Friuli Venezia-Giulia (ERSA), Via Sabbatini 5, 33050 Pozzuolo del Friuli (UD), Italy

Frank Möller
Faculty of Agriculture, University of Milan, Milan, Italy

Bernd Neidhart
Philipps-Universität Marburg, Hans-Meerwein-Strasse, 35032 Marburg, Germany

Seppo I. Niemelä
Finnish Environment Institute, P. O. Box 140, 00251 Helsinki, Finland

Riitta Maarit Niemi
Finnish Environment Institute, P. O. Box 140, 00251 Helsinki, Finland

Alberto Noriega-Guerra
European Reference Laboratory of Air Pollution (ERLAP), Commission of the European Communities, Joint Research Centre, I-21020 Ispra, Italy

Viliam Pätoprstý
Slovak Institute of Metrology, Karloveská 63, 842 55 Bratislava, Slovak Republic

Jean Pauwels
European Commission, Joint Research Centre Institute for Reference Materials and Measurements (IRMM) B-2440 Geel, Belgium

Inge M. Petersen
Department of Metrology, Novo Nordisk, A/S, Krogshøjvej 51, DK-2880 Bagsværd, Denmark

Mark J.Q. Rafferty
Laboratory of the Government Chemist, Queens Road, Teddington, Middlesex, TW11 0LY, UK

Michael H. Ramsey
Centre for Environmental Research, School of Chemistry, Physics and Environmental Science, University of Sussex, Falmer, Brighton BN1 9QJ, UK

Wolfgang Riepe
University of Salzburg, Institute of Chemistry and Biochemistry, Hellbrunnerstrasse 34, 5020 Salzburg, Austria

Angel Ríos
Department of Analytical Chemistry, Faculty of Sciences, University of Córdoba, E-14004 Córdoba, Spain

Jordi Riu
Department of Analytical and Organic Chemistry, Institute of Advanced Studies, Rovira i Virgili University of Tarragona, Pl. Imperial Tàrraco, 1, 43005 Tarragona, Catalonia, Spain

F. Xavier Rius
Department of Analytical and Organic Chemistry, Institute of Advanced Studies, Rovira i Virgili University of Tarragona, Pl. Imperial Tàrraco, 1, 43005 Tarragona, Catalonia, Spain

Matthias Rösslein
EMPA St. Gallen, Department of Chemistry/Metrology, Lerchenfeldstrasse, 5, 9014 St. Gallen, Switzerland

Heinz Schimmel
European Commission, Joint Research Centre, Institute for Reference Materials and Measurements (IRMM), B-2440 Geel, Belgium

Petras Serapinas
Plasma Spectroscopy Laboratory, Institute of Theoretical Physics and Astronomy, A. Goštauto 12, 2600 Vilnius, Lithuania

Avinoam Shenhar
The National Physical Laboratory of Israel, Danciger A Building, Givat Ram, Jerusalem 91904, Israel

Felix Sherman
The National Physical Laboratory of Israel, Givat Ram, Jerusalem 91904, Israel

Naijie Shi
National Research Center for Certified Reference Material, No.18 Bei San Huan Dong Lu, Chaoyang Qu, Beijing, 100013, P. R. China

Gino Stringari
Agrarian Institute, I-38010 San Michele all'Adige, Trento, Italy

Pavol Tarapcik
Department of Analytical Chemistry, CHTF STU, Radlinského 9, 812 37 Bratislava, Slovak Republic

Christoph Tausch
Philipps-Universität Marburg, Hans-Meerwein-Strasse, 35032 Marburg, Germany

Michael Thompson
Department of Chemistry, Birkbeck College, Gordon House, 29 Gordon Square, London WC1H 0PP, UK

Guanghui Tian
National Research Center for Certified Reference Material, No.18 Bei San Huan Dong Lu, Chaoyang Qu, Beijing, 100013, P. R. China

Yakov I. Tur'yan
The National Physical Laboratory of Israel, Danciger A Building, Givat Ram, Jerusalem 91904, Israel

Miguel Valcárcel
Department of Analytical Chemistry, Faculty of Sciences, University of Córdoba, E-14004 Córdoba, Spain

Adriaan van der Veen
Nederlands Meetinstituut, P.O. Box 654, 2600 AR Delft, The Netherlands

Wolfhard Wegscheider
Montanuniversität Leoben, Franz-Josef-Strasse 18, A-8700 Leoben, Austria

Rainer Wennrich
UFZ Centre for Environmental Research Leipzig-Halle, Department of Analytical Chemistry, Permoserstrasse 15, 04318 Leipzig, Germany

Paul Willetts
Food Labelling and Standards Division, Ministry of Agriculture, Fisheries and Food, CSL Food Science Laboratory, Norwich Research Park, Colney, Norwich NR4 7UQ, UK

Alex Williams
19 Hamesmoor Way,Mytchett, Camberley, Surrey, GU16 6JG, UK

Carole Williams
LGC (Teddington) Ltd., Teddington, England TW11 0LY, UK

Roger Wood
Food Labelling and Standards Division, Ministry of Agriculture, Fisheries and Food, CSL Food Science Laboratory, Norwich Research Park, Colney, Norwich NR4 7UQ, UK

Accred Qual Assur (2002) 7:294–298
DOI 10.1007/s00769-002-0484-9

Rouvim Kadis

Analytical procedure in terms of measurement (quality) assurance

Abstract In the ordinary sense the term "analytical procedure" means a description of what has to be done while performing an analysis without reference to quality of the measurement. A more sound definition of "analytical procedure" can be given in terms of measurement (quality) assurance, in which a specified procedure to be followed is explicitly associated with an established accuracy of the results produced. The logic and consequences of such an approach are discussed, with background definitions and terminology as a starting point. Close attention is paid to the concept of measurement uncertainty as providing a single-number index of accuracy inherent in the procedure. The appropriateness of the uncertainty-based approach to analytical measurement is stressed in view of specific inaccuracy sources such as sampling and matrix effects. And methods for their evaluation are outlined. The question of a clear criterion for analytical procedure validation is also addressed from the standpoint of the quality requirement which measurement results need to meet as an end-product.

Keywords Accuracy · Analytical procedure · Measurement assurance · Measurement uncertainty · Quality assurance · Validation

R. Kadis (*)
D. I. Mendeleyev Institute for Metrology, 19 Moskovsky pr., 198005 St. Petersburg, Russia
e-mail: rkadis@mail.rcom.ru
Tel.: +7-812-3239644
Fax: +7-812-3279776

Introduction

There are different ways in which quality assurance concepts play a role in analytical chemistry. Most of them such as stipulating requirements for the competence and acceptance of analytical laboratories, writing quality manuals, performing systems audits, etc. can be viewed as something foreign to common analytical thinking forced upon analysts by external authorities. Perhaps another possible way is to try to integrate quality assurance aspects into common analytical concepts and (re)define them in such a way as to explicitly include the quality matters required. This may facilitate an effective quality assurance strategy in analytical chemistry. In ordinary usage the term "analytical procedure" hardly needs a referential definition and may be for this reason there are few official definitions of the term. The only definition quoted in the references [1] is rather diffuse:

"The analytical procedure refers to the way of performing the analysis. It should describe in detail the steps necessary to perform each analytical test. This may include but is not limited to: the sample, the reference standard and the reagents preparations, use of the apparatus, generation of the calibration curve, use of the formulae for the calculation, etc." [2].

In brief, this simply means a description of all that should be done in order to perform the analysis.

Leaving aside some prolixity in the definition above, the main thing that is lacking is the goal requirement needed in considering quality matters. As it is shown in this paper, a sound definition of an analytical procedure can be given in terms of measurement (quality) assurance. The case in point is not simply "the way of performing the analysis" but that which ensures obtaining the results of a specified quality. What this eventually means is a prescribed procedure to follow in producing results with a known uncertainty.

If we have indeed recognized chemical analysis to be measurement, though possessing its own peculiarities, we can apply the principles and techniques of quality assurance developed in measurement to analytical work. These principles and techniques constitute the field of measurement assurance [3], a system affording a confidence that all the measurements produced in a measurement process maintained in statistical control are good enough for their intended use. "Good enough" implies here nothing more than having an allowable uncertainty. Although measurement assurance was originally developed for instrument calibration, i.e. with emphasis on measurement traceability, it is reasonable to treat it more generally. One can say that a fixed measurement procedure is a means of assigning an uncertainty to a single measurement, and this is the essence of measurement assurance. This also reveals the role a prescribed (analytical) procedure plays in routine analytical measurement. We will focus on different aspects involved in the concept of an analytical procedure defined in terms of measurement assurance such as terminology, content, evaluation, and validation.

Starting point

Chemical analysis generally consists of several operational stages beginning with taking a sample representative of the whole mass of the material to be analysed and ending with calculation and reporting of results. In this sequence the measurement proper usually makes a relatively small contribution to the overall variability involved in the entire chemical measurement process (CMP) [4], the largest portion of which being concerned with "non-measurement" operations such as isolation, separation, and so on. Because of this, everything in the chain that affects the chemical measurement result must be predetermined as far as practically possible: the experimental operations, the apparatus and equipment, the materials and reagents, the calibration and data handling. Thus, a "*complete analytical procedure*, which is specified in every detail, by fixed working directions (order of analysis) and which is used for a particular analytical task" [5] – a concept presented by Kaiser and Specker as far back as in the 1950s [6] – becomes a point of critical importance in obtaining meaningful and reproducible results. We use the term "analytical procedure" or merely "procedure" for short, in the sense outlined above.

"Method", "procedure", or "protocol"

The importance of the correct usage of relevant terms, in particular, the term "procedure" rather than "method" is noteworthy. The terms actually correspond to different levels in the hierarchy of analytical methodology [7] expressed as a sequence from the general to the specific:

technique ⇨ method ⇨ procedure ⇨ protocol

Indeed, the procedure level provides the specific directions necessary to utilize a method, which is in line with the definition of *measurement procedure* given in the International Vocabulary of Basic and General Terms in Metrology (VIM): "set of operations, described specifically, used in the performance of particular measurements according to a given method" [8].

This nomenclature is however not always adhered to. In many cases, i.e. scientific publications, codes of practice, or official directives, an analytical procedure is virtually implied when an analytical method is spoken about. Commonly used expressions such as "validation of analytical methods" or "performance characteristics of analytical methods" are typical examples of incorrect usage. Such confusion appears even in the definition suggested by Wilson in 1970 for the term "analytical method" [9]. As he then put it, "an analytical method is to be regarded as the set of written instructions completely defining the procedure to be adopted by the analyst in order to obtain the required analytical result". It is actually difficult to make a distinction between the two notions when one of them is defined in terms of the other. On the other hand, there is normally no reason to differentiate the two most specific levels in the hierarchy above, carrying the term "procedure" over to the designated "protocol". The latter was defined [7] as "a set of definitive directions that must be followed, without exceptions, if the analytical results are to be accepted for a given purpose". So, written directions have to be faithfully followed in both cases. In many instances the term "procedure" actually signifies a document in which the procedure is recorded – this is specifically noted in VIM in respect to "measurement procedure". Besides, the term "standard operating procedure" (SOP), especially applied to a procedure intended for repetitive use, is popular in quality assurance terminology.

A clear distinction needs to be drawn between analytical procedure as a generalized concept and its particular realization, i.e. an individual version of the procedure arising in specific circumstances. In practice, an analytical procedure exists as a variety of realizations, differing in terms of specimens, equipment, reagents, environmental conditions, and even the analyst's own routine. Not distinguishing between these concepts can lead to a misinterpretation embodied, for instance, in the viewpoint that with a detailed specification the procedure will change "each time the analyst, the laboratory, the reagents or the apparatus changed" [9]. What will actually change is realizations of the procedure, only provided that all specified variables remain within the specification. Also one cannot but note that the hierarchy of methodology above concerns, in fact, a level of specificity rather than the extent to which the entire CMP may be covered. Although sampling is the first (and even the most critical) step of the process, it is often treated as a separate issue when addressing analytical methodology. A "complete analytical procedure" may or may not include sampling, depending on the particular analytical problem to be solved and the scope of the procedure.

An analytical procedure yields the results of established accuracy

In line with Doerffel's statement which refers to analytical science as "a discipline between chemistry and metrology" [10], one may define analytical service – as a sort of analytical industry, that is practical activities directed to meeting customer needs – as based upon concepts of chemistry, metrology, and industrial quality control. The intention of any analytical methodology in service is to produce data of appropriate quality, i.e. those that are fit for their intended purpose. The question to answer is what kind of criteria should be addressed in characterizing fitness-for-purpose.

From the viewpoint of objective of measurement, which is to estimate the true value of the quantity measured, and its applicability for decision-making, closeness of the result to the true value, no matter how it is expressed, should be such a criterion. If a measurement serves any practical need, it is to meet an adequate level of accuracy. It is compliance with an accuracy requirement that fundamentally defines the suitability of measurement results for a specific use, and hence corresponding demands are to be made on a measurement process that produces the results. Next it is assumed that the process is generated by the application of a measurement procedure and thus, the accuracy requirements should be finally referred to in the procedure itself. (The "requirements sequence" first implies substantiation of the demands on accuracy in a particular application area, the problem that needs special consideration in chemical analysis [11].)

Following this basic pattern, it is reasonable to re-define Kaiser's "complete analytical procedure", so that the fitnessfor- purpose criterion is explicitly allowed for. There must be an accuracy constraint built in the definition so as to give a determining aspect of the notion. It is probably unknown to most analytical chemists worldwide that such a definition has long since been adopted in analytical terminology in Russia. This was formulated in 1975 by the Scientific Council on Analytical Chemistry of the Russian Academy of Sciences. As defined by the latter an *analytical procedure* is the: " a detailed description of all the conditions and operations that ensure established characteristics of trueness and precision" [12]. This wording which goes beyond the scope of analytical chemistry specifically differs from the VIM definition of measurement procedure quoted above by including an accuracy requirement as a goal function. It clearly points out that adhering to such a fixed procedure ensures (at least conceptually) that the results obtained are of a guaranteed degree of accuracy.

Two basic statements underlie the definition above. First, an analytical procedure when performed as prescribed, with the chemical measurement process operating in a state of control, has an inherent accuracy to be evaluated. Second, a measure of the accuracy can be transferred to the results produced, providing a degree of their confidence. In essence, the measure of accuracy typical of a given procedure is being assigned to future results generated by the application of the procedure under specified conditions. The justification for both the propositions was given by Youden in his work on analytical performance [13, 14] where methods for determining accuracy in laboratories were discussed in detail.

As a prerequisite for practical implementation of the analytical procedure concept it is assumed that the chemical measurement process remains in a state of statistical control, being operated within the specifications. To furnish evidence for this and to avoid the reporting of invalid data the analytical system needs to be tested for continuing performance. A number of control tests may be used with this aim, for instance, testing the difference between parallel determinations when they are prescribed to be carried out, duplicating complete analysis of a current test material, and analysis of a reference material. An important point is that the control tests are to be performed in such a manner and in such a proportion as a given measurement requires, and are an integral part of the whole analytical procedure. The control tests may be more specific in this case and relate to critical points of the measurement process. As examples, calibration stability control with even one reference sample or interference control by spiking provide useful means of expeditious control in an analytical procedure.

A principal point in this scheme is that accuracy characteristics should be estimated before an analytical procedure is regularly used and should be the characteristics of any future result obtained by application of the procedure under specified conditions. Measurements of this type are most commonly performed (by technicians, not measurement scientists) in control engineering and are sometimes called "technical measurements". It is such measurements that are usually referred to as routine in chemical analysis. In fact, the problems of evaluation of routine analyses faced by chemists are treated more generally in the "technical measurements" theory [15].

Uncertainty as an index of accuracy of an analytical procedure

It is generally accepted that accuracy as a qualitative descriptor can be quantified only if described in terms of precision and trueness corresponding to random and systematic errors, respectively. Accordingly, the two measures of accuracy, the estimated standard deviation and the (bounds for) bias, taken separately, have to be generally evaluated and reported [16]. As the traditional theory of measurement errors holds, the two figures cannot be rigorously combined in any way to give an overall index of (in)accuracy. Notice that accuracy, as such, ("closeness of the agreement between the result of a measurement and a true value of the measurand" [8]) by no means involves any measurement error categorization.

On the other hand, it has long been recognized that the confidence to be placed in a measurement result is conveniently expressed by its uncertainty that was thought, from the outset, to mean an estimate of the likely limits to the error of measurement. So, uncertainty has traditionally been treated as "the range of values about the final result within which the true value of the measured quantity is believed to lie" [17]. However, there was no agreement on the best method for assessing uncertainty. Consistent with the traditional subdivision, the "random uncertainty" and the "systematic uncertainty" each arising from corresponding sources should be kept separate in the evaluation of a measurement, and the question of how to combine them was an issue of debate for decades.

Now a unified and widely applicable approach to the uncertainty statement set out in ISO Guide (GUM) [18] is being accepted in many fields of measurement, particularly in analytical measurements due to the helpful adaptation in the EURACHEM Guide [19]. Some peculiarities of the new approach can be intimated, specifically, the abandonment of the previous distinction between random and systematic uncertainties, treating all of them as standard-deviation- like quantities (after the corrections for known systematic effects have been made), and their possible estimation by other than statistical means. Fundamental, however, is that any numerical measurement is not thought of in isolation, but in relation to the process which generates the measurements. All the factors operative in the process being defined, they virtually determine the relevant uncertainty sources, so making practicable their quantification to finally derive the value of total uncertainty. One can say that the measurement uncertainty methodology fits neatly the starting idea of a procedure specified in every detail, since the procedure itself defines the context which the uncertainty statement refers to.

This is true of the component-by-component ("bottom-up") method for evaluating uncertainty that is directly in line with GUM. Also this is true for the "top-down" approach [20] that provides a valuable alternative when poorly understood steps are involved in the CMP and a full mathematical model is lacking. An important point is that the top-down methodology implies a reconciliation of information available with the required one that is based on a detailed analysis of the factors which affect the result. For both approaches to work advantageously a clear specification of the analytical procedure is evidently a necessary condition.

The break with the traditional subdivision of measurement errors has a crucial impact on the way accuracy may be quantified and expressed. In 1961, Youden wrote [14]: "There is no solution to the problem of devising a single number to represent the accuracy of a procedure". He was indeed right in the sense that a strict probability statement cannot be made about a combination of random and systematic errors. Today, thanks to the present uncertainty concept, we maintain the other opinion that such a solution does exist. It is measurement uncertainty that can be regarded as a single-number index of accuracy inherent in the procedure. In doing so we must not be confused by the fact that the operational definition of measurement uncertainty that GUM presents does not use the unknown "true value" of the measured quantity following pragmatic philosophy. The old definitions and, in particular, that cited above are equally valid and are now considered ideal.

Consequently, we can define an analytical procedure as leading to results with a known uncertainty, as in Fig. 1 in which typical "constituents" to be specified in an analytical procedure are shown.

Specific inaccuracy sources in an analytical procedure

What has been said in the previous section generally refers to specified measurement procedures used in many fields of measurement. There are, however, some special reasons, specific to chemical analysis, that make the uncertainty methodology particularly appealing in analytical measurements. This is because of specific inaccuracy sources in an analytical procedure which are difficult to be allowed for otherwise. Two such sources, sampling and matrix effects, will be mentioned here, with an outline of the methods for their evaluation.

Sampling

Where sampling forms part of the analytical procedure, all operations in producing the laboratory sample such as sampling proper, sample pre-treatment, carriage, and sub-sampling require examination in order to be taken into account as possible sources contributing to the total uncertainty.

It is generally accepted that a reliable estimate of this uncertainty can be obtained empirically rather than theoretically. Accordingly, an appropriate methodology has being developed [e.g. 21, 22] aimed at separating the sampling contribution from the total variability of the measurement results in a specially designed experiment. This is not, however, the only way of quantifying uncertainty in sampling. Explicit use of scientific judgement is now equally approved when experimental data are unavailable. An illustrative example from the EURACHEM Guide (Ref. 19, Example A4) clearly demonstrates the potential of mathematical modelling inhomogeneity as an alternative to the sampling assessment experiment.

It is significant that with the uncertainty methodology both the major analytical properties, "accuracy" and "representativeness" [23], which quality of analytical data relies on, can be quantified and properly taken into account to give a single index of accuracy. This index expresses consistency between the measurement results and the true value that refers to a bulk sample of the material rather than the test portion analysed.

Matrix effects

The problem of matrix mismatch is always attendant when one analyses an unknown sample "with the same matrix" using a fixed, previously determined, calibration function. Not uncommonly, an analytical procedure is developed to cover a range of sample matrices in such a way that an "overall" calibration function can be used. An error due to matrix mismatch is therefore inevitable if not necessary significant. Commonly regarded as systematic for a sample with a particular matrix, the error becomes random when a population of samples to which the procedure applies is considered; this in fact constitutes an inherent part of the total variability associated with the analytical procedure. Meanwhile, these effects are in no way included in the usual measures of accuracy as they result from a "method-performance study" in accordance with the accepted protocols [24, 25]. The *accuracy experiment* defined by ISO 5725 (Ref. 24, Part 1, Section 4) does not presuppose any variable matrix-dependent contribution, being confined to *identical test items*. The underlying statistical model assumes that solely laboratory components of bias and their distribution must be considered.

It is notable that such kinds of error sources are fairly treated using the concept of measurement uncertainty which makes no difference between "random" and "systematic". When simulated samples with known analyte content can be prepared, the effect of the matrix is a matter of direct investigation in respect of its chemical composition as well as physical properties that influence the result and may be at different levels for analytical samples and a calibration standard. It has long since been suggested in examination of matrix effects [26, 27] that the influence of matrix factors be varied (at least) at two levels corresponding to their upper and lower limits in accordance with an appropriate experimental design. The results from such an experiment enable the main effects of the factors and also interaction effects to be estimated as coefficients in a polynomial regression model, with the variance of matrix-induced error found by statistical analysis. This variance is simply the (squared) standard uncertainty we seek for the matrix effects.

In many ways, this approach is similar to ruggedness testing aimed at the identification of operational (not matrix-related) conditions that are critical to the analytical performance.

"Method validation" in terms of measurement assurance

The presented concept of analytical procedure offers a clear perspective on the problem of "method validation" which is an issue of great concern in quality matters. Validation is generally taken to mean a process of demonstration that a methodology is suitable for its intended application. The question is how should suitability be assessed, based on customer needs? It is commonly recommended [e.g. 2, 28–30] that a number of characteristics such as selectivity/specificity, limits of detection and quantitation, precision and bias, linearity and working ranges be considered as criteria for analytical performance and evaluated in the course of an validation study. In principle, they need to be compared to some standard; based on this, judgement is made as to whether the *procedure* under issue is capable of meeting the specified analytical requirements, that is to say, whether a "method is fit-for-purpose" [28]. However, from the perspective of endusers of analytical results, it is important that the *data* be only of the required quality and thus appropriate for their intended purpose. In other words, the matter of primary concern is quality of analytical results as an end-product. In this respect, a procedure will be deemed suitable when the data produced are fit-for-purpose.

It follows that the common criteria of validation should be made more specific in terms of measurement assurance. It is (the index of) accuracy that requires overriding consideration among the characteristics of analytical performance if quality of the results is primarily kept in mind. Other performance characteristics are desirable to ensure that a methodology is well-established and fully understood, but *validation* of an analytical procedure on those criteria seems impractical also in view of the lack of corresponding requirements as is commonly the case.

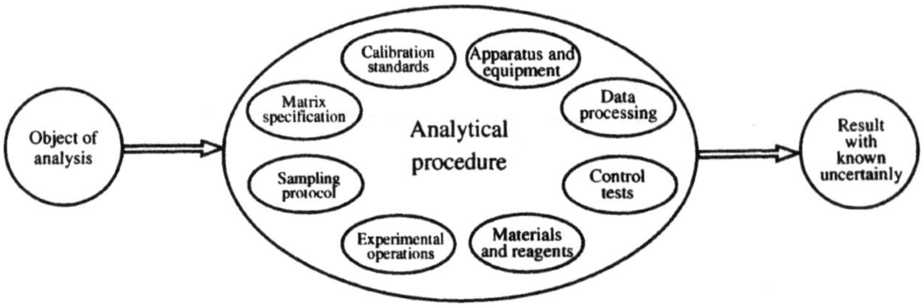

Fig. 1 Typical "constituents" to be specified within analytical procedure, which ensures obtaining the results with a known uncertainty

(Strictly speaking, there is no validation unless a particular requirement has been set.)

We have every reason to consider the estimation of measurement uncertainty in an analytical procedure followed by the judgement of compliance with a target uncertainty value as a kind of validation. This is in full agreement with ISO 17025 that points to several ways of validation, among them "systematic assessment of the factors influencing the result" and "assessment of the uncertainty of the results..." [31]. In line with this is also a statistical modelling approach to the validation process that has recently been developed and exemplified as applied to in-house [32] and interlaboratory [33] validation studies.

A concrete example of such validation is worthy of notice. *Certification* (*attestation*) of analytical procedures used in regulated fields such as environmental control and safety is operative in the Russian state measurement assurance system as a process of establishing metrological properties and confirming their compliance with relevant requirements. (By metrological properties we mean herein the assigned measurement error characteristics, i.e. measurement uncertainty.) This is introduced by the Russian Federation state standard GOST R 8.563 [34] which also covers procedures for quantitative chemical analysis. This certification is, in fact, a legal metrology measure similar, to some extent, to pattern evaluation and approval of measuring instruments. Some scepticism concerning the efficiency of legal metrology practice in ensuring the quality of analytical measurements may be in order. Nevertheless, the conceptual (measurement assurance) basis of this approach to validation deserves attention beyond doubt.

Conclusions

This debate allows the following propositions to be made:

1. The term "analytical procedure" commonly used without reference to the quality of data is best defined in terms of measurement (quality) assurance to explicitly include quality matters. This means a specified procedure which ensures results with an established accuracy.
2. The measurement uncertainty methodology neatly fits the idea of a specified measurement procedure and furthermore provides a tool for covering specific inaccuracy sources peculiar to analytical measurement. Uncertainty can be regarded as a single-number index of accuracy of an analytical procedure.
3. When an analytical procedure is so defined, uncertainty becomes the performance parameter that needs overriding consideration over and above all the others assessed during validation studies. This kind of validation gives a direct answer to the question whether the data produced are of required quality and thus appropriate for their intended use.

References

1. Holcombe D (1999) Accred Qual Assur 4: 525–530
2. International Conference on Harmonization of Technical Requirements for Registration of Pharmaceuticals for Human Use (1994) Text on validation of analytical procedures. ICH Quality topic Q2A: Definitions and terminology (http://www.ifpma.org/ich5q.html)
3. Cameron JM (1976) J Qual Technol 8: 53–55
4. Currie LA (1978) Sources of error and the approach to accuracy in analytical chemistry. In: Kolthoff IM, Elving PE (eds) Treatise on analytical chemistry. Part I. Theory and practice, vol.1, 2nd edn. Wiley, New York, pp 95–242
5. Kaiser H (1978) Spectrochim Acta 33B: 551–576
6. Kaiser H, Specker H (1956) Fresenius Z Anal Chem 149: 46–66
7. Taylor JK (1983) Anal Chem 55: 600A-604A, 608A
8. BIPM, IEC, IFCC, ISO, IUPAC, IUPAP, OIML (1993) International vocabulary of basic and general terms in metrology, 2nd edn. International Organization for Standardization (ISO), Geneva
9. Wilson AL (1970) Talanta 17: 21–29

10. Doerffel K (1998) Fresenius J Anal Chem 361: 393–394
11. Shaevich AB (1989) Fresenius Z Anal Chem 335: 9–14
12. Terms, definitions, and symbols for metrological characteristics in analysis of substance (1975) Zh Anal Chim 30: 2058–2063 (in Russian)
13. Youden WJ (1960) Anal Chem 32 (13): 23A-37A
14. Youden WJ (1961) Mat Res Stand 1: 268–271
15. Zemelman MA (1991) Metrological foundations of technical measurements. Izdatelstvo standartov, Moscow (in Russian)
16. Eisenhart C (1963) J Res Nat Bur Stand 67C: 161–187
17. Campion PJ, Burns JE, Williams A (1973) A code of practice for the detailed statement of accuracy. National Physical Laboratory, Her Majesty's Stationery Office, London
18. BIPM, IEC, IFCC, ISO, IUPAC, IUPAP, OIML (1993) Guide to the expression of uncertainty in measurement. ISO, Geneva
19. EURACHEM/CITAC Guide (2000) Quantifying uncertainty in analytical measurement, 2nd edn (http://www.eurachem.bam.de/guides/quam2.pdf)
20. Ellison SLR, Barwick VJ (1998) Analyst 123: 1387–1392
21. Ramsey MH (1998) J Anal Atom Spectr 13: 97–104
22. van der Veen AMH, Alink A (1998) Accred Qual Assur 3: 20–26
23. Valcárcel M, Ríos A (1993) Anal Chem 65: 781A-787A
24. ISO 5725 (1994) Accuracy (trueness and precision) of measurement methods and results. Parts 1–6. International Organization for Standardization, Geneva
25. IUPAC (1995) Protocol for the design, conduct and interpretation of methodperformance studies. Pure Appl Chem 67: 331–343
26. Makulov NA (1976) Zavod Lab 42: 1457–1464 (in Russian)
27. Parczewski A, Rokosz A (1978) Chem Analityczna 23: 225–230
28. EURACHEM (1998) The fitness for purpose of analytical methods. A laboratory guide to method validation and related topics. LGC, Teddington
29. Wegsheider W (1996) Validation of analytical methods. In: Günzler H (ed.) Accreditation and quality assurance in analytical chemistry. Springer, Berlin, etc., pp 135–158
30. Bruce P, Minkkinen P, Riekkola M-L (1998) Mikrochim Acta 128: 93–106
31. ISO/IEC 17025 (1999) General requirements for the competence of testing and calibration laboratories. International Organization for Standardization, Geneva
32. Jülicher B, Gowik P, Uhlig S (1999) Analyst 124: 537–545
33. van der Voet H, van Rhijn JA, van de Wiel HJ (1999) Anal Chim Acta 391: 159–171
34. GOST R 8.563–96 State system for ensuring the uniformity of measurements. Procedures of measurements. Gosstandart of Russia, Moscow (in Russian)

Accred Qual Assur (2001) 6:3–7

Gunther Dube

Metrology in chemistry – a public task

Presented at Analytica Conference 2000, 11–14 April 2000, Munich, Germany

G. Dube
Physikalisch-Technische Bundesanstalt, Bundesallee 100, 38116 Braunschweig, Germany
e-mail: gunther.dube@ptb.de
Tel.: +49-531-592 3210
Fax: +49-531-592 3015

Abstract The importance of analytical chemistry is increasing in many public fields, and the demand for reliable measurement results is growing accordingly. A measurement result will be reliable only if its uncertainty has been quantified. This can be achieved only by tracing the result back to a standard realizing the unit in which the measurement result is expressed. The National Metrology Institutes (NMIs) can contribute to the reliability of the measurement results by developing measuring methods, and by providing reference materials and standard measuring devices. In fields in which the comparability of measurement results is of particular importance, they establish traceability structures. Responding to the globalization of trade and industry the International Committee for Weights and Measures (CIPM) agreed on an arrangement on the mutual recognition of calibration certificates (CIPM MRA) issued by the NMIs.

Keywords Reliability · Uncertainty · Traceability · National Metrology Institutes · CIPM MRA

Introduction

In many fields in which quantitative analyses are called for, analytical chemistry is confronted with new challenges. Particularly in such spheres, on which countries spend a considerable part of their revenue like health care, environmental protection and nutrition, reliable measurement results are of great importance. "To judge analytical methods and results critically, this belongs at all times to the analyst's tasks" [1]. Therefore, analytical chemists also use metrological ways of thinking and terms like traceability and uncertainty of measurement results. On the other hand, metrologists understand their responsibility in determining amount-of-substance measurements, and try to reach uniformity of measurement to achieve comparability of the results. In doing so, they make use of the worldwide metrological infrastructure. The demand of the public for reliability of the measurement results in analytical chemistry can be best achieved by cooperation between the two parties.

The need for reliability of measurement results

The need for reliability of measurement results is demonstrated by the following examples. In Germany the expenditure for health care in 1994 amounted to DM 344.6 billion [2]. This was 10% of the gross national product. One-third of this went into medical services: 10% of these was spent on laboratory services, i.e. for the most part on measurements, and it is well known that 30% went into repeating measurements [3]. Repeat measurements are carried out only if the results do not seem reliable. In this particular case, the financial loss due to repetitions amounted to DM 3 billion.

The second example comes from the area of natural gas. In Germany the import of natural gas has increased over the last 30 years and in 1998 amounted to

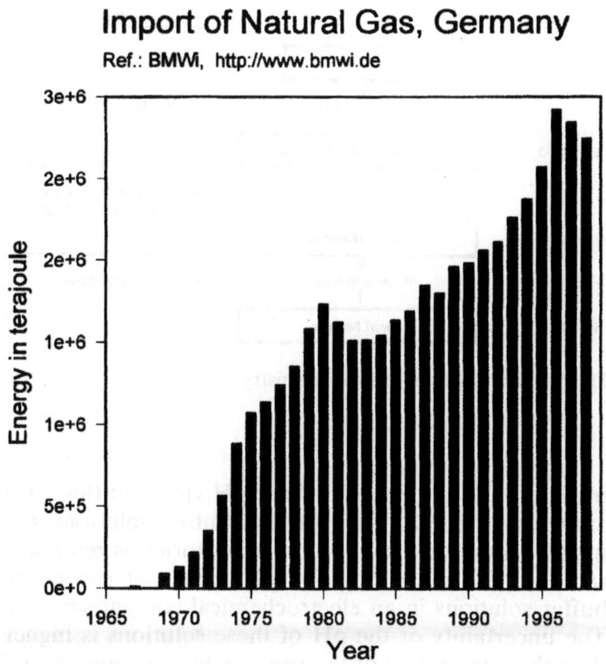

Fig. 1 Import of natural gas, Germany (Ref.: Bundesministerium für Wirtschaft und Technologie, BMWI)

nearly 3 million terajoule (TJ) [4]. The data shown in Fig. 1 are given in TJ, and that is why the calculation of the natural gas price is based on the energy consumed. This is the product of calorific value and volume. For the determination of the energy, the calorific value $\tilde{H}$ of the natural gas must be known. The well-known method for the calorimetric determination of calorific values is increasingly replaced by a new method. The main feature of this method is the determination of the mole fractions of the gas components using gas chromatography. The mole fractions x_j are multiplied by the molar calorific values $\bar{H}^\circ(t_1)$ of the gas components. These products are summarized and multiplied by p_2/RT_2 according to Eq. (1).

$$\tilde{H}^\circ[t_1 V(t_2 p_2)] = \sum_{j=1}^{N} x_j \times \bar{H}^\circ(t_1) \frac{p_2}{RT_2} \qquad (1)$$

where t_1 is temperature of combustion, V is volume, p_2, t_2 is the temperature and pressure at measurement, T is the absolute temperature, $\bar{H}^\circ$ is the molar calorific value, x is the mole fraction, R is the gas constant.

The effects that a wrong gas chromatographic result has on the natural gas price should be considered. If it is assumed that one component accounting for 10% of the natural gas was determined incorrectly by 1%. The calculated energy is affected by an error of 0.1%. If the price of the natural gas amounts to DM 20 billion per year, this error of 0.1% leads to a price difference of DM 20 million.

Uncertainty and traceability of measurement results

It follows from these examples that the reliability of the measurement results is of great public interest. Measurement results are reliable only if their uncertainty is known and quantified. Uncertainty is a metrological term which is defined as follows: Uncertainty: parameter, associated with the result of a measurement, that characterizes the dispersion of the values that could reasonably be attributed to the measurand [5].

The uncertainty can be stated only if the traceability of the measurement result to a system of units is guaranteed. Traceability is defined as follows [5]: Traceability: property of a result of a measurement or the value of a standard whereby it can be related to stated references, usually national or international standards, through an unbroken chain of comparisons, all having stated uncertainties.

Such a traceability system is demonstrated in Fig. 2. The International System of Units (SI) is at the top of the system. Its units are realized by standards. A measurement is a process, in the course of which the measurand is compared to a standard. For practical measurements, usually a working standard not a primary standard is used. To state the uncertainty of the measurement result, the uncertainty of the value assigned to the working standard must be known. It results from the uncertainty of the comparison measurement of the working standard with the reference standard. The uncertainty of the value assigned to the reference standard results from the uncertainty of the comparison measurement of the reference standard with the primary standard. This chain of comparison measurements is exactly what the definition of the term "traceability" means. If the traceability of a measurement result is

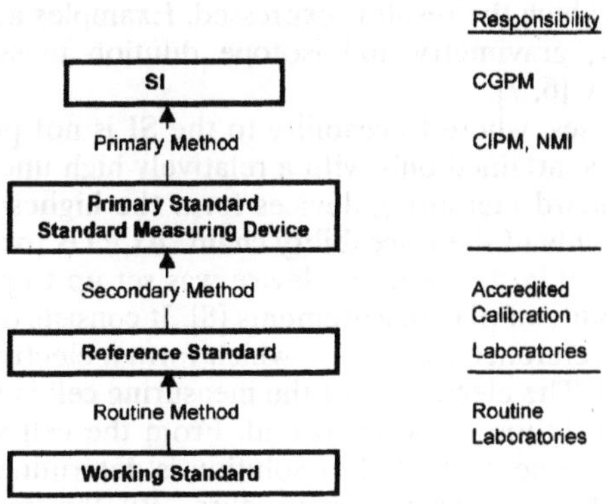

Fig. 2 Traceability scheme

guaranteed, its uncertainty can be stated. From this considerations it follows that metrology can provide the tools, necessary to get reliable measurement results.

In analytical chemistry, traceability of measurement results to SI units is not always possible and the traceability hierarchy ends below the level of the SI units. For example, in the case of standard measuring devices, or reference materials, the values are fixed by mutual agreement or in the case of methods are generally agreed upon. In these cases, the comparability of measurement results is limited.

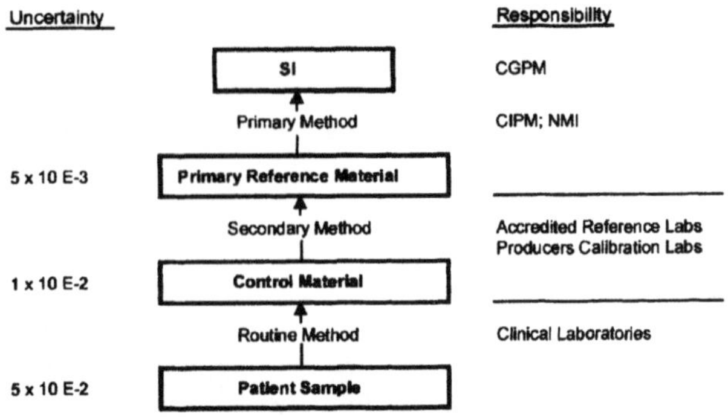

Fig. 3 Traceability in clinical chemistry

Tasks of the National Metrology Institutes (NMIs)

The tasks of the NMIs are :

- Realisation, maintenance and dissemination of the units
- Development and application of primary measurement methods
- Establishment of traceability structures
- Guaranteeing the equivalence of measurement standards.

In Germany, the Physikalisch-Technische Bundesanstalt (PTB) is responsible for the national standards. In chemistry, reference materials and standard measuring devices are the national standards. To cover the huge demand for reference materials, the PTB cooperates with other competent national institutions on the basis of agreements, first of all, with the Federal Institute of Material Research and Testing (BAM) and also with companies which produce and distribute reference materials. PTB also cooperates with other NMIs within the scope of joint projects aiming at the development of new reference materials not yet available on the market.

To trace back the values assigned to reference materials to the SI, the NMIs develop and apply "primary methods". It is the main feature of these methods which are sometimes called "absolute methods" that they do not make reference to standards of the same unit in which the result is expressed. Examples are coulometry, gravimetry and isotope dilution mass spectrometry [6, 7].

In cases, where traceability to the SI is not possible or can be attained only with a relatively high uncertainty, standard measuring devices form the highest reference points of the traceability chain. At PTB for example, a standard measuring device was set up to provide traceability of pH measurements [8]. It consists of a system of reference electrodes, which form a electrochemical cell. The electrolyte of the measuring cell is the pH buffer solution to be measured. From the cell voltage the pH value of the buffer solution is determined by a well-defined measuring procedure. By the measurement this solution gains the rank of a primary reference material. The uncertainty of the pH value on this level amounts to 0.002. The primary buffer solutions are used by accredited calibration laboratories as reference solutions for measuring the pH values of secondary buffer solutions in an electrochemical comparison cell. The uncertainty of the pH of these solutions is higher than the uncertainty of the primary buffer solution and amounts to 0.003. The secondary buffer solutions are used in the routine laboratories for calibrating commercial pH meters. The pH values measured by commercial pH meters show an uncertainty of 0.01. This traceability chain guarantees that the uncertainty of the pH values measured in the routine laboratories are correctly stated. They thus are reliable. Other NMIs also keep such pH measuring devices, and it was possible to demonstrate by comparison measurements that the standard measuring devices of different countries can provide measurement results which agree very well [9].

The establishment and the support of traceability structures is one of the most essential tasks of the NMIs. Traceability is of particular importance also in clinical chemistry. It must be ensured that the results obtained in the measurement of patient samples, are reliable. Therefore, in 1988, the Federal Medical Association (Bundesärztekammer) in Germany issued "*Guidelines for Quality Assurance in Medical Laboratories*" [10]. Although at present under revision, they prescribe the use of quality control samples to check the measurement results obtained in medical laboratories. For control materials used in internal quality control the producer and for control materials used in external quality control a reference laboratory must make sure that an uncertainty is stated for the assigned value. This is possible only if the control materials are linked to primary reference materials and to SI units. The link to the SI can be provided by reference laboratories as well as by the laboratories of control material producers, provided they have been accredited as calibration laboratories. In Germany this accreditation can be provided by PTB and is carried out by the German Calibration Service (DKD).

The International Committee for Weights and Measures Mutual Recognition Arrangement (CIPM MRA)

The NMIs are obliged by national law to realize, maintain and disseminate the national standards. However, they also take care of the uniformity of measurement worldwide. The first activity for this was the signing of the Meter Convention in 1875. On the basis of this treaty, the General Conference of Weights and Measures (CGPM) and CIPM work today. However, the NMIs are confronted today with new challenges [11]. The calibration certificates issued by them are generally valid only in the country of issue and are not accepted worldwide. This turned out to be a barrier to the international trade. So, different activities have been launched by various institutions to overcome these obstacles. The contribution of the NMIs to these efforts is the "Mutual Recognition Arrangement" (CIPM MRA), which was signed by the presidents of 38 NMIs in October 1999 during the twenty-first session of the CGPM [12]. Its objectives are:

- To establish the degree of equivalence of national measurement standards maintained by NMIs
- To provide for the mutual recognition of calibration and measurement certificates issued by the NMIs
- Thereby to provide governments and other parties with a safe technical foundation for wider agreements related to international trade, commerce and regulatory affairs.

The technical basis of the CIPM MRA is a system of key intercomparisons. Furthermore, the NMIs have to prove that they work in accordance with a quality system. The key comparisons are international comparison measurements. The Consultative Committee for Amount of Substance (CCQM) of CIPM is responsible for the comparisons in the field of chemistry. It selects the substance systems, organizes the realization of the measurements and the evaluation of the measurement results. The substance systems are chosen from areas of public interest in which traceability is necessary. Priority areas are:

Table 1 CCQM international comparisons, clinical diagnostic markers. NIST: National Institute of Standards and Technology, USA; LGC: Laboratory of the Government Chemist, UK; IRMM: Institute for Reference Materials and Measurements, Belgium; SP: Sveriges Provnings- och Forskningsinstitut, Sweden

	Reference No.	Pilot lab	Date
Cholesterol in serum	CCQM-P6	NIST	1998
	CCQM-K6	NIST	1999
Glucose in serum	CCQM-P8	NIST	1999
Creatinine in serum	CCQM-P9	NIST	1999
Creatinine in serum	CCQM-K12	NIST	2000
Ca in serum	CCQM-P14	IRMM/SP	2001
Anabolic steroids in urine	–	In preparation	–
Hormones in serum	–	In preparation	–

- Health
- Food
- Environment
- Advanced materials
- Commodities
- Forensic matters
- Pharmaceuticals
- Biotechnology

In the field of amount-of-substance measurements, 70 comparisons have been planned. Some of them have already been started. Up to now, the measurement programme for clinical chemistry has carried out the comparisons given in (Table 1) [13].

The results of the key comparisons – including the uncertainty statement – will be stored in an Internet-accessible database. This will enable companies, accrediting bodies, and institutions to evaluate the equivalence of the measurement results performed by the NMIs. The database will make it easier for businesses and organizations relying on these services to prove compliance with the measurement-related requirements of regulations and standards. The database will be an integral part of the infrastructure necessary to expand free trade and to eliminate technical barriers to export.

Acknowledgements Stimulating discussions with Mrs. P. Spitzer and Dr. P. Ulbig are thankfully acknowledged.

References

1. Doerffel K (1987) Preface to: Statistik in der analytischen Chemie, 4th edn. VCH, Weinheim, Germany
2. Bundesministerium für Bildung und Forschung, Bundesministerium für Gesundheit, Statistisches Bundesamt (2000) Die Gesundheitsberichterstattung des Bundes; http://www.gbe-bund.de
3. Semerjian HG (1998) Metrology: Impact on national economy and international trade. In: Seiler E (ed) The role of metrology in economic and social development. PTB-Texte, Band 9, Braunschweig, pp 99–133

4. Bundesministerium für Wirtschaft und Technologie (1999) Entwicklung der Einfuhr Naturgas in die Bundesrepublik; HYPERLINK http://www.bmwi.de
5. Deutsches Institut für Normung (1994) Internationales Wörterbuch der Metrologie, 2nd edn. Beuth, Berlin Wien Zürich
6. Quinn T (1997) Metrologia 34:61–65
7. Richter W (1997) Accred Qual Assur 2:354–359
8. Spitzer P, Eberhardt E, Schmidt I, Sudmeier U (1996) Fresenius J Anal Chem 356:178–181
9. Spitzer P (1997) Metrologia 34:375–376
10. Bundesärztekammer (1988) Dt Ärzteblatt 85:699–706;(1994) 91:211–212
11. Richter W (1999) Fresenius J Anal Chem 365:569–573
12. Comité international des poids et mesures (CIPM) (1999) Mutual recognition of national measurement standards and of calibration and measurement certificates issued by national metrology institutes. Bureau international des poids et mesures (BIPM), Sèvres, France
13. BIPM (1999) Comité consultatif pour la quantité de matière (CCQM). Report of the 5th Meeting (February 1999). Bureau international des poids et mesures (BIPM), Sèvres, France

DOI 10.1007/s00769-001-0438-7

John L. Love

Chemical metrology, chemistry and the uncertainty of chemical measurements

Abstract Chemical results normally involve traceability to two reference points, the specific chemical entity and the quantity of this entity. Results must also be traceable back to the original sample. As a consequence, any useful estimation of uncertainty in results must include components arising from any lack of specificity of the method, the variation between repeats of the measurement and the relationship of the result to the original sample. Chemical metrology does not yet incorporate uncertainty arising from any lack of specificity from the method selected or the traceability of the result to the original sample. These sources of uncertainty may however have much more impact on the reliability of the result than will any uncertainty associated with the repeatability of the measurement. Uncertainty associated with sampling may amount to 50–1000% of the reported result. Chemical metrology must be expanded to include estimations of uncertainty associated with lack of specificity and sampling.

Keywords Metrology · Sampling · Chemical · Specificity · Uncertainty

J.L. Love
Institute of Environmental Sciences and Research,
P.O. Box 29 181
Christchurch, New Zealand
e-mail: john.love@esr.cri.nz
Tel.: +64-3-351 0017
Fax: +64-3-351 0010

Introduction

Dependable measurement is critical to both science and trade. Without a common understanding of the meaning of results of measurements, science would not function and systems of trade would become inefficient. Trade requires reliable measurements for quantity, quality and safety of goods and without these, delivery slows and disagreements as to their compliance with specifications proliferate. Reliable measurements in science and trade depend on having defined standards for analytes, demonstrable traceability of results to the defined standards and an understanding of the uncertainties of these processes.

International trade agreements under the World Trade Organization are now emphasizing the current, less than satisfactory, state of chemical measurements. In trading relationships, both the buyer and seller usually repeat tests and often, regulatory agencies require their own independent check. This replication of effort is obviously inefficient [1] but reflects the current inability of chemical measurement to produce consistent results over distance and time.

The situation with respect to physical measurements is in complete contrast to results from different sources generally accepted as being comparable. Metrology, the science of measurement, has been developed from physical measurement and emphasizes results traceable to defined reference points, normally the International System of Units (SI), and fully analysed uncertainty budgets based on the processes set out in the Guide to the Expression of the Uncertainty of Measurement (GUM) [2]. This process involves identifying each component of the measurement that contributes to uncertainty, estimating the contribution of each component of uncertainty, then combining these estimations to calculate the total uncertainty. Much of the improvement in consistency of physical measurements has been achieved by use of the uncertainty budget to better define and control the test environment.

In the last ten years much effort has been applied to introduce these same concepts of physical measurement into chemical measurement. For example:

- The Bureau International des Poids et Mesures (BIPM) has put in place a consultative committee, the Consultative Committee on the Quality of Material (CCQM) [3], to strengthen the relationship of chemical measurements to its SI unit, the mole.
- EURACHEM and CITAC [4] have developed a guide for quantifying uncertainty in chemical analysis based on metrological principles and GUM [2] to quantifying uncertainty of measurement.
- ISO/IEC 17025:1999 [5] is replacing ISO Guide 25 [6] as the standard against which laboratories are accredited and supports these moves by having an increased emphasis on this metrological approach.

Incorporating traceability to the mole and uncertainty budgets into chemical analysis is more complex than is their application to physical measurement. Normally a chemical measurement depends on a combination of physical measurements, chemical separation of the compounds of interest and the selection of the test portion from the bulk material. An understanding of the chemistry involved in these separation processes is vital before reliable results can be achieved and chemical analysts have tended to concentrate on this area of analysis. It is however a part of the measurement that is tending to be ignored in moves to align chemical measurement with the traditional physical metrological process. The sampling process, both in the laboratory and outside in the field also contributes to the uncertainty of the measurement but has tended to be ignored by analysts. Understanding of the uncertainty of chemical measurements will not be achieved without an understanding of the whole process.

Discussion

Chemical measurement has a fundamental difference from physical measurement in that it does not take place under controlled and defined conditions. Almost always, the primary objective of a chemical measurement is to determine the amount of components of interest, not the total composition of the sample. Total composition will almost always remain unknown and therefore the total environment under which the measurement is taking place cannot be defined or controlled. Unknowns will always increase the uncertainty associated with any measurement.

Three components can be considered as contributing to uncertainty in chemical measurement. These are sources of uncertainty associated with the sampling process, the underlying chemistry of the chosen method, including its selectivity, and the more readily quantifiable aspects of uncertainty associated with the repeatability of the measurement. The cause and effect diagram in Fig 1 represents this situation. Each of these components is important. Get one wrong and the result is unlikely to be 'fit for purpose'.

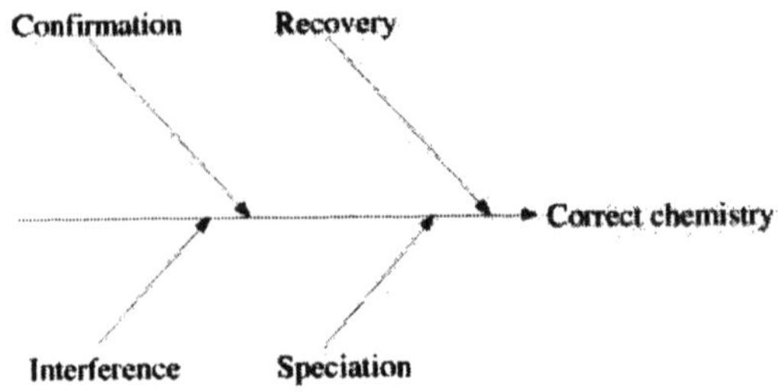

Fig. 1 Cause and effect diagram showing sources of uncertainty associated with chemical measurements

For many years, analytical chemists have used reference methods as a means of limiting the numbers of unknowns by removing those associated with traceability of the measurement to the defined chemical entity. Although reference methods remove uncertainty associated with traceability of the result to the named chemical entity and thereby eliminate most chemical unknowns as an issue, they always have the disadvantage that they redefine the analyte in terms of a method rather than as a chemical species. Amongst the best reference methods are those published by the Association of Official Analytical Chemists International (AOAC) [7]. These methods will have been validated within a number of laboratories from a collaborative study and will have associated estimations of uncertainty based on repeatability and reproducibility results from the study.

In reference methods, uncertainty in the result will directly relate to the measured repeatability. Defining the analyte as the method result eliminates any uncertainty related to the underlying chemistry. It may also define the procedure for taking the test portion in the laboratory and thereby include some of the uncertainty associated with sampling.

Modern methods of analytical chemistry are less conducive than traditional methods to the reference method approach. Instrument and equipment combinations are much more variable between laboratories and change over time as manufacturers add technical improvements. Reference methods also cause problems between countries unless they have international acceptance and they limit the adoption of new analytical methodology and equipment.

As a consequence of the problems associated with reference methods, there is now more emphasis on an absolute measure of analytes of interest where these are distinct chemical entities. For instance, the Codex Committee on Sampling and Analysis is presently debating whether analytical requirements for discrete chemical components in foods can be defined by method performance criteria or whether a prescribed method is also required in dispute situations [8]. Reference methods must still remain for those analytes not readily definable as a distinct chemical entity [8].

Moves away from reference methods towards performance criteria will make problems with the underlying chemistry of the selected method or in taking the test portion more significant. A problem in either of these areas may have the consequence of making the result meaningless. However, much of the recent work on the analyses of uncertainty in chemical measurement has neglected these issues and instead has tended to concentrate on alternatives, based on GUM [2], to repeatability estimated from collaborative trials, examples being the work of EURACHEM [4] and the survey of King [9].

Uncertainty arising from the repeatability of chemical measurement is a characteristic of the method. Its calculation has similarities to the calculation of uncertainty in physical measurement and many components are identical to those involved in physical measurement, components such as uncertainty in mass and volume. However, others such as purity of reference materials and recovery are rather more unique to chemistry, but once determined, can still be incorporated into the uncertainty budget using the standard techniques developed for physical metrology and described in the publication, GUM [2]. Once determined, this estimation of uncertainty can be applied to future tests using the same method in the same laboratory with the same equipment, reagents and staff.

Uncertainty due to the underlying chemistry and sampling is much more difficult to estimate. Both may vary with changes of sample. Realistic estimations of uncertainty as to the specificity of the method when testing samples from one source may not be applicable to samples from other sources. However, laboratories must be able to incorporate uncertainty arising from any lack of specificity into their estimation of total uncertainty if they are to be able to judge if results will be "fit for purpose" and give reliable information on uncertainty as required by ISO 17025 [5].

This same issue that uncertainty in analytical chemistry involves more than the uncertainty in the reported result has been highlighted previously in a small number of papers including those of Wells and Smith [10] and Alexandrov [11].

Uncertainty in the underlying chemistry

The underlying chemistry involved in the test method is obviously important. A simple cause and effect diagram for the underlying chemistry is shown in Fig. 2.

Almost without exception, modern methods of trace analysis require separation of the analyte of interest from the sample matrix, then estimation of the amount of analyte present using some unique characteristic, or combination of characteristics, of the molecules involved. Examples are systems involving chromatography, with or without, mass spectrometric detection, atomic absorption spectrometry and emission spectrometry. These methods do not work unless the measured characteristic, or combination of characteristics, is unique to the compound of interest or the impact of known interferences can be removed by calculation. The analyst should have used enough properties of the compound to make it unlikely responses from interfering substances could be incorporated into the reported result.

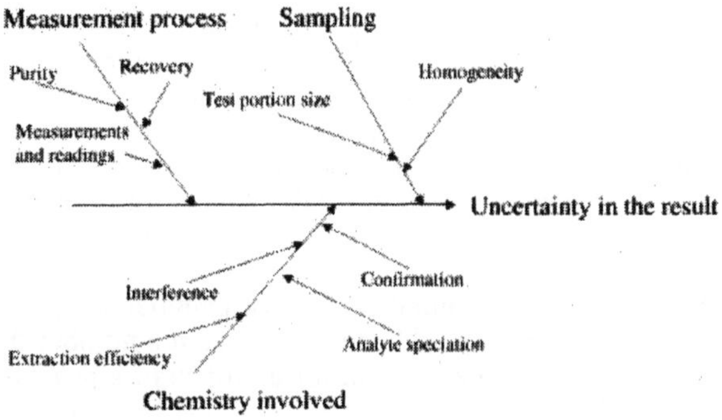

Fig. 2 Cause and effect diagram showing the issues around understanding the chemistry

Interference has been a concern of analysts using chromatography for a long time. All positive results from analyses based on chromatographic methods must have an appropriate level of confirmation if they are to be specific to the analyte of interest and results are to be reliable. Methods of confirmation have included:

- Independent analysis of the sample by a different method.
- Re-analysis of the sample on a column of different polarity known to separate compounds in a significantly different order.
- Re-analysis of the sample using a different wavelength on the detector. The second wavelength must be chosen so that it will give a good indication of the shape of the absorption curve.
- Re-analysis of the sample using a detector that operates using a different principle. An example could be an FID and nitrogen specific detector but the different sensitivities of detectors will limit this option.
- A statement from the client of the expected level of the analyte in the sample. Normally, it could be considered that the test has an appropriate degree of specificity if the result is similar to this expected value.
- Use of a detector such as a mass spectrometer that gives additional informational on the nature of the compound detected.
- Recording of the spectrum of the detected compound using for instance a diode array detector.

- Use of a spike of the compound of interest at a level that gives a similar response to the measured compound. A spike gives good evidence that a compound differs from the compound of interest when retention times are close but it does not provide good confirmation if retention times coincide. Spikes also reveal modifications to responses that can arise with new matrixes.
- Experience from a number of known similar samples. This is unlikely for ad hoc samples from a number of sources but may be relevant for a project or a process control laboratory.

Many of these approaches to confirmation involve an additional check procedure following that used to generate the analytical result. A change in the method of confirmation may have a major impact on the specificity of the result and its potential "fitness for purpose" without having any impact on the metrological estimation of uncertainty. In fact, a laboratory could ignore confirmation to get a commercial advantage and still be able to demonstrate the same uncertainty using the current metrological approach. Confirmation will normally have no effect on the components of uncertainty included in the metrological approach but lack of confirmation is likely to make the specificity of the result so uncertain that it will be "unfit for any purpose".

Suspect chemistry may not only apply to chromatography, it can make any measurement meaningless. The December 1999 report from the Canadian Food Inspection Agency on their Histamine Quality Assurance Programme [12] shows major differences between results obtained by high performance liquid chromatography (HPLC) methods, mainly with fluorescence detection, and some of those obtained by immunoassay based systems. Using immunoassay methods for the sample of fish sauce (code H16), 2 of the 3 results at 48.00 and 105.83 mg/100 g were considerably higher than all except 1 of the 21 HPLC based results, which had a mean of 11.83 mg/100 g. It seems unlikely these laboratories or the manufacturers of the immunoassay kits did not understand or had failed to validate the method. Obviously something is missing from these validations and the present understandings of these chemical measurements, and is unlikely to have been included in any uncertainty associated with the method. Alternatively, but less likely, an interference with the same retention time as histamine has quenched the observed fluorescence and resulted in low results for HPLC based methods.

Uncertainty arising from any lack of specificity of the underlying chemistry is not going to go away. It is also not a new concern. Interference has always dominated thinking and precautions in analytical chemistry and is well discussed in traditional reference textbooks such as Vogel [13]. Uncertainty may not be decreased by the new methods analysts continue to introduce or by manufacturers who continue to improve the ease of operation of existing equipment. Improvements can be very good for the expert analyst who understands the limitations of all measurements but can also allow laboratories to downgrade the skill level of equipment operators. Automated equipment can also lead to a decreased level of appraisal of individual results with an increase in uncertainty arising from overlooking problems with the underlying chemistry.

Automated equipment often allows replacement of expert technicians with less skilled staff, which will decrease reliability of judgements made during the analytical process. A decrease in the skill level of laboratory staff seems to be a problem throughout the world [14]. It is partly addressed by accreditation authorities requiring a minimum level of expertise within laboratories accredited to ISO/IEC 17025:1999 [5] but it also requires support of laboratory owners. Too often this is not forthcoming. Accreditation authorities can probably control major reductions in skill levels within a laboratory over a short-time frame but may have less ability to resist longer-term incremental shifts that have the same overall consequence. Skill levels in the laboratory certainly has a major impact on quality of analytical judgements and analytical reliability without necessarily any effect on the calculated uncertainty of the recorded numerical result.

It is often stated, by for instance King, that a significant number of reported chemical measurements are wrong [9]. This is almost certainly true and likely arises from problems in the chemistry underlying the method resulting in lack of specificity and/or loss of the analyte. Uncertainty arising from problems with the chemistry will be relevant to most chemical measurements aimed at measuring discrete chemical entities. In addition, lower limits of detection are often associated with increased uncertainty unless significantly improved equipment is used.

The sampling process

Sampling is the process used to obtain a portion of the bulk material for testing. It may take place in the field or in the laboratory. Most chemical analyses will involve at least two stages of sampling, primary sampling in the field and secondary sampling in the laboratory to give the final test portion. All samples are heterogeneous if looked at on a small enough scale [15]. The uncertainty arising from sampling will depend on the degree of heterogeneity in that sample.

Chemical laboratories are usually not responsible for primary sampling in the field but this must be appropriate or the result will be meaningless. They will however have to prepare the sample delivered to the laboratory and take a representative secondary sample for testing. The uncertainty associated with sampling is dependent on the sample and may vary between nominally identical samples although procedures used are identical. This is

Table 1 Protein and moisture content of rice flour

	Sample 1				Sample 2			
	Moisture		Protein		Moisture		Protein	
	1	2	1	2	1	2	1	2
	13.38	13.36	8.69	8.71	12.36	12.41	7.44	7.47
	13.58	13.66	8.24	8.24	12.50	12.50	7.45	7.44
	13.60	13.57	8.26	8.25	12.54	12.38	7.46	7.45
	13.36	13.36	8.74	8.69	12.51	12.53	7.44	7.45
	13.46	13.43	8.62	8.68	12.50	12.50	7.44	7.46
	13.48	13.45	8.24	8.26	12.48	12.56	7.64	7.69
	13.28	13.29	8.64	8.65	12.56	12.56	7.63	7.69
Mean	13.45		8.49		12.49		7.52	
SD Overall	0.12		0.21		0.07		0.10	
SD Replicate	0.026		0.022		0.050		0.023	

amply demonstrated by some test data on rice flour that are shown in Table 1.

The test data in Table 1 are taken from homogeneity data from two sets of samples each produced from two bags of rice flour and intended for use in a proficiency system. Seven replicates for each set of samples is too limited to produce a reliable estimate of the standard deviation. However, it is clear even from these limited data that for sample 1, the uncertainty associated with the location of the sample in the bag is about twice that of the second sample for both analyses. Variation between replicates of the same sample from within the batch is similar within both sets of samples for the protein determination and possibly slightly higher in batch 2 for the moisture determination.

An even more extreme example of the effect of lack of information on sample heterogeneity relates to the analysis of infant formula for iodine content [16]. A number of standard methods of analysis for the iodine content in dairy based products take 1 g test portions for the analysis and this size test portion is completely satisfactory for most formulations. One product on the New Zealand market when this testing was carried out was however formulated by dry mixing and 1 g test portions were completely inappropriate. Replicates from 1 g test portions of this sample were repeatable with no indication of an analytical problem although the sampling procedure meant results were well below the expected level and unfit for purpose. Test portions of 25 g gave more variable results but at least these were centered on the expected level. Test portions of 100 g gave results within a tight range still centered at the expected level. These results are summarized in Fig. 3 [16].

Uncertainty associated with sampling depends on three main attributes, the heterogeneity of the sample, the size of the sample and the method of taking this sample from the material in the field or in taking the test portion from the sample in the laboratory. If these attributes are all known then the uncertainty can be calculated

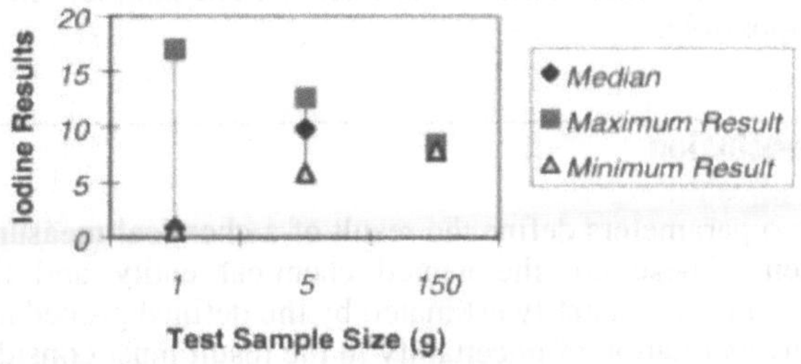

Fig. 3 Iodine levels in a dry mix infant formula

from statistical relationships such as have been developed by Gy [15] and incorporated into the metrological approach to estimating the uncertainty in the defined quality of the measurand.

Without knowledge of the uncertainty associated with sampling, any estimation of uncertainty in the result is meaningless. Gy [15] notes that the bias at the primary stage of sampling can be as high as 1000% of the result and at the secondary stage, as high as 50%. Any analytical uncertainty of a few percent is then essentially irrelevant to the uncertainty budget while the potential biases in sampling are this large.

Estimating uncertainty associated with the sampling process requires that test portions be representative of the parent material, that is, they have to be selected by a method that is both accurate and reproducible [15]. The term for "representative sampling" (r_0^2) has been defined by Gy [15] as a composite quantity dependent on both the maximum variance allowed (s_0^2) and the square of the maximum bias allowed (m_0^2), that is:

$$(r_0^2)=(m_0^2)+(s_0^2)$$

Thus, representative sampling is characterized by the absence of bias together with an acceptable variance [15]. Sub-samples that are not representative of the whole will make the final measurements irrelevant to any decision making.

Analysts must understand and include the uncertainty associated with sampling if they are to realistically estimate uncertainty of chemical measurements for clients. They must understand what is meant by a representative sample and have appropriate data to ensure samples tested are representative. However, as Gy [15] has pointed out this is not a traditional part of the education or training of analysts although it is critical to the use of measurements and their "fitness for purpose". This must change and sampling processes must be included as part of the metrological approach to chemical measurement.

In the past laboratories may have excluded primary sampling from their concern but ISO/IEC 17025:1999 [5] now imposes a requirement that the test and calibration methods selected are capable of meeting the client's requirements. Without knowledge of the traceability of the measurement back to the client's bulk sample, this is impossible.

Conclusion

Two parameters define the result of a chemical measurement. These are the named chemical entity and the amount of this entity estimated by the defined procedure. Any estimation of uncertainty in the result must consider traceability of the measurement to both these reference points. To be useful, the result must also be traceable back to the original sample. The uncertainty in chemical measurements must include:

- Uncertainty arising from assumptions made in the chemistry on which the method is based and the possibility that the measured result does not solely represent the chemical entity of interest and/or has not measured all of this entity present.
- Uncertainty arising from variation between repeats of the measurement.
- Uncertainty arising from the sampling procedure and the likelihood that the test portion does not containing a representative amount of the analyte of interest.

Present approaches to uncertainty based on metrological principles developed for physical measurement concentrate on estimating the uncertainty inherent in the traceability of the quantity of measurand to its reference point. Incorporation of metrological practices of physical measurement into chemical measurement is certainly an improvement but more effort is needed to incorporate uncertainty associated with the sampling process. The metrological practices of physical measurement do not address uncertainty arising from the choice of chemistry and the traceability of the result to the defined chemical entity, that is, the specificity of the method or uncertainty associated with sampling. At present chemists use judgement as to when uncertainty associated with sampling and specificity can be ignored. However, judgements are subjective and incompatible with formal methods to estimate uncertainty. Effort must be spent in developing a metrological approach for chemical measurements of discrete chemical entities that will allow realistic estimations of the total uncertainty associated with the reported result, not just the uncertainty associated with the numerical value.

Acknowledgements The author thanks Dr. Don Ferry from International Accreditation New Zealand (IANZ) for the supply of the data on rice flour replicates. This paper arose from some discussions between the author and the late Dr. John Nicholas at New Zealand Measurements Standards Laboratory (MSL).

References

1. CITAC (2000) Traceability in chemical measurement. CITAC web page at http://www.vtt.fi/ket/citac/traceability.pdf
2. BIPM, IEC, IFCC, ISO, IUPAC, IUPAP, OIML (1995) Guide to the Expression of the uncertainty in measurement. ISO, Geneva
3. Consultative Committee for Amount of Substance (Bureau International des Poids et Mesures) http://www.bipm.org/enus/2_Committees/CCQM.shtml
4. EURACHEM/CITAC Guide (2000) Qualifying uncertainty in analytical measurement, 2nd edn., Final Draft April 2000. EURACHEM
5. ISO/IEC 17025 (1999) General requirements for the competence of testing and calibration laboratories. ISO, Geneva
6. ISO 25 (1990) General requirements for the competence of calibration and testing laboratories. ISO, Geneva
7. Horwitz W (2000) Official methods of analysis of AOAC International. AOAC International, Gaithersburg, Md., USA
8. Codex Committee on Methods of Analysis and Sampling 23rd Session, Budapest, Hungary 26 February–2 March, 2001. Proposed draft guidelines for the application of the criteria approach by the committee on methods of analysis and sampling. Agenda item 4a (CX/MAS 01/4)
9. King B (2000) Accred Qual Assur 5: 173–179
10. Wells RJ, Smith RJ (1996) Chem Aust April 1996: 167–168
11. Alexandrov YI (1997) Fresenius J Anal Chem 357: 563–571
12. Burns-Flett E (2000) Report by the Histamine Quality Assurance Co-ordinator, Canadian Food Inspection Agency dated 31 March 2000. Canadian Food Inspection Agency, 501 University Crescent, Winnipeg, Manitoba R3T 2N6
13. Vogel AI (1961) A textbook of quantitative inorganic analysis including elementary instrumental analysis, 3rd edn. Longmans, London, UK
14. Clapp S (2000) Professional qualifications: How close should we look? Inside Laboratory Management June 2000: 18–20. AOAC International, Gaithersburg, Md., USA
15. Gy P (1998) Sampling for analytical purposes. Wiley, Chichester, UK
16. Love JL (2000) Sampling – What should analytical chemists learn from microbiologists? Inside Laboratory Management February 2000: 17–18. AOAC International, Gaithersburg, Md., USA

Accred Qual Assur (1999) 4:401–405

René Dybkaer

From total allowable error via metrological traceability to uncertainty of measurement of the unbiased result

Presented at: 4th Conference on Quality [R]evolution in Clinical Laboratories, Antwerp, Belgium 29–30 October 1998

R. Dybkaer
Copenhagen Hospital Corporation, Department of Standardization in Laboratory Medicine, H:S Kommunehospitalet, Øster Farimagsgade 5, DK-1399 Copenhagen K, Denmark
Tel.: +45-33-38-37 85/86
Fax: +45-33-38-37-89

Abstract The concept of "total allowable error", investigated by Westgard and co-workers over a quarter of a century for use in laboratory medicine, comprises bias as well as random elements. Yet, to minimize diagnostic misclassifications, it is necessary to have spatio-temporal comparability of results. This requires trueness obtained through metrological traceability based on a calibration hierarchy. Hereby, the result is associated with a final uncertainty of measurement purged of known biases of procedure and laboratory. The sources of bias are discussed and the importance of commutability of calibrators and analytical specificity of the measurement procedure is stressed. The practicability of traceability to various levels and the advantages of the GUM approach for estimating uncertainty are shown.

Key words Metrological traceability · Total allowable error · Trueness · Unbiased result · Uncertainty of measurement

Introduction

The important contributions of Prof. James O. Westgard to quality assurance in laboratory medicine have spanned a quarter of a century. His initial interest in statistical comparison of measurement procedures [1] soon led to criteria for judging precision and accuracy in a procedure [2]. Based on the concept "total analytic error", comprising constant systematic error, proportional systematic error, and random error, the concept "allowable total error" (originally called total allowable error) was defined with respect to clinical requirements, usually as a 95% limit. This measure has been maintained during all later developments by Westgard and his co-workers and has been recently applied to the "analytical model" used in a paper on the Validator® 2.0 which is a computer programme for automatic selection of statistical quality control procedures [3]. In worded form, the following equation is said to apply (where QC = internal quality control rules):

allowable total error
= constant inaccuracy of procedure
+ varying inaccuracy due to sample matrix
+ unstable inaccuracy detectable by QC
+ z (unstable imprecision detectable by QC)

where $z = 1.65$ yields a maximum allowable number fraction of defects of 5%.

The present discussion is about the constant bias of the measurement procedure (the first term, called constant inaccuracy, in the equation above). This component of overall bias is, in principle, a known detriment to trueness of measurement (defined as average closeness to a reference value).

Trueness and consequences of procedure-dependent bias

It is relevant to ask whether trueness is important or whether the sometimes heard pronouncement "precision is better than accuracy [meaning trueness]" relegates trueness to a lower priority. The reliance on precision is repeatedly seen in the results from external quality assessment (or proficiency testing) schemes all over the world, where method-dependent groupings of results for a given measurand are abundant.

Bias always impairs the comparability over space and time of the results for a given type of quantity and distorts the relationships between different types of quantity. Biological reference intervals are changed in comparison with a true distribution [e.g. 4, 5]. Harris even suggested a new term for such intervals, "medical indifference ranges" [6]. Whereas serial monitoring for change can sometimes live with a constant bias, this is not the case with screening, initial diagnosis, and movement towards a fixed discriminatory true limit, where diagnostic misclassifications are the outcome [e.g. 6–10]. A positive or negative bias of, say, 1 mmol/l in the amount-of-substance concentration of cholesterol or glucose in blood plasma has enormous effects on population health and economy.

Reduction of bias

Several approaches to the elimination of known bias should be considered when selecting, describing and operating a measurement procedure for a given type of quantity:

1. The type of quantity that is to be measured must be defined sufficiently well. This is particularly demanding when analyte isomorphs or speciation are involved.
2. The principle and method of measurement must be carefully selected for analytical specificity.
3. A practicable measurement procedure including sampling must be exhaustively described.
4. A calibration hierarchy must be defined to allow metrological traceability, preferably to a unit of the International System of Units (SI). Traceability involves plugging into a reference measurement system of reference procedures and commutable calibration materials.
5. An internal quality control system must be devised to reveal increases in bias.
6. Any correction procedures must be defined and validated.
7. Where possible, there should be participation in external quality assessment ("proficiency testing") using material with reference measurement values.

Metrological traceability

The necessary anchor for the trueness of a measurement procedure is obtained by strict metrological traceability of result, based on a calibration hierarchy. The official definition of traceability in metrology is: "property of the result of a measurement or the value of a measurement standard whereby it can be related to stated references, usually national or international measurement standards, through an unbroken chain of comparisons all having stated uncertainty" [11]. As stressed in the first resolution of the 20th General Conference on Weights and Measures (CGPM) in 1995 [12], the top of the calibration hierarchy, when possible, should be the definition of an SI unit.

The physical calibration hierarchy

In physics, the use of calibration hierarchies is well established and is used in any laboratory, e.g. for balances, volumetric equipment, spectrometer wavelengths, cuvette light path lengths, thermometers, barometers and clocks.

The chemical calibration hierarchy

For chemical quantities, involving the SI base unit for amount of substance, the "mole", its definition demands specification of the elementary entities of the component under consideration. According to the physical calibration hierarchy, a primary standard would be needed for each of the huge number of different compounds that are defined in the measurements. To circumvent this obstacle, the Consultative Committee for Amount of Substance of the International Committee on Weights and Measures (CIPM-CCQM) defines a primary reference method, which is claimed directly to give amount of substance in moles without prior calibration by a primary standard [13, 14]. Current examples of primary reference methods are isotope dilution-mass spectrometry and gravimetry. It should be realized, however, that establishing the more complicated measurement procedures based on such primary methods is by no means simple [15] and may require the expertise of the International Bureau of Weights and Measures (BIPM) or a national metrology institute (NMI). A primary reference measurement procedure (prim. RMP) assigns a value with uncertainty of measurement to a primary reference material [13], usually purified and stable, used as a primary calibrator (prim. C). The steps of the calibration series may be as follows with the responsible bodies in parentheses (accr. CL = accredited calibration laboratory; mf. = manufacturer).

SI unit (definition)	(CGPM)
prim. RMP	(BIPM, NMI)
prim. C	(BIPM, NMI)
sec. RMP	(NMI, accr. CL)

sec. C	(NMI→accr. CL→mf.'s lab.)
mf.'s selected MP	(mf.'s lab.)
mf.'s working C	(mf.'s lab.)
mf.'s standing MP	(mf.'s lab.)
mf.'s product C	(mf.→user)
routine MP	(mf., user)
routine sample	(user)
result	(user)

The length of the hierarchy can be reduced by eliminating pairs of consecutive steps, thereby reducing uncertainty.

Commutability and analytical specificity

There are two major reasons why a traceability chain may be broken and trueness lost due to the introduction of bias: insufficient commutability of a calibration material and non-specificity of a measurement procedure. The effect of these separate properties are often indiscriminately lumped together as "matrix effect". Commutability refers to the ability of a material, here a calibrator, to show the same relationships between results from a set of procedures as given by routine samples [16, 17]. Analytical specificity refers to the ability of a measurement procedure to measure solely that quantity which it purports to examine [16, 18]. Discrepancies between results of a reference procedure and a routine procedure applied to routine samples are often caused by non-specificity of the routine procedure. The use of a set of human samples as a manufacturer's calibrator to eliminate so-called matrix effects should only be accepted if the relationship between the results from reference and routine procedures is sufficiently constant to allow explicit correction with consequent increased uncertainty of assigned values.

Traceability in practice

It is relevant to ask how often the routine measurement procedures currently used in laboratory medicine provide results that are traceable to high-level calibrators and reference measurement procedures (Lequin: personal communication). It turns out that primary reference measurement procedures and primary calibrators are only available for about 30 types of quantity such as blood plasma concentration of bilirubins, cholesterols and sodium ion. International reference measurement procedures from the International Federation of Clinical Chemistry and Laboratory Medicine (IFCC) and corresponding certified reference material from BCR are available for the catalytic activity concentration of a few enzymes such as alkaline phosphatase and creatine kinase in plasma. For another 25 types of quantity, such as the catalytic activity concentration of aspartate aminotransferase in plasma and number concentration of erythrocytes in blood, no high-level calibrators exist. International calibrators, e.g. from WHO, but no high-level in vitro procedures characterize a couple of hundred types of quantity involving, for example, choriogonadotropin. An overwhelming number of types of quantity have no high-level ending of the traceability chain, but rely on the internal best-measurement procedure and calibrator of the reagent set manufacturer or individual laboratory. The end-user, as a rule, cannot be expected to establish the entire traceability chain if that goes above an in-house procedure. The laboratorian usually has to rely on the manufacturer which, in turn, may claim traceability of its product calibrators to the highest available level, preferably provided by a national metrology institute, an accredited calibration laboratory, or a reference measurement laboratory. In fact, this responsibility of the manufacturer is now enshrined in the EU Directive on in vitro diagnostic medical devices [19], which will be supported by four EN/ISO standards under development. The laboratorian should, however, bolster his or her belief in trueness and comparability – especially if the traceability chain does not reach high – by recovery experiments [20], comparison with a selected procedure [21], and interlaboratory parallel measurements [22], including external quality assessment [23], preferably on material with reference measurement procedure assigned values [24]. The internal quality control system finally checks, with a given probability, whether the current measurements are in statistical control with no sign of change in the assumed zero bias.

Uncertainty of measurement

The definition of metrological traceability (see above) stipulates that each link in the chain has a known uncertainty. Nowadays, this concept and its application have been reformulated by the BIPM and recently detailed in the "Guide to the expression of uncertainty in measurement" (GUM) [26]: "parameter, associated with the result of a measurement, that characterizes the dispersion of the values that could reasonably be attributed to the measurand". Useful explanations are provided in several other guides [26–30] as well as commentaries [e.g. 31–33]. The philosophy is to apply a bottom-up approach by formulating a function of all input quantities giving the measurand as output. An uncertainty budget of all sources of uncertainty is established. Important items to consider are:

- definition of the measurand
- realization of the measurand
- sampling
- speciation and matrix

- instability
- environment and contamination
- measuring system
- published reference data
- calibrator values
- commutability
- algorithms and software
- corrections and correction factor.

Each contribution is assessed as a standard uncertainty, either by statistical procedure on experimental data in the form of an a posteriori distribution, the so-called Type A evaluation, or by scientific judgement based on an a priori chosen distribution, Type B evaluation. The few standard uncertainties of important magnitude are combined quadratically, including any covariances, and the combined uncertainty, u_c, is obtained as the positive square root.

The advantages of this approach are important:

- The transparent budget invites improvement where major contributions are identified in the total sequence from definition onwards.
- There is no known significant bias allowing one, usually symmetric, measure of uncertainty.
- The combined uncertainty is comparable with that of other results.
- The combined uncertainty can be quadratically added to those of other results as demanded for traceability.
- The combined uncertainty can be compared with the classical top-down approach of calculating an uncertainty directly from replicate final results to reveal any discrepancy requiring further investigation.

The role of certified reference materials (with assigned value and uncertainty) in obtaining traceability and avoiding bias is obvious.

The GUM approach to uncertainty is rapidly gaining acceptance in metrological institutes and industry, and must be applied in ISO and CEN standards. It should be used in accredited laboratory work but chemists often find the implementation difficult and therefore hesitate [34]. Additionally, sometimes, there is a fear that honest GUM uncertainty intervals, which may be wider than classical precision intervals, are bad for business. Also, the perceived psychological effect on the customer of the term "uncertainty" seems to have led the food industry – naturally concerned about palatability – to propose the substitute term "reliability". Although it would be possible to define a concept with a "comforting" term inversely related to the measures of uncertainty – analogously to accuracy, trueness, and precision – the term reliability is already used for a more comprehensive concept covering several analytical performance criteria. There should be no doubt, however, that, as the GUM says, "The evaluation of uncertainty is neither a routine task nor a purely mathematical one; it depends on detailed knowledge of the nature of the measurand and of the measurement" [25]. To alleviate the calculations involved, commercial EDP programmes are being offered.

Conclusions

The upshot of these considerations is that one should cease to define a so-called allowable total error of result, with assessable biases of procedure and laboratory included. Instead, it is necessary to provide corrected results with a defined allowable maximum uncertainty at an agreed level of confidence. Likewise, a manufacturer may be asked to specify an expected uncertainty for a measuring system performing according to a measurement procedure under statistical control. Finally, the laboratorian can provide the customer with a corrected result and an accompanying uncertainty interval comprising a stated proportion of values that could reasonably be attributed to the measurand. This view is not in conflict with the 25-year-old statement by Westgard and co-workers – using classical terminology – that "In principle, only random error need be tolerated. Systematic errors can be eliminated by appropriate improvements in methodology" [1].

Acknowledgements Ms Inger Danielsen is gratefully thanked for her excellent secretarial assistance.

References

1. Westgard JO, Hunt MR (1973) Clin Chem 19:49–57
2. Westgard JO, Carey RN, Wold S (1974) Clin Chem 20:825–833
3. Westgard JO, Stein B, Westgard SA, Kennedy R (1997) Comput Method Programs Biomed 53:175–186
4. Gowans EMS, Hyltoft Petersen P, Blaabjerg O, Hørder M (1988) Scand J Clin Lab Invest 48:757–764
5. Hyltoft Petersen P, Gowans EMS, Blaabjerg O, Hørder M (1989) Scand J Clin Lab Invest 49:727–737
6. Harris EK (1988) Arch Pathol Lab Med 112:416–420
7. Ehrmeyer SS, Laessig RH (1988) Am J Clin Path 89:14–18
8. Hyltoft Petersen P, Lytken Larsen M, Hörder M, Blaabjerg O (1990) Scand J Clin Lab Invest 50 (Suppl 198):66–72
9. Hyltoft Petersen P, Hørder M (1992) Scand J Clin Lab Invest 52 (Suppl 208):65–87
10. Hyltoft Petersen P, de Verdier C-H, Groth T, Fraser CG, Blaabjerg O, Hørder M (1997) Clin Chim Acta 260:189–206
11. BIPM, IEC, IFCC, ISO, IUPAC, IUPAP, OIML (1993) International vocabulary of basic and general terms in metrology. ISO, Geneva

12. Comité International des Poids et Mesures (1998) National and international needs relating to metrology. Bureau International des Poids et Mesures, Sèvres
13. Kaarls R, Quinn TJ (1997) Metrologia 34:1–5
14. Quinn TJ (1997) Metrologia 34:61–65
15. Adams F (1998) Accred Qual Assur 3:308–316
16. Dybkaer R (1997) Eur J Clin Chem Clin Biochem 35:141–173
17. Fasce CF, Rej R, Copeland WH, Vanderlinde RE (1973) Clin Chem 19:5–9
18. Kaiser H (1972) Z Anal Chem 260:252–260
19. EU Directive 98/79/EC (1998) Off J Eur Comm L 331:1–37
20. Willets P, Wood R (1998) Accred Qual Assur 3:231–236
21. Hyltoft Petersen P, Stöckl D, Blaabjerg O, Pedersen B, Birkemose E, Thienpont L, Flensted Lassen J, Kjeldsen J (1997) Clin Chem 43:2039–2046
22. Groth T, de Verdier C-H (1993) Upsala J Med Sci 98:259–274
23. Hirst AD (1998) Ann Clin Biochem 35:12–18
24. Stamm D (1982) J Clin Chem Clin Biochem 20:817–824
25. BIPM, IEC, IFCC, ISO, IUPAC, IUPAP, OIML (1993) Guide to the expression of uncertainty in measurement. ISO, Geneva
26. Taylor BN, Kuyatt CE (1994) NIST Technical Note 1297. National Institute of Standards and Technology, Washington
27. Eurachem (1995) Quantifying uncertainty in analytical measurement.
28. EAL-G23 (1996) The expression of uncertainty in quantitative testing.
29. EAL-R2 (1997) Expression of the uncertainty of measurement in calibration.
30. ISO TR 14253-2 (1998) Geometrical product specifications (GPS) – Inspection by measurement of workpieces and measuring equipment – Part 2: Guide to the estimation of uncertainty in GPS measurement, in calibration of measuring equipment and in product verification. ISO, Geneva
31. Kadis R (1998) Accred Qual Assur 3:237–241
32. Bremser W (1998) Accred Qual Assur 3:398–402
33. Hässelbarth W (1998) Accred Qual Assur 3:418–422
34. Golze M (1998) Accred Qual Assur 3:227–230

Accred Qual Assur (1998) 3:180–184

Jean Pauwels
Andrée Lamberty
Heinz Schimmel

The determination of the uncertainty of reference materials certified by laboratory intercomparison

Jean Pauwels (✉) · Andrée Lamberty
Heinz Schimmel
European Commission,
Joint Research Centre
Institute for Reference Materials and Measurements (IRMM)
B-2440 Geel, Belgium
Tel.: +32-14-571722
Fax: +32-14-590406
e-mail: pauwels@irmm.jrc.be

Abstract A pragmatic method is proposed for the implementation of the Guide to the expression of uncertainty in measurement in the certification of reference materials by laboratory intercomparison. It is based on the establishment of a full uncertainty budget for each laboratory result and the estimation of the impact of various laboratory standard uncertainties and of between-units variability on the certified reference material (CRM) uncertainty.

Key words Reference material · Laboratory intercomparison · Certified value · Uncertainty

Introduction

Many reference materials, produced worldwide, are certified by laboratory intercomparison, involving a large number of independent and, if possible, equally competent laboratories [1]. Normally, methods used are based on a variety of chemical and/or physical principles. It is then assumed that the differences between individual results, both within and between laboratories, are all of a statistical nature regardless of their causes. Each laboratory mean is considered as an unbiased estimate of the property of the material to be certified, and usually an unweighted mean of the laboratory means is assumed to be the best estimate of that property. In general, a reference material certification involves different laboratories, each of which measures the requisite property on different samples, with each sample measurement consisting of a number of independent repeated observations. The certified value and its uncertainty are then estimated on the basis of an analysis of variance, after verification that all data belong to the same normally distributed population.

If this is the case, the mean value of all individual data is taken as the certified value, and the half-width of the 95% confidence interval of the mean value of all individual data as its uncertainty. If, on the contrary, pooling is not allowed because individual data do not belong to the same normally distributed population, the mean value of the laboratory means is taken as the certified value and the half-width of the 95% confidence interval of the mean value of the laboratory means as its uncertainty.

The limitation of such procedures is that the distribution of the considered values should be normal and that no other sources of uncertainty than "random experimental uncertainties" should exist [1].

The above procedure finds its justification in the fact that one presumes that, if a large variety of independent laboratories and methods is used, possible systematic effects in the individual laboratory results will be "randomized" and that, eventually, both the residual systematic error and its uncertainty are reduced to zero.

Determination of an uncertainty according to the Guide to the expression of uncertainty in measurement

According to the Guide to the expression of uncertainty in measurement (GUM)[2], the result of a measurement corresponds to the estimate of the value of a measurand and should, therefore, always be accompanied by an uncertainty statement. It is, generally, determined on the basis of a series of observations obtained under repeatability conditions; its standard uncertainty is expressed as a standard deviation. It is assumed that measurement results are corrected for recognized significant systematic effects and that every effort has been made to identify and quantify such effects. Moreover, any other sources of uncertainty should be estimated and taken into account.

Uncertainty components are of two different types based on the method used for their evaluation: *type A* uncertainties are evaluated statistically on the basis of a series of observations, and *type B* uncertainties on the basis of all means other than statistical ones (e.g. previous experimental data, knowledge or experience, manufacturer's specifications, data from certificates, published reference data, etc). Both *type A* and *type B* uncertainties can be of a *"random"* as well as of a *"systematic"* nature.

A measurand Y is, however, generally not measured directly, but determined from N other quantities X_1, X_2, ..., X_N through a functional relationship f:

$$Y = f(X_1, X_2, ..., X_N) \quad (1)$$

The set of input quantities $X_1, X_2, ..., X_N$ may be categorized as

– quantities whose values and uncertainties are directly determined in the current measurement; they may then be obtained from a single observation, repeated observations or judgement based on experience; they may involve the determination of corrections to instrument readings and corrections for influence quantities

– quantities whose values and uncertainties are brought into the measurement from external sources, such as quantities associated with calibrated measurement standards, certified reference materials, reference data obtained from handbooks, etc.

The estimated standard deviation associated with the output estimate y of Y, termed *combined standard uncertainty* and denoted $u_c(y)$ is determined from the estimated standard deviation associated with each input estimate x_i of X_i, termed *standard uncertainty* and denoted $u(x_i)$. In its second recommendation, the Comité International des Poids et Mesures (CIPM) requested that this combined standard uncertainty be used "by all participants in giving results of all international comparisons or other work done under the auspices of the CIPM and Comités Consultatifs" [3].

Although $u_c(y)$ can be universally used to express the uncertainty of the result of a measurement, it may be required to give a measure of uncertainty that defines an interval about the measurement result that may be expected to encompass a large fraction of the distribution of values that could reasonably be attributed to the measurand. This additional measure is termed the *expanded uncertainty* and is denoted U. It is obtained by multiplying $u_c(y)$ by a *coverage factor k*:

$$U = k \cdot u_c(y) \quad (2)$$

The generally chosen value of the coverage factor k is 2 or 3. If the probability distribution characterized by y and $u_c(y)$ is approximately normal and the effective degrees of freedom of $u_c(y)$ of significant size, $k=2$ or 3 corresponds to a level of confidence of approximately 95 or 99%.

The result of a measurement is conveniently expressed as:

$$Y = y \pm U \quad (3)$$

which means that the best estimate of the value attributable to the measurand Y is y, and that

$$y - U < Y < y + U \quad (4)$$

is the interval that may be expected to encompass a large fraction (p) of the distribution of values that could reasonably be attributed to Y. The fraction p of the probability distribution is named *coverage probability* or *level of confidence.*

The Eurachem document "Quantifying Uncertainty in Analytical Measurement"[4] shows how the GUM concept should be applied in chemical measurement and illustrates this by four worked examples. These examples are however limited to simple analytical determinations, and the document discusses neither the problem of laboratory intercomparisons nor their use for the certification of reference materials.

Application of the GUM to the determination of the uncertainty of CRMs by laboratory intercomparison

A typical example of a certification exercise by laboratory intercomparison (e.g. for BCR CRMs) is shown in Fig. 1:

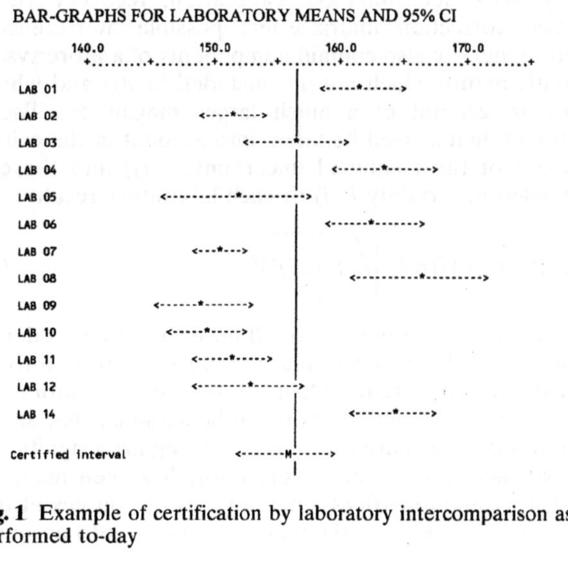

Fig. 1 Example of certification by laboratory intercomparison as performed to-day

– Between 6 and 15 laboratories carry out each six measurements spread on two different units.
– Samples of each of both units are measured on two different days.
– The measurement is (e.g. for BCR CRMs) carried out *under reproducibility conditions*, i.e. such that each replicate has its own calibration, dissolution, extraction, blank determination, etc.

The comparison of the results is, however, limited to the bare values of the six replicates carried out by each laboratory, with the immediate consequence that laboratories very often do not overlap between each other. Frequently, it is observed that the results of several laboratories participating in the certification do not even overlap with the value which is certified. The reason for this is not, as is generally believed on the basis of routine statistical tests, that there are significant differences between the results of the different laboratories, but because only the standard uncertainty on the six replicates is considered and because calculation of a combined standard uncertainty for each participating laboratory result is omitted. As already indicated, each analyst carrying out a measurement should always make up a complete uncertainty budget considering all *recognized* components of standard uncertainty affecting his measurement result. This should a fortiori apply to any laboratory which is invited to contribute to the certification of a reference material. The standard deviation $s(j)$ of the six replicates carried out by laboratory j, further denoted as $u_1(j)$ already includes part of the uncertainties of a purely statistical nature due to day-to-day variation, calibration (at least if each replicate has its own calibration), recovery yield (same remark), etc, as the measurements are in principle executed under *reproducibility conditions*. However, the standard uncertainties $u_i(j)$ (for i ranging from 2 to n) due to sampling, dry mass determination, calibration, recovery yield, blank correction, matrix effect, possible interferences, etc, generally also contain components of a more systematic nature which are not included in $s(j)$ and which are in general of a much larger magnitude. These should then as well be taken into account in the calculation of the combined uncertainty $u_c(j)$ and the expanded uncertainty $U(j)$ of each laboratory result:

$$U(j) = k \cdot u_c(j) = k \cdot \sqrt{\sum_{i=1}^{n} [u_i(j)]^2} \tag{5}$$

i = identification number of all uncertainties considered in each individual laboratory j, varying from 1 to n, with n not necessarily identical for each laboratory

From this moment on, it can be assumed that all laboratory results are corrected for recognized significant systematic effects, that every effort has been made to identify and quantify them, and that all sources of uncertainty have been estimated and taken into account.

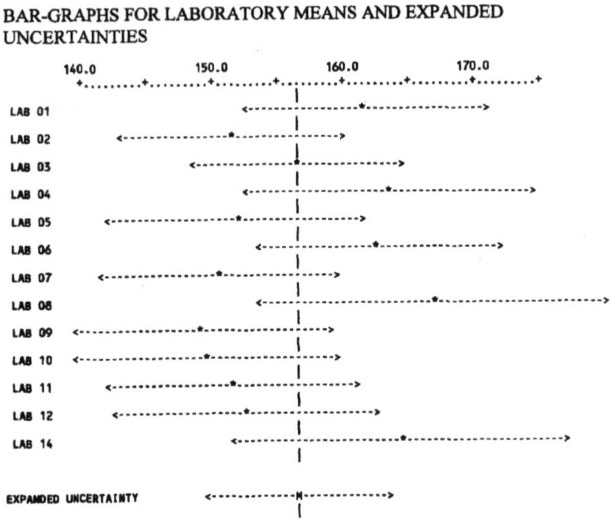

Fig. 2 Example of certification by laboratory intercomparison with consideration of combined standard uncertainties

Therefore, all results should in principle overlap and any discrepancies as shown in Fig. 1 should no longer exist (see Fig. 2). At this point, it should however be noted that components of standard uncertainty which are not laboratory specific but which are common to all or to part of the participating laboratories (e.g. those using identical methods) should be considered separately. For this reason it is essential that each laboratory supplies the project leader with a fully detailed uncertainty budget and that these uncertainty budgets are extensively discussed with the experts of all participating laboratories.

The certified value can then be calculated as either the unweighted or as the weighted mean of the laboratory means. In principle the former should be preferred, but in practice it may be unfair towards some laboratories, as especially type B components of uncertainty may have been evaluated differently from one laboratory to another. The certified uncertainty can be calculated after deconvolution (and later recombination) of all laboratory standard uncertainties in distinct categories of (combined) standard uncertainties, which may be evaluated as type A and/or as type B:

1. uncertainties which are *exclusively laboratory-dependent* $[u_c(l)]$
These affect the certified uncertainty interval in such a way that the more laboratories are involved in the intercomparison the smaller their contribution becomes:

$$u_c(l) = \frac{\sqrt{\sum_{j=1}^{l} [u_c(j)]^2}}{l} \tag{6}$$

j=laboratory identification number, varying from 1 to l
l=total number of laboratories

2. Uncertainties which are *common to all laboratories* participating in the certification $[u_c(II)]$
These affect the certified uncertainty interval in such a way that their contribution is independent of the number of participating laboratories:

$$u_c(II) = \sqrt{\sum_{i=1}^{n} [u_i(II)]^2} \qquad (7)$$

i=category II uncertainty identification number, varying from 1 to n
Typical examples of this category are the use of a common calibrant by all laboratories or material-related effects such as between-units variation (see "Effect of possible inhomogeneity and instability on the certified uncertainty").

3. Uncertainties in between the two above categories $[u_c(III)]$
These are *common to groups of limited numbers of laboratories* $\sum_{q=1}^{g} h_q = l$, such as those using an identical analysis procedure:

$$u_c(III) = \sqrt{\frac{\sum_{q=1}^{g} h_q \cdot [u_c(q)]^2}{g \cdot l}} \qquad (8)$$

with:

$$u_c(q) = \sqrt{\sum_{i=1}^{n} [u_i(q)]^2} \qquad (9)$$

q=group identification number, varying from 1 to g
g=total number of groups
l=total number of laboratories
h_q=number of laboratories in group q
i=category III uncertainty identification number in group q, varying from 1 to n

4. Moreover, as all laboratory means are not completely identical, a *residual component* $[u(R)]$ corresponding to the standard uncertainty of the average of the laboratory means should be considered as well:

$$u(R) = \frac{s_{betw}}{\sqrt{l}} \qquad (10)$$

s_{betw}=standard deviation of the laboratory means
l=total number of laboratories

As already indicated, if the expanded uncertainty of each laboratory is correctly estimated, all laboratory results should overlap. More specifically, one can state that in fact laboratories within the same group should have mean values $x(j)$ differing from the mean group value $\bar{x}(q)$ by less than:

$$|\bar{x}(q) - x(j)| \leq k \cdot u_c(I,j) \qquad (11)$$

whereby $u_c(I,j)$ corresponds to the combined category I uncertainty of laboratory j, whereas all laboratory mean values $x(j)$ should differ from the overall mean value $\bar{x}$ by less than:

$$|\bar{x} - x(j)| \leq k \cdot \sqrt{[u_c(I,j)]^2 + [u_c(III,q)]^2} \qquad (12)$$

whereby $u_c(III,q)$ corresponds to the combined category III uncertainty of the group to which laboratory j belongs.

Laboratories whose results do not overlap within these limits are either affected by unrecognized systematic errors and/or by uncertainties that have been underestimated or omitted. Their results should therefore not be considered for certification.

The final uncertainty of the laboratory intercomparison can then be calculated as:

$$U = k \cdot \sqrt{[u_c(I)]^2 + [u_c(II)]^2 + [u_c(III)]^2 + [u(R)]^2} \qquad (13)$$

Effect of possible inhomogeneity and instability on the certified uncertainty

Most frequently the between-units variability resulting from a homogeneity study is not insignificant compared to the uncertainty of the mean value. In addition it is generally preferred to assign a single certified value to all units of the entire CRM batch. Therefore, the uncertainty associated with the (possible) between-units inhomogeneity of the material should be included in the total uncertainty of the CRM. As indicated in [5], this can be done either by basing the CRM uncertainty on the statistical tolerance interval of the homogeneity study or by including the between-units standard uncertainty in the "category II" combined uncertainty $[u_c(II)]$ calculated according to Eq. 7.

The within-unit inhomogeneity, on the contrary, should in general not be included in the CRM uncertainty, except if such small sample intakes are used (e.g. in microanalysis techniques) that the sample inhomogeneity becomes significant compared to the certified uncertainty of the CRM. The main difference with between-units homogeneity testings is that if the observed within-unit inhomogeneity is significantly larger than the CRM uncertainty, it is sufficient to recommend the use of a larger sample intake on the basis of the fact that the uncertainty due to material inhomogeneity is inversely proportional to the square root of the mass of the analysed sample [6]. It is on the basis of this property that microanalysis was effectively proposed to determine experimentally the minimum sample mass down to which CRM certificates remain valid [7].

Linear regression and correlation can be used for the prediction of the possible instability of CRMs [8]. Quantitative characteristics expected to decrease (or increase) with time are determined by calculating the time at which the 95% lower (or higher) confidence limit intersects the acceptable lower (or higher) specification limit, i.e. the lower or higher limit of the certified interval. The time so determined may then be considered as the expiration date, as one may be 95% confident that the average value of the batch characteristic will remain within specification until that date. As was the case for the within-unit variation, this possible instability should, in general, not be included in the CRM uncertainty, except if the degradation is significant compared to the certified uncertainty of the CRM. In such cases it might be preferred, rather than to reject the material as CRM, to certify an arbitrarily chosen interval within which the material can be expected to remain stable during a significant period of time, i.e. until the expiry date of the certificate.

Conclusion

The Guide to the expression of uncertainty in measurement provides a framework for assessing uncertainty which can and should be used for the certification of reference materials by laboratory intercomparison. However, as is stated in its paragraph 3.4.8., the following should be noted:

– *It cannot substitute for critical thinking, intellectual honesty, and professional skill.*

– *The evaluation of uncertainty is neither a routine task nor a purely mathematical one and depends on detailed knowledge of the nature of the measurand and of the measurement.*

– *The quality and utility of the uncertainty quoted for the result of a measurement therefore ultimately depend on the understanding, critical analysis, and integrity of those who contribute to the assignment of its value.*

This is particularly the case for the certification of reference materials. The above procedures can be used to obtain an estimation of both the certified value of a reference material and its uncertainty. However, there must be room for critical evaluation of the results by the people and organizations taking up responsibility for the values assigned to a CRM. Therefore it may be common practice in some organizations to increase the calculated uncertainty as it is felt to be optimistic. One should however be careful not to give lower uncertainties just on the basis of the fact that large uncertainty intervals may be interpreted as being the consequence of e.g. an analytical artefact.

References

1. Guidelines for the production and certification of BCR reference materials (1997) - document BCR/01/97, European Commission, Dg XII-5-C (SMT Programme).
2. Guide to the expression of uncertainty in measurement (1995) ISO, Geneva, ISBN 92-67-10188-9
3. Giacomo P (1987) Metrologia 24:49–50
4. Quantifying uncertainty in analytical measurement, 1st edn (1995) Eurachem, ISBN 0-948926-08-2
5. Pauwels J, Lamberty A, Schimmel H, Homogeneity testing of reference materials, Accred Qual Assur 2:51–55
6. Ingamells CO, Switzer P (1973) Talanta 20:547–568
7. Pauwels J, Vandecasteele C (1993) Fres J Anal Chem 345:121–123
8. Pauwels J, Lamberty A, Schimmel H, Quantification of the expected shelf-life of certified reference materials, Fres J Anal Chem (accepted)

Accred Qual Assur (2000) 5:95–99

Jean Pauwels
Adriaan van der Veen
Andrée Lamberty
Heinz Schimmel

Evaluation of uncertainty of reference materials

Presented at: EURACHEM Workshop on Efficient Methodology for the Evaluation of Uncertainty in Analytical Chemistry, Helsinki, Finland 14–15 June 1999

J. Pauwels (✉) · A. Lamberty
H. Schimmel
Institute for Reference Materials and Measurements, EC-JRC-IRMM,
2440 Geel, Belgium
e-mail: jean.pauwels@irmm.jrc.be
Tel.: +32-14-571722
Fax: +32-14-590406

A. van der Veen
Nederlands Meetinstituut, P.O. Box 654,
2600 AR Delft, The Netherlands

Abstract Certification of reference materials is far more than just characterisation of a selected homogeneous batch of material. From the perspective of the ISO Guide on the Expression of Uncertainty in Measurement (GUM) all uncertainty sources relevant to the user of an individual certified reference material (CRM) sample at a moment in time should be part of the CRM uncertainty. This not only includes the full uncertainty of the batch characterisation (rather than the statistical variation), but also all uncertainties related to possible between-bottle variation, instability upon long-term storage and instability during transport to the customer.

Key words Certified reference materials · Uncertainty · Characterisation · Uncertainty analysis

Introduction

The accurate and traceable determination of a mean value of a quantity (content, amount) in a sample or a batch of material can be obtained in various ways, such as carrying out a number of independent repetitions using a primary method of analysis [1], comparing the results of a limited number of reference methods, or comparing the results of various independent methods applied in a series of laboratories. These three different methods are used by various producers to certify the values assigned to their reference materials (RMs), whereby this assignment is done using quite *similar statements*, but these statements may sometimes have very *different meanings*. Moreover, it must also be realised that the certification of a RM is much more than just carrying out a series of precise and accurate measurements traceable to the SI or to any other system of units, to written or agreed standards or to an artefact, such as, e.g. the primary WHO materials to which several clinical RMs are traceable. The certification of a RM involves, in the first instance, the preparation of a larger number of homogeneous, stable and adequately packaged samples which are all representative of the complete batch, as well as the proper assessment of their homogeneity and stability. Ignoring this is not only one of the main reasons why problems occur with certified reference materials (CRMs), but also why they are the subject of needless discussions about primary, secondary, consensus, working, etc. RMs. This *distiction* in classes of RMs mainly exists in the mind of some metrologists, but is fully absent in the existing ISO-REMCO Guides. The latter only differentiate between (just) RMs and CRMs, whereby a RM is defined as “a material or substance one or more of whose property values are sufficiently homogeneous and well established to be used for the calibration of an apparatus, the assessment of a measurement method, or for assigning values to materials", and a CRM is just "a RM with a certificate in which the certified values are accompanied by an uncertainty at a stated level of confidence" [2].

What is a CRM user interested in?

CRMs are sometimes forced into a hierarchical system depending on the fact that there certified values were determined using a primary method of analysis or are based on "less traceable" measurements obtained in a laboratory intercomparison. In reality, such a differentiation is meaningless, considering that very often the uncertainty component which originates from the characterisation of the RM is dominated by uncertainty components originating from several other sources such as insufficient guarantee of absence of inhomogeneity and/or instability. Therefore, it is not correct when producers certify their RMs just considering the results of their accurate and traceable determinations of the mean value of the content of the *CRM batch*, knowing that their customers (users) are only interested in the mean value of *the single bottle* they ordered on condition that it is received on the day of dispatch.

The ongoing revision of the ISO-Guide 35 [3] – which constitutes a complete rewriting – is therefore a unique opportunity to reconsider the production of CRMs. It will consider production as an integrated process of correct preparation, *positive demonstration* of homogeneity and stability, and accurate and traceable characterisation, and thus of full implementation of the principles laid down in the Guide to the Expression of Uncertainty in Measurement (GUM) [4]. This means that *all* components of uncertainty of "the sample on the desk of the user" should be properly evaluated and accounted for. Thereby, it must be strongly emphasised that the inability to demonstrate between bottle variation or instability during storage or transportation, as well as confining the uncertainty of the batch characterisation to the statistical between-laboratory variation is no longer acceptable. Ignoring this is one of the major causes of the so-called "Jorhem paradox" discussed at BERM-7 [5] where it was (rightly) found unacceptable that "results found to be unacceptable for user laboratories are good enough to be used in the certification of the CRM", even if it is statistically just logic [6]! The consequence is, however, that one will have to accept – just as was the case for testing laboratories introducing GUM – that uncertainties of CRMs will increase "from fiction to reality": an idea which is apparently difficult for many analysts to become accustom to, and which, moreover, may confuse those who tend to compare the quality of the CRMs of various producers just on the basis of the *quoted* uncertainty.

Uncertainty analysis in the preparation of a CRM

From the reasoning given above, it becomes apparent that the certification of a RM includes far more than just the characterisation of the material. This step, often carried out as a collaborative study between multiple laboratories, is crucial for the quality of the material as a CRM, but it is generally insufficient.

From the perspective of EURACHEM Guide [7] as well as from GUM [4], a producer should include all uncertainty sources that are relevant to the package sold to the customer. Internal consistency of the uncertainty analysis requires the inclusion of the (residual) uncertainty from the experiments carried our for homogeneity and stability testing. So, even if the producer cannot demonstrate any inhomogeneity or instability, there is still a (small) uncertainty budget to be included. Usually, this budget will be small, but in cases where only poorly repeatable methods of measurement are available, this contribution may be of significance.

A further consequence of this is that it really "pays off" in terms of uncertainty if a sufficient number of replicate measurements is carried out in homogeneity and stability testing. The use of methods with good linearity, selectivity and repeatability will also greatly contribute to reducing the uncertainty from these experiments. These factors are all in the hands of the producer. Implementing them correctly and consistently will reduce the costs of "after sales" of a CRM producer, not to speak of the subsequent damage due to wrongly certified RMs.

This way of thinking may seem new, but those who have already gained experience with inhomogeneous and/or instable RMs have already developed ways to deal with these aspects. A CRM producer should include in an uncertainty statement everything that "reasonably attributes" (GUM) to the uncertainty of the measurand, i.e. the property value to be certified. This ends where accidents and incidents start: if something happens to a CRM during transport that goes beyond what can be foreseen, it is not part of an uncertainty statement, as the information on the certificate will stipulate under what conditions the certificate (and the CRM) are valid.

What is important in the preparation of a CRM?

Good measurements carried out on bad quality candidate RMs are a nonsense and a complete waste of time and money! Therefore, extreme care should be taken not only to prepare a stable and homogeneous base material, but also to sample it in a tight and inert containment [8]. Matrix CRMs require in general to be clean and dry, to be transformed into an optimal physical and chemical form, and to be stored at the correct temperature from a very early stage in the production process. In general, microbiological degradation can be minimised by reducing the water content of the material to a level between 1 and 3%. Packaging is best carried out in an atmosphere of argon – not under vacuum as this may become a source of leaks – whereby all precautions must be taken to guarantee absolute tightness. This can be achieved using bottles with inserts, penicillin vials or ampoules, whereby it must be stressed that all three solutions have failed in the past: bottles and vials due to insufficiently tight or retracting inserts (e.g. due to ageing or freeze-temperature effects) or ampoules due to cracks appearing during storage as a consequence of stresses present in the glass.

What is important in homogeneity testing?

Homogeneity testing addresses a double problem: What is the variation in mean value which exists between the various units of a batch of candidate RM? And, how inhomogeneous is the material contained in a bottle?

The first problem is of utmost importance to the user as he/she will, in general, buy just one bottle, and will not care about the other ones! Therefore, between-units variation is an important component of uncertainty which *must* be included in the certified value of the CRM. The determination of the between-units variation is carried out by measuring the value of a significant number of units. As the result of such measurements is a combination of two effects, the between-bottle variability [s_{bb}] and the measurement repeatability [s_{meas}]

$$u_{bb}^2 = s_{meas}^2 + s_{bb}^2 \quad (1)$$

the variation between the mean value of the bottles can only be obtained from *measurements* carried out with the highest repeatability: i.e. that each bottle must be analysed, using a highly repeatable method, on sample intakes of optimal size and carrying out a number of repetitions which is sufficient to obtain a measurement uncertainty which is negligible compared to the variation between the bottles, i.e. $s_{meas}^2 < < s_{bb}^2$. Usually, this is however not the case. Then, u_{bb}^2 should, as far as possible, be corrected for s_{meas}^2 to obtain the best estimate of s_{bb}^2 [9].

To evaluate the inhomogeneity of the material contained in a bottle, within-bottle measurements have to be carried out. Also here, the result of such measurements is a combination of two effects, the within-bottle inhomogeneity [s_{inh}] and the *method* repeatability [s_{meth}]

$$u_{inh}^2 = s_{meth}^2 + s_{inh}^2 \quad (2)$$

and the variation between the different samples within a bottle can only be obtained from measurements carried out using a highly repeatable method so that the method repeatability is negligible compared to the variation between the samples in a bottle, i.e. $s_{meth}^2 \ll s_{inh}^2$. In this case, sample intakes must however be minimal, as the contribution of s_{meth}^2 to u_{inh}^2 becomes negligible when extrapolating s_{inh}^2 from smaller (m) to larger (M) sample sizes according to:

$$[s_{inh}^2]_M = [s_{inh}^2]_m \cdot m/M \cong [u_{inh}^2]_m \cdot m/M \quad (3)$$

It must be emphasised that s_{inh} is irrelevant for the CRM uncertainty, provided the *minimum representative sample intake* is properly determined. The value of s_{inh} is, however, of prime importance to estimate this minimum representative sample intake correctly [10].

In both cases it should be noted that:

- Not correcting u_{bb} or u_{inh} for s_{meas} or s_{meth} is not really a problem, but leads to (too) conservative CRM uncertainty estimates.
- Corrected s_{bb} or s_{inh} values may never be taken smaller than their respective combined uncertainties, i.e. $u(s_{bb})$ and $u(s_{inh})$ [9].

What to do with stability data?

Stability testing at higher temperatures simulating possible transport conditions and conditions of long-term storage are often part of procedures describing the production of CRMs [11]. In most cases they do, however, not give quantitative information on presumed instability, mainly as a consequence of insufficient measurement reproducibility and of an insufficient number of replicates. With the upcoming requirements of fixing expiry dates [12], it will be mandatory that not only quantitative data be available, but that their quality is such that high precision extrapolations can be made. This requires however that data are produced with measurement reproducibilities (or repeatabilities when isochronous measurements are carried out [13]) which are negligible compared to the certified uncertainty. An extrapolation method was recently proposed by Pauwels et al. [14] to determine the time for which the certified value of a CRM remains valid, based on the determination of the intersection of the lower 95% confidence bound with the lower limit of the certified confidence interval (see Fig. 1). Such calculations show however that, with the levels of uncertainty presently certifed, either unrealistically high precisions are required, or that shelf-lifes must be reduced to unrealistically short periods of time, even if one considers that further stability monitoring during the lifetime of the CRM makes regular re-evaluation and updating of the shelf-life possible. Therefore, in many cases, it may become necessary to re-evaluate the certified uncertainties of

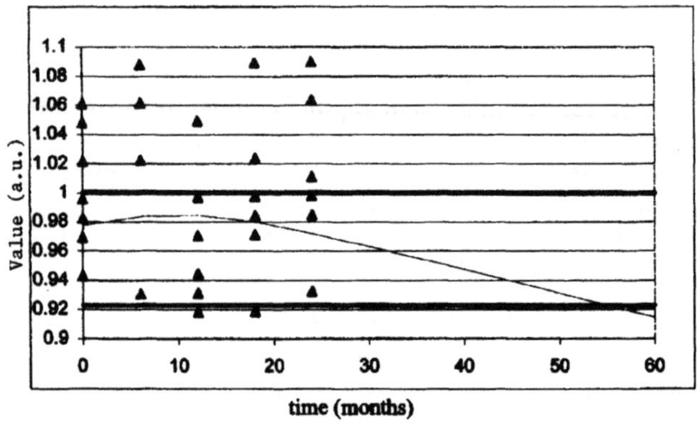

Fig. 1 Example of determination of the long-term stability of certified reference material (CRM): Cr in CRM 278R (mussel tissue)

RMs taking into account a realistic *stability uncertainty*. Possibly, other approaches may be found to solve this extremely important problem, such as the one proposed by a group of experts working in the framework of a "Standards, Measurements and Testing Accompanying Measure" under the co-ordination of LGC (s. Burke, personal communication), consisting in extrapolating the certified value to mid-way of an arbitrarily chosen life-time and calculating the associated supplemental uncertainty.

A similar reasoning may be appropriate for possible degradation of the CRM during transportation to the customer.

The characterisation of a homogenous batch of material

The estimation of the mean value of a quantity of a CRM batch using: (1) a primary method of analysis, or (2) by comparing the results of a limited number of reference methods, or (3) the results of various independent methods applied in a series of laboratories should, in fact, only be variants of one and the same philosophy. The third characterisation method, however, requires that a number of analyses are carried out by one or more techniques in one or more laboratories, whereby *each series of measurements is carried out with maximal guarantees of accuracy and traceability, and must be documented by a full uncertainty budget*. For each set of determinations an expanded standard uncertainty according to GUM should then be calculated. The final estimation of the uncertainty of the characterisation of the batch (u_{char}) should then take into account all these standard uncertainties, considering that those uncertainties which have been repeatedly determined in an independent way, decrease proportionally with the square root of the number of degrees of freedom. A proposal to handle this problem was published by Pauwels et al. [15]. It is based on a separate consideration of three types of standard uncertainties:

- Those which are exclusively laboratory dependent.
- Those which are common to all laboratories, such as the effect of between-bottle variation or the use of a common calibrant.
- Those which are common to groups of laboratories, e.g. those using the same measurement procedure.

In this context it should be noted that matrix CRMs are generally certified for mass fractions related to dry matter, i.e. that not only the amount of substance but also the dry sample mass has to be assessed and its uncertainty evaluated: a problem that is ignored and/or underestimated by many analytical chemists and a potential source of significant errors and unaccounted uncertainties in CRMs.

The CRM uncertainty according to GUM

The final uncertainty of a CRM according to GUM should consider all sources of uncertainty described above:

$$u_{\mathrm{CRM}} = [u_{\mathrm{char}}^2 + u_{\mathrm{bb}}^2 + u_{\mathrm{lts}}^2 + u_{\mathrm{sts}}^2]^{1/2}, \quad (4)$$

whereby lts and sts refer to long-term stability (upon storage) and short-term stability (during transport), respectively.

It is good practice to quantitatively determine *all* sources of uncertainty, be they significant or not. In the latter case they will anyhow disappear in the rounding-off of the calculation, but it will:

- Avoid the risk of overlooking sources of uncertainty due to ignorance.
- Demonstrate to users that they have been considered and what is their magnitude.

References

1. Quinn TJ (1997) Metrologia 34:61–65
2. ISO Guide 30 (1981) Terms and definitions used in connection with reference materials. ISO, Geneva, Switzerland
3. ISO Guide 35 (1989) Certification of reference materials – General and statistical principles. ISO, Geneva, Switzerland
4. ISO (1995) Guide to the expression of uncertainty in measurement. ISO, Geneva, ISBN 92-67-10188-9
5. Jorhem L (1998) Fresenius J Anal Chem 306:370 373

6. Pauwels J (1999) In: Fajgeli A, Parkany M (eds) The use of matrix reference materials in environmental analytical processes. The Royal Chemical Society, London, pp 31–45
7. EURACHEM (1995) Quantifying uncertainty in analytical measurement. EURACHEM, London, ISBN 0-948926-08-2
8. Kramer GN, Pauwels J (1996) Mikrochim Acta 123:87 –93
9. Pauwels J, Lamberty A, Schimmel H (1998) Accred Qual Assur 3:51–55
10. Pauwels J, Vandecasteele C (1993) Fresenius J Anal Chem 345:121–123
11. European Commission: DG XII-C-5 – document BCR/01/97 (1997) Guidelines for the production and certification of BCR reference materials. European Commission, Brussels
12. ISO Guide 31 (1998) Reference materials – Contents of certificates and labels (draft). ISO, Geneva, Switzerland
13. Lamberty A, Schimmel H, Pauwels J (1998) Fresenius J Anal Chem 360:359–361
14. Pauwels J, Lamberty A, Schimmel H (1998) Fresenius J Anal Chem 361:395–399
15. Pauwels J, Lamberty A, Schimmel H (1998) Accred Qual Assur 3:180–184

Accred Qual Assur (2002) 7:90–94
DOI 10.1007/s00769-001-0434-y

Alicia Maroto
Ricard Boqué
Jordi Riu
F. Xavier Rius

Should non-significant bias be included in the uncertainty budget?

A. Maroto (✉) · R.Boqué · J. Riu
F. X. Rius
Department of Analytical
and Organic Chemistry,
Institute of Advanced Studies,
Rovira i Virgili University of Tarragona,
Pl. Imperial Tàrraco, 1, 43005 Tarragona,
Catalonia, Spain.
e-mail: maroto@quimica.urv.es
Tel.: +34-977-558187
Fax: +34-977-559563

Abstract The bias of an analytical procedure is calculated in the assessment of trueness. If this experimental bias is not significant, we assume that the procedure is unbiased and, consequently, the results obtained with this procedure are not corrected for this bias. However, when assessing trueness there is always a probability of incorrectly concluding that the experimental bias is not significant. Therefore, non-significant experimental bias should be included as a component of uncertainty. In this paper, we have studied if it is always necessary to include this term and which is the best approach to include this bias in the uncertainty budget. To answer these questions, we have used the Monte-Carlo method to simulate the assessment of trueness of biased procedures and the future results these procedures provide. The results show that non-significant experimental bias should be included as a component of uncertainty when the uncertainty of this bias represents at least a 30% of the overall uncertainty.

Keywords Bias · Uncertainty · Assessment of trueness

Introduction

One of the most important steps in the validation of an analytical procedure is the assessment of trueness. In this process, the experimental bias of the analytical procedure is estimated. If this bias is statistically not significant, we assume that the procedure is unbiased and, consequently, results are not corrected for the experimental bias. However, can we be sure that the procedure does not have any bias? In fact, when assessing trueness, there is always a probability of incorrectly concluding that the experimental bias is statistically not significant. As a result, this probability should be included (expressed as a quantity related to the experimental bias) as a component of the uncertainty of the results obtained with the analytical procedure. However, several questions arise, i.e. is it always necessary to include this component of the uncertainty? Moreover, if it is necessary, how should this non-significant experimental bias be included?

Non-significant experimental bias has not been included so far as a component of uncertainty in chemical measurements. However, different approaches have been proposed to include bias as a component of uncertainty when physical measurements are not corrected for systematic errors [1]. In this paper we study whether these approaches can be applied to include non-significant experimental bias in the uncertainty budget of chemical measurements and whether it is always necessary to include this term. To answer these questions, we have simulated the process of assessment of trueness of biased analytical procedures and, subsequently, the future results these procedures provide. We simulated these results covering most of the possible situations that may happen in practice.

Assessment of trueness

Checking the trueness of an analytical procedure involves estimating its experimental bias. If the routine samples have similar levels of concentration, we can assume that we have the same bias in the whole concentration range and, consequently, the experimental bias can be estimated using one reference sample with a concentration similar to the routine samples. If this is the case, the experimental bias is calculated as the difference between the reference value, c_{ref}, and the mean value, $bias = c_{ref} - \bar{c}_{found}$. The experimental bias is not significant if:

$$bias \leq t_{\alpha/2,eff} \cdot u(bias) \qquad (1)$$

where $t_{\alpha/2,\ eff}$ is the two-sided t tabulated value for the effective degrees of freedom, ν_{eff}, [2] associated with $u(bias)$, and can be replaced by the coverage factor k if the effective degrees of freedom are large enough [3, 4]. The uncertainty of the experimental bias, $u(bias)$, depends on the reference used to assess trueness. If a certified reference material (CRM) is used, this uncertainty is calculated as:

$$u(bias) = \sqrt{\frac{s_I^2}{p} + u(c_{ref})^2} \qquad (2)$$

where s_I is the standard deviation of the p results obtained when analysing the CRM and $u(c_{ref})$ is the standard uncertainty of the CRM (i.e. $U(c_{ref})/k$, where k is normally equal to 2 and $U(c_{ref})$ is the uncertainty of the CRM provided by the manufacturer).

If the experimental bias is significant, the procedure should subsequently be revised in order to identify and eliminate the systematic errors which produced the bias. Otherwise, we assume that the procedure is unbiased and, consequently, we do not correct results for the experimental bias. However, several questions arise in this latter case because, from a chemical point of view, some bias is always to be expected in an analytical procedure.

Calculation of uncertainty

Uncertainty can be obtained either by calculating all the sources of uncertainty individually [3, 4] or by grouping different sources of uncertainty whenever possible [5–9]. In this paper, the latter strategy is followed to calculate uncertainty using information obtained in the process of assessment of trueness [5, 6]:

$$u = \sqrt{u(\text{proc})^2 + u(\text{trueness})^2 + u(\text{pret})^2 + u(\text{other therms})^2} \qquad (3)$$

where u is the standard uncertainty [3], u(proc) is the uncertainty of the procedure and corresponds to the intermediate standard deviation of the procedure, i.e. $u(\text{proc}) = s_I$. u(trueness) is the uncertainty of the experimental bias and corresponds to $u(bias)$. u(pret) is the uncertainty associated to subsampling and to sample pretreatments not considered in the assessment of trueness. Finally, u(other terms) considers other terms of uncertainty due to factors not representatively varied when estimating precision. In this paper, this latter term will be considered to be negligible. The overall expanded uncertainty, U, is then calculated by multiplying the standard uncertainty, u, by the two-sided t tabulated value, $t_{\alpha/2,\ eff}$, for the effective degrees of freedom, ν_{eff} [2], i.e. $U = t_{\alpha/2,\ eff} \cdot u$. A coverage factor of $k=2$ is recommended for most purposes when the effective degrees of freedom, ν_{eff}, are large enough. This value represents a level of confidence of approximately 95%. Strictly, the uncertainty calculated in Eq. 3 corresponds to results of future samples obtained after correcting the concentration found for the experimental bias. However, analytical results are never corrected for non-significant experimental bias. As a result, this bias should be included as a component of uncertainty because the procedure may have a true bias.

Approaches for including non-significant experimental bias in the uncertainty budget

Different approaches have been proposed in the field of physical measurements to include bias as a component of uncertainty when results are not corrected for systematic errors [1]. In this paper, we will study whether these approaches can be applied to include non-significant experimental bias in chemical measurements. The first approach consists of including this bias as another component of uncertainty and simply to add it in the usual root-sum-of-squares (RSS) manner, i.e.

$$U(RSSu) = t_{\alpha/2,eff} \cdot \sqrt{u^2 + bias^2}\,.$$

The second approach sums this bias in a RSS manner with the expanded uncertainty, U, i.e.

$$U(RSSu) = \sqrt{U^2 + bias^2}\,.$$

The third procedure consists of adding this bias to the expanded uncertainty. This approach is denoted as SUMU [1] and is equivalent to correcting the results:

$$U_+ = U + bias; \quad U_- = U - bias. \qquad (4)$$

Finally, the last procedure to be studied consists of adding the absolute value of the experimental bias to the expanded uncertainty, i.e.

$$U(bias) = U + |bias|.$$

Numerical example: bias in the assessment of trueness

To investigate the effects of including the non-significant experimental bias as a component of uncertainty, the

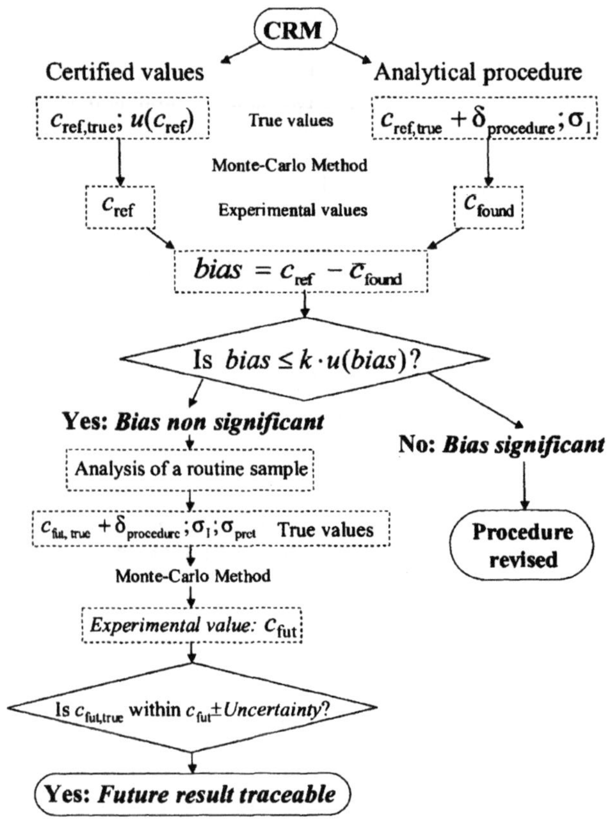

Fig. 1 Scheme of the process of assessment of trueness simulated with the Monte-Carlo method. Future results of routine sample were simulated if the experimental bias was identified as non significant

process of assessment of trueness was simulated with the Monte-Carlo method (see Fig. 1). A true reference value, $c_{ref,\ true}$, together with its standard uncertainty, $u(c_{ref})$, was assigned to a "hypothetical CRM". A true bias, $\delta_{procedure}$, together with a true intermediate standard deviation, σ_I, was associated to the analytical procedure. Moreover, the possibility of having other steps in the analytical procedure (i.e. pretreatments and/or subsampling) not carried out when analysing the CRM was also studied. In this case, an additional true standard deviation, σ_{pret}, was also added to the future results obtained with the analytical procedure.

In the assessment of trueness, we simulated the certified reference value, c_{ref}, and the *p* results obtained when analysing the CRM. The true value and the standard deviation used to simulate the *p* results were, respectively, $c_{ref,\ true}+\delta_{procedure}$ and σ_I. Results were simulated assuming that they followed a normal distribution. After this, we calculated the mean of these results, $\bar{c}_{found}$, and the experimental bias. The uncertainty of this bias, *u*(*bias*), was calculated with Eq. 2. Afterwards, it was checked

Table 1 Three different situations simulated in the assessment of trueness

	p	$u(c_{ref})$	σ_I	σ_{pret}	u	$u(bias)$	$\frac{u(bias)}{u} \cdot 100\%$
Case 1	5	0.2	1.5	0	1.64	0.70	43
Case 2	10	0.2	1.5	0	1.59	0.51	32
Case 3	10	0.2	1.5	3	3.39	0.51	15

with Eq. 1 whether this bias was statistically significant or not. If it was not significant, a future concentration obtained for a routine sample, c_{fut}, was simulated with the Monte-Carlo method. The true value and the true variance used to simulate this result were, respectively, $c_{fut,\ true}$ and σ^2_{fut}. The variance, σ^2_{fut}, corresponded to $\sigma^2_I+\sigma^2_{pret}$. After this, we calculated the uncertainty of this future result following the section on calculating uncertainty. The experimental bias was included as a component of uncertainty using the four approaches explained in the section following that one. Then we checked whether these approaches included the true concentration of the routine sample, $c_{fut,\ true}$, within the interval $c_{fut} \pm Uncertainty$.

This process was simulated 300,000 times for 25 different values of $\delta_{procedure}$. After this, we calculated the percentage of times that the experimental bias was found to be non-significant. This corresponded to the probability of β error (or probability of false negative) because in the assessment of trueness we state that the method is unbiased when in fact is biased. We also calculated the percentage of times that, once the experimental bias was identified as non-significant, the different approaches studied included the true concentration of the routine sample, $c_{fut,\ true}$, within the interval $c_{fut} \pm Uncertainty$. We simulated this process for three different cases which cover different situations that may happen in practice (i.e. presence/absence of pretreatment steps and different number of replicated analysis of the CRM). Table 1 shows, for the three cases, the values of σ_I, σ_{pret}, $u(c_{ref})$ and *p* used for simulating the results.

Results and discussion

Table 2 shows the probability of β error committed for different values of the true bias of the analytical procedure, $\delta_{procedure}$, for the three cases studied. We can see that the higher is $\delta_{procedure}$, the lower is the probability of β error. This is because the higher is the true bias, the more likely is to detect that the procedure is biased. The probability of β error depends also on σ_I, $u(c_{ref})$ and *p*. We can see that for the same values of σ_I, $u(c_{ref})$ and $\delta_{procedure}$, the lower is *p*, the higher is the probability of β error.

We also studied the percentage of times that, once the experimental bias was identified as being non-significant, the different ways of calculating uncertainty includ-

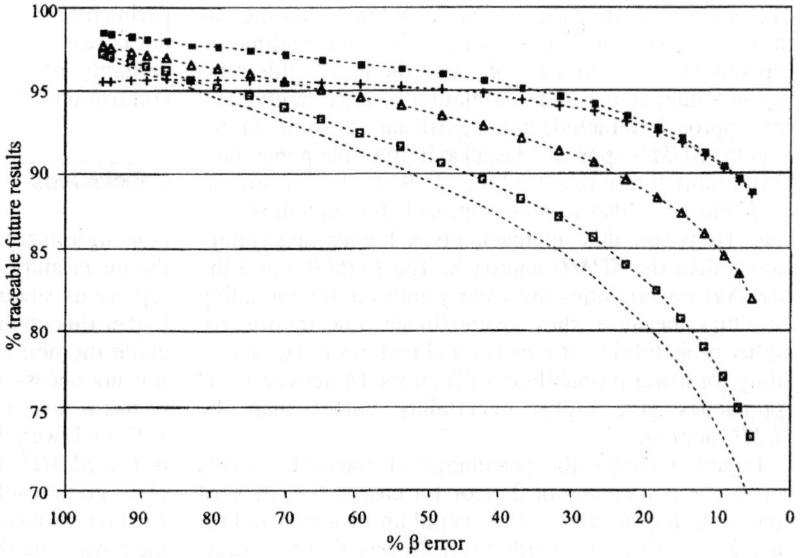

Fig. 2 Percentage of traceable future results for case 1 versus the probability of β error. Uncertainty is calculated without including non-significant experimental bias: ---- U and its inclusion using the four approaches described for including non-significant experimental bias in the uncertainty budget: --- △ $U(RSSu)$; --- ⊟ $U(RSSU)$; ---+---*SUMU* and ---■--- $U(bias)$

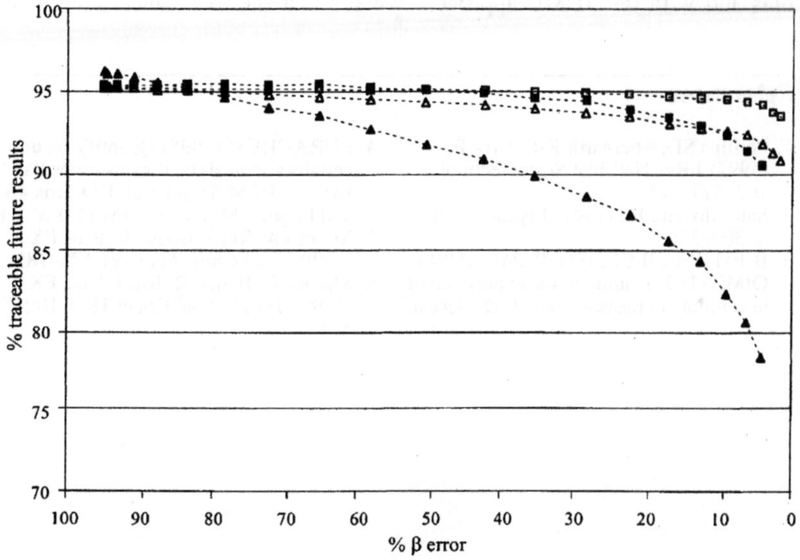

Fig. 3 Percentage of traceable future results for cases 2 and 3 versus the probability of β error. Uncertainty is calculated without including the experimental bias, U, (case 2: ---▲--- and case 3 --- △) and with the *SUMU* approach (case 2 ---■--- and case 3 --- ⊟)

Table 2 True bias of the analytical procedure, $\delta_{procedure}$, and percentage of times that the experimental bias is identified as non-significant, i.e. % β error, for the three cases described in Table 1

Case 1		Cases 2 and 3	
$\delta_{procedure}$	% β error	$\delta_{procedure}$	% β error
0	95	0	95
0.4	91	0.4	88
0.8	79	0.8	66
1.2	60	1.2	36
1.6	37	1.4	22
2	18	1.6	13
2.4	7	1.8	6

ed the true concentration of the routine sample, $c_{fut,\,true}$, within the interval c_{fut}±*Uncertainty*. Uncertainty was calculated using the z-value for a level of significance α=5%. Therefore, if the uncertainty is correctly calculated, this percentage, i.e. % traceable future results, should be 95% (i.e. 100-α%). If uncertainty is underestimated, this percentage is lower than 95% and, if it is overestimated, it is higher than 95%. This percentage was calculated and plotted as a function of the β error committed in the assessment of trueness.

Figure 2 shows these results for case 1. In this case the contribution of $u(bias)$ to the overall uncertainty, $u(bias)/u$, is 43%. We see that uncertainty can be greatly underestimated when the experimental bias is not included as a component of uncertainty. This underestimation depends on the ratio *u*(*bias*)/*u*, i.e.: the higher this ratio is, the higher is the underestimation of uncertainty. The best approach to include non-significant experimental bias is the *SUMU* approach because it gives the percentage of traceable future results closest to 95%. The uncertainty, *U*(*bias*), is also a good approach for including this bias. However, this approach gives higher uncertainty values than the *SUMU* approach. The *U*(*RSSU*) and the *U*(*RSSu*) uncertainties are clearly inferior for including this bias because they overestimate uncertainty for higher probabilities of β errors and underestimate uncertainty for lower probabilities of β errors. Moreover, these approaches give higher uncertainty values than the *SUMU* approach.

Figure 3 shows the percentage of traceable results versus the percentage of β error for case 2 (i.e. 32% of contribution of *u*(*bias*) to the overall uncertainty) and for case 3 (i.e. 15% of contribution). In this Figure, uncertainty is calculated without including the experimental bias and with the *SUMU* approach. We see that the experimental bias should be included in case 2 but this is not necessary in case 3. This is because in case 3 the uncertainty of the experimental bias is negligible when compared to the overall uncertainty.

Conclusions

Non-significant experimental bias should be included in the uncertainty budget when the uncertainty of this bias represents about 30% of the overall uncertainty. The higher this contribution is, the more important it is to include the non-significant experimental bias. In contrast, it is not necessary to include this bias when its uncertainty has a low contribution to the overall uncertainty, i.e. 15% or lower. The best approach for including this bias is the *SUMU* approach. The uncertainty, *U*(*bias*), also gives good results. Otherwise, we can use the uncertainty *U*(*bias*) because, opposite to the *SUMU* approach, it has the advantage that it gives a symmetric confidence interval around the estimated result. However, it gives higher uncertainty values than the *SUMU* approach.

References

1. Phillips SD, Eberhardt KR, Parry B (1997) J Res Natl Inst Stand Technol 102: 577–585
2. Satterthwaite FE (1941) Psychometrika 6: 309–316
3. BIPM, IEC, IFCC, ISO, IUPAC, IUPAP, OIML (1993) Guide to the expression of uncertainty in measurement, ISO, Geneva
4. EURACHEM (1995) Quantifying uncertainty in analytical measurements, EURACHEM Secretariat, P.O. Box 46, Teddington, Middlesex, TW11 0LY, UK
5. Maroto A, Riu J, Boqué R, Rius FX (1999) Anal Chim Acta 391:173–185
6. Maroto A, Boqué R, Riu J, Rius FX (1999) Trends Anal Chem 18/9–10:577–584
7. Ellison SLR, Williams A (1998) Accred Qual Assur 3:6–10
8. Barwick VJ, Ellison SLR (2000) Accred Qual Assur 5:47–53
9. EURACHEM/CITAC Guide (2000) Quantifying uncertainty in analytical measurement, EURACHEM, 2nd Edition. Helsinki

Accred Qual Assur (2002) 7:269–273
DOI 10.1007/s00769-002-0485-8

Lutz Brüggemann
Rainer Wennrich

Evaluation of measurement uncertainty for analytical procedures using a linear calibration function

L. Brüggemann (✉) · R. Wennrich
UFZ Centre for Environmental Research Leipzig-Halle,
Department of Analytical Chemistry,
Permoserstrasse 15, 04318 Leipzig,
Germany
e-mail: bruegge@ana.ufz.de
Tel.: +49-341-2352512
Fax: +49-341-2352625

Abstract In the EURACHEM/CITAC draft "Quantifying uncertainty in analytical measurement" estimations of measurement uncertainty in analytical results for linear calibration are given. In this work these estimations are compared, i.e. the uncertainty deduced from repeated observations of the sample vs. the uncertainty deduced from the standard residual deviation of the regression. As a result of this study it is shown that an uncertainty estimation based on repeated observations can give more realistic values if the condition of variance homogeneity is not correctly fulfilled in the calibration range. The complete calculation of measurement uncertainty including assessment of trueness is represented by an example concerning the determination of zinc in sediment samples using ICP-atomic emission spectrometry.

Keywords Measurement uncertainty · Linear calibration · Method validation

Introduction

For some years international guidelines and recommendations [1, 2] have existed for the evaluation of measurement uncertainty of analytical procedures. In the draft of the EURACHEM/CITAC Guide [2] two formulas for estimating uncertainty are indicated relating to analytical procedures with linear calibration. Application of these different formulas is only briefly described in the Guide. The aim of this study was to discuss the use of these estimations for the purpose of method validation. Moreover, the calculation of measurement uncertainty in a closed form using an analytical task is demonstrated.

The linear regression model

$$y = b_0 + b_1 x \tag{1}$$

with the response y of the measuring system to the analyte content x of the measured component is applied. The regression coefficients b_0 and b_1 are estimated from the calibration data set $\{x_i, y_i\}$ and the (predicted) content is calculated from the observed response with the help of the inverse function

$$x_{pred} = \frac{y_{obs} - b_0}{b_1}. \tag{2}$$

The analytical procedure (sampling preparation, measuring process) is regarded as a whole. On condition that the calibration is statistically justified [3] (in particular, the variance homogeneity must be within the calibration range and the regression model must be suitable) and that the trueness (referring to the ISO 5725 standard [4]) of the process can be checked from recovery of the analyte content of a reference material, the measurement uncertainty belonging to x_{pred} can be determined.

The two estimations of the variance of x_{pred}, indicated in [2], are calculated from different inputs. For given variances/covariances of the inputs y_{obs}, b_0 and b_1, the variance of x_{pred} is estimated by

$$\mathrm{var}(x_{pred}) = \frac{1}{b_1^2}(\mathrm{var}(y_{obs}) + x_{pred}^2 \,\mathrm{var}(b_1) + 2 \cdot x_{pred} \,\mathrm{cov}(b_0, b_1) + \mathrm{var}(b_1)) \tag{3}$$

(Eq. 3 marked in [2] as E3.3). A second estimation

$$\operatorname{var}(x_{pred}) = \frac{s_{y,x}^2}{b_1^2}\left(\frac{1}{p}+\frac{1}{n}+\frac{\left(x_{pred}-\overline{x_{cal}}\right)}{\sum_i\left(x_i-\overline{x_{cal}}\right)^2}\right) \tag{4}$$

is based on the calibration data (Eq. 4, E3.5 in [2]). It concerns the well-known formula for the variance estimation of an average value predicted when p repetitions (i number of calibration levels, *cal* symbolizes here that the average values belong to the calibration data set, and $s_{y.x}^2$ defines the residual variance of the regression model).

Comparison of the estimations of measurement uncertainty

With the help of the relations $\operatorname{var}(b_1) = \frac{s_{y.x}^2}{\sum_i\left(x_i-\overline{x_{cal}}\right)^2}$ and $\operatorname{var}(b_0) = s_{y,x}^2\left(\frac{1}{n}+\frac{\overline{x_{cal}}^2}{\sum_i\left(x_i-\overline{x_{cal}}\right)^2}\right)$ Eqs. (3) and (4) are identical if $s_{y.x}^2/p$ from Eq. (4) is used as an estimate of $\operatorname{var}(y_{obs})$ in Eq. (3) or vice versa.

Sometimes the available analytical quality assurance software gives different estimates of the input values of Eqs. (3) or (4). If no estimate for the covariance term in Eq. (3) is available, this term is replaced using the expression $\operatorname{cov}(b_0, b_1) = \overline{x_{cal}}\operatorname{var}(b_1)$ valid for this linear regression (derived from the rules of covariance algebra, see for instance [5]), thus

$$\operatorname{var}\left(x_{pred}\right) = \frac{1}{b_1^2}\Big(\operatorname{var}(y_{obs}) + \left(x_{pred}^2 - 2\overline{x_{cal}}x_{pred}\right)\operatorname{var}(b_1) + \operatorname{var}(b_0)\Big). \tag{5}$$

The relation

$$b_0 = \overline{y_{cal}} - b_1 \cdot \overline{x_{cal}} \tag{6}$$

applies to the regression parameters. On the assumption that the variance of calibration contents can be neglected, according to Eq. (6) one obtains

$$\operatorname{var}(b_0) = \operatorname{var}\left(\overline{y_{cal}}\right) + \overline{x_{cal}}^2 \operatorname{var}(b_1) \tag{7}$$

as well as the following equations for Eqs. (1), (2) and (3), respectively:

$$y = \overline{y_{cal}} + b_1\left(x - \overline{x_{cal}}\right) \tag{8}$$

$$x_{pred} = \overline{x_{cal}} + \frac{y_{obs} - \overline{y_{cal}}}{b_1} \tag{9}$$

and

$$\operatorname{var}(x_{pred}) = \frac{1}{b_1^2}\left(\operatorname{var}(y_{obs}) + \left(\frac{y_{obs}-\overline{y_{cal}}}{b_1}\right)^2 \operatorname{var}(b_1) + \operatorname{var}\left(\overline{y_{cal}}\right)\right) \tag{10}$$

According to [1] now the variance estimations of the measurement uncertainty are marked by u^2 and their standard measurement uncertainty by $u=\sqrt{u^2}$. Thus for the estimation of the standard measurement uncertainty of a content x_{pred} one obtains the expressions

$$u(x_{pred}) = \frac{s_{y,x}^2}{b_1}\sqrt{\frac{1}{p}+\frac{1}{n}+\frac{\left(x_{pred}-\overline{x_{cal}}\right)^2}{\sum_i\left(x_i-\overline{x_{cal}}\right)^2}} \tag{11}$$

$$u\left(x_{pred}\right) = \frac{1}{b_1}\sqrt{u^2(y_{obs}) + \left(x_{pred}^2 - 2\overline{x_{cal}}x_{pred}\right)u^2(b_1) + u^2(b_0)} \tag{12}$$

$$u(x_{pred}) = \frac{1}{b_1}\sqrt{u^2(y_{obs}) + \left(\frac{y_{obs}-\overline{y_{cal}}}{b_1}\right)^2 u^2(b_1) + u^2\left(\overline{y_{cal}}\right)} \tag{13}$$

following from Eqs. (4), (5) and (10) (Eq. 11 corresponds to E3.5; Eq. 12 and 13 correspond to E3.3 [2]).

Equations (12) and (13) only formally differ concerning the used regression parameters and supply identical results. Equation (13) has an advantage in comparison to Eq. (12). The individual uncertainty contributions for the sample measurement, the calibration model and the calibration measurement, given by $u_{obs}(x_{pred}) = |1/b_1| u(y_{obs})$; $u_{b_1}(x_{pred}) = |-(y_{obs} - \overline{y_{cal}})/b_1| u(b_1)$ and $u_{cal}(x_{pred}) = |-1/b_1| u(\overline{y_{cal}})$, respectively, referring to the formal representation $u(x_{pred}) = \sqrt{u_{obs}^2(x_{pred}) + u_{b_1}^2(x_{pred}) + u_{cal}^2(x_{pred})}$ of (13), are easy to interpret.

In contrast to Eqs. (12) and (13), Eq. (11) contains the residual standard deviation $s_{y.x}$ of the regression instead of the standard uncertainty $u(y_{obs})$ derived from repeated observations. For $x_{pred} = \overline{x_{cal}}$ Eq. (11) gives a minimum. If $\operatorname{var}(y_{obs}) = s_{y.x}^2/p$, Eqs. (12) and (13) give the same results as Eq. (11). Normally, Eq. (11) is applied for the estimation of measurement uncertainty in the case of linear least squares calibration. In the case of special applications, it can be used as an actual estimate $u(y_{obs})$, or an estimate $u(x_{pred})$ calculated according to Eq. (12) or (13) which gives a more realistic result.

Assessment of trueness

In order to check the trueness of an analytical procedure concerning the analysis of a sample with the content x_s of one component, an additional analytical quality control (AQC) measurement of a reference material with the certified content x_r is to be executed. The determined content x_q (observed from the reference material) is compared with x_r. Two further uncertainties have to be considered: the uncertainty of the AQC measurement $u(x_q)$, based on the standard uncertainty $u(y_q)$ of the appropriate response values, and the uncertainty $u(x_r)$ concerning the content specification of the reference material [6]. With the help of a correction factor f_r, defined by

$$f_r = x_q / x_r \quad (14)$$

and the model equation for the corrected content

$$x_{corr} = x_s / f_r \quad (15)$$

$u(x_q)$ and $u(x_r)$ can be included in the calculation of combined measurement uncertainty.

The standard uncertainties of f_r and x_{corr} are estimated by

$$u(f_r) = f_r \cdot \sqrt{(u(x_q)/x_q)^2 + (u(x_r)/x_r)^2} \quad (16)$$

and

$$u(x_{corr}) = x_{corr} \cdot \sqrt{(u(x_s)/x_s)^2 + (u(f_r)/f_r)^2}, \quad (17)$$

respectively. By multiplication of the combined measurement uncertainty $u(x_{corr})$ with a coverage factor (k=2) the expanded measurement uncertainty

$$U = k \cdot u(x_{corr}) \quad (18)$$

is obtained and reported in the result of the analysis.

Using the test statistic

$$T = \frac{|1 - f_r|}{u(f_r)} \quad (19)$$

based on the t-distribution, it can be proven whether f_r significantly differs from 1. If T is larger than 2 (according to the size of k), an existent method bias is suggested [7]. This method bias should be eliminated in the context of further investigations. If this is not possible, then it must be considered with the calculation of the sample content.

Example

The evaluation of measurement uncertainty is presented on the basis of a calibration data set (Table 1) for the determination of zinc in *aqua regia* extracts of polluted sediment samples. The *aqua regia* extracts were prepared according to DIN ISO 11466 [8]. The concentration of zinc was determined in the diluted (deionized water) extracts by ICP-atomic emission spectrometry with pneumatic nebulization (Spectroflame M/P, Spectro A.I.). The aim of this work was to estimate the applied calibration procedure based on a set of diluted ICP multielemental standards (Merck IV) in 0.1 mol l^{-1} nitric acid for "true" results in the *aqua regia* extracts. The trueness of this analytical procedure should be proved on the basis of SRM 2709, Montana soil (NIST), which was handled as a sample within the procedure. The certified value for zinc (*aqua regia* soluble) is reported to be 100 mg kg^{-1} (range 87–120 mg kg^{-1}). If one considers the conversion factor from the sample preparation, the mean concentration of zinc in the solution amounts to 0.285 mg l^{-1}.

Spread-sheet programs and special software solutions can be used for the necessary calculations. For example the program tool "SQS98" [9] supplies the standard uncertainties $u(b_0)$, $u(b_1)$, and $u(\overline{y_{cal}})$ needed in Eqs. (12) and (13). However, a small side-calculation is necessary (viz. Table 2).

Table 2 Standard uncertainty in the parameters of linear regression

Quantity	Standard uncertainty
$y=b_0+b_1x$	$s_{y.x}=48.33$ Residual standard deviation
$b_0=-42.5$ Intercept	$CI(b_0)=71.16$ $u(b_0)=CI(b_0)/t_{4;5\%}$ $u(b_0)=71.16/2.776=25.63$
$b_1=2.224.9$ Slope (Sensitivity)	$CI(b_1)=31.68$ $u(b_1)=CI(b_1)/t_{4;5\%}$ $u(b_1)=31.68/2.776=11.41$
$\overline{x_{cal}}=1.433$ Abscissa centroid	$u(\overline{x_{cal}})=0$
$\overline{y_{cal}}=3146.6$ Ordinate centroid	$u(\overline{y_{cal}})=\sqrt{u^2(b_2)-\overline{x_{cal}}^2 u^2(b_1)}$ $u(\overline{y_{cal}})=\sqrt{25.63^2-1.433^2 \cdot 11.41^2}$ $u(\overline{y_{cal}})=19.74$

Table 1 Calibration data set for n=6 calibration levels (p=5 repetitions)

Analyte Content[a]	Response from repeated measurements					Mean	SD
x	y1	y2	y3	y4	y5		
0	2.3	6.3	0	4.4	11.8	4.96	4.49
0.1	210.6	216.4	233.7	224.3	216	220.2	8.99
0.5	1053	1042	1070	1024	1033	1044.4	17.90
1.0	2144	2142	2126	2106	2125	2128.6	15.39
2.0	4288	4380	4387	4431	4376	4372.4	52.06
5.0	11155	11050	11061	11150	11127	11108.6	49.76

[a] content in mg l^{-1}

Table 3 Measurement uncertainty concerning analyte contents in calibration range (without AQC measurement)

Analyte Content[a] x_i	Response y_{obs}	Stand. unc. $u(y_{obs})$	Analyte Content[a] x_{pred}	Unc.-contr. u_{obs}	Unc.-contr. u_{bl}	Unc.-contr. u_{cal}	Measurement uncertainty[b] Eq. (13) $u(x_{pred})$	Eq. (11) $u(x_{pred})$
0	4.96	4.49	0.021	0.0020	0.0072	0.0089	0.011629	0.015015
0.1	220.2	8.99	0.118	0.0040	0.0067	0.0089	0.011856	0.014782
0.5	1044.4	17.90	0.488	0.0080	0.0048	0.0089	0.012920	0.014017
1	2128.6	15.39	0.975	0.0069	0.0023	0.0089	0.011492	0.013361
2	4372.4	52.06	1.984	0.0234	0.0028	0.0089	0.025183	0.013454
5	11108.6	49.76	5.012	0.0224	0.0184	0.0089	0.030264	0.022585

[a] content in mg l^{-1}
[b] The Eqs. (13) and (11) correspond to EURACHEM Guide [2], E3.3 and E3.5, respectively.

Table 4 Evaluation of measurement uncertainty for the special application (analyte content x=2) including the assessment of trueness (contents given in mg l^{-1})

Quantity	Standard uncertainty
Response sample measurement: $y_s=4372.4$	$u(y_s)=52.06$
Response AQC measurement: $y_q=679.3$	$u(y_q)=30.95$
Sample content: $x_s=1.433+\frac{4372.4-3146.5}{2224.9}$	$u(x_s)=\frac{1}{2224.9}\sqrt{52.06^2+\left(\frac{4372.4-3146.5}{2224.9}\right)^2 11.41^2+19.74^2}$
$x_s=1.984$	$u(x_s)=0.0252$
RM Content from AQC measurement: $x_q=1.433+\frac{679.3-3146.5}{2224.9}$	$u(x_q)=\frac{1}{2224.9}\sqrt{30.95^2+\left(\frac{679.3-3146.5}{2224.9}\right)^2 11.41^2+19.74^2}$
$x_q=0.324$	$u(x_q)=0.0175$
RM Content, certified: $x_r=0.285$	$u(x_r)=0.0157$
Correction factor: $f_r=x_q/x_r$	$u(f_r)=f_r\cdot\sqrt{\left(\frac{u(x_q)}{x_q}\right)^2+\left(\frac{u(x_r)}{x_r}\right)^2}$
$f_r=0.324/0.285=1.137$	$u(f_r)=1.137\cdot\sqrt{\left(\frac{0.0175}{0.324}\right)^2+\left(\frac{0.0157}{0.285}\right)^2}=0.0877$
Corrected content: $x_{corr}=x_s/f_r$	$u(x_{corr})=x_{corr}\cdot\sqrt{\left(\frac{u(x_s)}{x_s}\right)^2+\left(\frac{u(f_r)}{f_r}\right)^2}$
$x_{corr}=1.984/1.137=1.745$	$u(x_{corr})=1.745\cdot\sqrt{\left(\frac{0.0252}{1.984}\right)^2+\left(\frac{0.0877}{1.137}\right)^2}=0.136$

In Table 3, according to Eqs. (13) and (11) and without consideration of the AQC measurement, the calculated standard uncertainties are arranged for all calibration levels. One can see that the measurement uncertainties in the part of the calibration range, within which the condition of the variance homogeneity is correctly fulfilled (see the third column in Table 3 and Fig. 1), are nearly equal. For $u(y_{obs})=\sqrt{s_{y.x}^2/p}=\sqrt{48.33^2/5}=21.6$ the curves in Fig. 1 have an intersection point.

In Table 4 the determination of the combined measurement uncertainty, including the assessment of trueness, for a special analytical application (analyte con-

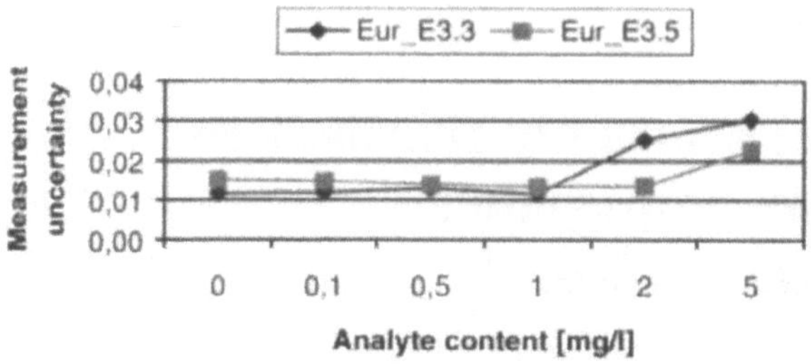

Fig. 1 Comparison of the measurement uncertainty estimations (Eqs. (13) and (11) correspond to EURACHEM Guide [2], E3.3 and E3.5, respectively)

tent x=2 mg l^{-1}), is represented. Here f_r=1.137 and $u(f_r)$=0.088 for the test statistic T=1.56≤2, so that a significant method bias cannot be proven, although the relatively large difference between x_r and x_y suggests a biased error. Because of the non-significance of the method bias the corrected sample content x_{corr}=1.75 is not used. Only its measurement uncertainty $u(x_{corr})$=0.136 is used, thus for the analysis result x_s=1.98 the associated expanded measurement uncertainty is U=2·0.136=0.27.

Conclusion

The application of unweighted ordinary least squares regression for linear calibration can lead to a slightly underestimated uncertainty value, if the condition of variance homogeneity in the calibration range is not correctly fulfilled and the uncertainty calculation is based on the residual standard deviation of the regression (Eq. E3.5 in [2]). In this case the other indicated possibility of uncertainty calculation (uncertainty deduced from repeated measurements, Eq. E3.3 in [2]) can result in more realistic estimations.

References

1. ISO (1995) Guide to the expression of uncertainty in measurement. ISO, Geneva
2. EURACHEM/CITAC Guide (2000) Quantifying uncertainty in analytical measurement, 2nd edn., Final Draft April 2000. EURACHEM: http://www.measurementuncertainty.org
3. IUPAC Recommendations (1998) Guidelines for calibration in analytical chemistry: http://www.iupac.org
4. ISO 5725-2 (2000) Accuracy (trueness and precision) of measurements methods and results, Draft May 2000. ISO, Geneva
5. Rawlings JO (1988) Applied regression analysis. Wadsworth and Brooks, California, USA
6. Kurfürst U (1998) Accred Qual Assur 3:406–411
7. Barwick VJ, Ellison SLR (2000) Accred Qual Assur 5:47–53
8. DIN ISO 11466: 06.97 Soil quality – extraction of trace elements soluble in *aqua regia*. Beuth, Berlin
9. Kleiner J, Lernhardt U (1998) Program SQS98. Perkin Elmer GmbH, Überlingen, Germany

Accred Qual Assur (2001) 6:352–359

Pavol Tarapčík
Ján Labuda
Blandine Fourest
Viliam Pätoprstý

Measurement uncertainty distributions and uncertainty propagation by the simulation approach

Paper based on a talk given at the 3rd EURACHEM Workshop "Status of Traceability in Chemical Measurement", 6–8 September 1999, Bratislava, Slovak Republic

P. Tarapčík (✉) · J. Labuda
Department of Analytical Chemistry, CHTF STU, Radlinského 9, 812 37 Bratislava, Slovak Republic
e-mail: tarapcik@chtf.stuba.sk
Tel.: +4217-59325, ext. 311 or 302
Fax: +4217-52926043 or 52493198

B. Fourest
Institut de Physique Nucléaire, 91406 Orsay Cedex, France

V. Pätoprstý
Slovak Institute of Metrology, Karloveská 63, 842 55 Bratislava, Slovak Republic

Abstract A complete and accurate evaluation of measurement uncertainty requires the knowledge of the uncertainty distributions. The latter are rarely determined or verified experimentally, and hence up to now only crude estimates or assumptions based on intuition have been used. The simulation of experimental results is readily accessible and provides a more reliable solution to this problem. When using an appropriate model of measurement and after determination of input value parameters by present state-of-the-art techniques, simulation data supply reliable information about the distribution of the output results of a complex measurement. The method permits simple variation of preposition and therefore ready analysis of various features influencing the measurement of uncertainty intervals. In the paper we described examples of such evaluations related to the preparation of certified reference materials, where there is excellent agreement between the traditional and simulation approaches. And evaluation of more complex measurements of diffusion coefficients by the open capillary method, where uncertainty of the simulated result is more realistic than the result from the traditional error method due to non-linearity and probably Cauchy distribution in some steps.

Keywords Measurement · Uncertainties · Chemical analysis · Distribution law · Monte Carlo simulation

Introduction

Chemical measurement uncertainty expressed according to the latest metrological requirements [1–3] is a clear improvement compared to the traditional "error approach." Up to now, the results of measurement of the same parameter would often differ widely from one paper to another, although both papers claimed very narrow confidence intervals (see for example stability constant of complexes). This is because some significant sources of uncertainty had been ignored or precessions were overestimated. In the process of uncertainty evaluation it is essential to make the most exhaustive assessment of all involved effects. All partial sources of uncertainty are scouted, then their single contributions are assessed and subsequently combined to provide the estimate of the uncertainty of the result. The combination of single constituents of uncertainty is made by common procedures using standard uncertainties; the standard uncertainty result is used to derive the uncertainty interval by applying a probability distribution law.

Traditional, applied statistics in chemistry text books is based on Fisher statistics [4, 5]. However, application of the latter is now considerably less strenuous due to the ability of computing devices to treat large data sets automatically, and hence their use is more widely spread. Unfortunately users are often too unaware of the limitations of the corresponding models and tools.

Mathematical statistics is developing continuously and a number of new procedures (e.g. robust procedures) [6] are built in new specialized statistics software. Improved software to calculate precision has been published [7, 8]. These new tools provide more correct results compared to traditional ones. It is therefore reasonable to expect that their use becomes standard.

Complete characterisation of analytical results and uncertainty distribution law

The use of standard uncertainty has one important limitation (but often forgotten) – results can be compared only in the case of the same uncertainty distribution law. The other interval measures of precision and derived values, such as confidence intervals, detection limit or determination limit, cannot be reliably determined without knowledge of the distribution, and likewise further statistical deductions are not valid without this information.

Gaussian distribution is often supposed. The analytical determinations are, however, generally complex problems – the measurements are indirect, many single operation are carried out, each of them with its own uncertainty (considered as direct measurement or more or less complex, indirect measurement). The uncertainty of a result is the composition of these elemental uncertainties. There is widely accepted opinion that due to the large number of uncertainty sources, result distribution is often Gaussian. This is valid only in the case of addition or subtraction of the individual constituents, in other cases solution of the distribution law problem is rather difficult – simple application of Central limiting theorem without careful consideration of its limitation leads to overestimation of Normal distribution applicability.

The identification of the distribution law is reliable only for very large data sets [9], which are practically impossible in real chemical measurements. Verification using common statistical tests requires, for good reliability, large data sets too. In a real situation often only an assumption can be made. As this assumption about the distribution law is not well-founded, there is a risk of false information about uncertainty. The assumption of a domination of Gaussian distribution was not confirmed in the study, where a number of large data sets of various types [10] were assessed. Only about 25% of the studied measurements could be considered as a Gaussian distribution.

In recommendations on uncertainty evaluation [1–3]. Gaussian distribution is said to be fundamental, but some other distributions are possible to use. The application of assumed distribution is also made less strict by using a less firm relation between probability and standard deviation as is used in traditional statistics.

There is no consideration about the relation between distribution of input and output values – while the result of the measurement is obtained as a function of more values with their own uncertainties, the standard deviation of result is determined using error propagation and the interval estimate is based on the assumption that the distribution is Gaussian. But the shape of the distribution depends on the input values distribution and their function relation too.

The applied function relation can be relatively simple or more complex. The measurement of individual input values is often a complex process and one input number is the result of many single operations. In the evaluation of chemical analytical results these relations are of various complexity, but they can be mostly expressed as an explicit function and their evaluation using the propagation rule is simple. But some important types of measurement are more difficult to treat. Two examples are described below.

Example 1. A simple case: preparation of a standard solution of zinc

When a certified reference material (CRM) is prepared from a weighted amount of Zn-metal by dissolving in acid and adjusting to the desired volume, the obtained concentration from this process is influenced by several factors and a relatively complete description can be obtained as an extensive equation comprising 16 parameters, the values of which carry some uncertainty [11]. For comparison: the simple but not complete relationship in this case is:

$$c = m/V.M, \tag{1}$$

where m is the amount weighed using analytical balances, V is the volume measured using a volumetric flask and M is the molar weight. The more complete form is as follows:

$$c = \frac{\left(\left(m_0 + m_{1read} + m_{1cert} + m_{1oper} + m_{2read} + m_{2cert} + m_{2oper}\right)/1 - r_{pair}\cdot\left(1/r_{probj} - 1/r_{pweidth}\right)\right)\times\left(1 + K_1 t_1 - K_1^2\right)}{\left(V_{decl} + V_{read} + V_{decl.\varepsilon.}(t_2 - 20)\right)\cdot M} \tag{2}$$

where all used parameters (except m_0, K_1 and K_2) are considered as uncertain values and their symbol meanings are described in [11]. The actual values and origin of non-experimental values (molar mass, densities, etc.) are given in [11]. This relationship is even more complex if the contribution of impurities of the materials used is included.

From the relationship above, standard uncertainty can be obtained, but one has to do relatively extensive work – the error propagation law must be applied, that means one needs 16 sensitivity coefficients (square of derivation) and one has to estimate the standard uncertainties of all 16 constituents. The derivation can be obtained without deep knowledge of mathematical analysis using the appropriate software, or using a similar numerical approach as described in [12] using a common spreadsheet. This procedure, however, after the large amount of work, gives no information about the result distribution law, even if the input distributions are known, and so the result interval estimates are usually based on a non-verified assumption.

Example 2. A more complex case: measurement of diffusion coefficient

The situation is more difficult if an analytical expression is not possible in the explicit form $y = f(x_i)$. This differs only in the more complicated method to obtain the sensitivity coefficient. The method for this is described, for example, in [2]. Even more complex was our particular measurement of diffusion coefficients using the open capillary method, where this process is described by infinite order [13]:

$$y = c_{str}/c_0 = \sum 8/\pi^2(2n+1)^2 \cdot \exp[-\pi^2(2n+1)^2 Dt/4l^2] \tag{3}$$

the sum is for n integers (i.e. from 0 to infinity), where c_{str} and c_0 are the concentrations in the capillary before and after diffusion, l is the length of the capillary, D is the measured diffusion coefficient, important for example for characterisation of particle size, t is the diffusion time, π has its common meaning and n is the consecutive number in infinity order.

This order converges rapidly – for a simple mathematical solution, the time of diffusion is usually adjusted so as to obtain a second term of order (n=1) 1000 times lower than the first term (n=0).

The solution for the first term only (n=0) is:

$$D_{(n=0)} = (4 \cdot l^2/\pi^2 \cdot t) \cdot ln(8/\gamma \cdot \pi^2) \tag{4}$$

and the relative standard uncertainty from the error propagation rule:

$$\delta_D^2 = 4 \cdot \delta_l^2 + \delta_t^2 + \delta_\gamma^2 (1/ln(8/\pi^2 \cdot \gamma^2)) \tag{5}$$

If one needs to use two terms – due to somewhat higher values of γ (lower rate of diffusion), the equation can be designed in the form:

$$\gamma = 8/\pi^2 \cdot \exp(-\pi^2 Dt/4l^2) \cdot \{1 + 1/9 \cdot \exp(-8\pi^2 Dt/4l^2)\} \tag{6}$$

The solution for D is treated for example in [14]:

$$D = 4l^2/\pi^2 t \cdot \ln(8/\gamma\pi^2) + 4l^2\pi^{14}\gamma^8/9t8^8 \tag{7}$$

and for uncertainty after very awkward work

$$\delta_D^2 = 4\delta_l^2 + \delta_t^2 + \delta_\gamma^2 \{8 \cdot (\pi^{16} \cdot \gamma^8 - 9.8^7)/ (9.8^8 \cdot \ln(8/\pi^2 \cdot \gamma) + \pi^{16} \cdot \gamma^8)\} \tag{8}$$

The value γ is conveniently measured as the radioactivity ratio (after background corrections), hence it is also a typical indirect measurement, which is a combination of four individual measurements of radioactivity decay and four measurements of time.

If one needs a more complete equation ($n = 2$ or more), due to slow diffusion of larger particles, when γ values are higher over a reasonable time period, the solution is more complex – D can be calculated by an iterative procedure, but determination of uncertainly according to the usual procedure [2] is very awkward. The general relationship for relative standard uncertainty is as follows:

$$\delta_D^2 = 4\delta_l^2 + \delta_t^2 + \delta_g^2 \cdot \{l^2 \cdot \gamma/[2tD(\sum \exp[-\pi^2(2n+l)^2 Dt/4l^2])]\} \tag{9}$$

and nothing is known about the result distribution.

The examples above show excessive work is required using the traditional approach without achieving completely reliable output information from well-measured input data.

Looking at the distribution law for composition of more values

More complete output information requires knowledge about the distribution law. The composition of distribution can be considered as two values (e.g. a ratio of two Gaussians gives a Cauchy distribution, multiplication gives Laplace distribution in this case). More complex relations are very awkward and for most analysts also technically impossible. By solving the problem of uncertainty distribution in the above way one has two values and can divide the problem and work in single steps, dealing with one problem at a time.

The distribution of results is obtained as a combination of two uncertain values

The procedures to obtain the probability of density functions in this case can be found in advanced mathematical statistics [5], here we provide some examples. Table 1 presents convolutions of some distributions

Table 1 Combined uncertainty distributions of results obtained by arithmetic combination of two measurements with known and common uncertainty distributions

f(A,B)	Distributions of constituents		Resulting distribution type, probability density		Graph type
	A	B			
A+B	R(–a,a)	R(–b,b)	trapezoid (a>b) p(x)=0 =(a+b+x)/4b(a+b) =1/2(a+b) =a+b–x)/4b(a+b)	 $\lvert x\rvert \geq a+b$ $-a-b \leq x \leq b-a$ $b-a \leq x \leq a-b$ $a-b \leq x \leq a+b$	1
A·B	R(–a,a)	R(–b,b)	p(x)=ln(ab/|x|)/2ab =0	$\lvert x\rvert \leq ab$ $\lvert x\rvert > ab$	2
A/B	R(–a,a)	R(–b,b)	p(x)=a/4b $=b/4ax^2$	$\lvert x\rvert \leq b/a$ $\lvert x\rvert > b/a$	3
A+B	N(0,σ_A)	N(0,σ_B)	$p(x)=\exp\left(-\frac{x^2}{2(\sigma_A^2+\sigma_B^2)}\right)/\sqrt{2\pi(\sigma_A^2+\sigma_B^2)}$ (Gauss)		4
A·B	N(0,σ_A)	N(0,σ_B)	$p(x)=\exp\left(-\frac{\lvert x\rvert}{\sigma_A\sigma_B}\right)/\sqrt{4\pi\lvert x\rvert\sigma_A\sigma_B}$		5
A/B	N(0,σ_A)	N(0,σ_B)	$p(x)=\sigma_B/(\pi\sigma_A(1+x^2\sigma_B^2/\sigma_A^2))$ (Cauchy)		6
A·B	L(0,σ_A)	L(0,σ_B)	$p(x)=\sqrt[4]{\pi^2\sigma_A\sigma_B/2\lvert x\rvert}\cdot\exp\left(-8\sqrt{\lvert x\rvert/\sigma_A\sigma_B}\right)/\sigma_A\sigma_B$		5*
A/B	L/0,σ_A)	L(0,σ_B)	$p(x)=\sigma_B/(2\sigma_A(1+x2\sigma_B^2/\sigma_A^2))$		6

generally known and often used. The Gaussian and Laplace distributions are signed N(μ, σ) and L (μ, σ), respectively, (where μ is the true value and σ is the standard deviation), R(–a, a) means rectangular distribution on interval <–a,a>. The graphic presentations of convolutions are shown in Fig. 1: graph type 5* is similar to graph type 5 with the exception that the values of high deviation have higher probability.

The traditional suggestion (ISO) in such cases recommends as a first approximation a Gaussian distribution (graph type 4 in Table 1 and Fig. 1), where standard uncertainty of the result is obtained by a combination of standard uncertainties of the constituents according to the rules of differential calculus. The other case is, of coarse, relatively simple and can be found in the basic text books or derived by a person with moderate skills.

The illustrative example can be given as a comparison of distributions obtained by the addition of two rectangular distributed values – graph type 1 in Table 1 and Fig. 1. According to ISO one can obtain the probability density function by:

$$p(x)=\exp\left(-\frac{(x-\mu_A-\mu_B)^2}{2(\sigma_A^2+\sigma_B^2)}\right)/\sqrt{2\pi(\sigma_A^2+\sigma_B^2)} \qquad (10)$$

and a comparison of the same input values in Fig. 2 emphasises a sharp difference in this case. The example presented is, however, quite academic. Real measurements are not so simple and the distribution laws and their parameters are only known from small data sets, i.e. in most cases only an estimation is used.

Some remarks can be made on this basis:

- The distribution of the value obtained by dividing two uncertainty values is often calculated without defining statistical moments other than the zero moment (e.g. Cauchy), and there is no sense in determining such characteristics as standard deviation because there is skewness based on the higher moments.
- When multiplication of uncertainty values is used, the resulting distribution has a large value of kurtosis and the usual precision characteristics have small statistical efficiency. More suitable are estimates based on median or robust methods.

Simulation approach for uncertainties

The possible solution is an exploitation of the ability of standard software to generate data with defined statistical properties – with defined distribution and its parameters, e.g. mean value, standard deviation and the like. In this way, sets of data can be obtained each representing a single "experiment". The number of these simulated data sets can be very large and a set of results can be treated traditionally – that means by determining statistical moments and derived values, construing histograms, polygons and the like. Or interval estimates can be obtained simply from percentiles of the set of results (this is really more correct than the moment method where the assumption about distribution often plays a negative role). This procedure requires no new

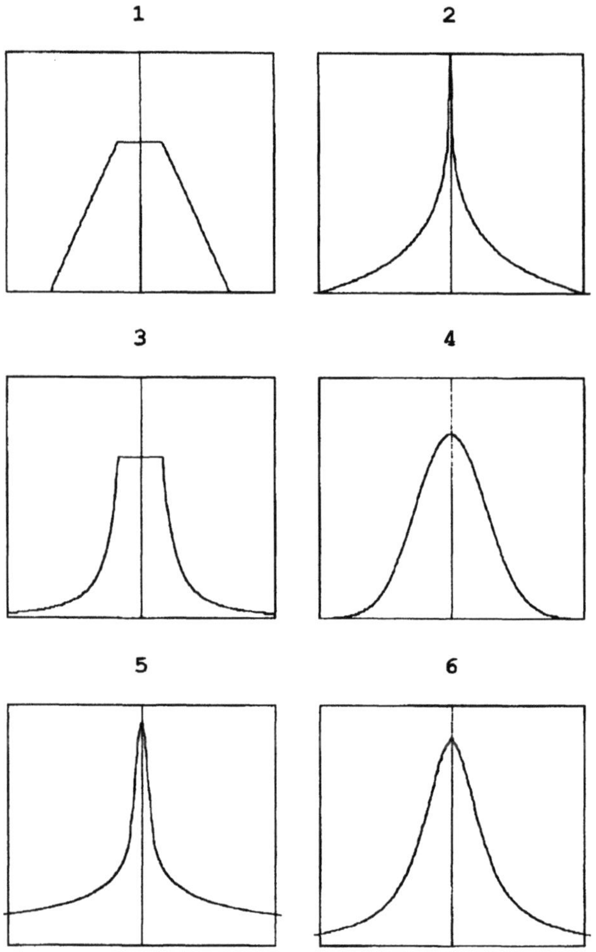

Fig. 1 Combined uncertainty distributions corresponding to the cases given in Table 1

mathematical knowledge or abilities, and so does not change the analyst into a theoretician. Only moderate skills in applying standard software are required.

Simulating procedure

The steps of identifying the partial uncertainty sources and the design of the measurement model represent the basis of the uncertaintly conception and cannot be changed. The main goal of the simulation procedure is to "calculate the combined uncertainty". The previous steps are conducted as usual and provide the input values for the next simulation procedure. The assumed distribution is an important parameter for each individual source of uncertainty. The treatment differs from the traditional one as follows:

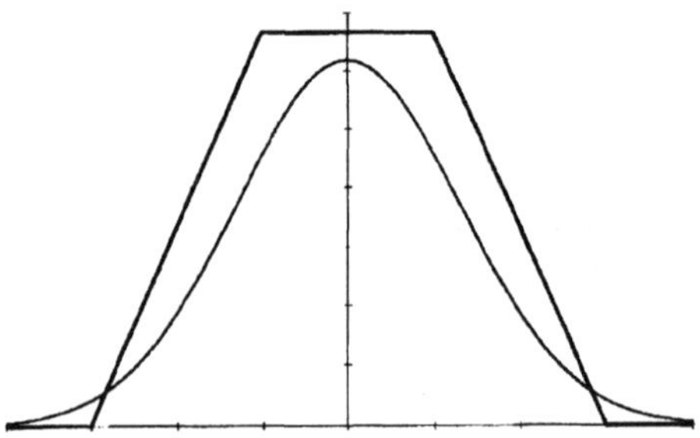

Fig. 2 Comparison of true distribution in the case of type 1 in Table 1 and the shape obtained according to traditional approximation by Gaussian distribution for the same input values R(–0.2;0.2) and R(–0.1;0.1)

- There is no need to calculate sensitivity coefficients (calculation of sensitivity coefficient is sometimes very awkward).
- By simulating on a PC, large set of individual values influencing the final uncertainty are made. Each set of values has its mean value, standard deviation and originates from a distribution which is known or assumed – in this case can be readily changed. (To generate these, a built-in function in the standard software can be used or some simple procedure can be written by a skilled user).
- Using a known measurement model (relation between input data and result), a set of results is calculated from simulated sets of individual input data.
- The number of simulated results can be so large that traditional statistical treatment will give correct interval estimates. The assumptions about input data can be readily controlled and so one can simply estimate the influence of changes in assumptions, by recalculations. Such calculation can also provide similar information useful for the optimization of measurement conditions, such as the sensitivity coefficient analysis common in traditional methods. These calculations do not require more skill than simple result calculation – no new mathematical equation, no derivation.

Solution using a spreadsheet

The working sheet includes, for each input value, the mean and standard deviation (or other data concerning uncertainty and the distribution law, from which the standard deviation can be calculated). The result is calculated using a known measurement model:

- In the 1st line, the input measured values are written and from these the result of the measurement is calculated.
- The next line(lines) includes further parameter(s) of the distribution – for two parameters distribution, as

in a Gaussian, it is for example standard uncertainty.
- Further, in the next line there is information about the type of distribution law.
- In a new line a simulated set of input data is calculated using the parameters above and known generators of random values.
- The last line is then realised (simple copy) n times (e.g. 1000×), each new line includes new simulated data automatically.
- Sets of results are then in one column, and are treated by the usual method (common integral part of spreadsheet). For example: the mean value is in cell A1 and its standard deviation is in cell A2:, cell A3 contains information about the distribution law, then if in cells A11–A1010 we want obtain 1000 values from a Normal distribution with the parameters above, we input into each cell term:

= NORMINV(RAND(); A1; A2) (11)

or if the distribution law has to be rectangular

= A1–A2*SQRT(3) +
RAND()*2*A2*SQRT(3) (12)

(NORMINV, SQRT and RAND are the names of appropriate built-in functions of spreadsheet).

How to generate numbers of many other distributions is a well known and solved problem [6]:

- If cells of column B have other values than in column C the result of the calculation for the functional relation of A and B (for example the ratio) one obtains a 1000 lines with simulated sets of input data and a set of 1000 calculated results.
- Statistical analysis of the last set gives standard statistical outputs, e.g. interval estimates (standard deviation, percentiles) or better still a more complete picture using a graph (polygon or histogram) of the simulated distribution, which can be then readily compared with an a priori assumed distribution – simply visually or by application of standard statistical tests. The result distribution can be described by an empirical equation and used in further result utilisation.

Examples of the simulation approach

Example 1. Preparation of standard solution of zinc

We used parameters and assumption of individual influencing values and model of measurement as stated in the procedure according to the usual recommended method [2]. The above mentioned procedure was used, only more complex.

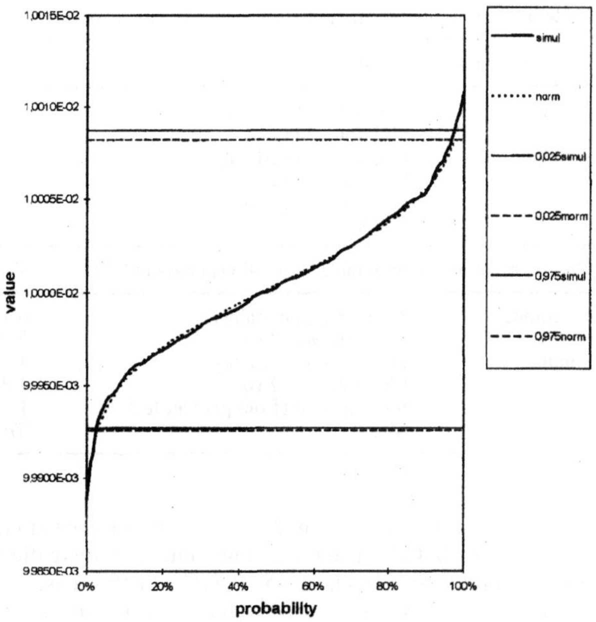

Fig. 3 Integral distribution function (polygon) of simulated certified reference material preparations

Table 2 Statistical evaluation of simulated certified reference material "preparations"

Mean	0.0100004 mol/l
Median	0.0100002 mol/l
Standard uncertainty	0.000004 mol/l
Asymmetry	0.0902
Kurtosis	–0.2527
Relative standard deviation	0.03996%

The results are given in Table 2 and in graphical form of integral distribution function (polygon) in Fig. 3. In this graph there is also a curve for the Gaussian distribution using the moment method and the 95% interval obtained from the percentiles, and the same estimate from parameters of the Gaussian distribution. It is evident, that in this case the assumptions were adequate. The simulating procedure gave results in agreement with the results of the standard procedure but without the element of vagueness due to the unknown composition of the individual distributions.

Example 2. Measurement of diffusion coefficient

This more complex example requires only a few more skills. Each simulated data set was solved iteratively – but automatically by the macro "find solution" function of the spreadsheet.

Table 3 Statistical evaluation of simulated diffusion experiment

Measurement at "favourable" conditions (γ about 0.4)		Type 1	Type 2
Traditional	Stand. rel. uncertainty	1.18%	1.18%
	95% interval (2 σ)	8.241×10^{-6}–8.640×10^{-6} cm²/s	8.250×10^{-6}–8.649×10^{-6} cm²/s
Simulation	Stand. rel. uncertainty	1.14%	1.19%
	95% interval (2 σ)	8.249×10^{-6}–8.633×10^{-6} cm²/s	8.249×10^{-6}–8.651×10^{-6} cm²/s
	95% interval (from percentiles)	8.243×10^{-6}–8.639×10^{-6} cm²/s	8.263×10^{-6}–8.626×10^{-6} cm²/s
	Shape	Gaussian	Trapezoid
Measurement at "unfavourable" conditions (γ about 0.75)		**Type 1**	**Type 2**
Traditional	Stand. rel. uncertainty	6.71%	6.62%
	95% interval (2 σ)	2.180×10^{-6}–2.497×10^{-6} cm²/s	2.152×10^{-6}–2.457×10^{-6} cm²/s
Simulation	Stand. rel. uncertainty	11.0%	12.75%
	95% interval (2 σ)	1.824×10^{-6}–2.850×10^{-6} cm²/s	1.717×10^{-6}–2.892×10^{-6} cm²/s
	95% interval (from percentiles)	1.857×10^{-6}–2.851×10^{-6} cm²/s	1.715×10^{-6}–2.935×10^{-6} cm²/s
	Shape	Triangular	"skewed" Gaussian

For a rapid diffusion of Na^+, at a mean measured value of $\gamma = 0.424$, using the uncertainty propagation rule we obtained a value of δ_γ=0,71%. In this case simulation gave a value of 0.70% and the relative uncertainty of the calculated D was 1.15% by uncertainty propagation and 1.14% by the simulation. At the same time, the stimulation procedure gave details of the uncertainty distribution of the result (expressed, e.g., as a histogram, moments of distribution, etc.), and it was possible to show the interval limits directly from the simulated set of results, regardless of the distribution type. In activities measurements, assumptions were used usually for radiometric measurements – normal distribution for counts, measurement of time and length were also of a Normal distribution (type 1). The histogram and parameters of simulated set of results indicated a normal-like distribution, and this character was maintained also in the case of the rectangular distribution used for time and length measurements (type 2).

For slow diffusion and a more critical measurement of Eu^{3+}, and not a long enough diffusion time, γ was about 0.75, and the situation is different. The application of the uncertainty propagation rule leads to an overestimation of precision by about 40–50% of the standard deviation, as compared to the values from the simulation. Moreover, the simulation gave information about the shape of the distribution. It is also simple to judge the influence of an assumption; the simulation gave an objective result. Illustrative examples are shown in Table 3.

Favourable conditions means a measurement in which γ is about 0.4, so higher terms of diffusion equations are not important. Under unfavourable conditions the values of γ was about 0.75: about four members of the order are of importance. In both cases in an iterative calculation, the first eight members of a given order were applied.

Under favourable conditions there is no difference in a number of characteristics between the traditional and simulation procedure, and similarly there is no significant difference in relation to the assumption about the distribution. However, the simulation gave the shape of the distribution and for often used assumptions in type 2 a slightly narrower interval at 95% probability from percentiles technique is seen, which is most evident.

In the less favourable conditions, the traditional procedure overestimated precision. The influence of a given choice of assumptions about distributions even displayed a skewed distribution and an evident enlargement of interval, that shows the importance of well-founded assumptions.

For such measurements, more capillaries are used in the same experiment to obtain repeated values, although a very limited number (8 for example) and with risk of correlation's, but the details are known by a skilled person. The uncertainties thus obtained show that the result of the simulation is more realistic compared to the result of the traditional method.

Conclusions

Almost every analytical result is obtained as an indirect measurement. The distribution of the results is given by the distribution of direct measured values and their relations. There is no reason to suppose a Gaussian distribution of elementary measured values and results simultaneously. In some cases, the simple error propagation law is sufficient but in some cases could be in error and a simulation approach would be worth adopting. It is almost impossible to decide a priori which method to use.

If an assumption about the distribution is used in one step of the analysis, further conclusions in the other steps must be in harmony with this assumption, e.g.

if the parameters of calibration of a straight line were tested as Gaussian distributed, simultaneously the assumption of Cauchy distribution of the concentrations obtained by using this parameters is automatically made and further statistical deductions must agree with this fact.

If assumptions about the direct measured values are used, the result of the determination of the distribution law is also an assumption and it is not worth making a difficult exact analysis of distribution.

The determination of the distribution of an analytical result is rather a difficult problem. Even if distributions of direct measured values are known, it is reasonable to turn to statisticians for a solution, but a simulation procedure is available for the moderately skilled person.

The simulation approach for interval estimates of measured values does not solve this problem entirely, it cannot overcome the problem of small numbers of degrees of freedom, but in most cases it is a good tool:

- It can be used in cases where the distribution and distribution parameters of the constituent is known and in cases where only approximate values of parameters, derived from small data sets are used. This method is also only an approximation, but in many cases more correct and less strenuous than by the error propagation method.
- The result is easily variable as relate to the assumptions, this variability includes the distribution law too.
- The method is able to solve difficulties originating from complex calculations.
- It is easy to realise, and the procedure does not require any special mathematical knowledge or ability.
- Only a moderate knowledge of standard software is required.
- Result distribution is given in table and graphical form.

The simulated distribution, in many cases without known (assumed) input distributions, remains a model of frequency distribution to be expected. This model is particularly useful when it is compared to experimentally determined distributions, to check whether all significant sources of uncertainties have been identified and properly assessed.

References

1. EURACHEM (1995) Quantifying uncertainty in analytical measurement. EURACHEM, London, UK; ISBN 0-948926-08-2
2. TPM 0051-93 (1993) Stanovenie neistôt pri meraniach (*Determination of measurement uncertainties*). Slovak Institute of Metrology, Bratislava, Slovak Republic
3. ISO (1993) Guide to the expression of uncertainty in measurement. ISO, Geneva, Switzerland; ISBN 92-67-10188-9
4. Taylor JK (1990) Statistical techniques for data analysis. Lewis Publishers, Chelsea, Mich., USA
5. Koroliuk VS, Portenko NI, Skorokhod AV, Turbin AF (1985) Spravotchnik po teorii veroyatnostei i matematitcheskoi statistike (*Handbook of probability theory and mathematical statistics*). Nauka, Moscow, Russia
6. Antoch J, Vorlíčková D (1992) Vybrané metody statistické analýzy dat (*Selected statistical methods*). Academia, Prague, Czech Republic
7. Eckschlager K (1991) Collect Czech Chem Commun 56:505–559
8. Tarapčik P (1992) Chem Lett 86:648–652
9. Thompson M, Howarth RJ (1980) Analyst 105:1188–1195
10. Novitskii PV, Zograf IA (1985) Otsenka pogreshnostei rezultatov izmerenii (*Errors evaluation of results of measurements*). Energoatomizdat, Leningrad, Russia
11. Tarapčik P, Buzinkaiová T, Polonský J, Dlouhá M, Chromek F (1997) Metrológia a skúšobníctvo, 2:6–10; 3:10–14
12. Kragten J (1994) Analyst 119:2161–2166
13. Gosman A, Jech Č (1989) Jaderné metody v chemickém výskumu (*Nuclear method in chemical resarch*). Academia, Prague, Czech Republic
14. Fourest B (1983) Coefficients de diffusion limites et structure de quelques ion aquo d'elements 5f et 4f. PhD Thesis, Institute de Physique Nucleaire, Orsay, France

Accred Qual Assur (2000) 5:88–94

Matthias Rösslein

Evaluation of uncertainty utilising the component by component approach

Presented at: EURACHEM Workshop of Effecient Methodology for theEvaluation of Uncertainty in Analytical Chemistry, Helsinki, Finland, 14–15 June 1999

M. Rösslein
EMPA St. Gallen, Department of Chemistry/Metrology, Lerchenfeldstrasse 5, 9014 St. Gallen, Switzerland
e-mail: matthias.roesslein@empa.ch
www.empa.ch/metrology

Abstract This paper reviews the so-called "component by component approach" of evaluating measurement uncertainty. An overview of the evaluation process is given followed by an in-depth discussion of some of the differences between this approach and the approach of utilising validation data. Some of the advantages and disadvantages of using the component by component approach are outlined at the end.

Key words Uncertainty · Component by component approach · Evaluation of measurement uncetrtainty · Measurement uncertainty

Introduction to the evaluation of measurement uncertainty

The evaluation process of measurement uncertainty consist of four steps (see Fig. 1):

1. Specification

A clear statement of what is being measured is written down. This includes the relationship between the measurand and the parameters upon which it depends, and the scope of the measurement.

GUM makes an important remark about this step, which is quite often overlooked in the day-to-day business of calculating measurement uncertainty:

> The measurand cannot be specified by a value but only by a description of a quantity. However, in principle, a measurand cannot be completely described without an infinite amount of information. Thus, to the extent that it leaves room for interpretation, incomplete definition of the measurand introduces into the uncertainty of the result of a measurement a component of uncertainty that may or may not be significant relative to the accuracy required of the measurement [1].

2. Identify uncertainty sources

A feasible approach to identify the uncertainty sources is as follows [2]:

A. Write down the complete calculation involved in obtaining the result, including all intermediate measurements. List the parameters involved.
B. Study the method, step by step, and identify any other factors acting on the result. Add these to the list. For example, ambient conditions such as temperature and pressure affect many results.
C. Consider factors which will affect the parameters identified in the previous two paragraphs and add them to the list. Continue the process until the effects become to remote to be worth consideration.
D. Resolve any duplicate entries in the list. Listing uncertainty contributions separately for every input parameter might result in duplications in the list. Three cases arise and the following rules should be applied to resolve duplication:
   - Cancelling effects: remove both instances from the list.

Fig. 1 Procedure to evaluate the measurement uncertainty

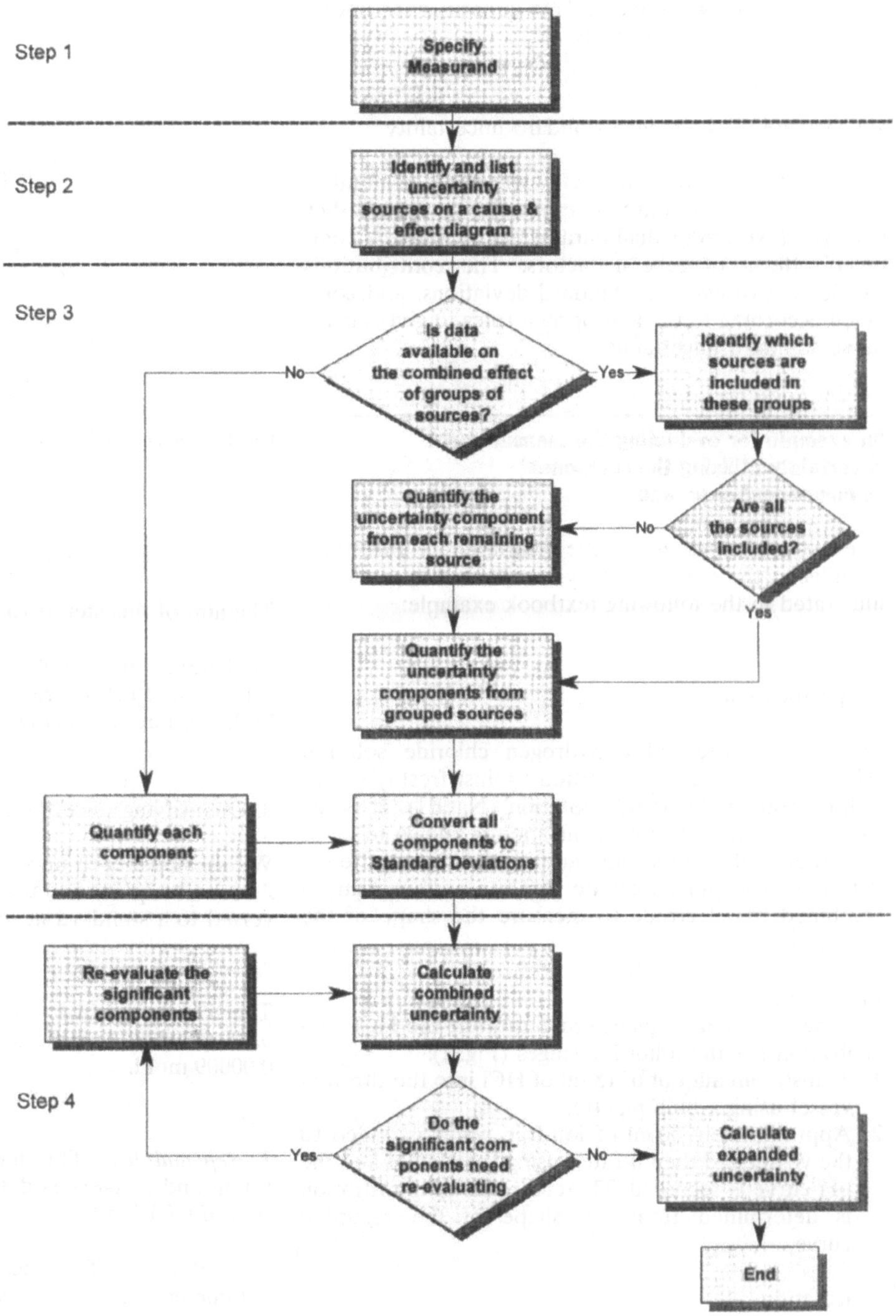

- Similar effects, same time: check carefully, if similar effects are accounted for twice. In that case resolve the duplication.
- Similar effects, different instances: re-label.

3. Quantifying uncertainty components

Measure or estimate the size of the uncertainty associated with each potential source of uncertainty identified. The goal of the component by component approach is to find ways and tools to quantify each uncertainty component individually.

4. Calculate the combined standard uncertainty

The information obtained will consist of a number quantified contributions to overall uncertainty, whether associated with individual parameters or with the combined effects of several factors. The contributions should be expressed as standard deviations, and combined according to the appropriate rules to give a combined standard uncertainty.

An example for evaluating the measurement uncertainty utilising the component by component approach

The evaluation process for the measurement uncertainty utilising the component by component approach is illustrated in the following textbook example:

1. Specification

The concentration of a hydrogen chloride solution (HCl) is determined by titration against freshly standardised sodium hydroxide solution (NaOH). It is assumed that the NaOH concentration is known to be of the order of 0.1 mol/l. The end-point of the titration is determined by an automatic titration system using a combined pH-electrode to measure the shape of the pH-curve.

Procedure
The measurement sequence to determine the HCl concentration has the following stages (Fig. 2):
1. Transfer an aliquot of 15 ml of HCl into the titration vessel using a bulb pipette.
2. Approximately 50 ml of ion-free water is added to the vessel and then titrated using the NaOH and the pH-curve is recorded. The end-point of the titration is determined from the shape of the recorded curve.

Calculation

$$c_{HCl}=\frac{c_{NaOH}\cdot V_{Tit}}{V_{HCl}} \quad \text{(mol/l)}$$

c_{HCl}: concentration of the HCl solution (mol/l)
c_{NaOH}: concentration of the NaOH solution (mol/l)
V_{Tit}: titration volume of NaOH solution (ml)
V_{HCl}: aliquot of HCl titrated with NaOH solution (ml)

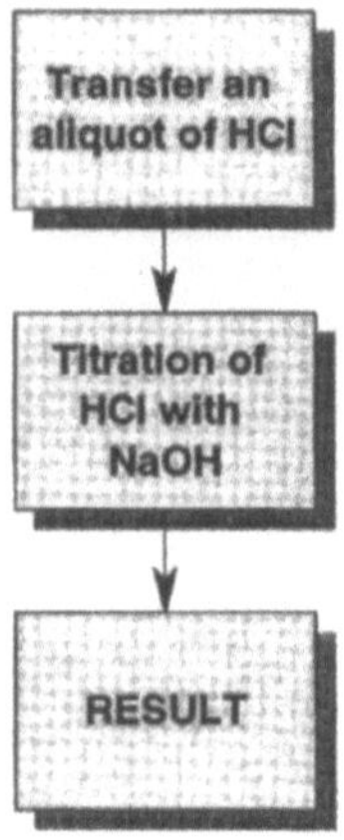

Fig. 2 Flow chart of the analytical procedure

2. Identifying and analysing uncertainty sources

The aim of this step is to identify all major uncertainty sources and to understand their effect on the measurand and its uncertainty. This is best done by drawing a cause and effect diagram using the procedure described by Ellison et al. [2] (Fig. 3).

3. Quantifying uncertainty component by component

Within step 3 each uncertainty source identified in step 2 has to be quantified using relevant data and then converted to a standard uncertainty.

c_{NaOH}

The concentration of the NaOH solution is 0.10215 mol/l with a standard uncertainty of 0.00009 mol/l.

V_{HCl}

1. Repeatability. The uncertainty due to variability in filling and delivery is determined as a standard deviation of 0.0037 ml.

2. Calibration. The uncertainty on the stated internal volume is given by the manufacturer as ±0.02 ml. This value is transformed into a standard uncertainty assuming a triangular distribution $0.02\sqrt{6}=0.0082$ ml.

3. Temperature. The effect of temperature difference from the pipette calibration temperature to the laboratory environment can be calculated from an estimated temperature range and the coefficient of volume expansion.

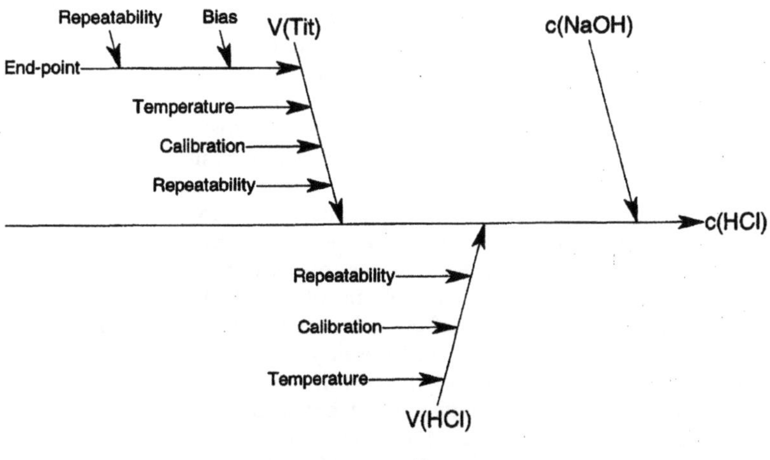

Fig. 3 Cause and effect diagram displaying the effect of the different components

$$\frac{15\ \text{ml}\cdot 2.1\cdot 10^{-4}{}^{\circ}\text{C}^{-1}\cdot 4{}^{\circ}\text{C}}{\sqrt{3}}=0.0073\ \text{ml}.$$

Combining the three contributions to the uncertainty $u(V_{HCl})$ of the volume V_{HCl} gives a value of

$$u(V_{HCl})=\sqrt{0.0037^2+0.0082^2+0.0072^2}=0.012\ \text{ml}.$$

1. Repeatability of the volume delivery. The variability of the delivered volume of the piston burette is determined as a standard deviation of 0.004 ml.

2. Calibration. The limits of accuracy of the delivered volume is indicated by the manufacturer as ± 0.03 ml for a 20-ml piston burette. The standard uncertainty is calculated assuming a triangular distribution $0.03/\sqrt{6}=$ 0.012 ml.

3. Temperature. The uncertainty due to the lack of temperature control is calculated in the same way as for V_{HCl}: 0.0073 ml.

4. Repeatability of the end-point detection. The repeatability of the end-point detection is thoroughly investigated during the method evaluation under the given conditions and a standard uncertainty of 0.004 ml is found appropriate.

5. Bias of the end-point detection. During the method evaluation no indication for any bias was found, because a strong base (NaOH) is used to titrate a strong acid (HCl). This leads to a large change of the pH-values around the end-point resulting in a very accurate determination of the value of the end-point.

V_{Tit} is determined to be 14.89 ml and combining the four remaining contributions to the uncertainty $u(V_{Tit})$ gives a value of:

$$u(V_{Tit})=\sqrt{0.004^2+0.012^2+0.0073^2+0.004^2}=0.015\ \text{ml}.$$

The above section shows that even for a relatively simple textbook example the process to calculate the standard uncertainties for each of the components is time-consuming.

4. Calculate the combined standard uncertainty

The intermediate values, their standard uncertainty and their relative standard uncertainty are shown in Table 1.

Using the values obtained above:

$$c_{HCl}=\frac{0.10215\cdot 14.89}{15}=0.10140\ \text{mol/l}$$

and

$$u(c_{HCl})=0.00016\ \text{mol/l}.$$

Table 1 The standard uncertainties and relative standard uncertainties of the components used to calculate the combined standard uncertainty

	Description	Value	Standard uncertainty	Relative standard uncertainty
c_{NaOH}	Concentration of NaOH	0.10215 mol/l	0.00009 mol/l	0.0009
V_{HCl}	HCl aliquot for NaOH titration	15 ml	0.012 ml	0.0008
V_{Tit}	Volume of NaOH for HCl titration	14.89 ml	0.015 ml	0.001

Within step 4 one also wants to find out which of the relevant contributions should be re-evaluated. To accomplish this the relevant contributions have first to be found. One possible way to do this is to compare the relative variance of the different components in a histogram. It is better to compare the relative variances instead of the relative standard uncertainties, because they reflect more accurately their influence on the final result, i.e. combined standard uncertainty (Figs. 4–6).

The contribution of V_{Tit} is the largest one according to the histograms in Fig. 4. The volume of NaOH for titration of HCl (V_{Tit}) itself is affected by four influencing quantities, which are the repeatability of the volume delivery, the calibration of the piston burette, the difference in temperature between the bench and the calibration of the burette, and the repeatability of the end-point detection. Checking the size of each contribution the calibration is by far the largest one. Therefore this contribution has to be investigated more thoroughly, because its size is based on a rough estimate of the shape of the distribution function.

The standard uncertainty of the calibration of V_{Tit} was calculated assuming a triangular distribution. The influence of the choice of shape of the distribution is shown in Table 2.

It is not surprising that the combined standard uncertainty $u_c(c_{HCl})$ shows little effect from the choice of the distribution function of the largest influencing quantity. The contribution of the different components found during the evaluation can also be combined in different ways. One possibility is shown in the following Fig. 7 which includes that the calibration of the experimental equipment (38.7%) is the largest contribution to the overall uncertainty. The uncertainty of the titer concentration of NaOH (33.1%) is around the same size whereas the effect of the repeatability (8.3%) and the temperature influence (19.9%) are considerably smaller.

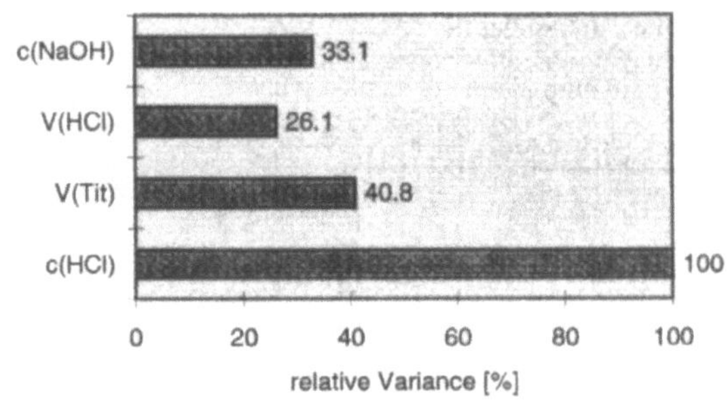

Fig. 4 Contribution of the main parameters to the overall uncertainty

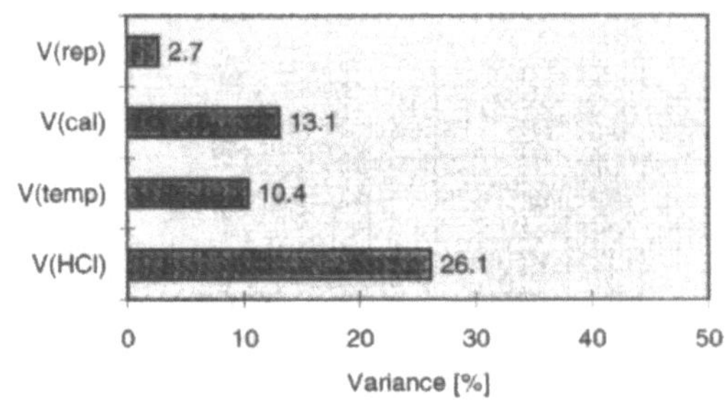

Fig. 5 Contribution of the components (V_{HCl}) to the overall uncertainty of a single parameter

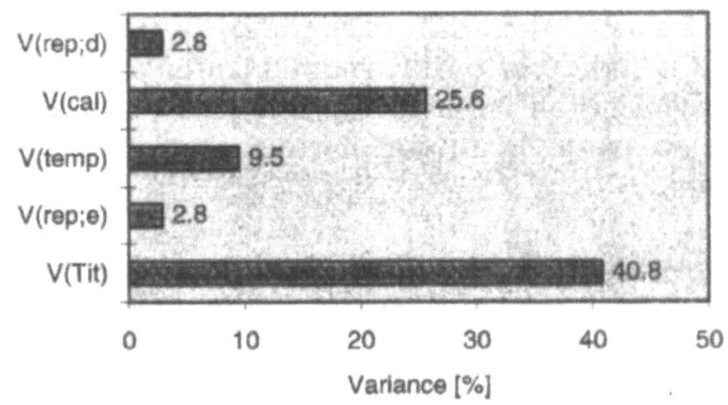

Fig. 6 Contribution of the components (V_{Tit}) to the overall uncertainty of a single parameter

Table 2 The influence of the choice of shape of the distribution on the combined standard uncertainty

Distribution	Factor	$u(V(T; cal))$	$u(V(T))$	$u_c(c_{HCl})$
Rectangular	$\sqrt{3}$	0.017 ml	0.019 ml	0.00018 mol/l
Triangular	$\sqrt{6}$	0.012 ml	0.015 ml	0.00016 mol/l
Normal	$\sqrt{9}$	0.010 ml	0.014 ml	0.00015 mol/l

Similarities and differences between the two approaches

Similar components, such as repeatability or temperature dependence, are combined in Fig. 7. These components influence different parameters of the equation of the measurand, but by combining them in this new way allows to visualise their overall effect on the measurement and measurement uncertainty. Employing this approach, the user of the data might gain a better understanding of the measurement process. This is another benefit of the evaluation of measurement uncertainty besides its major objective to obtain comparable results.

A closer look at the process of combining components shows that it is very similar to the grouping of influencing quantities in step 3 for the utilising of validation data approach (Figs. 8, 9). In this approach the grouping of influencing quantities has to be done to accommodate the sometimes limited amount of informa-

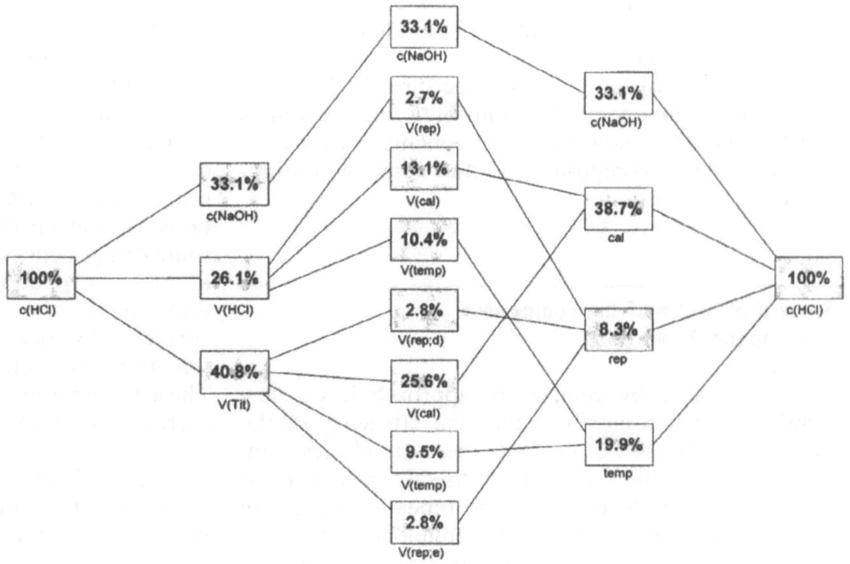

Fig. 7 Analysis of the effect of the different components

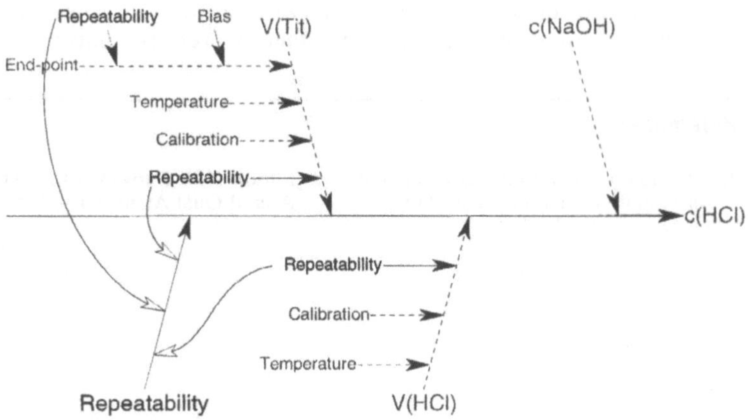

Fig. 8 Process of grouping the repeatability within step 3 of the evaluation process utilising validation data

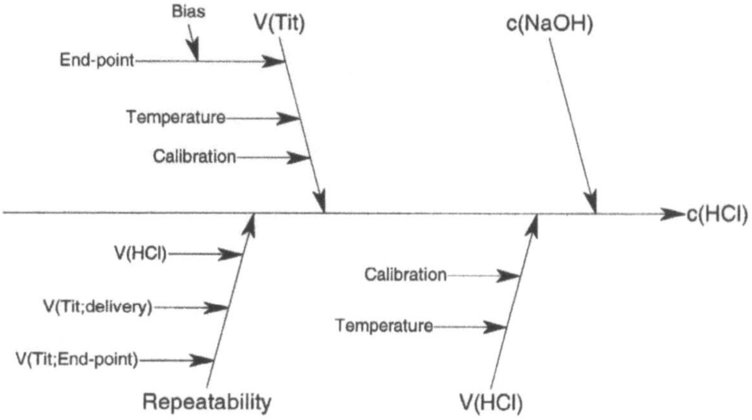

Fig. 9 Grouped repeatability components

tion obtained during a validation study. GUM comments about combining individual components in its introduction [1]: "The actual quantity used to express uncertainty should be:

– internally consistent: it should be directly derivable from the components that contribute to it, as well as independent of how these components are grouped and of the decomposition of the components into subcomponents".

Some drawbacks of the component by component approach

The component by component approach has a few drawbacks, which mostly reduce the efficiency of the evaluation process. For example there are always components whose effect cannot be directly measured. In the above textbook example, the repeatability of the end-point determination is such a component. The variation of the end-point detection is always part of the overall repeatability of the experiment and there is no other means to directly determine its size in an independent measurement. In contrast the other two repeatability components, the variation of the two volume deliveries, have been determined independently in a series of ten delivery and weight experiments.

The determination of the different components in independent experiments increases the risk of neglecting correlation between these components in a given analytical procedure. The aim of experiments to determine the size of a single component is to reduce any other influences, especially one tries to avoid correlations. In a given routine analysis correlations between some components might occur. Therefore the relevance of these correlations have to be investigated if one wants to use the component by component approach correctly. In contrast overall method performance parameters are determined during validation studies. These parameters already take into account most of the correlations between different components.

The drawbacks outlined above show that the utiliation of validation data can be more efficient than the component by component approach to evaluate measurement uncertainty. However the former approach does not provide information about the relative size of the components, this information is importnat when it is necessary to develop the method further to reduce its uncertainty.

References

1. ISO (1993) Guide to the expression of uncertainty in measurement. ISO, Geneva, Switzerland
2. Ellison SLR, Barwick VJ (1998) Accred Qual Assur 3:101–105

Accred Qual Assur (1998) 3:145–149

Petras Serapinas

Uncertainty – statistical approach, 1/f noise and chaos

Presented at: 2nd EURACHEM Workshop on Measurement Uncertainty in Chemical Analysis, Berlin, 29–30 September 1997

P. Serapinas (✉)
Plasma Spectroscopy Laboratory, Institute of Theoretical Physics and Astronomy,
A. Goštauto 12, 2600 Vilnius, Lithuania
Tel.: +370-2-616018
Fax: +370-2-225361
e-mail: pserapin@itpa.lt

Abstract Some general reasons for poor applicability of the statistical approach based on approximation of normal data distribution to interlaboratory test results and analytical measurements at high data dispersion are considered. They include a symmetry of the concentration scale, low-frequency noise, and nonlinear phenomena in atomization processes and chemical reactions. The relationship of 1/f noise and nonlinear phenomena to uncertainty balance, experimental verification of the assigned uncertainty value, ruggedness tests and statistical data distribution are briefly discussed.

Key words Uncertainty · Interlaboratory test · Noise · Data distribution · Nonlinear phenomena

Introduction

Attention to uncertainty in chemical analysis is mainly for two reasons: (a) uncertainty is an indispensable part of any test results; and (b) the dispersion between measurement results of different analytical laboratories is usually much greater than that between the results of experiments reapeated by one laboratory. There is a tendency to explain this difference as inherent in the uncertainties of the individual results, and the need to consider the full list of uncertainty sources and their contributions is usually stressed [1–3]. Nevertheless, the interlaboratory character of the final experimental uncertainty test remains a problem, and gaps exist not only between repeatability and reproducibility of RSD figures but also in links between the two concepts. Repeatability and reproducibility definitions [4] concentrate on the difference between these, but more detailed understanding of the connections between them is also needed. The relationship of uncertainty to 1/f noise, oscillations and chaos in chemical reactions seems to be of importance in the context and is considered in the paper.

Gaps also exist in the preparation of methods and the study of correspondence between the assessed contributions of the identified sources of uncertainty and their direct manifestation in real measurement sets. Moreover, as at least some comparatively large uncertainty sources seem to be the typical situation (if only one main source is identified it usually can be reduced to approximate to the group of comparatively intensive contributions). So, essential tasks seem to include the development of means to obtain information on uncertainty in the measurement process itself as well as the application of this information for uncertainty reduction.

On asymmetry of the data distribution

Applicability of statistical control is a general characteristic of analytical methods. For this, analytical data distributions ought to be normal. Large data sets are needed for detailed examination of the data distributions. Often, an obvious asymmetric character of the distribution manifests itself, indicating deviations from the normal distribution. This is almost the usual state of affairs for large data dispersions, e.g. in interlaboratory tests. Nevertheless, the very definition of uncertainty implies at least the nearly symmetric distribution of values about the measured result (see e.g. [5]).

The applicability of other data distributions, such as the logarithmically normal, for analytical data is also discussed, but the procedures currently used in analytical data analysis (e.g., homogeneity, outlier tests etc. [6, 7]) have mainly been developed for normal data distributions. Asymmetry can be due to contamination and other errors, but one general reason is asymmetry of the concentration scale – zero acts as the lower limit and there is no upper limit. According to our analysis, the distribution peak position in relation to the mean value and symmetry of the data scatter of the interlaboratory test data can be improved essentially if the symmetric scale of logarithms of concentration (– infinity, + infinity) instead of the concentration itself is used. The difference is important at large data scatters only, and all the data scales are almost symmetric if the data dispersion is small enough. So the problem can be formulated as follows: can scales be found in which the applicability of normal distribution would be wider than in the linear scale?

In some cases, the applicability of the log-normal distribution has been strongly contested (e.g. at the detection limit [8]). Even in these cases, some arguments seem to be reformulated. Of course, no distribution is universal (the normal one as well), and if more suitable types of distribution (or data scales) exist, these ought to be applicable only to some definite types of measurement data. Concerning data distributions in interlaboratory tests, more extensive analysis of the interlaboratory data distributions from different test schemes ought to be carried out. Better applicability of the normal distribution could enable wider fitness of the statistical approach and wider possibilities for validation of the analytical methods. Besides this, possibly some of the results at the large concentration "tail" could be regarded as normal. Estimation of uncertainty in the case of other distributions, including the asymmetric ones, also needs attention.

The 1/f noise approach to understanding bias, error functions and uncertainty reduction possibilities

Repeatability standard deviation of the measurements with the same apparatus at identical conditions is usually limited by low-frequency noise. This noise is $1/f^{\alpha}$ frequency dependent. Often α is close to 1, and the noise is known as 1/f noise (in parallel, the term flicker noise is used). The phenomenon, while not fully understood (see e.g. [9]), takes part in many natural, physical, chemical, social, economical etc. phenomena and is of special interest when accuracy of the measurements is considered. In analytical measurements, the measurement time usually exceeds the time at which the 1/f noise, which increases with time, becomes comparable to other noises. Usually this takes place in the order of seconds.

Long-lasting measurement sets seem to be characteristic of chemical analysis. Quality control charting [10], interlaboratory studies [6, 7, 11], and applications of the certified reference materials are typical examples. In spite of this, the concept of 1/f noise is hardly ever applied to analytical measurements with large time intervals. The following question in relation to 1/f noise is considered in this paper: is the 1/f noise concept applicable to the long-term variations of analytical data, where the information on the characteristic variation time or variation time dependences can be of importance?

The result of an attempt to measure the low-frequency noise spectrum from AAS soil analysis quality control data lasting for almost two years is presented in Fig. 1. Fourier analysis of the traditional quality control data time series was carried out. It follows that 1/f noise seems to be a useful approximation for such a low-frequency spectrum. In addition to the common 1/f character of the noise some characteristic frequencies at about 1/(25 days) and 1/(2 months) are observed.

The example presented seems to be of interest from different points of view. First of all, characteristic noise frequencies or time constants of the main slowly varying uncertainty sources are displayed. This is an additional, parallel source of information compared to the ruggedness tests and uncertainty evaluation, and can be used to confirm, support or supplement the obtained conclusions or to control the effectiveness of uncertainty reduction procedures. Not only the characteristic frequencies but the contributions of the corresponding uncertainty sources to the combined uncertainty can be obtained as integrals of the noise spectral density in the characteristic frequency ranges.

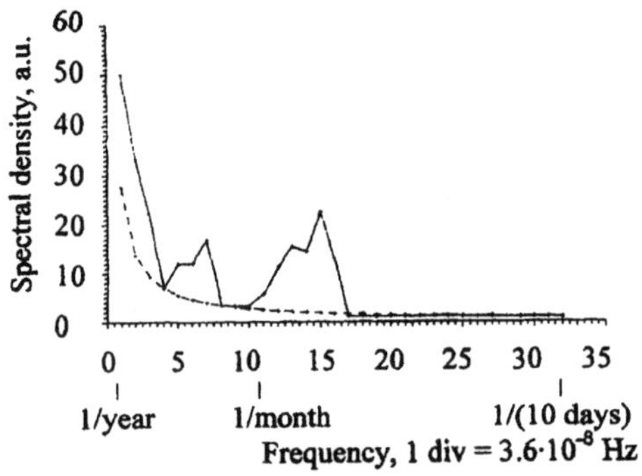

Fig. 1 Noise spectrum of the AAS quality control analytical results: *solid line* from measured results for the period 1995.01–1996.09, *dashed line* 1/f dependence

It is easy to understand that the number of such long measurement data sets available from individual laboratory is restricted. More extensive and profound studies can hardly be carried out without interlaboratory collaboration. Comparison of similar data from different laboratories would allow one to distinguish experimentally between the uncertainty sources characteristic of the method and those of individual laboratories.

The integral of the noise spectrum up to the frequencies comparable to the inverse characteristic time of the individual measurement would contain the full uncertainty due to all sources, representing the transition from one measurement to the long time scale. Of course, the integration frequency range is a problem. The high-frequency limit of the order of the inverse time of the individual measurement seems to be natural. The low-frequency limit, longest time scale, depends on the practical problems being considered. In any case, we ought to take into account that if the measurement time being regarded is comparable with the characteristic time of some uncertainty source, such a source will reveal itself only as a tendency, trend, or bias. If the measurement lasts for an essentially longer time, such a source will manifest itself as an error function. These circumstances represent a more rigid, mathematical, basis to the bias-error relativity [5] problem and could be important when the collaborative trial repeatability and reproducibility data are subjected to uncertainty and bias estimates [12].

If the noise spectrum is regarded as information for assessment of the integral uncertainty in spite of the spectral density increment at low frequencies, an accurate value of the lowest frequency is not important. As an example, a time interval of between 1 and 100 years covers a frequency range of only about $3 \cdot 10^{-8}$ Hz, while time intervals from 1 year to 1 min cover about $2 \cdot 10^{-2}$ Hz. So, for example, contributions to uncertainty for pure 1/f noise from 1 s to 1 h and 1-h to half-year time intervals would be comparable.

Probably the low-frequency noise studies need too much data for routine uncertainty estimates. It is almost accepted that the total uncertainty can be assessed as the combined effect of the identified uncertainty sources. But the situation is possible when it is mainly due to a large number of comparatively weak sources, and experimental uncertainty measurement from data dispersion or a noise spectrum would be highly desirable in such cases.

Another problem is how far the uncertainty can be reduced. Many methods can be used for such reduction. The most straightforward procedure seems to be the "bottom-up" assessment of the contributions from all the uncertainty sources and modification of the analytical method to eliminate the largest of them. As was discussed earlier, the noise spectrum studies can aid such identification and assessment as well. If further elimination of the uncertainty sources is not possible, multichannel measurements can be used to monitor the variations of the important conditions of the experiment (see e.g. [13]). If such information is available, many methods can be used to exclude the effect of the varying experimental conditions from the measurement results through determination of the relations between those variations and variations of the analytical signal. Cross-correlation measurements between the final signal and the variations of the experimental conditions is a very sensitive tool to ascertain how complete the exclusion was. How far can we go in such a process? If the 1/f noise is, as we understand till now, really a general phenomenon due to a large number of effects, this component can hardly be reduced significantly while the characteristic (excess) noise components (see Fig. 1) ought to be reducible. For the case represented in the figure (the accuracy of the spectrum itself is still under study), contributions of the two noises to the uncertainty seem to be comparable, and uncertainty reduction by a factor about 2 can be expected, and only fundamental changes to the system (including preparation of extremely low-noise methods as a special case) could result in a different 1/f noise level.

Nonlinear phenomena – oscillations and chaos

Hitherto, stochastic (or random) fluctuations have been considered. Sometimes special, quasi-oscillatory variations of the analytical signal are observed. The amplitude of the oscillations is comparable with the analytical signal itself, approaching two or more possible values of the signal. Variations of the integral analytical signal depend on both the signal intensity dynamics and the number of the spikes observed. Because of the limited number of such spikes, some special values of the integral analytical signal are preferable, and the distribution of this quantity would not be normal in this case. The well-known example of such variations are spikes in graphite furnace atomic absorption measurements (e.g. [14]). In the still-continuing discussion on the detailed mechanism of this phenomenon, processes being considered usually include autocatalytic, essentially nonlinear, reactions.

More or less similar signal-time dependences seem to follow analytical sample transformations more often than was traditionally expected. Arc discharge is well known as a radiation source susceptible to instabilities. An example of the quasi-oscillatory radiation intensity time dependence from original measurements in a carbon arc is presented in Fig. 2. Two repetitive intensity peaks were very often observed and were traditionally attributed to different volatilities of the sample components or reaction products or different volatilization mechanisms. The studies reported in [15] were possibly

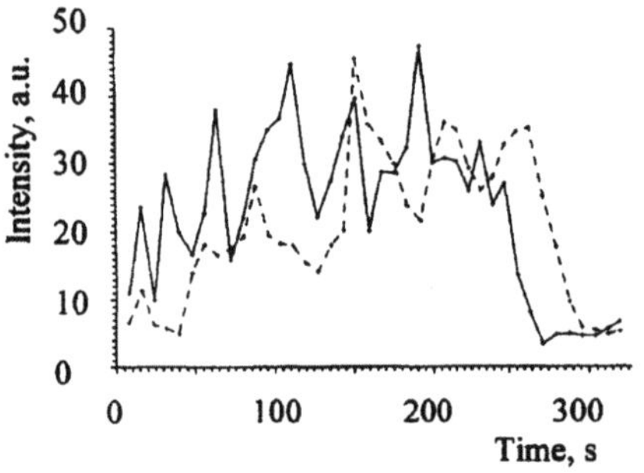

Fig. 2 Two individual Cu 261.8-nm spectral line intensity time dependences. Spiked multielemental oxide sample atomization in carbon arc: current 20 A, sample 15 mg, analyte concentrations 1 mg/g. Data points are means from 14 measured time signal values

the first to take notice of the oscillatory character of the process. It seems characteristic that often the quantity of the sample material involved is too small (or the observation time is too short) for the dynamical structure of the sample transformations to be explicitly displayed, and only one or two peaks are observed. In the atomization of solids and slurries such a situation can be expected more often now than hitherto [16].

The spike phenomenon, while interesting in itself, reduces repeatability and needs special attention in organization of measurements, but can be considered in the context of the common uncertainty assessment. The essential feature seems to be the limited reproducibility or irreproducibility of the time scenario or even the characteristic time constants observed in measurements similar to those of Fig. 2. From the mathematical point of view, if at least three reactions or processes, one of them being nonlinear, are involved, it can result in chaotic dynamics of the phenomenon in the sense that very small variations of the initial conditions can change the character of the phenomenon as a whole. Such numbers of reactions may well be the rule than the exception in the atomization of multielement samples. Thus, competition between different reaction mechanisms, autocatalytic and other nonlinear atomization processes, seem to be possible causes of oscillations and chaos in the analytical chain.

One of the early basic principles of analytical chemistry was to allow the reactions providing the analysis results to proceed to completion. In the fast modern methods of sample volatilization, this principle seems hardly to be maintained. Also, observation of only the initial stages of the process cannot improve the situation: the observed analytical signals depend on the character and reproducibility of the process.

Because of the high sensitivity of the causes of nonlinear phenomena to the initial and dynamic characteristics of the system, such phenomena, especially in interlaboratory tests, can be appreciable sources of uncertainty that are difficult to assess. Of course, even in chaotic phenomena, limits of variation of the signal exist, but the phenomenon as it affects analytical applications is hardly studied and can hardly be properly accounted for in certain cases. Understanding of the problem as such is in progress: "To date, chemometrics has dealt with systems as deterministic or random, yet many chemical systems behave chaotically" [17]. Studies of nonlinear phenomena should help in understanding atomization mechanisms, revealing links between the reproducibility, repeatability and ruggedness tests, and should present additional information for the assessment of balance and reduction of uncertainty.

Conclusions

The problems of uncertainty understanding and assessment in the range of small (a few per cent) and larger uncertainties seem to be different. Experimental online uncertainty analysis should be of interest in both cases. Comprehensive information inherent in interlaboratory tests and quality control measurements could (and possibly should) be more effectively gained, summarized and used in method development. Preliminary results show both the 1/f and characteristic noises in the quality control data. Attention to the nonlinear phenomena including chaos seems to be essential in method preparation and performance studies. From this point of view, problems in uncertainty understanding and reduction remain, and more intensive and in-depth interlaboratory collaboration would be highly desirable for achieving faster progress.

Acknowledgements The author thanks Dr. J. Lubyte, Agrochemical Research Center, Kaunas, for presenting quality control data and the Regional Programme on Quality Assurance PRAG-III for financial support for the participation at the 2nd Workshop "Measurement Uncertainty in Chemical Analysis – Current Practice and Future Directions".

References

1. Eurachem (1995) Quantifying Uncertainty in Analytical Measurement (1st edn). Laboratory of the Government Chemist, London
2. Horwitz W (1997) VAM Bulletin no. 16:5–6
3. Williams A (1997) VAM Bulletin no. 16:6–7
4. ISO (1993) International Vocabulary of Basic and General Terms in Metrology. International Organization for Standardization, Geneva

5. Analytical Methods Committee (1995) Analyst 120:2303–2308
6. Horwitz W (1988) Appl Chem 60: 855–864
7. Sutarno R (1993) Procedure for statistical evaluation of analytical data resulting from international tests. ISO TC 102 N 458
8. Thompson M, Howarth RJ (1980) Analyst 105:1188–1195
9. Hooge FN (1997) In: Claeys C, Simoen E (eds) Noise in physical systems and 1/f fluctuations, Proc 14th Int Conf, World Scientific. Singapore New Jersey London Hong Kong, pp 3–10
10. Howarth RJ (1995) Analyst 120:1851–1873
11. Thompson M, Wood R (1993) Pure Appl Chem 65:2123–2144
12. Ellison S (1997) In: Measurement uncertainty in chemical analysis – current practice and future directions, Proc 2nd Workshop. BAM, Berlin
13. Oberauskas J, Serapinas P, Šalkauskas J, Švedas V (1981) Spectrochim Acta B 36:799–807
14. L'vov BV (1996) Spectrochim Acta B 51:533–541
15. Katasus Portuondo MR, Petrov AA, Sheinina GA (1980) Zh Prikl Spektr 33:19–24
16. Jackson KW, Chen G (1996) Anal Chem 68:243R–244R
17. Brown SD, Sum ST, Despagne F, Lavine K (1996) Anal Chem 68:23R

Accred Qual Assur (2002) 7:153–158
DOI 10.1007/s00769-002-0440-8

Kaj Heydorn
Thomas Anglov

Calibration uncertainty

Presented at the 10th International Metrology Congress, 22–25 October 2001, Saint Louis, France

K. Heydorn
Department of Chemistry,
Technical University of Denmark,
2800 Lyngby, Denmark
e-mail; heydorn@kemi.dtu.dk
Tel.: +45-4525-2342
Fax: +45-4588-3136

T. Anglov
Department of Metrology,
Novo Nordisk A/S, Krogshøjvej 51,
2880 Bagsværd, Denmark

Abstract Methods recommended by the International Standardization Organisation and Eurachem are not satisfactory for the correct estimation of calibration uncertainty. A novel approach is introduced and tested on actual calibration data for the determination of Pb by ICP-AES. The improved calibration uncertainty was verified from independent measurements of the same sample by demonstrating statistical control of analytical results and the absence of bias. The proposed method takes into account uncertainties of the measurement, as well as of the amount of calibrant. It is applicable to all types of calibration data, including cases where linearity can be assumed only over a limited range.

Keywords Metrology in chemistry · Calibration · Uncertainty · Traceability · Verification

Introduction

In chemical analysis, the first and most essential link in the traceability chain is the calibration of the measurement system with known calibrants. The uncertainty from this calibration is therefore a most important uncertainty component, which is sometimes the largest contribution to the combined uncertainty of the analytical result.

Both classical [1] and contemporary [2] guidelines present methods intended for the determination of calibration uncertainty, and many of their recommendations complement each other. The application of these methods to actual calibration data should lead to a correct estimation of the contribution of calibration to the uncertainty budget. However, neither the classical nor the contemporary approach are capable of extracting the maximum information from the data, corresponding to the minimum uncertainty of the calibration; a new approach is therefore warranted.

We have applied a novel combination of classical and contemporary methods to actual calibration data for the determination of Pb in aqueous solution by ICP-AES in the concentration range of 0–10 mg/L. The resulting uncertainty budget was used to test the linearity of calibration using the T-statistic originally developed for the Analysis of Precision [3]; when the deviations of the calibration points from the straight line relationship are completely attributable to the uncertainty of the measurement, the value of T is closely approximated by a chi-square distribution with n-2 degrees of freedom.

Verification of the resulting uncertainty budget can be carried out by the analysis of a particular sample by a number of independent operators, each using their own calibration data. With these data in statistical control, the new method for estimating calibration uncertainty is well suited for propagation of uncertainty in accordance with the BIPM philosophy [4].

Materials and methods

All measurements were carried out by atomic emission spectrometry with an inductively coupled plasma unit for excitation of the sample. The ICP-AES instrument was a Perkin-Elmer model *Plasma II emission spectrometer.*

Calibrants

All samples were prepared by dilution of a Merck certified lead standard with a nominal concentration of 1000 mg/L. Aliquots of the standard were taken with a calibrated pipette and diluted with a mixture of hydrochloric and nitric acids in calibrated volumetric flasks.

Pure acid mixture was used as a blank, and calibration samples were prepared with nominal concentrations of 0.25, 1.00, 2.00, 5.00, and 10.00 mg/L. A nominal 1.5 mg/L sample was chosen as an unknown. Samples were labelled A, B ... G in random order.

Uncertainty budgets prepared in accordance with Example A1 in Ref. [2] provided an estimated relative standard uncertainty of 0.22% for the dilution process.

Measurements

Ten people working together two and two carried out all measurements in one afternoon. Each participant measured all samples in alphabetic order, and the instrument was zero-adjusted at the beginning of each series.

Readings in arbitrary units – counts – are presented in Table 1 together with the mean and standard deviation for each level.

The calibration data – Levels A to F – were tested in accordance with the procedure recommended by ISO [1] as Example 3. No outliers were detected. In addition we used ANOVA to detect any influence from the order of measurement or from a possible difference between the teams; no such effects were found.

The variability of measurements expressed as standard deviations – SD – are therefore assumed to depend only on the measurement level.

Results and discussion

The first step in predicting the uncertainty of calibration is to establish the relationship between the standard deviation and the level of the signal, and this is dealt with in paragraph 7.5 in [1]. Three types of functional relationship are considered for the reproducibility standard deviation, s_r, and the mean level, m

I $s_r = bm$, where b is the slope
II $s_r = a + bm$, where a is the intercept
III $s_r = Cm^d$ $d \leq 1$, where d is the logarithmic slope

Simple linear regression analysis may be used for the determination of the parameters a and b, and from the logarithm of expression III

$\log s_r = \log C + d \cdot \log m$

we can determine in the same way log C as the intercept a, and d as the slope b.

These expressions are mathematically simple, but make no physical or chemical sense. Much more sense is found in Ref. [2] Appendix E4, which reflects the normal situation that the standard deviation is independent of the result, x, at low signal levels and proportional at high levels

$$E4 \quad u(x) = \sqrt{s_0^2 + (x \cdot s_1)^2} \qquad (1)$$

As proposed in paragraph E.4.5.2 in Ref. [2] linear regression of $u(x)^2$ on x^2 can be used to determine s_0 as the square root of the intercept, and s_1 as the square root of the slope.

Relationships found by simple linear regression

The quality of these representations can be judged by the statistic T [3], which compares the deviations of the observed values from the calculated values with the uncertainty of the observations

$$T = \sum_{1}^{n} \frac{(observed - expected)^2}{(standard.uncertainty)^2} \qquad (2)$$

Table 1 Readings obtained by 10 participants for 6 reference solutions of Pb by ICP-AES

Team	Level A	Level B	Level C	Level D	Level E	Level F	Level G
1	1074.0	159.6	2855.6	573.5	13.1	5780.5	830.3
1	1177.9	157.7	2946.1	564.7	9.3	5668.6	888.7
2	1109.3	146.6	2824.1	567.4	9.0	5736.9	875.4
2	1180.3	139.7	2829.7	568.1	7.7	5824.4	883.0
3	1091.9	149.6	2828.4	556.6	13.1	5741.9	860.8
3	1158.1	129.0	2861.0	573.5	–3.3	5667.9	864.4
4	1084.8	145.4	2837.4	568.7	14.9	5673.4	864.9
4	1137.0	151.3	2875.5	555.3	26.2	5794.3	861.2
5	1115.4	147.9	2864.5	560.9	–1.9	5709.5	861.7
5	1145.4	155.5	2920.7	547.4	6.9	5628.7	848.0
Mean	1127.4	148.2	2864.3	563.6	9.5	5722.6	863.8
SD	38.2	9.0	40.7	8.5	8.4	63.9	16.7

Table 2 Test for adequacy of simple functional relationships for representation of the uncertainty of a reading

Level	SD counts	u(SD) counts	Linear Regression		Log/log Regression		sqr/sqr Regression	
			Type I or II	Sq residual	Type III	Sq residual	E 4.5.2	Sq residual
A	38.2	9.0	21.7	3.35	24.8	2.20	23.0	2.85
B	9.0	2.1	12.1	2.01	12.8	3.18	19.6	24.29
C	40.7	9.6	38.8	0.04	33.6	0.55	36.5	0.19
D	8.5	2.0	16.2	14.82	19.8	32.33	20.4	35.81
E	8.4	2.0	10.7	1.32	5.3	2.53	19.5	31.10
F	63.9	15.1	66.9	0.04	42.1	2.10	64.7	0.00
G	16.7	3.9	19.1	0.37	22.8	2.35	21.6	1.53
				T=21.94		T=45.24		T=95.77

and which closely follows a chi-squared distribution with n-2 degrees of freedom, when two parameters are estimated from the observations.

In the set of data in Table 1 we have n = 7 levels, and the uncertainty of each SD from m = 10 replicates is calculated from

$$u(SD) = \frac{SD}{\sqrt{2(m-1)}} \tag{3}$$

in good agreement with the value of 24% presented in Table E.1 in Ref. [4].

In Table 2, results for SD from Table 1 are presented together with their uncertainties at each level, and calculated values are shown for expressions I to III and E4 together with the normalized, squared residuals, contributing to the value of T. The intercept a is not significantly different from 0, which means that expression II is not significantly better than I, although it makes more physical sense.

However, at 5 degrees of freedom all values of T are significant at the 0.1% level, which means that none of the expressions are capable of predicting a correct standard deviation from the mean level.

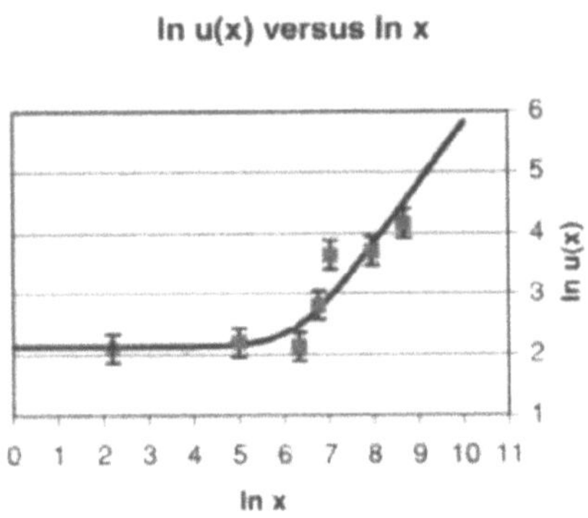

Fig. 1 Standard deviations ± their standard uncertainties according to Eq. 3 as a function of the level. The curve is drawn according to Eq. 4 with the parameters from Table 4 obtained by non-linear regression.

Relationships found by alternative regression methods

Conditions for estimating parameters by simple linear regression are not strictly fulfilled. First of all the independent variable is not without uncertainty; in fact its uncertainty is approximately √2 times larger than that of the dependent variable. However, b is usually below 0.1, which means that variations in x have very little influence on $u(x)$.

More important is that the uncertainty of the dependent variable SD is *not* independent of the value of the variable: in fact it is proportional to this variable, as indicated in Eq. 3.

There are two methods for alleviating this problem:

1. to resort to weighted linear regression to determine slope and intercept
2. to use log $u(x)$ as the dependent variable.

1. Weighted linear regression is carried out as described in paragraph 7.5 in Ref. [1] using initial weights, w_i, equal to $u(SD)^{-2}$ for relationships I and II. Three iterations were needed to obtain a stable solution. For relationship III, w_i equals $u(\log SD)^{-2}$, which is independent of SD, and therefore degenerates to simple linear regression.

   In the linear regression of $u(x)^2$ on x^2 the initial weight of $u(x)^2$ is taken as the squared reciprocal of

   $$u(u(x)^2) = u(x)^2 \cdot \sqrt{\frac{2}{n-1}}$$

   In this case we needed 4 iterations to obtain the stable solution shown in Table 3.

2. With log SD = log u(x) as the dependent variable all results have the same weight in the transformed Eq. (1)

   $$\log u(x) = 0.5 \log(s_0^2 + (s_1 \cdot x)^2) \tag{4}$$

   but the parameters have to be found by non-linear regression.[5].

   This may be carried out by means of the *Solver* program that can be accessed from the usual Microsoft Excel spreadsheet; the starting values were taken from the results obtained by weighted quadratic regression.

The quality of the resulting functional approximations is shown in Table 3, which presents the value of T calculat-

Table 3 Test for goodness of fit of functional relationships for representation of the uncertainty of a reading

Level	SD counts	u(SD) counts	Linear weighted		sqr/sqr weighted		Non-linear weighted	
			Type II	Sq residual	E 4.5.2	Sq residual	E 4.2.1	Sq residual
10	8.4	2.0	7.9	0.08	8.4	0.00	8.4	0.00
148	9.0	2.1	9.6	0.06	8.9	0.01	8.7	0.02
564	8.5	2.0	14.7	9.88	13.5	6.38	12.1	3.28
864	16.7	3.9	18.5	0.20	18.2	0.15	15.7	0.06
1127	38.2	9.0	21.7	3.33	22.7	2.94	19.3	4.41
2864	40.7	9.6	43.3	0.07	54.3	2.02	44.9	0.19
5723	63.9	15.1	78.8	0.97	107.6	8.41	88.4	2.65
				T=14.60		T=19.90		T=10.61

Table 4 Absolute and relative contributions to the standard deviation determined by different types of weighted regression

Type of regression	linear/linear	squared/squared	non-linear
Absolute value s_0	7.75	8.42	8.43
Relative value s_1	0.0124	0.0187	0.0154

ed according to Eq, (2). With 5 degrees of freedom the critical values for T are 15.1 for p <1% and 11.1 for p <5%, which means that neither of the weighted regressions give satisfactory agreement with the observations. Only the non-linear – unweighted – fit is acceptable at the 5% level of significance.

Moreover, the actual deviations seen in the last column of Table 3 are much more evenly distributed over the range than in the other two cases, which confirms that the non-linear case is the best overall representation of the data from Table 1. This is confirmed by the logarithmic plot of the standard deviation as a function of the level shown in Fig. 1.

With the ready availability of non-linear regression to users of Microsoft Excel the computational effort is probably less than that associated with the iterative, weighted regression. Consequently we are going to use the non-linear results presented in Table 4 for estimating the uncertainty of our calibration.

Linearity of calibration function

The calibration function expresses the relationship between the known concentrations of Pb in the reference solutions A to F and the readings of the instrument; to the extent that the superposition principle is observed this function is a straight line. However, deviations are expected to occur at higher levels of the indicator, so that the linearity range becomes restricted.

Instrument calibration is found by weighted linear regression of the mean readings using the reciprocal square of their respective uncertainties as the initial weight and Eq. (1) to calculate weights for subsequent iterations. The linearity is tested by the value of T from Eq. (2) with the standard uncertainties calculated from Eq. (1) using the original data, and results are presented in Table 5. Simple linear regression gave in all cases a less satisfactory fit.

It is concluded that a perfectly satisfactory linear calibration is maintained up to approx. 2 mg/L, which means that the observed deviations from the linear relationship are fully accounted for by the assumed uncertainty of the readings. Thus no additional uncertainty contribution needs to be included in the uncertainty budget.

The uncertainty budget

The uncertainty budget for the determination of Pb by ICP-AES is reduced to

$$u(y) = \sqrt{s_0^2 + (y \cdot s_1)^2}$$

with s_0=8.43 and s_1=0.0154, where y is the reading. The conversion of a reading to a concentration is linear up to at least 2 mg/L, and the additional uncertainty of 0.22% is negligible in comparison with u(y), which is always larger than 1.5%.

Experimental verification of the budget is based on the separate determination by each of the participants of the concentration of Pb in the unknown sample G from Table 1. With the uncertainty budget above each participant will calculate a result and its uncertainty from his own data alone.

The participant may choose to assume a linear calibration and use all his observations to arrive at a result, or the participant may calculate the result by interpolation only between the two calibration points bracketing the unknown.

Results based on linear calibration

Each participant determines his own calibration data from his readings in Table 1 of the reference solutions A

Table 5 Tests for linearity of calibration based on weighted linear regression over a decreasing range

Reference mg/L	Mean counts	Uncertainly u(counts)	Linear weighted		Linear weighted		Linear weighted	
			calibration	Sq. residual	calibration	Sq. residual	calibration	Sq. residual
0	9.5	2.7	5.0	2.89	6.3	1.47	8.6	0.11
0.248	148.2	2.9	147.1	0.18	147.6	0.05	148.3	0.00
0.992	563.6	2.7	573.4	6.54	571.5	4.31	567.2	0.87
1.979	1127.4	12.1	1138.9	3.57	1134.0	1.16	1122.9	0.54
4.956	2864.3	12.9	2844.7	1.90	2830.3	5.73	–	–
9.88	5722.6	20.2	5666.1	4.08	–	–	–	–
			4 d.f.	*** T=19.15	3 d.f.	** T=12.72	2 d.f.	n.s. T=1.52

Table 6 Results obtained by using a linear calibration function over a range from zero to at least 2 mg/L

Participant Initials	Calibration Points	Linear calibration		T statistic	Sample G Reading	Result mg/L	Uncertainty u(x)
		a	b				
TAng	4	20.42	542.45	3.69	830.3	1.493	0.039
HkW	6	8.56	581.20	6.51	888.7	1.514	0.032
EIS	6	5.00	569.12	3.11	875.4	1.529	0.030
RiX	6	1.03	581.93	5.90	883.0	1.516	0.032
LonM	6	6.39	565.10	7.73	860.8	1.512	0.033
FinC	6	–6.58	580.02	2.37	864.4	1.502	0.029
IMP	6	7.90	563.65	6.91	864.9	1.520	0.033
DB	5	14.90	565.60	6.61	861.2	1.496	0.036
JNoS	6	–1.04	573.21	2.39	861.7	1.505	0.029
Dorman	6	4.88	573.28	8.79	848.0	1.471	0.033
					Mean	1.507	0.010 T=2.29

to F with the known concentrations given in Table 5. The zero-offset and the slope are determined by weighted regression assigning to each reading y a weight of $u(y)^{-2}$, as calculated from the uncertainty budget above. The linear range that may be used is determined in exactly the same way as in Table 5.

With a 95% confidence level of T =9.49 at 4 degrees of freedom 8 out of 10 participants could use all 6 calibration points to determine their calibration data. One participant had to eliminate the highest level and one the two highest levels; only the last case was therefore restricted to the range found in Table 5.

Calibration data and results for the concentration of Pb in sample G are shown for individual participants in Table 6 together with the estimated uncertainty of their predicted result, x_{pred}. Its uncertainty is calculated in accordance with the approach presented in Appendix E.3 of Ref. [2], while using the T-values calculated for the weighted regression from paragraph 7.5 in Ref. *[1]*

$$u(x_{pred})^2 = \frac{1}{b^2}\left[u(y_{obs})^2 + \frac{T}{T_1(n-2)}\left(1+\frac{(x_{pred}T_1 - T_2)^2}{T_1T_3 - T_2^2}\right)\right] \quad (5)$$

The agreement between individual results is actually better than expected from their estimated uncertainties, and the calculated T-statistic for 9 degrees of freedom is far below the 95% level of significance of 16.9.

The calibration uncertainty as expressed by the second term in the expression (5) is based on 4–6 observations and therefore a minor contribution compared to the measurement uncertainty of the reading for sample G based on only one observation. Thus the variability among final results is probably less than the variability of the readings on which these results are based, which indicates that the readings obtained by a particular participant are not completely independent of each other.

Results based on interpolation

Calculation of results may also be carried out without assuming a linear calibration function over a particular range, but only between two calibration points bracketing the unknown. This is simply done as linear interpolation and has the further advantage that uncertainty in both x and y values are easily taken into account.

Results for the unknown sample G are obtained by each participant using simple interpolation between reference levels A and D; the interpolation formula is entered into a spreadsheet, and the procedure described by Kragten in Appendix E.2 in Ref. [2] calculates the com-

Table 7 Results obtained by interpolation between two references

Participant Initials	Reference samples		Sample G Reading	Result mg/L	Uncertainty u(x)
	1.979 mg/L	0.992 mg/L			
TAng	1074.0	573.5	830.3	1.498	0.037
HkW	1177.9	564.7	888.7	1.514	0.032
EIS	1109.3	567.4	875.4	1.553	0.036
RiX	1180.3	568.1	883.0	1.500	0.032
LonM	1091.9	556.6	860.8	1.553	0.036
FinC	1158.1	573.5	864.4	1.483	0.033
IMP	1084.8	568.7	864.9	1.558	0.038
DB	1137.0	555.3	861.2	1.511	0.033
JNoS	1115.4	560.9	861.7	1.527	0.035
Dorman	1145.4	547.4	848.0	1.488	0.032
			Mean	1.516	0.011 T=5.71

bined uncertainty from the uncertainties of the readings, as well as from the assigned value of the reference samples.

Table 7 lists results and their uncertainties, which show that also in this case the variability may be less than expected from the uncertainties with a value of the statistic T that is well below the 95% level.

Bias

The unknown sample G was prepared by dilution of the same lead standard as the calibrants, and its concentration was 1.485 mg/L with a standard uncertainty of 0.008 mg/L.

Weighted mean values of the analytical results obtained by the participants are presented in Tables 6 and 7 together with their standard uncertainty. In either case the bias is positive, but barely significant at the 5% level of confidence.

For the results based on linear calibration the bias is 0.022 mg/L with a standard uncertainty of 0.013 mg/L, which is not significantly different from zero. For the interpolated results the bias is 0.031 mg/L with a standard uncertainty of 0.014 mg/L which slightly exceeds the critical value of z=1.96.

Conclusion

Neither simple nor weighted linear regression yielded satisfactory results for the functional representation of standard deviation as a function of level with any of the alternatives proposed in Ref. [1]. Only the use of non-linear regression of the logarithmic standard deviation on the expression recommended in Ref. [2], gave acceptable agreement with experimental data.

An uncertainty budget based on this representation for the determination of Pb in an aqueous solution yielded experimental results in statistical control and without bias.

Acknowledgements The authors are indebted to the Danish Institute of Occupational Health for making their ICP-AES available to the first course in "Metrology in Chemistry" held at Novo Nordisk in the year 2000. The original readings made by the participants in the course are used in this study exactly as they were made on 2000-10-09. In particular we acknowledge Miss Dorrit Meincke for supervising the ICP-AES instrument and for preparing the calibration standards under careful statistical control.

References

1. ISO 5725–2, Accuracy of Measurement Methods and Results, 1st Edition, 1994
2. EURACHEM/CITAC Guide, Quantifying Uncertainty in Analytical Measurement, 2nd edn. 2000
3. Heydorn K (1991) Mikrochim Acta (Wien) III: 1–10
4. ISO Guide to the Expression of Uncertainty in Measurement, Geneva, 1993
5. Heydorn K, Griepink B (1990) Fresenius Z Anal Chem 338: 287–292

Accred Qual Assur (2001) 6:372–375

Riitta Maarit Niemi
Seppo I. Niemelä

Measurement uncertainty in microbiological cultivation methods

R.M. Niemi (✉) · S.I. Niemelä
Finnish Environment Institute,
P. O. Box 140, 00251 Helsinki, Finland
e-mail: maarit.niemi@vyh.fi
Tel.: +358-9-40300853,
Fax: +358-9-40300890

Abstract Microbiological analyses are carried out on clinical, food, feed and environmental samples. The aims of the analyses are diagnostic or estimation of the safety or the quality of the sample. Important decisions are made on the basis of microbiological analyses. Little attention, however, is paid to the uncertainty of measurement of microbiological analyses. In microbiological cultivation techniques the result is obtained by counting individual objects. The normally low number of counted objects strongly affects the result of the analysis and its uncertainty. Because of the importance of the particle statistical variation to the uncertainty, the approaches developed for chemical analyses are not directly applicable to microbiology. This paper discusses microbial analyses and describes a novel guidance document for the estimation of measurement uncertainty in culturing methods [1].

Keywords Microbiology · Cultivation methods · Measurement uncertainty

Introduction

It has been asked whether microbiology is more like the art of cooking than science. Cooking, and doing it correctly (e.g. utilising gravimetric, volumetric, temperature- and time-related metrological information) is an integral part of microbiological cultivation methods. Science is the ever present basis for analytical work in microbiology.

The aim in microbiological analyses is usually to detect and enumerate a known species or group of microorganisms in a measured amount of sample. If measurement means counting and identification, quantitative microbiological analyses belong to the sphere of metrology. Traceability to primary measurement standards can hardly be achieved, but the measurement units are the numbers per gravimetric or volumetric units.

Internationally available reference strains and materials, and international performance tests for microbiology are becoming increasingly available to aid traceability and comparability evaluation of analyses carried out in different laboratories.

In microbiological analysis, culture techniques are important because it is often relevant to detect viable microorganisms. Often the target microorganisms in human, animal, food or environmental samples constitute only a minor fraction of the microorganisms present. Different selection principles and indicator systems are applied in order to facilitate the growth of the target microorganism yielding characteristic reactions, while suppressing the growth of other microorganisms. The primary cultivation result is usually not sufficiently reliable, but necessitates the use of further tests to confirm the identity of the target microorganism. In practice, confirmation tests in routine work do not provide taxonomically valid identification, but rely on a limited set of tests confirming the identity of the target with high probability. The valid identification of microorganisms can be regarded as a rather complex measurement and the uncertainties inherent to microbial taxonomy complicate evaluation of the uncertainty involved. In this paper, the uncertainty of taxonomically valid identification is not discussed. In routine work, identification is usually limited to the agreed confirmation tests. Uncertainty of confirmation is addressed only to the extent of estimating the binomial sampling variance of fractions confirmed.

The Finnish guidance document for measuring uncertainty is a novel approach for the estimation of uncertainties in microbiological measurements based on cultivation and including confirmation [1]. This document has been elaborated as an activity of the Advisory Committee for Metrology in Finland and the Centre for Metrology and Accreditation in Finland will produce a translation into English. The complex measurements such as taxonomically valid identification are not included in this document. It is hardly possible to know the exact number of atoms or molecules in chemical analysis. Similarly, it is not possible to know the actual number of viable target cells or spores in a sample.

Therefore, it is only possible to measure the relative uncertainty of measurement. In this paper the uncertainty measurements of microbiological cultivation methods are described according to Niemelä [1].

Principles

The instructions for the calculation of uncertainty of measurement elaborated for chemistry [2, 3] are not directly applicable to microbiology. In the guidance document [1] the principles of these chemistry documents are interpreted and adapted to microbiology.

In microbiological measurements sample pretreatment is normally limited to homogenisation and dilution (or concentration by filtration). The measured portion of the sample or its dilution is transferred to the "detector" and the result is obtained by counting individual objects. The sole principle of the measurement is reflected in the formula for calculation of the result:

$$y = F \cdot \frac{C}{V}$$

where

y = number of microorganisms per volume

F = dilution factor

V = volume of the portion of the final dilution

C = number of microbial particles in V.

It is evident from the above formula that the counted number of microbial particles strongly affects the result. The particle statistical variation can be estimated by applying the Poisson theory:

$$RSD_c = u_c = \sqrt{\frac{1}{C}}$$

Because the microbial detectors function optimally at particle numbers between 25 and 100 per test portion, the particle statistical variation often dominates the uncertainty of the measurement. Therefore uncertainty estimates such as reproducibility and repeatability determined in collaborative efforts are not generally applicable in microbiology and it is not worthwhile investing much effort in calculating values for these parameters (see Type A below). Instead, the uncertainty of the method can be expressed as a formula into which observed values of each measurement can be inserted.

Because the Type A uncertainty estimation (based on replicate measurements) is usually not economically feasible in microbiology, the emphasis in the guidance document [1] is on the Type B approach.

Type A: The standard uncertainty is calculated from n independent replicate measurements $x_1, x_2, ..., x_n$ as the experimental standard deviation [2]:

$$s_x = \sqrt{\frac{\sum_{i=1}^{n}(x_i - x)^2}{n-1}}$$

The number of replicates must be rather high because even 30 replicates from sample sources following normal distribution yield estimates of the standard deviation with only 13% relative uncertainty.

Type B: According to [2], Type B uncertainty is obtained using other approaches than replicate samples. The uncertainty variance u^2 or the standard uncertainty u are based on the whole body of scientific information available (with the exception of replicate measurements) on the possible variation of the measurand. Information of statistical theory, earlier measurements, experience or general beliefs on instruments and materials, specifications of the manufacturer, published reference values in calibration and certification reports, and uncertainty estimates in handbooks can be utilised.

The common sources of uncertainty in cultivation methods are sample stability, dilution, counting (including particle statistical variation and personal interpretation of the target), yield on the medium, crowding effect (coincidence error) and uncertainty of confirmation. The combined uncertainty can be calculated as the quadratic sum of different uncertainty components.

Compilation of uncertainty in cultivation methods

In microbiological cultivation methods only four types of detection systems are used: the one-plate instrument, the set of plates instrument, the one-tube detector (Presence/Absence) and the set of tubes instrument (Most Probable Number). All these classes necessitate different formulas for uncertainty estimates. Different media and incubation conditions, and confirmation and identification tests offer the versatility needed for the enumeration of different organisms. However, this versatility is not reflected in the principles of the estimation of uncertainty.

In microbiological cultivation methods the sample is homogenised, after which a measured amount is diluted or concentrated (e.g. by membrane filtration) to the proper measuring range of the analytical method. Suitable aliquots are transferred to the plates, tubes or wells and, after incubation, colonies or numbers of positive and negative reactions in tubes or wells are counted.

It is often necessary to confirm that typical colonies or tubes yielding a positive reaction actually show the presence of the target microorganism. When it is possible to confirm all the presumptive results, there is no uncertainty due to the random error caused by sampling for binomial attributes. On the other hand, when only a fraction of the presumptive positive reactions are tested for confirmation, a significant increase will occur. The binomial sampling variation in confirmation tests should be addressed in addition to the error due to Poisson distribution of presumptive target colonies counted. In the primary cultivation step, counting results of typical colonies or reactions are susceptible to subjective judgement that can cause significant uncertainty.

The components causing uncertainty can be seen from the formula used for calculating the measurement result. Usually it includes all the dilution steps, colony counts from different plates (or most probable numbers) and numbers of isolates tested further and those confirmed. The uncertainty of counting of colonies caused by differences between technicians should be included. It is possible to correct systematic errors, e.g. confirmation rate (often expressed as % of typical colonies or reactions confirmed).

In the calculation of the result multiplication is needed and, therefore, the combined uncertainty is composed of the sum of the relative uncertainties. Fortunately, the components of uncertainty tend to be independent in microbiological measurements, so that covariances need seldom be considered.

Corrections for systematic errors, e.g. confirmation rate and dilution error, are mostly multipliers. The corrected final result is therefore expressed as:

$$y = k_1 \cdot k_2 \cdots k_n \cdot F \cdot \frac{C}{V}$$

where

k = coefficients of systematic errors.

The uncertainty of each correction coefficient should be included in the calculation of uncertainty.

As an example, the relative total uncertainty for a one-plate instrument can be calculated with the following formula:

$$u_y = \sqrt{u_{k_1}^2 + u_{k_2}^2 + \cdots + u_{k_n}^2 + u_p^2 + u_F^2 + u_c^2 + u_V^2 + u_z^2}$$

where

u_k^2 = correction uncertainties (e.g. dilution correction, yield correction)

u_p^2 = uncertainty of the confirmation coefficient

u_F^2 = dilution factor uncertainty

u_c^2 = particle statistical uncertainty

u_V^2 = inoculum volume uncertainty

u_z^2 = individual counting uncertainty.

Instructions for the estimation of individual uncertainty components are described by Niemelä [1].

A shortcut to uncertainty estimates of the instrument with several plates

Consider a detection instrument consisting of several plates with colony counts c_i derived from the volumes v_i (i=1,2,..., n) of the final suspension. The average particle concentration x of the final suspension is estimated from

$$x = \frac{\sum c_i}{\sum v_i}$$

The microbial concentration of the original sample is obtained from the calculation

$$y = Fx$$

where F is the dilution factor.

The uncertainty of the average particle density of the final suspension consists of particle statistical variation (u_c), counting uncertainty (u_z) and the uncertainty of inoculum volume measurements (u_V) including possible dilution effects within the detection instrument. These components merge to form the uncertainty of x. It can be estimated by using the log likelihood ratio statistic G^2 calculated with the following formula:

$$G_{n-1}^2 = 2\left[\sum_{i=1}^{n} c_i \ln\left(\frac{c_i}{v_i}\right) - \left(\sum c_i\right)\ln\left(\frac{\sum c_i}{\sum v_i}\right)\right]$$

where

c_i = colony count of the plate i

v_i = inoculum size (in ml of final suspension) on the plate i

n = number of plates

The relative uncertainty of the microbial concentration x is calculated with the formula

$$u_x = \sqrt{\frac{G_{n-1}^2}{n-1} \cdot \frac{1}{\sum c_i}}$$

The uncertainty of confirmation and any other uncertainties and correction factors that are known to affect the result can be taken into account in the same way as before. The formula for the combined uncertainty in the shortcut systems is therefore:

$$u_y = \sqrt{u_{k_1}^2 + u_{k_2}^2 + \cdots + u_{k_n}^2 + u_p^2 + u_F^2 + u_x^2}$$

where

u_k^2 = correction coefficient uncertainties

u_p^2 = binomial sampling uncertainty of confirmation

u_F^2 = dilution factor uncertainty

u_x^2 = measurement uncertainty of the multi-plate "instrument".

Discussion

Traditionally, microbiological uncertainty estimates have been based on replicate measurements or on the Poisson theory only. Uncertainty estimates that include different factors contributing to the uncertainty of microbiological measurements [1] offer the possibility to identify main sources of error and, therefore, aid in identification of the weak points in the analytical procedure. Attention should be directed to these estimates in order to improve reliability of the measurement. The novelty, however, means that it is challenging to the microbiologist to embark on the calculation of uncertainty estimates.

The temporal and spatial variations in numbers of microorganisms in the sample sources may be vast. Frequent sampling and replicate samples yield some information about this variation and should be considered together with uncertainty estimates from laboratory analyses.

Acknowledgements We thank Mr. Michael Bailey for correction of the language.

References

1. Niemelä SI (2001) Uncertainty of quantitative microbiological culture methods. (In Finnish, to be translated into English and later available in electronic form at: www.mikes.fi). The Centre for Metrology and Accreditation in Finland (MIKES) J1/2001, 69p ISBN 952–5209–3
2. ISO (1995) Guide to the expression of uncertainty in measurement, 1st edn. International Organization for Standardization (ISO), Geneva
3. EURACHEM (2000) Quantifying uncertainty in analytical measurement, 2nd edn. Laboratory of the Government Chemist (LGC), Teddington, UK

Accred Qual Assur (2002) 7:228–233
DOI 10.1007/s00769-002-479-6

Hans Andersson

The use of uncertainty estimates of test results in comparisons with acceptance limits

Abstract When a test is performed in order to qualify a material or a product for a certain use, the result is generally compared with an acceptance limit. The test result has an uncertainty which should be estimated and stated (e.g. in accordance with GUM). Very often this is not the case. Further, discussions often arise on the issue of how the uncertainty shall be considered in relationship to the acceptance limit. The intention of this note is to describe, in simple terms, the statistical background and to give some recommendations. In short, there are two clean-cut, extreme situations. The first case is when the uncertainty of the testing procedure is the dominating factor. Here it is found that the estimates of single laboratories cannot, generally, be used for comparisons with acceptance limits. One should have standardised, well-verified estimates based on comprehensive investigations of the method. It can also be concluded that comparisons between test results and acceptance limits have to be made with regard to the actual circumstances, as, e.g. how the acceptance limit is related to the risk. In the second case, the variation in the property of the material or product dominates and the uncertainty of the testing procedure is negligible. When the results are non-quantitative (go – no go), statistical methods can be used to estimate the risk taken with a certain sampling and acceptance strategy that a certain proportion of the batch to be delivered does not qualify. This should be considered more often in standardisation of product test methods. When the results are quantitative, a statistical analysis should be performed and the uncertainty should be compared with the acceptance limit as before, from the actual circumstances. When effects of testing uncertainty and product variation are comparable a sound treatment requires extensive experimental work. No short cuts can be made without loss of confidence!

Keywords Uncertainty · Conformity assessment · Acceptance limit

H. Andersson (✉)
SP Swedish National Testing and Research Institute,
P.O. Box 857, 501 15 Borås, Sweden
e-mail: hans.andersson@sp.se
Tel.: +46-33-165000
Fax: +46-33-165010

Assessment of risk

Consider a measurable property of a product, e.g. the content of a poisonous substance. There is an uncertainty in the measurement, x, which may depend on variations in the product and on imperfections in the measurement (sampling, weighing, etc.). This uncertainty may be described by a probability function $P_E(x)$, i.e. the probability that the true content lies between x and x+Δx is $P_E(x)\Delta x$.

Further, assume that the risk for damage, e.g. illness, can be described by a function $P_s(x)$ such that the risk if

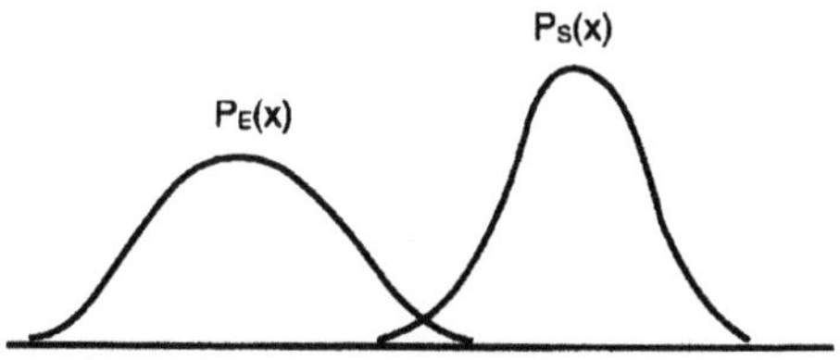

Fig. 1 Probability of property, $P_E(x)$, and risk, $P_s(x)$

the content lies between x and x+Δx is $P_s(x)\Delta x$. Then the probability for damage is described by the general relationship

$$P_{Damage} = \int_{-\infty}^{\infty} P_E(x)\left[\int_{-\infty}^{x} P_s(t)dt\right]dx \quad (1)$$

See Fig. 1. Note that a similar relationship may be set up for the case when the value of the property shall **exceed** the acceptance value, e.g. the strength of a structure.

If as well $P_E(x)$ as $P_s(x)$ have Gaussian distributions, not at all necessary but useful as an example, with means m_E and m_s, and standard deviations σ_E and σ_s, respectively, the risk, probability for damage is

$$\phi\left(\frac{\pm(m_E - m_s)}{\sqrt{\sigma_E^2 + \sigma_s^2}}\right) \quad (2)$$

where ϕ is the normalised Gaussian probability distribution. The plus and minus signs relate to the two cases above, respectively. It is readily seen that if the distribution of property and risk are well separated, the risk will be small. In practise, there are considerable difficulties to find $P_E(x)$ and $P_s(x)$, and therefore assumptions and approximations are used. The implications of this are treated next.

Assessment of $P_E(x)$

$P_E(x)$ comprises a description of how the property varies within the product, the "population". This is described under "Qualitative testing of products". Here, we consider the variation in the measured value of the property only due to the uncertainty in the measurement. This gives an analogous probability function $P_E(x)$. In cases where both the uncertainty in the measurement and the variation in product property are of the same magnitude, a special treatment is required, where the probability functions are combined.

The experimental determination of $P_E(x)$ is very costly. In theory, it would mean the performance of a "large" number of tests in each of a "large" number of laboratories. In practise one is confined to a limited number of tests in a small number (5–10) of laboratories. Then the degree of confidence becomes low, which has to be compensated for in the resulting estimation of the standard deviation.

The tests give, through statistical analysis, estimates of the mean and standard deviation of $P_E(x)$. A parameter related to the estimated standard deviation is the reproducibility. Further, statistical analysis can also yield the part of the uncertainty which is due to variations within the laboratories, which is related to the parameter repeatability. The statistical analysis, analysis of variances, and definitions of parameters are given in, e.g. ISO 5725 [1].

However, in most cases, and contrary to metrology, there are no means or time available for experiments at all, so estimates of the uncertainties are made through professional judgement, so called Type B estimates in [2].

Generally, one has to combine a number of uncertainty estimates for different measurements in order to obtain the uncertainty estimate for a complete test result or chemical analysis. It is then presumed that there exists some functional relationship, not always possible to define in mathematical terms, between the measurement, $t_i (i = 1,n)$, results and the test result (x)

$$x = f(t_1, t_2, \ldots, t_n) \quad (3)$$

If t_i are independent stochastic variables with standard deviations σ_{t_i} the following formula is generally valid

$$\sigma_x^2 = \sum_{i=1}^{N}\left(\frac{\partial f}{\partial t_i}\right)^2 \sigma_{t_i}^2, \quad (4)$$

If there is a correlation between some of the t_i:s, Eq. (4) becomes more complex, see [2]. In many cases f has the form

$$x = \pi_{i=1}^{n} t_i^{k_i}$$

then

$$\left(\frac{\sigma_x}{x}\right)^2 = \sum_{i=1}^{N} k_i^2 \left(\frac{\sigma_{t_i}}{t_i}\right)^2 \quad (5)$$

Obviously, if some t_i has a large standard deviation, or if some k_i is large, this component dominates.

According to GUM [2] one shall assess the components σ_{ti} in the following way and call them "standard uncertainties". If a σ_{ti} is based on experiments it is called a Type A standard uncertainty. It is simply the standard deviation estimate of the mean value. If it is not, it shall be estimated through professional judgement following certain, rather open, recommendations in GUM and be called a Type B standard uncertainty. These entities are then inserted in Eqs. (4) or (5), yielding a standard uncertainty, σ_x, for the test result x. The degree of confidence of σ_x depends on the experimental evidence and on the Type B estimates. A detailed description of this is given in GUM [2], Appendix G. See also [3]. The value of σ_x thus defined is then used for decision making, most often disregarding its degree of confidence.

It is worth noticing that Type A and Type B uncertainty estimates can both represent variations within as

	Type A estimate	Type B estimate
Variation between laboratories	X	X
Variations within laboratories	X	X

Fig. 2 Types of uncertainty components

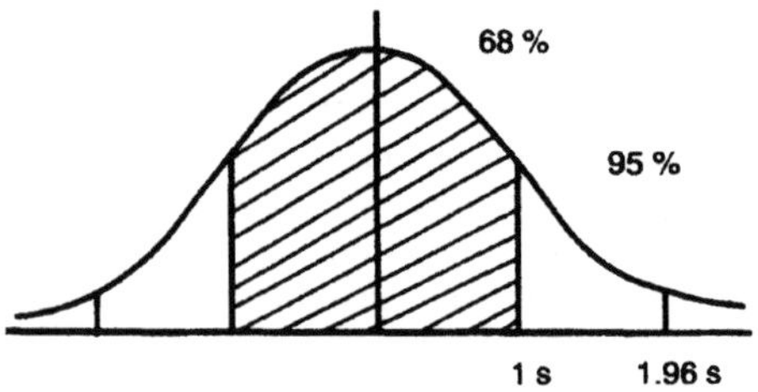

Fig. 3 Assumption of coverages of expanded standard uncertainties

well as between laboratories. Components of uncertainty may hence exist in all four of the positions of the matrix in Fig. 2.

It is often, erroneously, presumed that Type A uncertainties represent variations within a laboratory and that Type B uncertainties represent variations between laboratories. As an example, measurements with a volt meter of class 0.1% may be taken. Here, there is an uncertainty of each specimen, i.e. between "laboratories", within the nominal limits –0.1% to +0.1%, i.e. normally a Type B uncertainty. But there is also variations, uncertainties within each laboratory, due to operations, etc., which may be estimated as a Type A or Type B uncertainty.

As mentioned, the σ_x found, the standard uncertainty for the test results is now presumed to be useful as a measure of the standard deviation of $P_E(x)$. It is also presumed that $P_E(x)$ is approximately Gaussian so that "expanded" standard uncertainties may give coverages of $P_E(x)$ according to Fig. 3. This is based on the general relationship that the sum of a number of stochastic variables becomes approximately Gaussian, independently of the distributions of the various variables.

Very often, an expansion factor of K=2 is recommended (instead of 1.96) to represent a coverage of 95% (97.5% one-sided). Two reservations should be made. It is not quite true that Eq. (4) gives an approximately Gaussian distribution, e.g. if there is one or two dominating factors which are not Gaussian, but perhaps "rectangular". This gives errors, in particular for large K-values. Further, the degree of confidence becomes low if the experimental evidence is poor. This should be compensated for, through the Student (t) distribution, but this is very rarely done.

$P_E(x)$ estimates in the real world

The procedures described in "Assessment of $P_E(x)$" are best suited to metrology and certain areas of chemical analysis where, e.g. statistical analysis can be used extensively thanks to large numbers of repeated measurements. They are less easy to apply in product testing where only few, expensive experiments can be performed and where the varying properties of the products are of importance. See also "Qualitative testing of products".

Even in favourable situations there are problems, however. Extensive background data in chemical analysis has been collected through IMEP, large programmes for proficiency tests in "pure" chemistry. One example is shown in Fig. 4, which is typical for many investigations.

Here, ordered results from 168 laboratories are shown from analysis of lead in water together with their estimated, extended uncertainties (K = 2, meaning a coverage of approximately 95%). As can be seen, the estimates vary strongly and are generally much too small. Probably uncertainty components have been both overlooked and underestimated. This indicates that there is room for training and education and that guides as [4] should be widely spread and used. It can also be seen that there is no distinction between accredited and non-accredited laboratories in this investigation performed some years ago. This goes both for mean values and uncertainty estimates.

With Fig. 4 as a basis one can discuss what would happen in some real cases. Assume, for instance, that there is a legal regulation that the lead content may not be more than 0.12 μ mol/kg. Then, around 20 laboratories, to the left in the diagram, accredited and non-accredited, will approve the test sample although the content is well above the limit, even if one uses the whole stated uncertainty as a margin.

Hence, one should be careful to use the uncertainty estimate of a single laboratory for comparison with stated limiting values in testing, particularly when many Type B estimates have to be relied upon. There is a significant risk that the estimates made by a single person or a group in a single environment make omissions or misjudgments. If possible one should use an uncertainty estimate which includes the experience of several laboratories, e g from interlaboratory comparisons in a learning process.

If all the laboratories use the same method in the same way and make correct uncertainty estimates (K=2) one would end up with a diagram like Fig. 5, analogous to Fig. 4. All laboratories state approximately equal uncertainties and they cover the true value except in a few cases. (The S-shaped curve is an idealised, ordered sample of $P_E(x)$.)

So, if one could, anyhow, obtain a trustworthy estimate of $P_E(x)$ and express it with the uncertainty, how

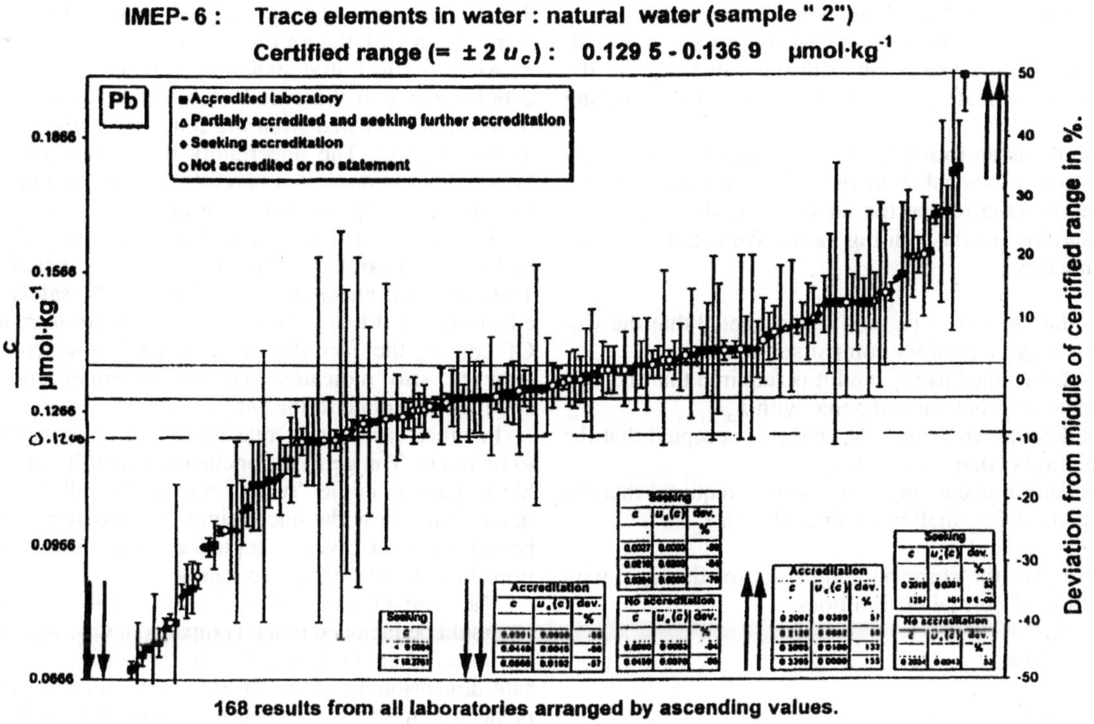

Fig. 4 Example of uncertainty estimates in a large interlaboratory comparison

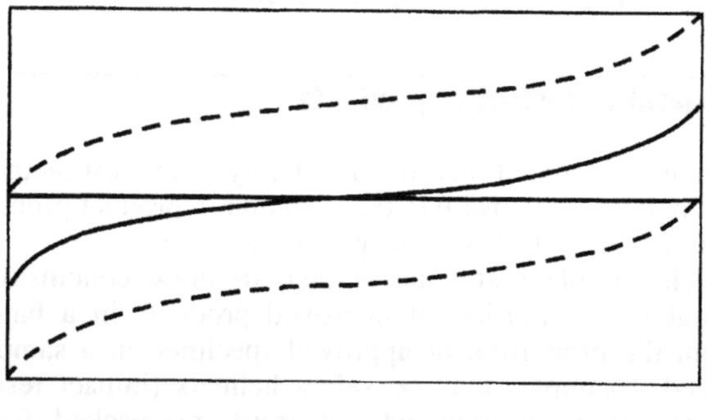

Fig. 5 Idealised diagram of results and uncertainties, cf. Fig. 4

can it be used for comparisons with an acceptance limit? This issue is treated next.

Replacing the risk distribution $P_s(x)$ with a limiting value

For various reasons authorities, and written standards, routinely replace the real distribution of risk with a limiting value which is intended to "safely" encompass all risks. In analogy with Fig. 1, this corresponds to a situation according to Fig. 6. The limiting distribution is formally a Dirac delta function, and one gets

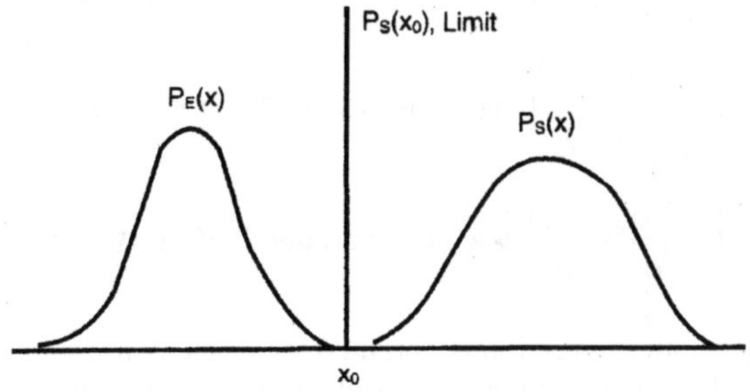

Fig. 6 Replacing real risk with a limiting value

$$P_{Limit'} = \int_{-\infty}^{\infty} P_E(x)\left[\int_{-\infty}^{\infty} \delta(t-x_o)dt\right]dx = \int_{-\infty}^{\infty} P_E(x)1(x-x_o)dx$$

where $1(x-x_o) = 1$ if $x>x_o$ and 0 else, i.e.

$$P_{Limit'} = \int_{x_o}^{\infty} P_E(x)dx$$

Assuming a Gaussian distribution yields

$$P_{Limit'} = \phi\left(\frac{m_E - x_o}{\sigma_E}\right)$$

This is in accordance with Eq. (3), if $x_o = m_s$ and $\sigma_s = 0$.

The result may seem trivial, but the reasoning gives some insight in the process of replacing a real risk distribution with a limiting value.

It is evident that the relationships between the parameters, m_s, σ_s and x_0 play an important part when it shall be decided how a test result with the value m_E and the uncertainty estimate $2\sigma_E$ shall be related to a limiting value x_0.

Assume, as an example, that the limiting value x_0 is set in some relationship to the risk distribution P_s (x), and that we require that the measure m_E shall have certain "margins" to the limiting value. We use the following four cases:

1) x_0 is determined as m_s, and it is accepted that the value m_E may be used for comparison with x_0.
2) x_0 is determined as m_s, and it is required that the value $m_E + 2\sigma_e$ shall be compared with x_0.
3) x_0 is determined as $m_s - 2\sigma_e$ and it is accepted that the value m_E is compared with x_0.
4) x_0 is determined as $m_s - 2\sigma_s$ and it is required that the value $m_E + 2\sigma_E$ shall be compared with x_0.

This will give the corresponding risks for damage, with assumption of normal distributions.

1) $\phi(o) = 50\%$

2) $\phi\left(\frac{-2\sigma_E}{\sqrt{\sigma_E^2+\sigma_s^2}}\right)$, which is around 8% if $\sigma_E = \sigma_s$

3) $\phi\left(\frac{-2\sigma_s}{\sqrt{\sigma_E^2+\sigma_s^2}}\right)$, which is around 8% if $\sigma_E = \sigma_s$

4) $\phi\left(\frac{-2(\sigma_E+\sigma_s)}{\sqrt{\sigma_E^2+\sigma_s^2}}\right)$, which is around 0,2% if $\sigma_E = \sigma_s$

Hence, an impression is obtained about the consequences of different strategies. If, for example, it is required that $x_0 = m_s - 3\sigma_s$ and $\sigma_E \ll \sigma_s$, which is a usual situation, the risk of accepting m_E for comparison with x_0 is only

$$\phi\left(\frac{-4\sigma_s}{\sqrt{\sigma_s^2+\sigma_E^2}}\right) \approx \phi(-3) \approx 0,1\%$$

It is noted that in most cases the risk distribution is not Gaussian, still the reasoning remains the same.

It is usual to set limits for contents of hazardous substances in food or in the occupational environment. Here, the determination of the risk distribution, $P_s(x)$, is very unsure for small probabilities of injury or health effects. For this reason, there is a tendency for responsible authorities to set the acceptance limit far below the lowest levels observed to cause effects, corresponding to large k-values (k = 4,5 or even 6). This may cause difficulties in performing the measurement at all since the acceptance level and detection levels become comparable. It is also clear that in many such situations the uncertainty of the measurement can be allowed to be omitted in the comparison with the limiting value.

Another illustrative example is the case where the alcohol content in the blood of a car driver is measured from the air exhaled from the lungs by a direct spectrometric method. The uncertainty of the method gives a $P_E(x)$ distribution with a carefully determined uncertainty approximating σ_E. The limiting value in Sweden, x_0 =0.2 ‰, is decided for political and pedagogical reasons and it is comparable to factors as fatigue, irritation, etc. The expanded uncertainty can therefore be safely credited to the car driver. There is still a margin to real risks. Of course, the fact that a sentence has severe consequences also indicates that the technical uncertainty should be credited to the car driver.

In other cases other reasoning and conclusions have to be made. The general conclusion is that it is not possible to have a single "rule of thumb" for all cases when determining how the uncertainty of a measurement shall be related to a given acceptance limit. Different situations have to be treated separately.

There are, of course, also situations where the uncertainty should be included in the comparison. One such is when the material strength, e.g. fatigue properties are used for safe dimensioning of critical structures, such as pipe lines or nuclear reactors. (Even here, though, the limits are set with safety factors far exceeding the testing uncertainty.)

Again, the important principle to follow is that the relationship between limiting values and measurements and their uncertainties should be technically well founded. This requires a case by case decision.

Qualitative testing of products

It is now assumed that the uncertainty in the test method is small in relationship to the variation in product property and that the test is of the go – no go type.

The problem at hand is then to draw conclusions about the proportion of approved products in a batch from the proportion of approved specimen in a sample tested. Examples may be safety helmets (impact resistance), iron ore (contents of iron), pre-packed food (weight). This type of test is described, e.g. in ISO 2859 [4] and may be described by so-called OC-diagrams (operating characteristics).

Such diagrams are based on the binomial distribution and the functional relationship is

$$P_{Accept}(x) = \sum_0^p \binom{n}{p} x^p (1-x)^{n-p}$$

where $P_{Accept}(x)$ is the probability to accept a batch with the proportions x of non-accepted units if a sample of size n is tested and if the batch is approved if at most p non-accepted units are found. The diagram has the general shape according to Fig. 7.

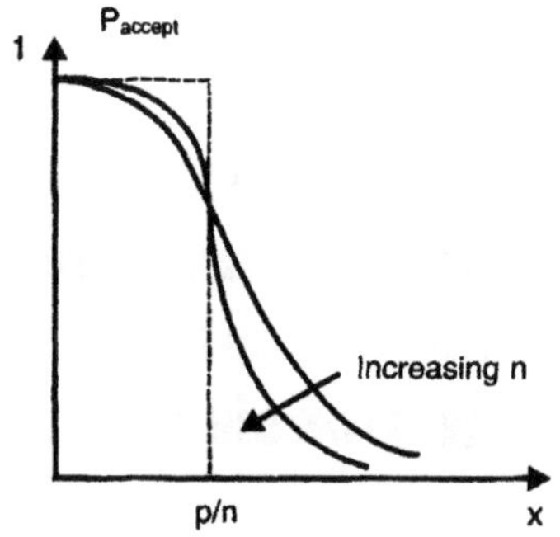

Fig. 7 General features of an OC-diagram

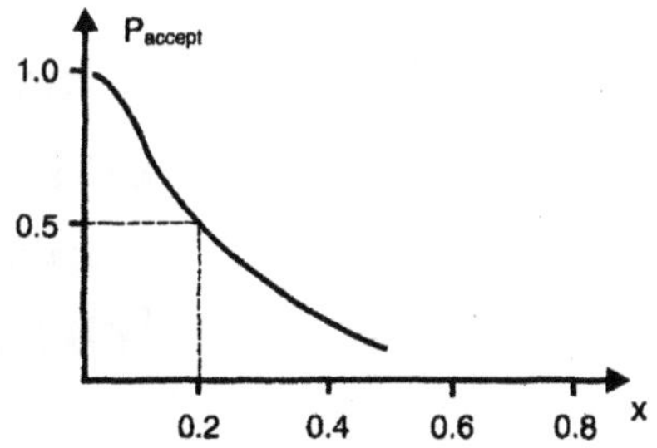

Fig. 8 OC-diagram for p=0, n =3

Of course it is desirable that the diagram shall be decisive, i.e. that it is close to the rectangular shape of the limiting curve obtained for n→∞. If the non-accepted proportion is smaller than p/n the batch should not be accepted and vice versa.

As an example may be taken testing of packaging for dangerous goods, where a drop test, from a certain height is approved if no leakage occurs. Three drops are made and the batch is approved if no leakage occurs in any of them.

If it is assumed that the batch has a proportion x which would not pass the test the probability to accept the batch will be

$$P_{Accept}(x) = \binom{3}{0} x^0 (1-x)^3 = (1-x)^3$$

The corresponding curve is shown in Fig. 8.

Hence, if the batch contains 20% (x = 0.2) erroneous units there is a 50% probability to accept it. This means that two laboratories may well get contradictory results or that the supplier may get different results from the same laboratory on two consecutive occasions!

It should be noted that when quantitative measurements are made one can determine in the usual fashion an approximate $P_E(x)$ function.

This type of product testing is very usual in written standards, and much confusion and debate occurs between customers and laboratories, due to lack of understanding of the properties of the testing procedure. Yet it is necessary to have these methods, for cost reasons, in product testing. What is needed is that the standards are produced with more care, and with clear explanations of the risks and properties of the test methods.

References

1. ISO 5725 (1994) Precision of test methods – Determination of repeatability and reproducibility by inter-laboratory tests. ISO, Geneva
2. ISO (1993) Guide to the expression of uncertainty in measurements. ISO, Geneva
3. Andersson H (1994)Assessment and practical use of uncertainty in test results. 2nd EUROLAB Symposium, Florence, 1994. EUROLAB,
4. EURACHEM Guide (2000) Quantifying uncertainty in analytical measurements, 2nd edn. EURACHEM, LGC, Teddington, UK
5. ISO 2859 (1974) Sampling procedures and tables for inspection by attributes. ISO, Geneva

Accred Qual Assur (2001) 6:493–500

Dean A. Flinchbaugh
Larry F. Crawford
David Bradley

A model to set measurement quality objectives and to establish measurement uncertainty expectations in analytical chemistry laboratories using ASTM proficiency test data

Electronic supplementary material to this paper can be obtained by using the Springer LINK server located at http://dx.doi.org/10.1007/s007690100398-y.

D.A. Flinchbaugh
Flinchbaugh Consulting
Bethlehem, Pa., USA
e-mail: DAFlinch@bellatlantic.net,
Tel.: +1–610–694 6473,

L. F. Crawford (✉)
Bethlehem Steel Corporation
Bethlehem, Pa., USA
e-mail: Larry.Crawford@Bethsteel.com,
Tel.: +1–610–694 6646,
Fax: +1–610–694 1739

D. Bradley
ASTM Headquarters
West Conshohocken, Pa., USA
e-mail: Dbradley@ASTM.org,
Tel.: +1–610–832 9681

Abstract A model is presented that correlates historical proficiency test data as the log of interlaboratory standard deviations versus the log of analyte concentrations, independent of analyte (measurand) or matrix. Analytical chemistry laboratories can use this model to set their internal measurement quality objectives and to apply the uncertainty budget process to assign the maximum allowable variation in each major step in their bias-free measurement systems. Laboratories that are compliant with this model are able to pass future proficiency tests and demonstrate competence to laboratory clients and ISO 17025 accreditation bodies. Electronic supplementary material to this paper can be obtained by using the Springer LINK server located at http://dx.doi.org/10.1007/s007690100398-y.

Keywords Proficiency test · Uncertainty · Measurement quality objectives · Accreditation · ISO 17025

Introduction

This paper describes one model for use by analytical chemistry laboratories that carry out compositional analyses to apply reliable interlaboratory test data, such as from the ASTM Proficiency Test (PT) programs, to determine intralaboratory measurement quality objectives (MQOs) and to establish and consistently meet individual-method uncertainties consistent with those objectives. In this work, we define intralaboratory MQOs to be the maximum measurement uncertainty to be allowed in any series of test methods. Intralaboratory MQOs should be set so that the data quality needs of clients are met in a way that allows the laboratory to operate effectively, efficiently, and competitively. Once MQOs are established, laboratory personnel are responsible for implementing test methods that have estimated uncertainties that comply with the MQOs.

ISO defines measurement uncertainty [1] as "A parameter associated with the result of a measurement that characterizes the dispersion of the values that could reasonably be attributed to the measurand." A report value with an estimated measurement uncertainty must be accompanied by a confidence interval to be meaningful, for example, 0.100+/–0.008% (m/m, 95% confidence). Successful application of the model requires that the laboratory is compliant with ISO 17025 and is capable of successfully participating in PT programs for the method(s) to be applied to the model. Laboratories that do not meet these basic requirements can use the model to help establish their overall Quality System program but should not expect to consistently receive acceptable ratings on PTs without being in compliance with ISO 17025.

The implementation of the model is simple. It makes extensive use of the logarithmic correlation between performance statistic of a test method and concentration, independent of analyte (measurand), matrix, or method that was initially developed by Horwitz [2–5], beginning in the early 1980s and further developed by Rocke and Lorenzato [6]. Horwitz et. al., used this model to analyze the performance characteristics of test methods used for regulatory purposes by the Food and Drug Administration. As discussed in the following representative examples, others use the concept to model the uncertainty in analytical measurement systems.

ISO 5725 [7] provides practical numerical definitions for the repeatability, r, and reproducibility, R, of a standard test method and describes the organization and analysis of interlaboratory experiments for the numerical determination of r and R. As part of its work program, ISO Technical Committee 17 (Iron and Steel), Subcommittee 1 (Chemical Analysis) applied these definitions and practices to the Horwitz model [8]. That work included 45 published and ready-to-publish national and international standard test methods (BSI, ECISS/TC20, and ISO/TC 17/SC 1) that employed 6 method principles (gravimetric, spectrophotometric, flame AAS, titrimetric-visual, titirimetric – potentiometric, and combustion/infra-red) to determine 21 elements in iron and steel. The work showed that a clear correlation existed between the log of both r and R vs. the log of the analyte concentration. Based on that work, ISO/TC 17/SC 1 set a policy [9] to use the log-log plots of r and R vs. concentration to evaluate test data of candidate ISO test methods. Any method that has data that exceeds specified limits beyond the historical log-log plots will not be submitted for international ballot to elevate to international standard status. ISO 15350 [10] is a recent example of a test method that met those requirements.

More recently, Flinchbaugh and Poholarz [11] described how the Horwitz model could be adapted to set MQOs for a Reference Materials (RM) Program. That program uses the ISO/TC 17/ SC1 plot in a manner similar to the model described in this paper to predict the MQOs (uncertainties) needed in certified homogeneity and concentrations. A laboratory using those RMs should be able to meet performance requirements consistent with the r and R limits set by the ISO/TC 17/ SC 1 plot. The program based on that model is now accredited to ISO Guide 34 [12, 13].

ASTM Committee E01 recently developed a Standard Guide [14] for managing uncertainties within a laboratory organization that uses multiple locations (workstations) to perform the same tests. The guide presumes that the laboratory organization has established data quality objectives and is committed to meeting them. The Annex to that Guide describes a procedure for setting laboratory-wide data quality objectives based on the ISO/ TC 17/SC 1 plot. Other recent ASTM Standards also make use of the logarithmic correlation between variation and concentration [15–17].

The referenced programs typically use classical statistical techniques, such as Dixon or Cochran, to reject individual data points and then use the log-log plot concept to identify data set outliers. Outliers are defined as points above or below a specified limit on the log-log plot. Root cause analysis of log-log plot outliers above the limit usually show possible deficiencies including: 1) heterogeneity of one element in the test material, 2) unanticipated test method interference, 3) attempted application of the test method above or below the optimum concentration range or 4) inadequate test method calibration and control protocols. Conversely, outliers below the line, although infrequent, appear to be "too good", and usually indicate that the experimental design did not include all of the normal sources of variation.

ASTM'S proficiency test programs as a source of interlaboratory data

ASTM currently sponsors PT programs in petroleum products and lubricants, stainless steel, plain carbon and low alloy steel, aluminum, gold in bullion; plastics: mechanical properties testing, plastics testing (polyethylene): melt index and ash; Textiles, and engine coolants. More information regarding ASTM PTs can be obtained on their website, www.astm.org. These programs provide laboratories ongoing, statistically sound objective evidence of their performance on common test materials as compared with other competent laboratories around the world. With the exception of the petroleum-related programs, the PT reports provide feedback to laboratories on the current PT samples only and ASTM does not monitor a laboratory's long-term performance. The petroleum-related programs publish 2-year summary reports of each laboratory's robust standard deviations. These reports provide indications as to whether a lab may have a persistent relative bias on a test or if their precision performance has been poorer than the majority of labs. This analysis also attempts to identify labs with especially tight performance. ASTM's Gold in Bullion Program offer guidance to laboratories on how to track and monitor their own PT performance [18].

ASTM Committee E01, on the Chemical Analysis of Metals, Ores, and Related Materials, conducts PT programs in plain carbon and low alloy steels, stainless steels, aluminum, and gold in bullion. These programs are conducted in compliance with ISO Guide 43 and

ASTM E2027 [19, 20]. This paper utilizes data from the first two of the E01 programs for its source of interlaboratory test data collected over a 2-year period. Samples used to generate the PT data incorporated in this paper were supplied by the Brammer Standard Company and tested by ASTM E-826 [21], Standard Practice for Testing Homogeneity of Materials for Development of RMs. Typically, carbon, manganese, phosphorous, sulfur, silicon, copper, nickel, chromium, molybdenum, aluminum, and tin are analyzed and reported by participants with additional elements included periodically.

Once the PT data are collected, ASTM calculates the "robust" mean and "robust" [22] standard deviation for each element at each concentration. Robust statistics are computed using all data, but unusually large or small observations have little effect on the mean and standard deviation estimates. Details on how ASTM calculates these parameters are available from ASTM. As a measure of laboratory performance, ASTM provides a Z-score for each laboratory on each element reported. Z-scores are calculated by subtracting the average laboratory result from the overall average and dividing the difference by the overall standard deviation. Laboratories obtaining Z-scores greater than 2 (i.e. reported test data is greater than 2 standard deviations from the overall average) are advised to investigate their test data and measurement systems for evidence of systematic errors. Under the present system, conscientious laboratories react to high Z-scores by internally auditing their test reports and measurements systems and taking corrective actions as indicated by root-cause analysis. However, in many cases, the root-cause analysis does not yield a convincing cause or solution because no apparent system failure can be identified. One cause might be that the test result was within the laboratory's normal variation, but that the laboratory's normal variation was large enough to allow it to occasionally report data that did not meet the robust standard deviations expected by the PT program.

Compilation of historical PT data

For this study, we combined data from ASTM Committee E01's Plain Carbon and Low Alloy Program and their Stainless Steel Program over a 2-year period, beginning in the third quarter of 1998. In 1999, over 120 laboratories participated in the Plain Carbon and Low Alloy Steel Program and over 40 laboratories participated in the Stainless Steel Program. These two programs were selected to incorporate different matrixes, different sets of participants, and to expand the overall concentration ranges of the analytes evaluated.

A plot of the log of 2 times the robust standard deviation (95% confidence) vs. the log of the robust mean concentration is shown in Fig. 1. For the purposes of this publication, references to robust standard deviations,

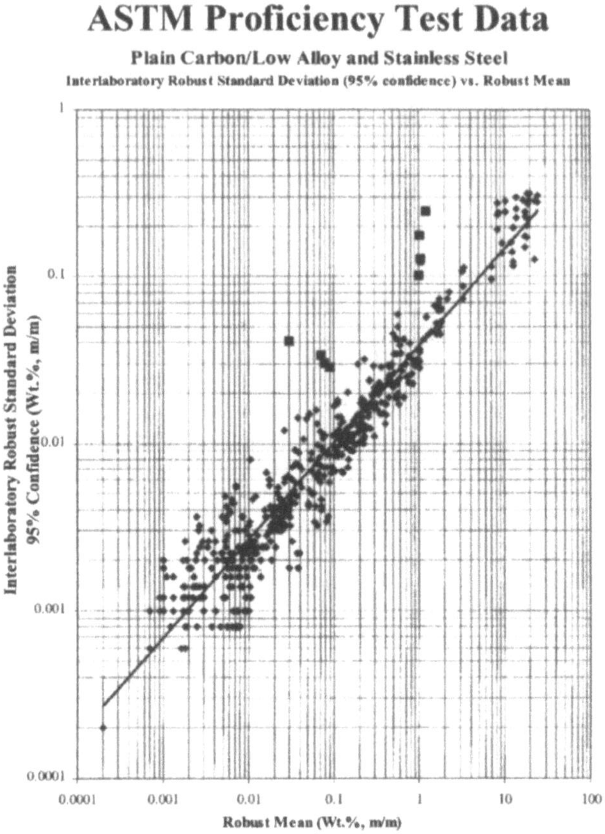

Fig. 1 ASTM Proficiency Test Data: plain carbon/low alloy and stainless steel. Interlaboratory robust standard deviation (95% confidence) vs. robust mean

method performance statistic, and other probability parameters are presented at the 95% confidence level, or 2 times the parameter, unless otherwise noted. These data cover more than 4 orders of magnitude for concentration and include over 20 elements. Data were generated using about six analytical test methods and from two different test matrixes. Figure 1 clearly shows that a well-defined correlation exists between the robust standard deviation and robust mean concentration. A "power" fit of the data in Fig. 1 yields the equation

$$y=0.0384x^{0.58} \tag{1}$$

which mathematically describes the correlation between the robust mean concentration and the interlaboratory robust standard deviation obtained by multiple labs/instruments/methods.

Visual inspection of Fig. 1 shows that most of the data points form a well-defined band around the best-fit line through all of the data points. The notable exceptions are on the positive side of the standard deviation line at about 1.0 and 0.1% concentrations and shown as

Table 1

Element, Conc.	Matrix	Method[a]	Method range
Si, 0.98%	Stainless	E1086/E572	0.01-0.9%/Not included in scope of method
Al, 1.02%	Stainless	[b]	Not applicable
Al, 1.04%	Stainless	[b]	Not applicable
Al, 1.04%	Stainless	[b]	Not applicable
Al, 1.18%	Stainless	[b]	Not applicable
Al, (0.08%	Stainless	[b]	Not applicable
Si, (0.07%	Stainless	E1086/E572	0.01–0.9%/Not included in scope of method
Si, 0.09%	Stainless	E1086/E572	0.01–0.9%/Not included in scope of method
Si, 0.03%	Stainless	E1086/E572	0.01–0.9%/Not included in scope of method

[a] Method(s) shown are those used most frequently in that test.
[b] Laboratories reported various methods of analysis for Al, including ASTM E572 (X-ray, Ref. [23]) and ASTM E1086 (optical emission (OES), Ref. [24]), neither of them include the determination of Al. Other laboratories reported using techniques not supported by ASTM Test Methods, such as inductively coupled plasma and glow discharge OES.

blocks on Fig. 1 in lieu of diamonds. Table 1 shows the source of those points. This visual analysis indicates that the most clearly outlying points were at or beyond the specification ranges of the alloys typically produced, were at or beyond the maximum concentrations for which the Standard Test Methods were validated, or were beyond the scope of the standard test method. In the case of Al, neither of the standard test methods reported (ASTM E572 and ASTM E1086) include Al in the method scope. For Si, ASTM E572 does not include Si in the method scope and the 1% silicon value is above the validated range of the test method. Laboratories should not use/report standard test methods for elements not included in the standard test method scope nor for samples above (or below) the analytical range. Based on these findings, there is a strong possibility that at least some of the laboratories provided the test data outside the demonstrated capability of ASTM test methods.

This brief overview illustrates how PT providers can use the log-log plot of two related programs under one provider to identify and improve performance on any element/concentration pairs that tend to generate unusually high standard deviations. As progressive PT providers improve their robust standard deviations, their participating laboratories will need to improve their performance to maintain consistent Z-scores over time. Clearly, laboratories will not benefit by participating in PT programs with unusually large robust standard deviations, because those laboratories will earn relatively low Z-scores when their measurement uncertainties are relatively large. Conversely, laboratories will benefit by participating in PT programs that consistently provide low robust standard deviations. The long-term goal is to have all compositional-based PT programs show the same, minimum standard deviations at each tested concentration.

We believe that as this model is used and overall data quality is improved the best-fit line will drop to some minimum standard deviation, which is undefined at this time. However, we can compare log-log plots of interlaboratory standard deviation data from other programs. Figure 4 (Figs. 4–7 are available as electronic supplementary material) compares the best-fit line from Fig. 1 with the best-fit lines from the ISO TC 17/SC 1 plot referred to in the Introduction. Note that the general agreement is good, especially considering that the two lines represent very large, totally independent data sets. Both measure interlaboratory standard deviations, but they were calculated from data derived under very different protocols. Many of the individual ISO data were from the final, most successful iteration of several optimization experiments. On the other hand, most of the ASTM data were generated under production conditions using many more pieces of equipment, each designed to handle a specific facility's fairly unique product mix. In many cases, the number of digits reported in test results are preset to meet local production requirements and may create rounding errors that make precision worse. Most of the ISO data resulted from calibrations with certified reference materials (CRMs) and pure chemicals that are highly reliable, while many of the ASTM data sets are from X-ray and optical emission instruments calibrated with secondary materials because appropriate calibrants are not always available. Also, the trace determinations of most residual elements in the 0.001 to 0.01 concentration range are not critical to routine spectrometric laboratories. Therefore, laboratories may not exercise as much diligence in controlling those elements.

Using data from two ASTM PT and ISO/TC 17/SC 1 test methods programs, we have shown that interlaboratory standard deviations from these diverse programs are very similar at any given concentration. We have also shown that opportunities exist for further improvement in the degree of curve-fit, once the respective organizations begin to evaluate and improve their procedures using the model. The similarities of these two curves tend to verify the model and should give laboratories the confidence they need to use this model to establish MQOs.

This correlation can be used to predict future performance of PT exercises. We will show how laboratories can use the predictability of future PT results to set MQOs and design control schemes to ensure satisfactory performance.

Establishing MQOs

Laboratories must understand client expectations and establish MQOs that are consistent with the agreed, client expectations. High-quality PT programs that attract large numbers of ISO 17025 [25] compliant laboratories are reliable sources of competitive performance data that can be used by laboratories to negotiate realistic performance goals with clients. This approach is also useful to laboratories in helping clients better understand the uncertainty in their test results by establishing performance goals where none previously existed.

Starting with the interlaboratory robust standard deviation (data shown in Fig. 1), we estimate the intralaboratory standard deviation of the participants by dividing interlaboratory data by $\sqrt{2}$, a standard practice for estimating a reduction in variation by removing one major source of error [7]. These data are modeled using a "power" fit, shown as a line in Fig. 5 (available as supplemental electronic material), to yield

$$y=0.0271x^{0.58} \tag{2}$$

The intralaboratory line represents the maximum uncertainty a single laboratory can have and still expect to receive Z-scores of less than 2 in 19 out of 20 PTs. This line can be used as the maximum allowable uncertainty for a laboratory. For example, to establish the maximum allowable uncertainty for a 0.3 wt.% Mn sample, Eq. (2) yields a result of 0.0135 wt.%.

Thus, if a laboratory's estimated uncertainty is 0.0135 wt.% (m/m) at 0.3wt.% (m/m) they can be assured that they will receive Z-scores of less than 2 in any ASTM PT 95% of the time and that their instrument is operating correctly at this concentration level. This also means that a laboratory will receive Z-scores of greater than 2 in a PT 5% of the time due to random causes when a 95% confidence level is employed. It may be in the laboratory's best interest to reduce the MQO to allow them to work with less than a 5% probability of receiving a Z-score of greater than 2 due to random error. For the determination of some chemical species, only one method may be available and it may not be optimized, resulting in higher than the desired variation. In these cases, efforts should be made to continue optimizing the test method to reduce the amount of uncertainty in the reported value.

Developing control limits to comply with MQOs

Having defined the maximum allowable error to meet MQOs, such as to pass 95% of all PTs, it is easy to use the uncertainty budget concept to establish the maximum error to be allowed at each major step in the subject method. In a bias free environment, measurement uncertainty can be described by pooling standard deviations as follows:

$$\sigma^2_{R,total} = \Sigma\sigma^2_{ith} \tag{3}$$

where:

$\sigma^2_{R,total}$ = combined uncertainty in a measurement system, $R\sigma^2_{ith}$ = each source of variance in measurement system R.

An essential step in utilizing the uncertainty budget is to identify all sources of uncertainty in the measurement system and then to quantify these sources to the extent possible. Individual sources of variation that contribute significantly to the combined uncertainty for a typical measurement system utilizing an analytical instrument are: instrument calibration (i.e. quantity and quality of CRMs used in the calibration protocol), instrument control, and field sampling. Other significant sources of variation may exist for specific situations and should be accounted for, as appropriate. Although field sampling is generally considered a significant source of uncertainty in an analytical measurement system, it will not be considered here because sampling variation is not significant in PT programs. It is recommended that accepted sampling techniques are used and documented when sampling is under the control of the laboratory organization to reduce this source of variation as much as possible.

The identified sources of variation can be used to expand Eq. 3 to better describe the measurement system:

$$\sigma^2_{R,\,total} = \sigma^2_{R,\,Control} + \sigma^2_{R,\,Calib} \tag{4}$$

or

$$\sigma_{R,total} = \sqrt{\sigma^2_{R,Control} + \sigma^2_{R,Calib}} \tag{5}$$

where:

$\sigma_{R,\,total}$ = combined uncertainty in measurement system R
$\sigma^2_{R,\,Control}$ = uncertainty due to instrumental control in measurement system R
$\sigma^2_{R,\,Calib}$ = uncertainty due to instrument calibration in measurement system R.

$\sigma_{R,\,total}$ from Eq. 5 can be set as equal to the MQO determined by Fig. 5. The laboratory must design and control its measurement systems so that the right-hand side of Eq. 5 is equal to or less than the combined uncertainty allowed in the system. This requires that the laboratory quantify the individual sources of variation contributing

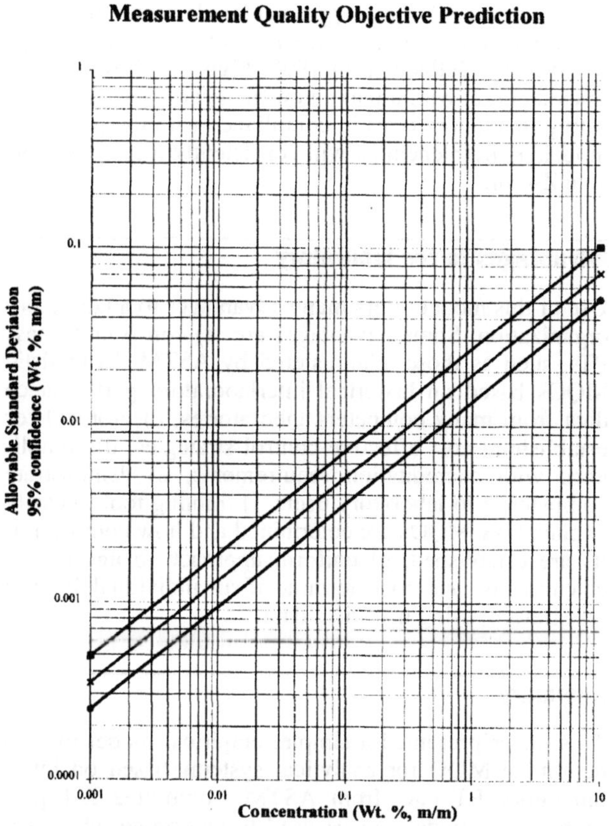

Fig. 2 Measurement quality objective (MQO) prediction: squares – overall (intralaboratory), crosses – control material, circles – calibration

to $\sigma_{R,\ total}$. It is best, although not always practical, to quantify each source of variation separately.

Quantitation of the first variable in Eq. 5, calibration uncertainty, ($\sigma^2_{R,\ Calib}$) can be done using published references, such as uncertainty statements on Certificates of Analysis for CRMs and the degree of fit of the calibration curve. It is of particular importance to use the appropriate type and quantity of high quality CRMs when calibrating the instrument to minimize this source of variation. Using CRMs produced by an ISO Guide 34 [12] compliant producer will help ensure that the material is adequately characterized and that estimated uncertainties are given. ISO Guide 34, *General Requirements for the Competence of Reference Material Producers*, details the necessary elements a RM producer must have in place to generate high quality RMs.

A general means for distributing the allowable variance between the two sources and still meeting the combined MQO (Eq. 2) is by consecutive divisions by $\sqrt{2}$ as each source of variation is removed. Starting with the combined MQO, the first division will yield the MQO for control sample variation, the second division will yield the MQO for calibration variation. The MQO for control sample variation is described by

$$y=0.0192x^{0.58} \quad (6)$$

and the MQO for calibration variation is described by

$$y=0.0136x^{0.58} \quad (7)$$

Equations 6 and 7 are shown graphically in Fig. 2.

Equations 6 and 7 give the maximum allowable variation for control and calibration and are used to design test methods in a cost-effective manner. If the amount of variation is less than the maximum allowed in either control or calibration, it reduces the probability of failing a PT due to random causes or allows the variation to be given to other parts of the measurement system, as needed.

Applying the model to an ISO 17025 accredited laboratory

We offer the following data from Bethlehem Steel's Research Analytical Laboratory to demonstrate the practical utility of this model. The laboratory is accredited [26] for the analysis of steel, slag, and coated sheet steel.

For most accredited methods, the uncertainty in calibration is very low due to the fairly good supply of CRMs. With a good supply of reliable CRMs, the curve-fit of calibration curves is usually "tight" and there are abundant numbers of CRMs to calibrate and independently verify calibration curves. In these cases, the calibration errors are very small compared to the combined method uncertainty and usually do not have to be addressed as part of the uncertainty budget. Difficulties arise when there is a shortage of reliable calibrants such that the calibration errors become relatively significant and make it impossible to meet MQOs. In Bethlehem's laboratory, we try to maintain the calibration contribution to the allowable variation in a measurement system to between 10–30%. We found this to be an achievable goal for calibration, in the majority of cases.

Figure 3 shows Bethlehem's two-sigma values for the statistical process control (SPC) samples associated with accredited methods. Note that most of the points are below the MQO for control, as reproduced from Fig. 2. The few points above the MQO for control samples are from 1.6% Al, 5.98% Mn, and 6.6% MgO in slag by X-ray fluorescence (XRF) and 0.022% S in steel by optical emission spectrometry (OES). We believe these slight problems are related to a minor homogeneity problem in the RMs used to control the OES and XRF.

As an ISO 17025 accredited laboratory, the Bethlehem Laboratory quantifies and reports to their clientele, the estimated uncertainties (95% confidence) associated with their measurement systems. To comply with the log-log model, Bethlehem's estimated uncertainty (95% confidence) must be below the predicted MQO. The best-fit line of the re-

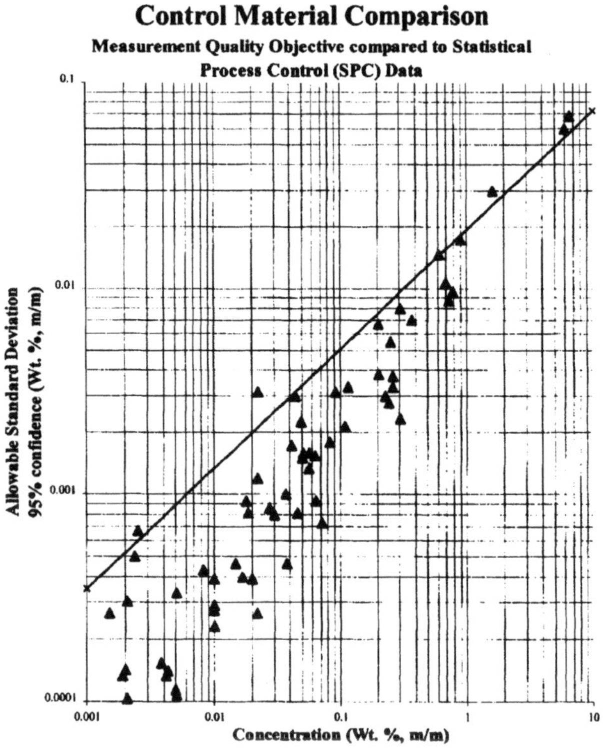

Fig. 3 Control material comparison. MQO compared to statistical process control (SPC) data: crosses – control material MQO, triangles – SPC data

ported estimated uncertainties for each accredited test method is shown in Fig. 6 (available as supplemental electronic material). Figure 6 shows that the best-fit line describing the estimated uncertainties (95% confidence) is very near the MQO, as reproduced from Fig. 2.

As final confirmation that the model will help laboratories pass PTs, we offer the Bethlehem Research Analytical Laboratory historical performance on ASTM's Plain Carbon/Low Alloy Steel PTs. In 1999, the Bethlehem Laboratory had an average Z-score (defined as the summation of Z-score value/the total number of Z-score values) of about 0.04, indicating that they were an average of 0.04 robust standard deviations from the robust mean of the sample being tested, an exemplary record. Figure 7 (available as supplemental electronic material) is a frequency diagram of the individual Z-score ratings received by Bethlehem in 1999. Figure 7 shows a bell-shaped curve, with a maximum frequency near 0.1, indicating that laboratories in compliance with the presented model, as Bethlehem is, will consistently receive acceptable Z-scores.

Advantages to the laboratory

Use of this model offers many advantages to a laboratory organization. Some advantages are: a) consistently pass PTs, such as those administered by ASTM, b) realistic MQOs based on historical interlaboratory performance data from many competent laboratories, c) protocol for establishing cost effective control strategies, d) compliance with accreditation requirements, e) demonstrate competence to laboratory users, f) training tool to easily explain how MQOs are determined and how control limits are established, g) uniform approach to uncertainty calculations, and h) determine if an individual PT run was flawed, using historical data.

Summary

This paper presented a simple, empirical model to help determine MQO for analytical systems based on interlaboratory PT data from ASTM Committee E-1 programs. The model can be used to determine what estimated uncertainty can be expected in a measurement result generated during routine analysis if the measurement system is optimized and bias free. Using the model and the principles of the uncertainty budget, compliant laboratories can set MQOs and ensure that they to consistently pass PTs and to establish control protocols in a cost-effective manner.

Acknowledgements The authors thank the management of Bethlehem Steel Corp. and ASTM for their support and for permission to publish this paper. In addition, we thank Bethlehem's Research Analytical Laboratory and support staff at ASTM for their valuable input and supporting data needed to develop this model. Finally, we acknowledge the tireless efforts of the volunteers which draft, debate, revise, and ballot the nationally and internationally accepted guides and standards used in developing this publication.

References

1. ISO VIM (1993) International vocabulary of basic and general terms in metrology, 2nd edn. ISO, Geneva
2. Horwitz W, Kamps LR, Boyer KW (1980) J Assoc Off Anal Chem 63: 1344–1354
3. Horwitz W (1982) Anal Chem 54: 67A-76A
4. Margosis M, Horwitz W, Albert R (1988) J. Assoc Off Anal Chem 71: 619–635
5. Horwitz W, Britton P, Chitrel SJ (1998) J Assoc Off Anal Chem 81: 1257–1265
6. Rocke DM, Lorenzato S (1995) Technometrics 37: 176–184
7. ISO 5725 (1994) Accuracy (trueness and precision) of measurement methods and results. ISO, Geneva
8. Hobson JD; ISO Document N 938 (1992) A survey of the precision of standard methods for the analysis of steel and iron, based on ISO, ECISS/EN and BSI Statistics, revised 1992-06-01. ISO, Geneva

9. ISO TC/17/SC1 N 1235 (1998) The procedures for activities of ISO/TC 17/SC1 (The 5 edn., version 2). ISO, Geneva
10. ISO 15350: Steel and iron: determination of total carbon and sulfur contents – Infrared absorption method after combustion in an induction furnace (routine method). ISO, Geneva
11. Flinchbaugh DA, Poholarz JM (1998) Accred Qual Assur 3: 367–372
12. ISO Guide 34 (2000) General requirements for the competence of reference material producers. ISO, Geneva
13. American Association of Laboratory Accreditation (A2LA) Certificate No. 300.03, presented 7 November, 2000 valid through 31 August, 2002
14. E2093–00: Standard guide for optimizing, controlling and reporting test method uncertainties from multiple workstations in the same laboratory organization
15. ASTM D6091–97: Standard practice for 99%/95% interlaboratory detection estimate (IDE) for analytical methods with negligible calibration error.
16. ASTM D6591–99: Standard practice for an interlaboratory quantitation estimate
17. ASTM E1763: Guide for the interpretation and use of results from the interlaboratory testing of chemical analytical methods
18. ASTM Proficiency Test Program Report: Determination of gold in bullion by cupellation (E1335) May/June 200 Appendix "Tracking and Monitoring your own Proficiency Test Performance"
19. ISO/IEC Guide 43–1 (1997) Proficiency testing by interlaboratory comparisons – Part 1: Development and operation of proficiency testing schemes. ISO, Geneva
20. ASTM E2027–99: Standard practice for conducting proficiency tests in the chemical analysis of metals, ores, and related materials
21. ASTM: E-826–85 (Reapproved 1996): Standard practice for testing homogeneity of materials for development of reference materials
22. Analytical Methods Committee of the Royal Society of Chemistry (1989) Analyst 114: 1693–1697
23. ASTM E572–94 (reapproved 2000): Standard test method for X-ray emission spectrometric analysis of stainless steel
24. ASTM E1086–94 (reapproved 2000): Standard test method for optical emission vacuum spectrometric analysis of stainless steel by the point-to-plane excitation technique
25. ISO 17025(1999) General requirements for the competence of calibration and testing laboratories. ISO, Geneva
26. A2LA Certificate No. 300.01, presented October 18, 2000 valid through August 31, 2002

Accred Qual Assur (2000) 5:464–469

Adriaan M.H. van der Veen
Jean Pauwels

Uncertainty calculations in the certification of reference materials. 1. Principles of analysis of variance

A.M.H. van der Veen (✉)
Nederlands Meetinstituut,
Schoemakerstraat 97, 2600 AR Delft,
The Netherlands
e-mail: avdveen@nmi.nl
Tel.: +31-15-2691 733
Fax: +31-15-261 2971

J. Pauwels
European Commission, Joint Research Centre, Institute for Reference Materials and Measurements, Retieseweg,
2440 Geel, Belgium
e-mail: Jean.Pauwels@irmm.jrc.be
Tel.: +32-14-571 722
Fax: +32-14-590 406

Abstract The preparation and certification of reference materials is a rapidly developing area. Many innovative reference materials have limited homogeneity and stability, and, additionally, the uncertainty estimation of the property values must be brought in agreement with the principles of the "*Guide to the expression of uncertainty in measurement*" (GUM). The results of the homogeneity and stability studies must be included to a certain extent in the uncertainty of the property values of the reference material, in order to comply with these requirements. The basic theory needed to accomplish this is essentially the theory of analysis of variance (ANOVA). As GUM also allows alternative evaluations other than Type A evaluations, a reinterpretation of the theory of ANOVA is necessary to establish a model for the certification of reference materials that is widely applicable. For this, analysis of variance can be used as a statistical technique to derive standard uncertainties from homogeneity, stability and characterisation data.

Keywords Reference materials · Measurement uncertainty · Analysis of variance · Homogeneity study · Stability study

Introduction

There are many experiments where two or more uncertainty components are involved, which are evaluated simultaneously by means of a Type A evaluation. In the certification of (batch) reference materials, this situation can be observed in the homogeneity study, stability study, and/or characterisation of the material. For instance, the homogeneity of a batch of subsamples is to be determined. The experiment involves two uncertainty components: the repeatability of the measurement (method) and the between-sample variation. In terms of analysis of variance (ANOVA), the samples are at the level of *groups*, and the repeatability of the measurement method is found within the groups. The basic model for this type of ANOVA reads as

$$Y_{ij} = \mu + A_i + \varepsilon_{ij} \quad (1)$$

where Y_{ij} is the result of a single measurement in the experiment. μ is the expectation of Y_{ij}, which is the value that Y_{ij} takes up when the number of repeated measurements tends to infinity. It should be noted that μ is not the true value. It even does not include any aspect of traceability to external references, unless special precautions have been taken. A_i is a bias term, due to the (random) differences in the extracts. The variable is assumed to be normally distributed, with mean zero and variance σ_A^2. Furthermore, it is assumed that A_i is independent of all ε_{ij} which are also normally distributed variables with mean zero and variance σ^2 [1]. There are *a* groups, and each of them contains n_i members. Ideally, the number of members in groups should be equal, but in practice this is often not the case. In order to make this paper useful to practitioners, the more complex formulae for incomplete data sets are given, rather than the simpler ones for complete data sets.

The objective of the uncertainty evaluation of the experiment sketched is to obtain estimates for σ_A^2 and σ^2, where the first refers to the variance of the A_i, and the second refers to the variance in ε_{ij}. The model for a two-way fully nested ANOVA reads as

$$Y_{ijk} = \mu + A_i + B_{ij} + \varepsilon_{ijk} \quad (2)$$

where a second bias term has been introduced: B_{ij}. For this bias term, which is at the level of subgroups, the same assumptions are made as for A_i, that it is normally distributed with mean zero, and that it is independent from both any A_i and any ε_{ijk}. The subscript "$B \subset A$" should be read as "among subgroups, within groups", as A represents the level of groups (for example: samples), and B represents the level of subgroups (for example: extracts). In a homogeneity study, a two-way ANOVA might be considered if additionally to the between-sample variation also the repeatability of subsampling and extraction is to be determined.

For the calculation of variances from these complex experiments, two things are needed. First a method is needed to partition the total scattering into contributions, attributed to the various levels in the ANOVA. In a second step, these contributions are converted into variances. These variances can directly be used in uncertainty evaluations that are compliant to the "*Guide to the expression of uncertainty in measurement*" (GUM) [2].

Partitioning sums of squares

Scattering of data can be represented in various ways. Probably the best known way is to express scattering of data in terms of variances, covariances, and standard deviations. In analysis of variance, the scattering is often expressed in terms of sums of squared differences, or in short "sums of squares". These sums of squares express the scattering at various (hierarchic) levels in the analysis of variance. At the top level, the total sum of squares (SS_{total}) is defined

$$SS_{total} = \sum_{i=1}^{a} \sum_{j=1}^{n_i} (Y_{ij} - \overline{\overline{Y}})^2 \quad (3)$$

where

$$\overline{\overline{Y}} = \frac{1}{\sum_{i=1}^{a} n_i} \sum_{i=1}^{a} \sum_{j=1}^{n_i} Y_{ij} \quad (4)$$

denotes the grand mean; a is the number of groups and n_i is the number of members in the group. SS_{total}, the total sum of squares, can be partitioned as follows. First, SS_{within} will be defined

$$SS_{within} = \sum_{i=1}^{a} \sum_{j=1}^{n_i} (Y_{ij} - \overline{Y}_i)^2 \quad (5)$$

SS_{within} is the part of the total sum of squares that can be attributed to the variation within groups. At the level of groups, the expression for the sum of squares reads

$$SS_{among} = \sum_{i=1}^{a} n_i (\overline{Y}_i - \overline{\overline{Y}})^2 \quad (6)$$

Without proof, the relationship between the three sums of squares reads as

$$SS_{total} = SS_{among} + SS_{within} \quad (7)$$

Each of these sums of squares has well-defined numbers of degrees of freedom. SS_{among} has $a-1$ degrees of freedom, SS_{within} has $\sum n_i - a$ degrees of freedom and SS_{total} has $\sum n_i - 1$ degrees of freedom. Dividing SS_{within} and SS_{among} by their respective number of degrees of freedom leads to the respective mean squares, abbreviated as MS. MS_{within} can thus be calculated as

$$MS_{within} = \frac{SS_{within}}{\sum_{i=1}^{a} n_i - a} \quad (8)$$

and MS_{among} is thus defined as

$$MS_{among} = \frac{S_{among}}{a-1} \quad (9)$$

The mean squares take up the form of variances, but they are, apart from MS_{within}, not equal to the variance at their specific level.

The main objective of this partitioning is that it enables the separation of different effects that contribute to the combined standard uncertainty of the measurand. This separation only makes sense if an uncertainty component, obtained from one experiment is used in another experiment, thus in the case of a kind of Type B evaluation [2].

Estimating uncertainties from ANOVA

In order to be able to calculate variances from mean squares, the first thing needed is the expectations of the mean squares, expressed in terms of variances σ_A^2 and σ^2. These relationships read as

$$MS_{within} = \sigma^2 \quad (10)$$

$$MS_{among} = \sigma^2 + n_0 \sigma_A^2 \quad (11)$$

where n_0 is a function of the number of degrees of freedom. For a complete data set, where for any value of i, $n_i = n$, then $n_0 = n$. It has been determined by mathematical statisticians that the appropriate value for n_0 for incomplete date sets reads as

$$n_0 = \frac{1}{a-1}\left[\sum_{i=1}^{a} n_i - \frac{\sum_{i=1}^{a} n_i^2}{\sum_{i=1}^{a} n_i}\right] \tag{12}$$

Using the expressions for the mathematical expectations of the mean squares, and considering the fact that only estimates are available, the following equations result

$$s_{within}^2 = MS_{within} \tag{13}$$

$$s_A^2 = \frac{MS_{among} - MS_{within}}{n_0} \tag{14}$$

Although s_A^2 is an unbiased estimator for σ_A^2, and can be employed as such, there is some aspect to keep in mind. In various references [3], the following expression for the confidence interval for this variance ratio can be found

$$F_{0.975} + n\sigma_A^2/\sigma^2 < \frac{s_A^2}{s^2} < F_{0.025} + n\sigma_A^2/\sigma^2 \tag{15}$$

where $F_{0.975}$ and $F_{0.025}$ are the lower and upper 2.5% one-tailed levels of F with numbers of degrees of freedom of $a-1$ and $a(n-1)$, respectively. This expression can be rearranged to

$$\frac{1}{n}\left(\frac{s_A^2}{s^2 F_{0.025}} - 1\right) < \frac{\sigma_A^2}{\sigma^2} < \frac{1}{n}\left(\frac{s_A^2}{s^2 F_{0.975}} - 1\right) \tag{16}$$

From this formula, it can be seen that s^2 has an impact on the estimator s_A^2; this can be understood by considering, that s^2/n defines the "resolution" of the method for obtaining s_A^2. The smaller s^2/n, the better the estimator for s_A^2. This fact plays an important role when transferring uncertainty components from one experiment to an other.

Two-way fully nested design

The expressions of the variances s_A^2, $s_{B\subset A}^2$, and s^2 from a two-way fully nested design are developed in a similar way as those for the one-way layout. As in uncertainty calculations for especially stability studies often a two-way design is necessary (1: time; 2: samples; 3: repeatability of measurement), the formulae are given below. The model for a two-way ANOVA has been given in Eq. (2). The grand mean is computed using

$$\overline{\overline{Y}} = \frac{1}{\sum_{i=1}^{a}\sum_{j=1}^{b_i} n_{ij}} \sum_{i=1}^{a}\sum_{j=1}^{b_i}\sum_{k=1}^{n_{ij}} Y_{ijk} \tag{17}$$

In this expression, a denotes the number of groups, b_i the number of subgroups within groups, and n_{ij} the number of observations in the subgroups. The formula for the grand mean takes into consideration that often the data set is not complete. The expression for the partitioned sum of squares read as [1]

$$SS_{among} = \sum_{i=1}^{a} n_i(\overline{Y}_A - \overline{\overline{Y}})^2 \tag{18}$$

$$SS_{B\subset A} = \sum_{i=1}^{a}\sum_{j=1}^{b_i} n_{ij}(\overline{Y}_B - \overline{Y}_A)^2 \tag{19}$$

$$SS_{within} = \sum_{i=1}^{a}\sum_{j=1}^{b_i}\sum_{k=1}^{n_{ij}} (Y_{ijk} - \overline{Y}_B)^2 \tag{20}$$

The double bar over Y denotes the grand mean; although one could argue that it should be a triple bar, a grand mean in the ANOVA literature is always denoted by a double bar. Likewise, group and subgroup means ($\overline{Y}_A$ *and* $\overline{Y}_B$) are always denoted by a single bar. The subscript denotes the level: A is the top level, B is the second level. The second step in developing expressions for the variances σ_A^2, $\sigma_{B\subset A}^2$, and σ^2 is to convert the sum of squares in their respective mean squares. The expressions for the three mean squares are

$$MS_{among} = \frac{SS_{among}}{a-1} \tag{21}$$

$$MS_{B\subset A} = \frac{SS_{B\subset A}}{\sum_{i=1}^{a} b_i - a} \tag{22}$$

$$MS_{within} = \frac{SS_{within}}{\sum_{i=1}^{a}\sum_{j=1}^{b_i} n_{ij} - \sum_{i=1}^{a} b_i} \tag{23}$$

which are, after the discussion of the one-way ANOVA self-explanatory. In the denominators, the expressions for the respective number of degrees of freedom are given. The number of degrees of freedom is equal to the number of observations, minus the number of parameters computed from them. Thus, the number of degrees of freedom among groups equals a-, as there are a group means and there is one parameter computed from them (in fact, the grand mean).

These mean squares can be converted into variances using the following expressions

$$\sigma_{method}^2 = MS_{within} \tag{24}$$

$$\sigma_{B\subset A}^2 = \frac{MS_{B\subset A} - MS_{within}}{n_0} \tag{25}$$

$$\sigma_A^2 = \frac{MS_{among} - n_0\sigma_{B\subset A}^2 - \sigma^2}{(nb)_0} \tag{26}$$

Again, MS_{within} equals σ^2. $\sigma_{B\subset A}^2$ is calculated by subtracting MS_{within} from $MS_{B\subset A}$. Likewise, σ_A^2 could be obtained by subtracting $MS_{B\subset A}$ from MS_{among}. For incomplete data sets, it is less effort to compute σ_A^2 as stated in Eq. (26) [1]. For higher-order ANOVAs, the pattern is similar. In [1], higher-order ANOVA as well as other designs of ANOVA are given. The expressions for the denominators of Eqs. (25 and 26) as well as in the nominator of Eq. (26) read as follows

$$n_0 = \frac{\sum_{i=1}^{a}\left(\frac{\sum_{j=1}^{b_i} n_{ij}^2}{\sum_{j=1}^{b_i} n_{ij}}\right) - \frac{\sum_{i=1}^{a}\sum_{j=1}^{b_i} n_{ij}^2}{\sum_{i=1}^{a}\sum_{j=1}^{b_i} n_{ij}}}{a-1} \tag{27}$$

$$n_0 = \frac{\sum_{i=1}^{a}\sum_{j=1}^{b_i} n_{ij} - \sum_{i=1}^{a}\left(\frac{\sum_{j=1}^{b_i} n_{ij}^2}{\sum_{j=1}^{b_i} n_{ij}}\right)}{\sum_{i=1}^{a} b_i - a} \tag{28}$$

$$(nb)_0 = \frac{\sum_{i=1}^{a}\sum_{j=1}^{b_i} n_{ij} - \frac{\sum_{i=1}^{a}\left(\sum_{j=1}^{b_i} n_{ij}\right)^2}{\sum_{i=1}^{a}\sum_{j=1}^{b_i} n_{ij}}}{a-1} \tag{29}$$

As already pointed out, for practical reasons, the formulae for incomplete data sets are given, deliberately. In most references, e.g. ISO 5725–3 [5] among others, usually only the formulae for complete data sets are given. From a theoretical point of view, this may find its justification in that ANOVA has been developed for complete data sets, but a few observations (or even subgroups) missing does not necessarily mean that the whole experimental set-up has become invalid. However, the formulae needed certain modifications in order to work with the (approximately) correct numbers of degrees of freedom. Obviously, the more "holes" in the data set, the poorer the method works, and the poorer the results.

Useful relationships and inferences

The significance of ANOVA goes beyond the applications sketched here. Traditionally, in chemistry ANOVA has always been associated with the F-test, testing mean square ratios for significance. Although there are cases where this becomes relevant, in uncertainty evaluations it is rarely needed. Often, it is sufficient to draw up the expression for the combined standard uncertainty, and if components coming from ANOVA are insignificant, then they will effectively drop out in the summation anyway. What matters is the significance of the uncertainty components in the ANOVA in relation to the combined standard uncertainty, not so much in relation to the repeatability of the measurement method used.

Returning to a one-way ANOVA, a useful relationship can be developed from the expression for MS_{among}. It is defined as

$$MS_{among} = \frac{\sum_{i=1}^{a} n_i(\overline{Y}_i - \overline{\overline{Y}})^2}{a-1} \tag{30}$$

The second relationship that is of interest is the expression for a sample variance

$$s^2 = \frac{\sum_{i=1}^{a} (Y_i - \overline{Y})^2}{a-1} \tag{31}$$

If, in the expression for MS_{among}, n_i is set to unity for all i, then this expression becomes identical with the one for a sample variance for a row of results. So, from a matrix of a one-way analysis of variance, MS_{among} can be computed directly from the variance of the group means.

Furthermore, returning to the model of a one-way ANOVA and the principle of propagation of uncertainties, the following expression can be developed

$$u^2(Y_{ij}) = u^2(A_i) + u^2(\varepsilon_{ij}) \tag{32}$$

For the group means, the following expression can be derived

$$u^2(Y_i) = u^2(A_i) + \frac{u^2(\varepsilon_{ij})}{n} \tag{33}$$

whereby it has been assumed that all groups are complete, i.e. $n_i = n$ for all i. Combining this result with Eq. (31) leads to the interesting result that

$$s^2 = s_A^2 + \frac{s_{within}^2}{n} \tag{34}$$

which is consistent. Furthermore, under the assumption that $n_i = n$ for all i,

$$MS_{among} = ns^2 = ns_A^2 + s_{within}^2 \tag{35}$$

These formulae also open up other options. According to GUM [2], there is no difference in nature and properties of a standard uncertainty coming from a Type A or a Type B evaluation. Accepting this principle, the formulae given also open up possibilities to work with a combination of Type A/Type B evaluation of uncertainty. Especially in cases, where a series of data is to be processed from which it is known that the values carry an uncertainty (apart from the variation inherent to the data set), the formulae given may provide a basis for developing a procedure for the uncertainty analysis, based on this ANOVA work. This actually is one of the reasons why the ANOVA theory is very suitable for the description of the certification of batch reference materials.

An illustration of this runs as follows. Suppose a series of data is obtained, with a certain degree of scattering, but from which it is known that each member of the series has some additional measurement uncertainty. This measurement uncertainty may be a combination of uncertainty from Type A and Type B analysis, but it is assumed that there is only one additional uncertainty component (u_{add}) to be considered, and it comes from a Type B evaluation. This assumption does not affect the general validity of this inference. Using Eq. (31), the variance can be computed. What about the uncertainty of the mean? It is known that each Y_i in the series has this component u_{add}. Given Eq. (33), it must be noted that it is impossible to determine $u^2(A_i)$ and $u^2(\varepsilon_{ij})$, separately. This is the "penalty" for not having more than one data point per "group".

Applying the principle of uncertainty propagation, two alternatives can now be developed, which represent extremes

$$u^2(m) = \frac{s^2}{a} + u_{add}^2 \tag{36}$$

and

$$u^2(m) = \frac{s^2 + u_{add}^2}{a} \tag{37}$$

The difference between the two is obvious. The first alternative leads to a greater value for $u^2(m)$ than the second. Under what conditions can the second be used, under what conditions must Eq. (36) be used? It depends on the nature of u_{add}, and at what level it affects the uncertainty in the mean m. If u_{add} is the same for all Y_i, it is clear that u_{add} affects m at its own level: the scattering of Y_i has no relationship with the value of u_{add}. If u_{add} is specific to each Y_i, then it affects the uncertainty of m at the level of Y_i, and in this case Eq. (37) can be used.

In terms of correlations, Eq. (36) represents the case where the system is fully correlated with respect to u_{add}, whereas Eq. (37) represents the fully independent case. In practice, usually it is not clear whether this additional uncertainty source is correlated or not. As a principle, in lack of information, the conservative alternative should be chosen, unless positive evidence is available that the assumption of independence holds. That is, in cases of doubt, Eq. (36) should be used instead of Eq. (37).

Underlying assumptions revisited

As with most mathematical and statistical techniques, the computational methods as presented are only valid within certain restrictions. The first equation setting restrictions is the model, as specified in Eq. (1) for a one-way ANOVA and in Eq. (2) for a fully nested two-way ANOVA. These assumptions have already been discussed. The requirement of independence of the variables on the right-hand sides of Eqs. (1) and (2) is probably the most critical one. This assumption seems to be met in many cases in analytical chemistry as well as in physical testing, but it cannot be taken for granted that this assumption is always valid. For instance, heterogeneity of a material will in a homogeneity study lead to a greater value for $\mathrm{Var}(A_i)$, but usually also $\mathrm{Var}(\varepsilon_{ij})$ increases with increasing heterogeneity.

Another important point related to the model is that the data obtained does not show any trend. This may seem obvious, but for some applications (e.g. computing uncertainty budgets from stability studies) it is something that should be carefully investigated. In absence of any kind of trend, the ANOVA approach is valid. Otherwise, some kind of trend analysis or regression technique is recommended. This requirement boils down to the usefulness of a grand mean: in absence of a trend, the grand mean is (from a theoretical point of view) a useful property. If there is a trend, the concept of calculating a grand mean, other than for internal purposes of the regression analysis, is of doubtful value.

A further assumption already mentioned is the normality of data. This assumption is quite notorious, and has lead to a variety of statistical tests for normality. Probably the best known test for this purpose is the Kolmogorov-Smirnov test [4] or one of its variants. Normality of data is often assumed, but not so often observed as desired. Statistical tests, which require normality of data, are known to be very sensitive with respect to deviations from normality. A skewed distribution may very well lead to completely wrong decisions.

In the applications discussed here, ANOVA is used as a method for obtaining values for uncertainty components. These values are variances, and their value is relatively insensitive with respect to the underlying distribution. This means that the evaluation method as discussed is quite robust with respect to the actual distribution of the data. This is an important aspect, as it makes testing for normality redundant. The variances thus obtained can be treated as any other variances from Type A or Type B evaluations [2].

The fact that non-normality of data does not play such a role as in the classical use of ANOVA is discussed in ISO 5725-1 and -3 [5, 6], and ISO Guide 35 [7]. In the classical approach, the F-test for testing significance of ratios of means of squares play a dominant role. This F-test is very sensitive with respect to non-normality of the underlying data, thus leading to false-positive or false-negative results.

In the whole process of uncertainty evaluation, testing on ratios of mean squares or testing on ratios of variance for significance has become superfluous, as in the establishment of the combined standard uncertainty any components that are insignificant drop out automatically. There is something more to add to the problem significance/insignificance. The classical *F*-test approach tests for significance within the context of the experiment. This may be relevant in some cases, but in most experiments carried out in this context, it is not. It is – in principle – not of interest whether an effect observed and quantified is significant with respect to some other uncertainty component present in the experiment.

Practical applications

The practical applications of partitioning of sums of squares are manifold, and many of them have been reported in the open literature the past few years. In analytical chemistry, steps like extraction, destruction, subsampling and sampling can only be evaluated using some kind of ANOVA technique. GUM [2] requires however to express the repeatability of these kinds of steps to be expressed as standard uncertainties. These components however only take into consideration the random effects of these steps. Effects systematic to the experiment, such as incomplete extraction, incomplete destruction, and biased (sub)sampling must be assessed separately if needed. The statistical techniques as discussed in this paper deal essentially with Type A evaluation of these experiments. However, this source of uncertainty is frequently overlooked in uncertainty evaluations.

In the process of preparation and certification of reference materials, ANOVA plays an important role in the evaluation of the uncertainty due to inhomogeneity, instability, and characterisation of the candidate reference material. In the parts to follow, the evaluation of data from these studies, as well as developing an expression for the combined standard uncertainty of the reference material will be discussed.

Concluding remarks

The use of ANOVA in analytical chemistry, and more general in physical, chemical and biological testing has gained renewed interest as a suitable technique to perform complex Type A-evaluations of experiments addressing multiple uncertainty components. These experiments include homogeneity and stability testing, as part of the process of producing (certified) reference materials. In routine laboratories, typical experiments that can be evaluated using ANOVA techniques are studies into the performance of destruction and extraction methods, methods for sample preparation and related studies.

The variances obtained from ANOVA can be used directly in uncertainty evaluations in compliance with GUM. The method of evaluation is less sensitive towards the influence of the actual distribution of data. Even if the distribution of data deviates considerably from the normal distribution, the method can be applied.

List of symbols

A_i	bias term at the group level
B_{ij}	bias term at the subgroup level (two-way ANOVA or higher)
ε_{ij}	random term at the within-group level (index from one-way ANOVA)
MS	mean of sum of squares
SS	sum of squares (sum of squared differences)
μ	expectation of Y_{ij}
s	standard deviation
s_A	standard deviation at the top level of ANOVA
$s_{B \subset A}$	standard deviation at the subgroup level (two-way ANOVA)
u_{add}	additional uncertainty component
Y_{ij}	result of a single measurement (index from one-way ANOVA)

References

1. Sokal RR, Rohlf FJ (1995) Biometry, 3rd edn. Freeman, New York
2. BIPM, IEC, IFCC, ISO, IUPAC, IUPAP, OIML (1995) Guide to the expression of uncertainty in measurement, 1st edn. ISO, Geneva
3. Snedecor GW, Cochran WG (1989) Statistical methods, 8th edn. Iowa State University Press, Iowa, USA, Chapter 13
4. Law AM, Kelton WD (1991) Simulation modeling and analysis, 2nd edn. McGraw Hill, New York, Chapter 6
5. ISO 5725-3:1994 (1994) Accuracy (trueness and precision) of measurement methods and results – Part 3: Intermediate measures of the precision of a standard measurement method in statistical methods for quality control. International Organization for Standardization (ISO), Geneva, pp 75–104
6. ISO 5725-1:1994 (1994) Accuracy (trueness and precision) of measurement methods and results – Part 1: General principles and definition of statistical methods for quality control, vol. 2. International Organization for Standardization (ISO), Geneva, pp 9–29
7. ISO Guide 35:1989 (1989)Certification of reference materials – General and statistical principles, 2 edn. International Organization for Standardization (ISO) Geneva

Accred Qual Assur (2001) 6:26–30

Adriaan M.H. van der Veen
Thomas Linsinger
Jean Pauwels

Uncertainty calculations in the certification of reference materials. 2. Homogeneity study

A.M.H. van der Veen (✉)
Nederlands Meetinstituut,
Schoemakerstraat 97, 2600 AR Delft,
The Netherlands
e-mail: avdveen@nmi.nl
Tel.: +31-15-2691 733
Fax: +31-15-261 2971

T.P. Linsinger · J. Pauwels
European Commission, Joint Research Centre, Institute for Reference Materials and Measurements, Retieseweg,
2440 Geel, Belgium

Abstract Many reference materials undergo a batch certification, which implies that a small number of samples is taken from a batch, characterised, and these results are then assumed to be representative of all remaining samples. An important aspect in this design is the translation of the characterisation data to a single sample, as usually the laboratory will be using only one sample of the batch. This form of homogeneity is very important and can be influenced to a certain extent by well-designed sample preparation procedures. Another subsampling problem associated with many reference materials is that only a small test portion is drawn from the sample to carry out the measurement. Obviously, this test portion must be representative of the sample, otherwise the certified value is still not applicable. Both kinds of homogeneity tests are examined in the paper and evaluated using practical examples.

Keywords Reference materials · Measurement uncertainty · Analysis of variance · Homogeneity study · Minimum sample intake

Introduction

Homogeneity testing is a well-known phenomenon in the preparation and certification of reference materials. It also finds its application in the preparation and checking of proficiency testing material. The design of the homogeneity study is the same for both between-bottle and within-bottle testing, as the same question needs to be answered. For a "between-bottle" homogeneity test, where the differences among samples are of interest, the problem can be stated as follows. What do these differences contribute to the uncertainty of the characterisation of material, taking into consideration that the user of the material will only be using one sample at a time? The name "between-bottle" has been chosen, as most reference materials (and proficiency testing materials) are shipped in bottles; however, the whole discussion is also valid for vials, gas cylinders, and test pieces, to give a few examples.

Within-bottle homogeneity plays a role if a subsampling step is required. For instance, for measuring a gas mixture shipped in a cylinder, a subsample is taken, so within-bottle homogeneity is an issue here. In this case, as in most other cases of solutions and mixtures, the heterogeneity in the bottle can effectively be removed by shaking, rolling or some other handling that allows the mixture to become homogeneous again. For particulate materials, the problem is different. Apart from reprocessing the material, there is no option of improving the within-bottle homogeneity. A certain minimum amount of the sample must be taken, so that the test portion represents the contents of the package. In terms of measurement uncertainty, the smaller the amount taken the higher the measurement uncertainty. Thus, this would lead to a situation where the uncertainty of the characterisation of the material is no longer valid for the test portion, even in the case where the between-bottle homogeneity has been taken into consideration.

Experimental set-up and the relationship with analysis of variance (ANOVA)

A typical experimental set-up for a between-bottle homogeneity study is visualised in Fig. 1. On the left-hand side, the case is given where subsampling is impossible, or just not done. With test pieces, often only one test is possible, so in this case, n, the number of replicates, equals 1. In those cases where the sample allows multiple measurements after transformation, n will generally be greater. In those cases where $n>1$, the data can be treated with ANOVA [1]. In this paper, a few examples will be given.

The effect of between-bottle homogeneity is in the variance "among groups" [1], as well as the effect of the transformation of the sample. The variance "within groups" [1] covers only the repeatability of the measurement. In contrast, if from each sample of the batch multiple test portions are taken, the variance "within groups" will cover the measurement, transformation, and subsampling. In this case, the variance "among groups" only covers the between-bottle heterogeneity. From the perspective of obtaining an unbiased estimate, this is the ideal situation.

For obvious reasons, it is not possible to obtain an exact estimate of the variance resulting from within-bottle heterogeneity. From a sample, multiple test portions must be drawn, which can obviously only be transformed once (Fig. 2).

This implies that the variance resulting from within-bottle heterogeneity will always be conservative, which is not really a problem. For the computation of the minimum sample intake [2], this implies in turn that the minimum sample intake will be overestimated, which is not a problem either. The amount of material stated will be greater than the critical amount, the amount actually needed to observe no effect of the sample intake on the repeatability of a measurement. However, the sample intake for the study to establish the minimum sample intake differs from the sample intake for a between-bottle study: in the latter case the optimal sample intake should be chosen to obtain a good repeatability, in the former, the minimum sample intake must be used. Doing so will increase influences of within-bottle heterogeneity and will thus reduce the influence of method repeatability on the estimated minimum sample intake.

Current practice versus uncertainty-based practice

In many cases, the well-known F-test [3–5] is still used to test for significance of a heterogeneity effect. Apart from the problems already stated in Part 1 [1], there is another big problem: this way of treating homogeneity data does not answer the question(s) raised in the introduction. In principle, it does not matter whether the heterogeneity observed is significant with respect to the repeatability of the test method. The repeatability of the test method is only of interest in the sense of the quality of what has been demonstrated. If the F-test does not indicate a significant result from the homogeneity test, but the repeatability of the test method used is poor, then it is still possible that the heterogeneity among samples (between-bottle homogeneity) is significant when compared to the uncertainty from characterisation.

As pointed out by Pauwels et. al. [6] and by Van der Veen and Alink [7], a method used for assessing sampling and/or subsampling performance, including homogeneity tests, should have a good repeatability. The repeatability of the test method, in conjunction with the number of replicates on each sample, defines the reso-

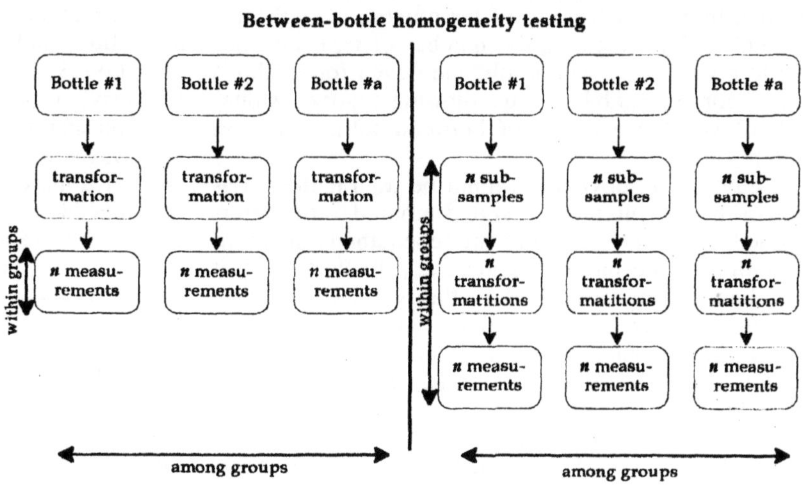

Fig. 1 Lay-out of a between-bottle homogeneity study

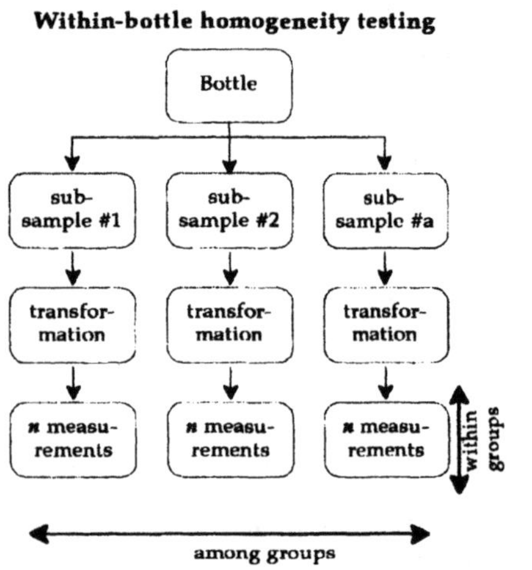

Fig. 2 Lay-out of a within-bottle homogeneity study

lution of the method. The better the resolution, the smaller the effects that can be estimated.

Looking from the perspective of the "*Guide to the expression of uncertainty in measurement*" (GUM) [8], the variation of the bottle averages ($u_{c(bb)}$) is a combined uncertainty consisting of the between-bottle heterogeneity (s_{bb}) and the measurement variation (s_{meas}). The latter comprises analytical variation and within-bottle heterogeneity, which should be pooled for the estimation of between-bottle heterogeneity anyway. The relationship between $u_{c(bb)}$, s_{bb} and s_{meas} can be expressed as [6]

$$u^2_{c(bb)} = s^2_{bb} + s^2_{meas} \tag{1}$$

which implies that

$$s^2_{bb} = u^2_{c(bb)} - s^2_{meas} \tag{2}$$

Note that $u_{c(bb)}$, s_{bb} and s_{meas} were named u_{exp}, u_{betw} and u_{meas}, respectively in [6]. s_{meas} is the analytical variation divided by the square root of the number of replicates per bottle. By increasing the number of replicates per bottle a small s_{meas} can be obtained even for methods with poor repeatability, thus allowing a good estimation of s_{bb}, the estimate of between-bottle variation sought for.

Equation (1) obviously cannot be used if the variation of the measurement is large compared to the heterogeneity, without looking at the repeatability standard deviation of the measurements, s_{meas}. In Part 1 [1], it has been demonstrated that the variance "among groups" is affected by the variance within groups. This means that for small values of s_{bb} a problem of quantification arises. There are two principle choices to deal with this

1. Acceptance of the value for s_{bb}, even if it is zero, because the samples are expected to be homogeneous (e.g. solutions) or known from previous experience to have negligible heterogeneity when prepared properly
2. Application of a more conservative estimation technique for this uncertainty source, based on s_{bb} as observed and s_{meas}.

Both options comply with GUM, and apart from the expectations about the heterogeneity aspect, s_{meas} should also fulfill the requirement, to be smaller than the repeatability standard deviation of the measurements in the characterisation. If this requirement is not clearly fulfilled, then in any case a more conservative estimator than the value of s_{bb} as observed from ANOVA is necessary. This is an effect of the existing correlation between both standard deviations. This topic will be covered in more detail in Part 4, about the certification process, as it also has consequences for the treatment of stability data.

For within-bottle homogeneity studies, a similar reasoning can be developed. The uncertainty from the experiment can be expressed as [6]

$$u^2_{c(wb)} = s^2_{wb} + s^2_{method} \tag{3}$$

which leads to

$$s^2_{wb} = u^2_{c(wb)} - s^2_{method} \tag{4}$$

with $u_{c(wb)}$ being the combined standard uncertainty of the experiment, s_{wb} the within-bottle variation and s_{method} the intrinsic variability of the method. These terms were named u_{exp}, u_{inh} and u_{meas}, respectively in [6]. The second term of this expression differs from that of Eq. (1) due to the difference in experimental design: Usually, s_{method} cannot be determined independently, as this would require a material of the same type with perfect within-unit homogeneity, which renders estimation of s_{wb} impossible. In these cases, $u_{c(wb)}$ must be used to estimate the minimum sample intake. To diminish the influence of s_{method} as much as possible, a sample intake should be chosen for which s_{wb} is much larger than s_{method}. Examples for this approach for (trace) elements are solid sampling (SS) techniques like solid sampling SS-ETAAS (Electrothermal Atomic Absorption Spectrometry) or solid sampling inorganic inductively coupled plasma-mass spectrometry (SS-ICP-MS). In any case, s_{wb} is not part of the uncertainty of the certified reference material, as will be explained in Part 4. It is only needed to establish the minimum sample intake for which the stated uncertainty is valid.

Evaluation of a between-bottle homogeneity study

For a clay soil sample, 18 samples were taken out of a batch for an homogeneity study on barium. The results, expressed in mg/kg on a dry basis are given in Table 1 [Van Son M, Van der Veen AMH, Verkuil D, unpub-

Table 1 Homogeneity study of barium in soil

Sample	Data #1	Data #2	Data #3	Mean	*s*	*n*
# 0118	323	301	310	311	11	3
# 0201	340	334	316	330	12	3
# 0383	320	321	309	317	7	3
# 0442	315	338	321	325	12	3
# 0557	326	338	325	330	7	3
# 0666	325	302	304	310	13	3
# 0791	324	331	317	324	7	3
# 0918	310	310	331	317	12	3
# 1026	336	321	328	328	8	3
# 1133	310	328	312	317	10	3
# 1249	314	314	302	310	7	3
# 1464	329	300	299	309	17	3
# 1581	320	329	311	320	9	3
# 1607	322	312	311	315	6	3
# 1799	332	317	299	316	17	3
# 1877	313	294	293	300	11	3
# 1996	324	314	335	324	10	3
# 2000	321	342	316	327	14	3

Table 2 Analysis of variance (ANOVA) table for barium in soil

Source of variation	*SS*	*df*	*MS*	*F*	*P-value*	F_{crit}
Between groups	3467	17	204	1.66	0.10	1.92
Within groups	4412	36	123			
Total	7880	53				

lished data]. The measurements were carried out on extracts obtained from aqua regia digestion using NEN 6465, and the measurements were carried out using ICP-MS. Using Excel, the following ANOVA table (one-way layout) can be computed (Table 2). The column "*SS*" provides the sums of squares, the column "*df*" the associated degrees of freedom, and the column "*MS*" the mean squares, which form the basis for the computation of variances as discussed in Part 1 [1]. The *F*-test indicates that the result of the homogeneity is insignificant ($F<F_{crit}$, the critical value of *F* for $\alpha=5\%$). The *P*-value gives the level for which the observed *F* equals $F_{crit.}$

The calculation of uncertainties is now very straightforward. The repeatability of the test method is just the square root of MS_{within}, equal to 11 mg/kg (=3,5%). For the variance among groups, the following expression can be used

$$s_A^2=s_{bb}^2=\frac{MS_{among}-MS_{within}}{n} \qquad (5)$$

This equation can be used instead of the formulas from Part 1, as in this case the data matrix is complete (all groups have the same number of members, $n=3$). The variance is 27 mg^2/kg^2; the standard deviation between bottles (s_{bb}) is 5.2 mg/kg (=1.6%), which is the contribution of heterogeneity to the uncertainty of one bottle. The link between Eqs. (1and 5) is as follows: MS_{among} is equal to *n* times $u^2_{c(bb)}$ and MS_{within} is equal to s^2_{method} (see part 1 of this paper). As $s_{meas}=s_{method}/\sqrt{n}$, the equivalence of Eqs. (2 and 4) follows. As it can be seen, the computation of the grand mean is not needed for this uncertainty evaluation, although Excel internally will calculate this parameter.

Evaluation of a between-bottle homogeneity study in an alternative format

In the experimental set-up described above, method repeatability and between-bottle heterogeneity were estimated by analysing several units *n*-times each. Frequently a different approach to obtain estimates for s_{meas} and s_{bb} is used: one unit is analysed several times to obtain an estimate for s_{method} and several units are then analysed in one replicate each to obtain an estimate of $u_{c(bb)}$. For the estimation of the between-bottle variability by this approach, it is vital that the results from one unit and from the different units are obtained by the same technique using the same sample intake. As $n=1$ in this case, $s_{meas}=s_{method}$ and s_{bb} can be estimated according to Eq. (1).

For the certification of a mussel-tissue material [9], ten units were used for the homogeneity study on selenium. Five determinations on one unit were performed, whereas the other nine units were analysed once. All analyses were done by k_0–NAA without sample pretreatment. The results of this study are given in Table 3. In this case, s_{bb} amounts to 2.84%, which is the uncertainty contribution of homogeneity to the uncertainty of one bottle.

Obviously, this method requires fewer measurements than performing the complete matrix of measurements for the ANOVA evaluation. This advantage is paid for with less significant results, as measurement re-

Table 3 Homogeneity study for selenium in mussel tissue

	5 replicates from one unit [mg/kg]	Results from 9 different units [mg/kg]
	1.907	1.872
	1.917	1.874
	1.961	1.928
	1.901	1.833
	1.834	1.944
		1.840
		1.726
		1.952
		1.861
Average	1.904	1.870
Standard deviation	0.046	0.070
Variation coefficient	2.40% (s_{meas})	3.72% ($u_{c(bb)}$)

peatability is not reduced by replication. The effect is that s_{meas} is more likely to be larger than $u_{c(bb)}$, leading to the problems already addressed about s_{bb} values tending to zero. This format may therefore lead to a greater uncertainty for the certified reference material. As in many cases a more conservative value for the uncertainty due to between-bottle variation must be inserted rather than the value of s_{bb} as obtained directly from ANOVA. On the other hand, this example also demonstrates how to conduct a homogeneity test in cases where repetition of measurements is impossible. In these cases, the results in the first column of Table 2 must be obtained from other sources ("Type B evaluation" [8]), such as the quality manual of the laboratory, validation data or some other source.

Conclusions

A general framework for the estimation of within- and between-bottle heterogeneity has been developed. The approach consists of correcting an estimation of heterogeneity, impaired by measurement variability, for measurement variability. Reliable estimates for between- and within-bottle heterogeneity can be obtained given that the measurement variability is small compared to the heterogeneity to be detected. If the requirement of low measurement variability compared to heterogeneity is not met, more conservative approaches should be employed.

The method as such works equally well on the fully nested ANOVA designs of experiments, and on other formats. The underlying theory and concepts are the same, but the implementation differs. This allows application of the method for homogeneity tests on test pieces in destructive testing as well, which has great benefits.

This work also shows that carrying out homogeneity tests cannot and must not be separated from other parts of the certification project (e.g. stability studies, characterisation measurements), as the accuracy of the measurements in the homogeneity study have important implications on the establishment of the combined standard uncertainty of the candidate reference material.

References

1. Van der Veen AMH, Pauwels J(2000) Accred Qual Assur 5:464–469
2. Pauwels J, Kurfürst U, Grobecker KH, Quevauviller P (1993) Fresenius J Anal Chem 345:478–481
3. ISO Guide 35:1989 (1989) Certification of reference materials – General and statistical principles, 2nd edn. International Organization for Standardization (ISO), Geneva, Switzerland
4. Schiller SB (1996) Statistical aspects of the certification of chemical batch SRMs. NIST Special Publication 260–125. NIST, Gaithersburg, USA
5. BCR Guidelines (1994) Standards, Measurement and Testing Programme, Brussels, Belgium
6. Pauwels J, Lamberty A, Schimmel H(1998) Accred Qual Assur 3:51–55
7. Van der Veen AMH, Alink A (1998) Accred Qual Assur 3:20–26
8. BIPM, IEC, IFCC, ISO, IUPAC, IUPAP, OIML (1995) Guide to the expression of uncertainty in measurement, 1st edn. ISO Geneva, Switzerland
9. Lamberty A, Muntau H (1999) The certification of the mass fractions of As, Cd, Cr. Cu, Hg, Mn, Pb, Se and Zn in mussel tissue *Mytilus edulis*. EUR 18840EN

Accred Qual Assur (2001) 6:257–263

Adriaan M.H. van der Veen
Thomas P.J. Linsinger
Andree Lamberty
Jean Pauwels

Uncertainty calculations in the certification of reference materials
3. Stability study

A.M.H. van der Veen (✉)
Nederlands Meetinstituut,
Schoemakerstraat 97, 2600 AR Delft,
The Netherlands
e-mail: avdveen@nmi.nl
Tel.: +31-15-2691 733
Fax: +31-15-261 29 71

T. P. J. Linsinger · A. Lamberty
J. Pauwels
European Commission, Joint Research Centre, Institute for Reference Materials and Measurements, Retieseweg,
2440 Geel, Belgium

Abstract To serve as a measurement standard, a (certified) reference material must be stable. For this purpose, the material should undergo stability testing after it has been prepared. This paper looks at the statistical aspects of stability testing. Essentially, these studies can be described with analysis of variance statistics, including variant regression analysis. The latter is used in practice for both trend analysis and for the development of expressions for extrapolations. Extrapolation of stability data is briefly touched upon, as far as the combined standard uncertainty of the reference material is concerned. There are different options to validate the extrapolations made from initial stability studies, and some of them might influence the uncertainty of the reference material and/or the shelf-life. The latter is the more commonly observed consequence of what is called 'stability monitoring'.

Keywords Uncertainty · Reference materials · Stability testing · Analysis of variance · Regression analysis

Introduction

Stability testing is, together with homogeneity testing, crucial in the process of certifying reference materials. As the lifetime of a typical certified reference material (CRM) is several years, the property value should be constant during the lifetime of the CRM. A further requirement is that during transport, under conditions to be specified, the stability of the material should be guaranteed by the producer. These aspects should be considered carefully in the design stage of the certification project, and the stability of reference materials is an aspect that should be demonstrable.

In this paper, the basic statistics for stability studies will be looked at, and it will be shown how these uncertainty components can be determined in order to develop a certification project in compliance with the "Guide to the expression of uncertainty in measurement" (GUM) [1]. The paper carries on from on Parts 1 and 2 [2, 3]. The statistics developed in this paper mainly concern cases where no statistically significant or otherwise relevant trend in the property value has been observed. In current practice, it is not acceptable to have a time-dependent property value, apart for radioisotope CRMs where a well-defined decay mechanism is present. This decay will lead to a trend, which should be included in the certification.

It should be noted however that the data should be assessed with trend analysis. Trend analysis, as will be shown, shares its basis with analysis of variance (ANOVA). A brief introduction to the assessment of trends in data is given, although it is considered that it is a subject that needs – depending on the application – more complete coverage. As the property value of a candidate CRM should be independent of time, it is sufficient to treat trend analysis in a simplified way.

Stability testing is often more complex than homogeneity testing. The reason for this is that at least one factor potentially influencing the measurement uncertainty has been added (time), but in some cases more (time, operator, calibration, etc.). It will be shown that these cases can be described with the same statistical theory, leading to different models for different cases.

Types of (in)stability

There are two types of (in)stability to be considered [4]:

1. Long-term stability of the material (e.g. shelf-life)
2. Short-term stability (e.g. stability of the material under "transport conditions")

The first kind of stability study is well known, and usually implemented in certification projects to a certain extent. The second kind of stability study is less common, but is possibly even more important than the first type. The behaviour of the sample during transport under the conditions specified may differ from that under the storage conditions of the producer/vendor due to external influences (e.g. temperature, daylight, etc.).

Often, the effect of short-term stability can be neglected. The producer will usually specify proper transport conditions, effectively reducing this uncertainty component. Usually, this specification of conditions leads to the case that the uncertainty due to short-term stability does not exceed that of long-term stability. In such cases, it is acceptable to set the (additional) effect of short-term stability, as obtained from stability testing to zero, as it has already been accounted for in the long-term stability study. Under certain conditions, it is equally important to know what might happen to the sample if proper transport conditions have not been maintained. In many cases, a simple verification of the CRM prior to first use might be sufficient, whereas in other cases it is evident that the CRM has become useless. This allows the producer to give better advice and, from the perspective of the user, supply a better product.

Monitoring should be envisaged during the lifetime of the CRM. A fundamental problem of stability studies is that they only account for the past, not necessarily for the present or future. Some kinds of degradation or other instability problems proceed very slowly and very gradually, but in many cases some abrupt change in properties takes place at some time, practically ending the lifetime of the CRM. As these mechanisms are highly unpredictable, monitoring of the stability is a necessity. As extrapolation of stability data is an absolute necessity to improve the marketability of the CRM, it is reasonable to make assumptions based on the 'past' (e.g. stability study) and verify them through stability monitoring.

Trend analysis

The key point in a stability study is to monitor the property value as a function of time. Often, such a study is conducted at different temperatures to determine the optimal conditions for storage and/or transport. The data thus obtained should first be investigated for a trend.

Before doing so, an assumption must be made about the kind of trend that might be observed. From an empirical point of view, the change of the property value (if any) is expected to be small anyway, so a simple linear model as a first-order approach may do. However, in some cases some physical, chemical or biological phenomenon might dictate instability, and in these cases such mechanisms may suggest the need for another approach. In this paper, the empirical approach using a linear model will be developed, along a paththat enables the use of other models.

The basic model for a simple linear regression can be expressed as [5]

$$Y=\beta_0+\beta_1 X+\varepsilon \quad (1)$$

where β_0 and β_1 are the *regression coefficients*, and ε denotes the random error component. The random error component, ε, may be composed of only random error, but it may also contain one or more systematic factors. The theory applicable for "decomposing" ε is given in [2]. In the case of stability studies, X denotes time and Y the property value of the candidate CRM. For a stable reference material, β_1 is expected to be zero. In those cases, the model can be simplified to

$$Y=\beta_0+\varepsilon \quad (2)$$

which is essentially the model underlying ANOVA. The development of expressions for estimates for the parameters β_0 and β_1, as well as the computation of variances of different kinds, follows the same paths as the development of the expressions for ANOVA, as shown in part 1 [2].

Given a set of n pair wise observations of Y versus X, for each Y_i the following expression can be developed

$$Y_i=\beta_0+\beta_1 X_i+\varepsilon_i \quad (3)$$

Often, more than one value of Y_i will be available for each X_i, due to repetition of measurement, the use of more than one bottle per point in time, etc. These aspects should be included in a real-life model of a particular stability study. Based on the theory of this paper and part 1 [2], these extensions can be developed quite straightforwardly.

The sum of squared deviations can now be expressed as

$$S=\sum_{i=1}^{n}\varepsilon_i^2=\sum_{i=1}^{n}(Y_i-\beta_0-\beta_i X_i)^2 \quad (4)$$

Frequently, the sum of squared deviations is expressed in terms of the χ^2-statistic. The only difference between χ^2

and S is a scaling factor, which equals $1/\sigma^2$. Equation (4) defines the regression problem: the objective is to minimise S (as function of β_0 and β_1). This can be accomplished by differentiating (4) with respect to β_0 and β_1 and setting the partial derivatives to zero. This is treated in detail in many statistical textbooks.

The regression parameters can be computed from the following expressions. For the estimator for the slope, solving Eq. (4), the following expression can be derived

$$b_i = \frac{\sum_{i=1}^{n}(X_i-\bar{X})(Y_i-\bar{Y})}{\sum_{i=1}^{n}(X_i-\bar{X})^2} = \frac{\sum_{i=1}^{n}X_iY_i - \frac{1}{n}\left(\sum_{i=1}^{n}X_i\right)\left(\sum_{i=1}^{n}Y_i\right)}{\sum_{i=1}^{n}X_i^2 - \frac{1}{n}\left(\sum_{i=1}^{n}X_i\right)^2} \quad (5)$$

whereby it should be noted that the first expression is more suitable for numerical work in computer programs, whereas the second is more suitable for pocket calculators, as the computations are less tedious. The first expression for the slope is less sensitive to round-off errors, so in most cases more accurate.

The estimate for the intercept can be computed from

$$b_0=\bar{Y}-b_i\bar{X} \quad (6)$$

Using the error propagation formula [1], the standard deviations in b_1 and b_0 can be computed. The *estimated standard deviation* of b_1 is given by

$$s(b_i) = \frac{s}{\sqrt{\sum_{i=1}^{n}(X_i-\bar{X})^2}} \quad (7)$$

whereby

$$s^2 = \frac{\sum_{i=1}^{n}(Y_1-b_0-b_1X_i)^2}{n-2} \quad (8)$$

The estimated variance of b_0 is given by

$$V(b_0) = V(\bar{Y}-b_1\bar{X}) = s^2\left[\frac{1}{n}+\frac{\bar{X}^2}{\sum_{i=1}^{n}(X_1-\bar{X})^2}\right] = \frac{s^2\sum_{i=1}^{n}X_i^2}{n\sum_{i=1}^{n}(X_1-\bar{X})^2} \quad (9)$$

whereby it should be noted that b_1 and <PIC 010> are uncorrelated.

Based on the standard deviation of b_1, a judgement can be made. Using Eq. (7), and an appropriate t-factor (number of degrees of freedom equals n-2), b_1 can be tested for significance. Although this method is quite uncomplicated, it requires the computation of $s(b_1)$, a parameter that is often not calculated by software. Most software does however compute an F-table, which can also be used for evaluating the significance of regression (Table 1).

The mean square due to regression is often denoted as $SS(b_1|b_0)$, to be read as "sum of squares for b_1 after allowance has been made for b_0". The mean square about regression (s^2) is an estimate for the property denoted by $\sigma^2_{Y.X}$ and called the variance about regression.

The ratio MS_{reg}:s^2 can be tested for significance using the F-tables. Table 1 provides the necessary information with respect to the degrees of freedom. The advantage of using the F-table instead of the method using the t-test is twofold:

1. The F-table is generated by most software systems by default.
2. The F-table can readily be extended to other regression models, which makes it more widely applicable.

Irrespective of what kind of test is used, it should be noted that the outcome is only meaningful if the repeatability standard deviation of measurement, possibly in conjunction with the between-bottle homogeneity is sufficiently small. It can be demonstrated that if the repeatability standard deviation is comparable to that of the homogeneity study and the characterisation of the material (e.g. the determination of the property value), this requirement is met.

Experimental layouts

There are two basic experimental layouts for stability studies:

1. Classical stability study
2. Isochronous stability study.

The isochronous stability study, introduced by Lamberty et. al. [6], has the great advantage that all measurements

Table 1 Analysis of variance table for linear regression

Source of variation	Degrees of freedom	Sum of squares (SS)	Mean square (MS)
Due to regression	1	$\sum_{i=1}^{n}(\hat{Y}_i-\bar{Y})^2$%	
About regression (residual)	n–2	$\sum_{i=1}^{n}(Y_i-\hat{Y}_i)^2$%	$s^2=\frac{SS}{n-2}$%
Total, corrected for mean	n–1	$\sum_{i=1}^{n}(Y_i-\bar{Y})^2$%	

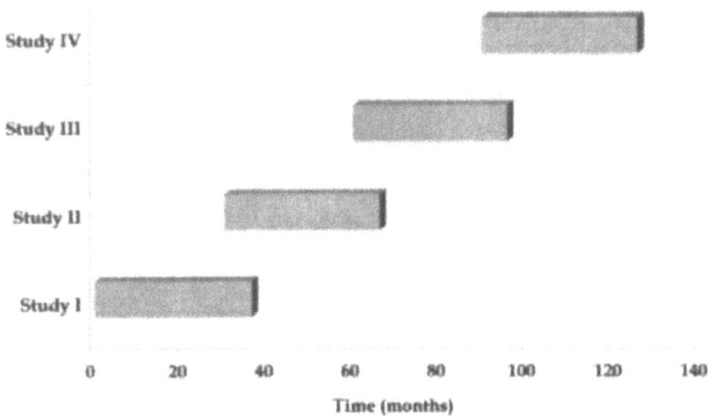

Fig. 1 Semi-continuous stability testing

can be carried out in one run, with one calibration. This reduces the scattering on each point in time, and thus improves the 'resolution' of the stability study. As a consequence, an isochronous stability study will lead to a smaller uncertainty than a classical one.

A classical stability study measures a sample as a function of time. In this case, the work is carried out under (within-laboratory) reproducibility conditions, which leads to a higher uncertainty, as the instability of the measurement system is now also included. When certifying a single package, such as a gas cylinder, it is impossible to use the isochronous layout. The isochronous layout is specifically designed for batch certifications.

Furthermore, monitoring typically takes place using the classical design. The problem here is that the isochronous design only provides data at the end of the stability study, whereas for monitoring it is essential that information becomes available during the lifetime of the CRM. This has no further consequences for the uncertainty of the CRM, in contrast to the other two stability studies, as monitoring only involves the demonstration that the uncertainty on the certificate is still valid. This should obviously be done with care, so that not too much uncertainty is added during the verification of the CRM, but there is no necessity to account for these results in the combined standard uncertainty of the CRM. It should be noted that monitoring data could be used to re-evaluate the uncertainty due to instability, and thus, affecting the uncertainty of the CRM. This topic is beyond the scope of this paper.

Another way of implementing the requirement of monitoring the CRM is to use a kind of semi-continuous stability testing. In essence, the first stability study of, for example, 36 months is succeeded by a second one, with some months of overlap. The principle is illustrated in Fig. 1. Measuring "reference samples" (i.e. samples that have been stored at reference temperature) at each monitoring point could be seen as a special case of this semi-continuous study with each study consisting of one point in time.

The reason that this kind of stability testing is not continuous has to do with the use of isochronous measurements: as these are done in a single run after the period of stability testing, it is necessary to make a cut in the stability testing. After such a cut, the uncertainty of the CRM can be reviewed as well, as these new stability data can be used as a renewed estimation of the uncertainty due to instability.

As with homogeneity testing, good repeatability of the test measurement method is an important prerequisite for stability testing as well. In Part 1 [2], it has been argued that the quality of the estimators for group and subgroup variances depends on the within-group variances. For a stability study, this means that the estimators for variance due to instability improve when the repeatability variance decreases. The within-bottle variance, as obtained from homogeneity testing [3], forms the basis for decreasing the repeatability variance to a level specific for the method. The number of repeated measurements (replicates) is the other variable that can be influenced to decrease the repeatability variance.

Uncertainty modelling

A stability study may include the following uncertainty components:

- Repeatability of measurement
- Instability of the material
- Instability of the measurement system (in the classical design)
- Between-bottle homogeneity (in batch certifications).

From this list, it can be seen that whenever possible, the isochronous design should be preferred over the classical one, as it reduces the number of components to look at. In a typical isochronous stability study, only three components of uncertainty are left, which can be separated through a fully nested two-way analysis of variance (see Part 1 [2] for details). The uncertainty for a single measurement in such an experiment can be expressed as

$$u^2\left(y_{ijk}\right) = s_{stab}^2 + s_{bb}^2 + s_r^2 \tag{10}$$

where s_{stab} is the standard deviation due to instability[1], s_{bb} denotes the between-bottle standard deviation, and s_r the repeatability standard deviation. As in the case of the homogeneity study, the quality of the estimator s_{stab} depends on s_{bb} (and s_r). Thus, the between-bottle homogeneity affects the quality of the estimator for instability! This is inevitable, as it is a property of analysis of variance, as discussed in Parts 1 and 2 [2, 3].

It should be noted that the model is only valid if the homogeneity and stability of the material are indepen-

[1] The subscript "stab" is used either to denote "lts" (long-term stability study) or to denote "sts" (short-term stability study).

dent. This might seem obvious, but it is not. If a material shows considerable between-bottle heterogeneity, it can also be expected that the stability of the material differ from bottle to bottle, as the stability of the material will depend (among others) on the composition of the material. However, as the preparation of a reference material involves the reduction of heterogeneity and the improvement of the stability, it is for most reference materials reasonable to assume independence between effects from heterogeneity and instability.

The model also raises another question: is it possible to estimate s_{bb} from a stability study? The answer is that statistically speakingit is possible. Especially in those cases where the effect of (in)stability is expected to be small anyway, it is certainly an option. A two-way fully nested ANOVA will do the job and provide the three standard deviations s_r, s_{bb}, and s_{stab}. As stability studies are often carried out at different temperatures, it is to be recommended to pick the value at one of the lowest temperatures.

Furthermore, it can be noted that if s_{stab} is zero or sufficiently close to zero, it is possible to scatter the bottles along the time axis, and to consider the set of data as a homogeneity study. For example, if at5 different points in time 3 bottles have been measured, then, provided that $s_{stab} \sim 0$, the experiment can be evaluated as a between-bottle homogeneity study with 15 bottles. If there is some instability left, that it, s_{stab} is not exactly zero, this will be accounted for in the estimate obtained for s_{bb} when evaluating the data as a homogeneity study.

In the classical design, the expression for the uncertainty reads as

$$u^2\left(y_{ijk}\right) = s_{stab}^2 + s_{lor}^2 + s_{bb}^2 + s_r^2 \tag{11}$$

whereby one term has been added, the variance due to lack of repeatability, s_{lor}^2. This term represents the stability of the measurement system. The measurements in a classical stability study take place under (within-laboratory) reproducibility conditions. The other terms are identical to the isochronous case. The problem with the classical stability study is that the separation between s_{stab} and s_{lor} is not possible; as a result, the model describing a typical analysis of variance layout (two-way, fully nested design) read as

$$u^2\left(y_{ijk}\right) = s_{stab'}^2 + s_{bb}^2 + s_r^2 \tag{12}$$

whereby $s_{stab'}$ now denotes the uncertainty component due to instability of the measurement system and the material. This is the case for both the short-term stability study as well as for the long-term stability study.

Uncertainty evaluation of stability monitoring

The uncertainty evaluation of stability monitoring is quite different from the long-term and short-term stability studies. First, it should be noted that stability monitoring does not affect the uncertainty statement of the reference material on the certificate, U_{CRM}. and it is unnecessary as will be demonstrated. The uncertainty from monitoring can be expressed as

$$u_{mon} = \sqrt{u_{CRM}^2 + u_{meas}^2} \tag{13}$$

whereby u_{CRM} denotes the combined standard uncertainty of the reference material, and u_{meas} the uncertainty from measurement, including calibration. Ideally, u_{CRM} is considerably greater than u_{meas}, but it should be considered that this is not always possible. Furthermore, the measurements should be carried out in such a way, that their validity must not be demonstrated from using the CRM. One cannot check two things at the same time in one experiment. The validity of the CRM is to be reconfirmed, which can only be valid if the measurement is demonstrably reliable.

If these experimental conditions are fulfilled, both the property value and its expanded uncertainty are reconfirmed. There is, under these conditions, no need to increase U_{CRM}, as the uncertainty from measurement is something which must be accounted for separately. This is true both for monitoring as well as for the normal use of the CRM. It should however be noted that for the sake of the validity of the monitoring measurement, u_{meas} should be as small as possible, and certainly not exceeding u_{meas} from a typical user of the CRM, who will use a similar approach for verifying her measurements.

An alternative approach is to consider the point obtained in monitoring as just the next point in the stability study, and from the complete set of data, a new estimate for s_{lts} can obtained. If necessary, the uncertainty of the CRM can be reviewed, but usually the evaluation will only reconfirm the value of s_{lts} already obtained and just extend the shelf-life (i.e. the time for which the certificate is considered to be valid).

Examples

An example of the results of an isochronous stability study is shown in Table 2, which lists the results of a 12-month isochronous measurement for the determination of total glucosinolate in rapeseed, BCR 190R. Two units were analysed in triplicate for each time.

A standard deviation between bottles of 0.28 µmol/kg (1.1%) was calculated, which corresponds well with the homogeneity study, in which a method repeatability of 3.7 % and a between unit variation of 1.4% was estimated [3]. Standard deviation between times was estimated 0.31 µmol/kg (1.4%). The detected instability was refuted by subsequent stability studies. Results of a classical stability study are shown in Table 3 [7].

Performing a one-way ANOVA gives a standard deviation within groups of 0.063 µg/kg (7.9%) and a standard

Table 2 Results of stability tests for total glucosinolate in rapeseed, BCR 190R. Concentrations are given in μmol/kg

	t=0 months	t=6 months	t=9 months	t=12 months
Bottle 1	21.84	23.26	23.25	22.25
	21.74	21.62	22.39	22.22
	22.45	22.55	22.51	22.22
Bottle 2	22.32	23.42	22.54	23.89
	21.89	23.16	22.61	22.78
	21.37	22.75	23.73	22.84

Table 3 Results for a stability study for aflatoxin M1 in milk powder in μg/kg [7]

	t=1 month	t=2 months	t=4 months	t=6 months	t=8 months	t=10 months	t=12 months
Value 1	0.72	0.63	0.83	0.85	0.89	0.73	0.74
Value 2	0.79	0.72	0.87	0.90	0.91	0.92	0.80

deviation between groups of 0.064 (8.0%). This result is obviously strongly influenced by measurement reproducibility. Looking at the results, the aflatoxin content seems to decrease initially, rise afterwards and decrease again. Such behaviour is very unlikely, but would nevertheless be included in an estimation of uncertainty of stability.

Discussion

Evaluating stability studies using ANOVA seems to neglect the information of the relative position of the measurement in time. As is shown in the Annex to this paper, the statistics from ANOVA and those of regression analysis are closely related. For obtaining the estimate of s_{lts}, it seems that a stability study with measurements after 0, 2, 4 and 6 months contains the same information as one with measurements after 0, 8, 16, 24 months. When using the appropriate expressions for extrapolating the data and the evaluation of measurement uncertainty, it will become clear that the 24-months stability study will be different from that of 6 months, as would be expected. The expressions for the uncertainty must appreciate the distance between the centre of gravity of the data.

In many cases, s^2_{stab} will be smaller than the sum of the other contributions to $u^2(y_{ijk})$ in Eqs. (1) and (3), which makes the estimation of s_{stab} impossible. This problem was already addressed in Part 2 for the homogeneity study [3]. The same options for treating these cases exist for the stability study:

1) Accept the low value and conclude that uncertainty of stability is negligible compared to the other uncertainty contributions.
2) Choose a more conservative approach like for example the ones outlined in [10].

If profound knowledge of the material and the production process is required to adopt possibility for the homogeneity case, it is even more true for stability testing. Homogeneity of a material can be assumed constant over time. Neglecting a small inhomogeneity will therefore result in only slight underestimation of uncertainty. On the contrary, instability exacerbates with time. Degradation between the monitoring measurements may therefore result in unrealistic uncertainty statements if no allowance for possible degradation is made.

All these points emphasise the importance of profound knowledge of the material and possible degradation pathways. Being able to predict degradation therefore allows one actively to counteract it, which is to be preferred over any statistical evaluation of the facts. Knowledge also allows a more reliable estimation of u_{lts} for those cases in which degradation cannot be prevented than statistical evaluation *a posteriori*.

Finally, it should be noted that the estimates s_{lts} and to a lesser extent s_{sts} form only the basis for the values for these uncertainty components in the expression of the measurement uncertainty of the property values. One of the aspects not covered here is the development of a recipe for a shelf-life, in conjunction with developing an appropriate estimate for u_{lts} at the shelf-life.

Conclusions

A framework for the estimation of uncertainty of stability from ANOVA has been developed. The approach separates between-bottle variation from measurement repeatability and variation in time. For isochronous measurements, this variation in time represents variation due to instability. For classical stability studies, this variation is confounded with the intralaboratory reproducibility.

Estimates of uncertainty of stability are therefore smaller when using isochronous schemes.

It has been shown that estimates for between-unit variation can be obtained from stability studies. These can be used to back up original homogeneity studies or may even serve as the sole homogeneity study.

The method requires that variation due to stability is not negligible compared to measurement and between-unit variation. If this requirement is not met, a more conservative approach should be employed. However, the decision about this should be made based on a profound knowledge of the material.

References

1. BIPM, IEC, IFCC, ISO, IUPAC, IUPAP, OIML (1995) Guide to the expression of uncertainty in measurement, 1st edn. ISO, Geneva, Switzerland
2. Van der Veen AMH., Pauwels J. (2000) Accred Qual Assur 5:464–469
3. Van der Veen AMH., Linsinger TPJ., Pauwels J. (2000) AccredQual Assur 6:26–30
4. Van der Veen AMH. (2000) "Determination of the certified value of a reference material appreciating the uncertainty statements obtained in the collaborative study", presented at AMCTM 2000 conference, Monte de Caparica, May 2000
5. Draper NR, Smith H (1981) "Applied regression analysis", 2nd edn. Wiley, New York, chapter 1
6. Lamberty A, Schimmel H, Pauwels J (1997) Fresenius J Anal Chem 360:359–361
7. Van Egmond HP, Wagstaffe PJ (1992) "The certification of aflatoxin M1 in four milk powder samples. CRM No's 282, 283, 284 and 285", European Commission, EUR 10412

Accred Qual Assur (2000) 5:231–237

Mirella Buzoianu

Some aspects of the evaluation of measurement uncertainty using reference materials

Invited paper presented at the 2nd Central European Conference on Reference Materials (CERM.2), 9–10 September 1999, Prague, Czech Republic

M. Buzoianu (✉)
National Institute of Metrology,
Reference Materials Group,
Sos. Vitan-Bârzesti No.11,
75669 Bucharest, Romania
e-mail: office@inm.ro
Tel.: +401-334-5060
Fax: +401-334-5345

Abstract In practice there are three aspects that need to be considered in order to achieve the required traceability according to its definition: the 'stated reference', the 'unbroken chain of calibrations' and the "stated uncertainty". For a certain chemical result, each of these aspects highly depends on the measurement uncertainty, both on its magnitude and how it was estimated. Therefore, the paper describes the experience of the Romanian National Institute of Metrology in estimating measurement uncertainty during the certification of reference materials (RMs), in metrological activities (calibration, pattern approval, periodical verification, etc.), as well as during the analytical measurement process. Practical examples of estimation of measurement uncertainty using RMs or certified reference materials are discussed for their applicability in spectrophotometric and turbidimetric analysis. Use of the analysis of variance to obtain some additional information on the components of measurement uncertainty and to identify the magnitude of individual random effects is described.

Key words Measurement uncertainty · Reference materials · Turbidimetry

Introduction

In Romania, increased attention is been paid to the accuracy of all measurements performed in trade, environmental monitoring, public health, research and industry. According to the Law of Metrology, issued in 1992 and amended in 1999, all legal measurements and instruments should be uniform, comparable and accurate, and the instruments or measurement standards should be traceable to national or international standards. Note that the experience and main tasks of the Reference Materials Group of the Romanian National Institute of Metrology (INM) related to metrological assurance of all measurements performed in chemical laboratories are described in Refs. [1–3]. It is widely accepted that the strength of the traceability chain (starting from the measurand in the sample being analysed in a chemical laboratory, up to a unit of the SI, or to the value of a recognized measurement scale) depends on the measurement uncertainty. Also, the need to estimate and report properly the measurement uncertainty during the constantly increasing accreditation activities results in the utmost attention to the application of this concept in routine level laboratories.

Within this framework, the paper describes the experience of the Reference Materials Group of INM in estimation of measurement uncertainty both in metrological-related activities and in routine chemical applications. Considering some representative spectrophotometric and turbidimetric analysis, the paper attempts to discuss practical problems on the estimation of measurement uncertainty using the reference material (RM) and certified reference material (CRM). approach.

A brief review of the experience of INM in estimating measurement uncertainty

In order to estimate the measurement uncertainty associated with a measurement result reported in a metrological issue (certificate of calibration, certificate of a RM, report of a highly accurate measurement, etc.) ISO GUM [4], adopted in the national area as the Romanian standard SR 13434:1999, should be followed according to a fully defined measurement process.

Since the purpose of a measurement is to determine the value of a measurand (the value of a particular quantity to be measured), the measurement begins with an appropriate specification of the measurand and of the method of measurement or measurement procedure. The required specification (definition) of the measurand is dictated by the necessary accuracy of measurement and it should be defined with sufficient completeness so that for all practical purposes associated with the measurement its value is unique. Further, an individually designed measurement method or procedure, taking into consideration the requirements of the data, is needed. Then, a measurement system, involving the coordinated interaction of a number of influencing factors, is chosen. This system is calibrated against physical and chemical standards traceable to national standards maintained by INM. Physical calibrations are needed for the measurement equipment itself and for ancillary measurements such as time, temperature, volume, mass, etc. Several kinds of chemical calibration, involving RMs and CRMs, are necessary to calibrate the system providing results expressed in concentration units.

Briefly presented in Fig. 1 are the necessary steps, recommended in ISO-GUM for measurement uncertainty estimation, referring to the expression of the mathematical relationship between the measurand and the input quantities, the identification of uncertainty sources, the quantification of the uncertainty components and the calculation of the expanded uncertainty.

According to its definition, the combined standard uncertainty in calibration includes the main components described in Table 1 (components related to the knowledge of the true value of the measurement standard, components introduced when using the standard or the CRMs for calibrating another instrument and the components introduced by the instrument being calibrated). Note that an example of how to evaluate the calibration uncertainty of an analytical spectro(photo)meter is discussed in Ref. [3].

As a rule, the certification of RMs in the Reference Material Group of INM follows the procedure described in ISO Guide 35:1989, adopted as the national standard SR 13252-2:1995. Mostly, certification based on interlaboratory testing and on a metrological approach are used for metrological purposes. Several types of CRMs developed and certified by INM are presented in Ref. [1], and examples of certification using the metrological approach are described in Ref. [2].

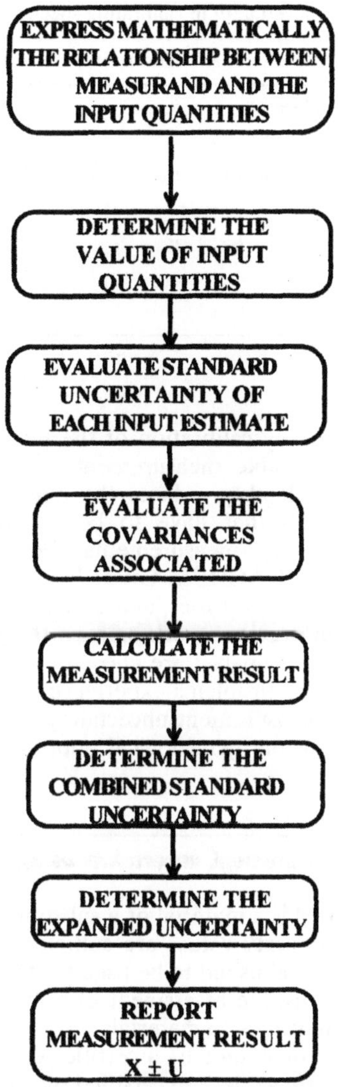

Fig. 1 Measurement uncertainty estimation process

The 'metrological approach' relies on assessing possible sources of error, then, by means of subsidiary experiments and/or theoretical analysis, determining the correction for each source and building an uncertainty budget by evaluating the uncertainty on the correction. Sometimes this approach is a tedious and time-consuming operation, although only few components of the uncertainty budget associated with a few corrections dominate. For routine measurements this approach can be

Table 1 Main sources of uncertainty in calibrating spectro(photo)meters

Main sources of uncertainty	Uncertainty	Some necessary characteristics	Method of evaluation
Due to CRMs used in calibration	u_{CRM}	Homogeneity, stability, certified value of concentration	As indicated in ISO Guide 35:1989, adopted as SR 13252-2:1995
Due to the use of the measurement standard	u_{use}	Method of certification, calibration, conditions. matrix mismatch	As indicated in ISO Guide 33:1989, adopted as SR 13252-4:1995
Due to the spectrophotometer	u_{sp}	Photometric linearity, photometric accuracy, wavelength accuracy, stray light	From technical specification of the producer, from calibration certificate or within the calibration process

considerably simplified by calibration of the measurement system with traceable measurement standards, since calibration considerably reduces the number of uncertainty components that have to be evaluated. Also, for some routine measurements, both SR 13434:1999 and the EURACHEM Guide [5] are used in accredited chemical laboratories, and RMs, or some data and results from previous work or even the judgment of the experienced analyst are chosen to do this evaluation. Some of the problems experience by INM in the evaluation of measurement uncertainty of spectrometric results using the genealogical approach are discussed in Ref. [6].

The evaluation of measurement uncertainty using RMs

By definition [7], a RM is a material or a substance one or more of whose property values are sufficiently homogeneous and well established to be used for the calibration of an apparatus, the assessment of a measurement or for assigning values to materials.

Further, a RM, accompanied by a certificate, one or more of whose property values are certified by a procedure which establishes its traceability to an accurate realization of the unit in which the property values are expressed, and for which each certified value is accompanied by an uncertainty at a stated level of confidence is a CRM.

Thus, according to the above definitions and if they are properly used, both RMs and CRMs may contribute to the evaluation of the performance of a measurement system and to the estimation of measurement uncertainty. Three general cases of how CRMs may be used to evaluate measurement uncertainty are illustrated in Fig. 2. When the matrix of the CRM matches the matrix of the sample being measured, and the homogeneity and stability of the sample have been proven (first situation), the uncertainty of the sample measurements can be equatable to that observed in measurement of the CRM. Further, if this match is not possible, but the CRM has a related matrix to the sample (second situation), the test sample uncertainty may be related by a factor of k to that observed when measuring the CRM. When the previous situation does not apply, the measurement of an appropriate CRM (third situation) can give an indication of the measurement uncertainty. Even so, it is recommended that isolated results on CRMs should be completed with the use of control charts if one consistently needs information on the measuring process.

A control chart is simply a graphical way to interpret test data. If a selected RM is measured periodically and the results are plotted sequentially on a graph (chart control type), a lot of information may be obtained on the combined effect of many potential sources of errors occurring in the measuring process. Limits for acceptable values are defined and the chemical measurement system is assumed to be in control as long as the results stay within these limits. The monitored precision of measurement and the accuracy of measurement of the reference material may be transferred, by inference, to all other appropriate measurements made by the system while it is in a state of control (i.e. repeated measurements over a period of time of standard samples processed right through the system are consistent with the measured variance of the system). Thus, the resulting judgment of uncertainty could be assigned to the sample data output of the process provided the following sources of uncertainty are taken into account:

- Uncertainty of the assigned value of the RM
- Reproducibility of the measurements made on the RM
- Any difference between the measured value of the RM and its assigned value
- Difference between the composition of RM and sample
- Difference in the response of measurement response due to interferences or matrix effects
- Operations that are carried out in the laboratory on samples but not on RMs due to subdivision of the original sample.

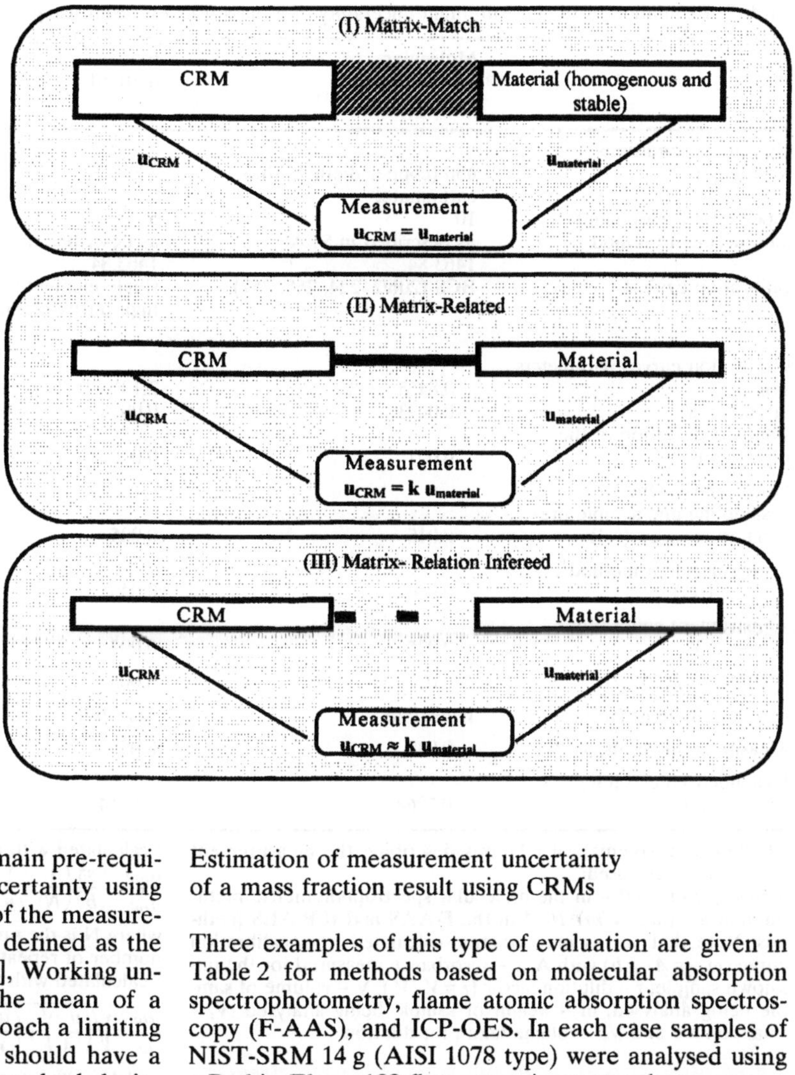

Fig. 2 Interpreting CRM measurements

One may easily note that one of the main pre-requisite for estimation of measurement uncertainty using RMs is the stability (statistical control) of the measurement system. Statistical control may be defined as the attainment of a state of predictability [8], Working under the above-mentioned conditions, the mean of a large number of measurements will approach a limiting value and the individual measurements should have a stable distribution, described by their standard deviation. The limits within any new measured value can be predicted with a specified probability. Confidence limits for a single measurement or for the mean of a set of measurements can be calculated, and the number of measurements required to obtain a mean value with a given confidence may be estimated.

Some outcomes on the estimation of measurement uncertainty using RMs and CRMs

Two examples of estimation of measurement uncertainty using both the CRM and RM approach are discussed below.

Estimation of measurement uncertainty of a mass fraction result using CRMs

Three examples of this type of evaluation are given in Table 2 for methods based on molecular absorption spectrophotometry, flame atomic absorption spectroscopy (F-AAS), and ICP-OES. In each case samples of NIST-SRM 14 g (AISI 1078 type) were analysed using a Perkin-Elmer 192 flame atomic spectrophotometer, a UV2-100 ATI Unicam spectrophotometer, and a Spectroflame ICP-P. Prior to this experiment, each measurement system was metrologically verified in accordance with the legal metrological norms (NML 9-12-97, NML 9-02-94 and OIML R116). Each system was calibrated against INM's own CRMs and internal quality procedure. The practical operation conditions and the parameters of the calibration curves are also indicated in the Table 2. Note that the uncertainty due to the sampling and sample preparation took into consideration [5] aspects regarding the homogeneity estimate, dissolution, dilution errors, chemical effects, etc. Therefore, several samples were taken and analysed separately according to the considered methods. The variability between the individual results was considered as a measure of the reproducibility of the specific analytical method and of the uncertainty of sampling and sam-

Table 2 Estimation of the measurement unceraintty of a mass fraction result using a CRM

	Molecular spectrometric method	Flame-AAS method	ICP-AES method
Operation conditions			
wavelength	450 (copper diethyldithio-carbamate)	324.8	324.75
repeated measurements	10	5	10
measurement method	as described in STAS 1463-84	as described in STAS 1463.84	as described in internal procedure
calibration against	BCS (206/2, 224, 255, 257)	single element solution from high purity elements	multielement solution from high purity metals (synthetic matrix form)
Calibration curves parameters			
slope, b	0.8455	2.1016	71517
intercept, a	−0.0093	0.0016	11
standard deviation of the regression, s_0	0.007	0.001	13.65
Uncertainties due to:			
– sampling and sample preparation, u_{sam}, (rel)	0.015	0.005	0.005
– measurement method[1], u_{meth}, (rel)	0.040	0.045	0.037
– instrument calibration[2], u_{cal}, (rel)	0.025	0.003	0.002
– CRM, u_{CRM}, (rel)	0.020	0.005	0.005
– data treatment (rel)	negligible	negligible	negligible
Measurement result, ω (%)	0.047	0.055	0.047
Expanded Uncertainty[3] (k=2), u_ω (rel)	0.1068	0.0914	0.0755

[1] Following mathematical relations describing the measurement procedures were used: $\omega=(\omega_{rec}\cdot r/m)\cdot 100$ – in the molecular spectrophotometric method, and $\omega=(\omega_{rec}\cdot V/m)\cdot 10^{-4}$ in the F-AAS and ICP-AES methods. Note that ω_{rec} is mass fraction calculated from calibration curve ($\omega_{rec}=A_{(x)}/b$) with $A_{(x)}$ – absorbance measured for the unknown sample; r – dilution factor ($r=V_f/V_i$); V – volume of sample being analysed; m – weight of sample being analysed; V_f – final volume and V_i – volume taken for dilution

[2] calculated with the equation:

$$\frac{u_{fcal}}{f_{cal}}=\frac{s_0^2}{b^2}\left(\frac{1}{N}+\frac{1}{n}+\frac{n(Ameas-\bar{A})^2}{b(n\Sigma c-(\Sigma c)^2)}\right);$$

where N is the number of CRMs used for calibration and n is the number of repeated measurements on each CRM

[3] calculated with the equations:

$$\frac{u\omega}{\omega}=\sqrt{\left(\frac{u_A}{A}\right)^2+\left(\frac{u_r}{r}\right)^2+\left(\frac{u_V}{V}\right)^2+\left(\frac{u_m}{m}\right)^2+\left(\frac{u_{CRM}}{c_{CRM}}\right)^2+\left(\frac{u_{cal}}{\omega_{rec}}\right)^2+\left(\frac{u_{sam}}{\omega_0}\right)^2}$$

and $U_\omega=2\cdot u_\omega/\omega$

ple preparation. Thus, the uncertainty due to sampling and sample preparation was determined as the difference between the above mentioned variabilities divided by the number of samples analysed.

As one may note, the absolute expanded uncertainty measurement (for $k=2$) evaluated for the mass fraction result obtain against the molecular spectrophotometric method (0.005%) was equal to the interlaboratory standard deviation of the standard method of analysis (0.005%). Using the F-AAS method, the absolute uncertainty of the measurement result (0.004%) exceeded the standard method accuracy (0.003%-abs). Comparing the two values, one may note that $\chi^2_{cal}=(0.004/0.003)^2=1.78$, is less than $\chi^2_{table(4;0.95)}=3.65$, and one may conclude that there is no evidence that measurement process is not as precise as required. For the ICP method, the expanded uncertainty agreed with the certified value of the CRM (0.003%) used in this experiment. Starting from the certified value of mass fraction of Cu in NIST-SRM (0.047%) one may also note that measurement results obtained both with molecular spectrophotometric and ICP methods were in good agreement. In the F-AAS method, a bias exceeding the prescribed limits was observed. Thus, the necessary optimization of the measurement process was concluded.

Table 3 Example of determination of measurement uncertainty of a turbidity result using ANOVA method

Instrument Mean turbidity measured	1	2	3	4	5
$\overline{T}_j$, FNU	100.42	97.98	103.60	99.80	103.34
$s(T_{ij})$, FNU	0.638	2.342	3.188	2.888	1.558
Instrument Mean turbidity measured	**6**	**7**	**8**	**9**	**10**
$\overline{T}_j$, FNU	101.50	102.00	100.22	99.88	99.16
$s(T_{ij})$, FNU	0.620	2.722	2.545	3.569	3.547

$\overline{T}=100.79$ $s_a^2=i\cdot s^2(\overline{T}_j)=5(1.801)^2=(4.027)^2$	$s(T_{ij})=1.801$ $s_b^2=\overline{s^2(T_{ij})}=(2.577)^2$ $F=\frac{s_a^2}{s_b^2}=2.44$ $F_{0.95(9,40)}=2.12$ $F_{0.975(9,40)}=2.45$
$F>F_{0.95(9,40)}$	$F<F_{0.975(9,40)}$
	$s^2(\overline{T}_j)=s_w^2/k+s_B^2$ $s_B^2=\frac{k\cdot s^2(\overline{T}_j)-\overline{s^2(T_{ij})}}{k}=(1.384)^2$ $s_w^2=\overline{s^2(T_{ij})}=(2.577)^2$
$s^2(\overline{T})=\frac{(J-1)\cdot s_a^2+J\cdot(k-1)\cdot s_b^2}{J\cdot k\cdot(J\cdot k-1)}$	$s^2(\overline{T})=s^2(\overline{T}_j)/J$
Measurement uncertainty: $(0.410)^2$	$(0.569)^2$
Reported result: $(100.79\pm k\cdot 0.410)$	$(100.79\pm k\cdot 0.569)$

On the use of analysis of variance (ANOVA) to estimate measurement uncertainty of a turbidity result

ANOVA methods are of special importance in measurements to identify and quantify individual random effects so that they be properly taken into account when the uncertainty of the result of the measurement is evaluated.

An example of the application of ANOVA to estimate turbidity measurement uncertainty is presented in the Table 3.

A standard of turbidity of (100.5±3.7) FNU, prepared from a (4000±20) primary standard of turbidity was measured against ten stable turbidimeters, denoted 1 to 10. On each instrument, five independent turbidity measurements were made, and the mean value ^{-}T and the experimental standard deviation $s(T_{ij})$ determined are presented in Table 3. Note that T_{ij} denotes the ith observation on the standard made on the jth turbidimeter. The experimental standard deviation of the mean, $s(T_{ij})$ is the measure of the uncertainty of measurement only if instrument-to-instrument variability of observations is the same as the variability of the observations made on a single instrument. If there is evidence that the between-instrument variability is significantly larger than the within-instrument variability the use of this expression can lead to a considerable understatement of the uncertainty of turbidity result. Therefore, the consistency of the within- and between-instrument variability of the observations was investigated by comparing the two independent estimates of the within-instrument component of variance, σ_w^2:

- The first one was denoted s_a^2 with (10-1) degrees of freedom. It was obtained from the variation of the measurements performed on the same instrument. The mean of measurements made on an instrument was the arithmetic average of five observations and its estimated variance was calculated;
- The second estimate of σ_w^2 was denoted s_b^2 and was calculated as the pooled estimate of variance obtained from the ten individual variances s_{ij}^2. Note that this estimate has 10(5-1) = 40 degrees of freedom.

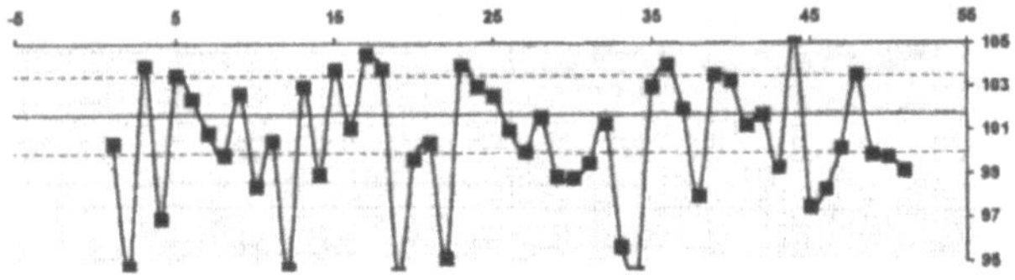

Fig. 3 Dispersion of turbidity measurement results on a RM of 100 FNU

The difference between s_a^2 and s_b^2 indicates the possible presence of an effect that varies from one instrument to another but that remains constant when measurements are made on any instrument. To test this possibility the *F*-test was used, and the value calculated was 2.44. Since $F_{0.95\ (9.40)}$ is 2.12 and $F_{0.975\ (9.40)}$ is 2.45, it was concluded that there is a statistical significance between instrument effect at the 5% level of significance but not at the 2.5 % level. Two further situations were considered. Presented in the left column of the table is the case in which the existence of the between-instrument effect was rejected because the difference between s_a^2 and s_b^2 was not viewed as statistically significant. Following the procedure indicated in [4], a standard measurement uncertainty of 0.410 FNU was obtained. Presented in the right column is the situation where the existence of a between-instrument effect was accepted. Assuming that this effect was random, the estimated variance of the mean value of turbidity was calculated taking into account both the within- and between-instrument random components of variance. Also, following the procedure indicated in [4] a standard measurement uncertainty of 0.569 FNU was obtained. Note that this value reported when a between-instrument variance was accepted, is a more prudent decision for practical purposes.

In the example considered above, the same RM was also tested on several turbidimeters within a short period of time, and the 48 turbidity values measured were plotted in the Fig. 3. Note that the upper limit of turbidity was of 105 FNU and the lower limit 95 FNU. An overall mean value of 102.3 FNU and a standard uncertainty of 2 FNU were determined. Over 35 results fell within the range of ±2 FNU.

Conclusions

This paper has examined the role of RMs and CRMs in estimation of the uncertainty of a measurement result. A clearly defined specification of the measurand, knowledge of the main sources of uncertainty and the correct use of RMs and CRMs are the main targets for the adequate application of the ISO-GUM. Different examples describing the use of CRMs and RMs emphasize the importance of the appropriate certification of RMs.

References

1. Buzoianu M, Duta S (1996) National System for Reference Materials in Romania. Proceedings of Central European Conference on Reference Materials, Slovacia
2. Buzoianu M (1998) Accred Qual Assur 3:270–277
3. Buzoianu M (1999) Metrological Calibration in Traceability. Proceedings of the EURACHEM Workshop on the Status of Traceability in chemistry, Bratislava
4. ISO (1993) Guide to the expression of uncertainty in measurements (GUM), 1st edn. ISO, Geneva
5. EURACHEM (1995) Guide to quantifying uncertainty in analytical measurements. EURACHEM, London
6. Buzoianu M, Aboul-Enein HY (1997) Accred Qual Assur 2:11–17
7. ISO (1993) International vocabulary of basic and general terms in metrology (VIM), 2nd edn. International Organization for Standardization (ISO), Geneva
8. ISO Guide 33 (1989) Uses of certified reference materials. ISO, Geneva

Accred Qual Assur (1998) 3:115–116

Werner Hässelbarth

Uncertainty – The key topic of metrology in chemistry

Presented at: 2nd EURACHEM Workshop on Measurement Uncertainty in Chemical Analysis, Berlin, 29–30 September 1997

W. Hässelbarth
Federal Institute for Materials Research and Testing (BAM)
D-12200 Berlin, Germany
Tel.: +49 30 6392 5861;
Fax: +49 30 6392 5972;
e-mail: werner.haesselbarth@bam.de

Abstract At the Second EURACHEM Workshop on Measurement Uncertainty in Chemical Analysis the author had the pleasure of chairing a working group on chemical metrology. This note presents some propositions arising from the preparation of, as well as from the discussion at and after, the working group session.

Key words Uncertainty · Metrology · Comparability · Traceability · Validation · Bias

There is no specific chemical metrology: Instead, as a truly horizontal discipline, metrology is applied in chemistry

Although "chemical metrology" is frequently used in chemical analysis, this term should be abandoned.

The main reason for this recommendation is that metrology is a truly horizontal discipline, operating on uniform principles, largely independently of the particular application field. Although first developed for physical measurements, the basic metrological terms, concepts and procedures are applicable throughout analytical chemistry. Nevertheless, there are specific challenges in chemical measurements such as the enormous diversity of measurands (i.e. analyte/level/matrix combinations) and the lack of direct measurement methods, which require specific strategies.

To promote the application of metrology in chemistry, its basic concepts and procedures have to be made crystal clear, emphasizing purposes instead of protocols.

The main objective of metrology in chemistry is known uncertainty of analytical results

By focussing on purposes instead of procedures, most of the current "buzz items" in measurement quality assurance such as comparability, traceability and validation are reduced to uncertainty as the primary performance characteristic, as follows.

Comparability

For comparing different measurement results on the same measurand, three basic requirements have to be fulfilled:

- The uncertainties of the measurement results have to be known.
- The units used to express the measurement results (including uncertainties) have to be the same, or at least convertible.
- The measures used to express the uncertainties have to be the same, or at least convertible.

Evidently, the last two requirements call for standardization. For the units of physical quantities the standardization problem was solved by establishing the International System of Units (SI) in 1960. The confusion about different uncertainty measures continued beyond that date and was solved only recently, with the appearance and world-wide acceptance of the "standard uncertainty" proposed by the Guide to the Expression of Uncertainty in Measurement (GUM) in 1993. The first requirement – known uncertainty – is still largely unsettled, although the concepts and methods of uncertainty evaluation proposed by the GUM have paved the way for substantial progress.

Traceability

Evaluation of measurement uncertainty according to the concept of the GUM basically includes two steps. In the first step, the measurement process is investigated for bias. If significant bias is found, a correction is applied. In the second step, the uncertainty on the bias correction is combined with the uncertainty due to random effects to yield the overall uncertainty of the corrected measurement.

Traceability serves the purpose of excluding, or of determining and correcting, significant measurement bias by comparison between measured values and corresponding reference values. Thus traceability, where applicable, provides a firm basis for valid uncertainty statements.

Validation

The performance characteristics considered in method validation serve the purpose of specifying, for a given analytical method, an application range with defined uncertainty. For example

- Specificity and selectivity are intended to specify the range of matrices where the uncertainty budget is valid. For other matrices, a correction and/or an additional uncertainty component are necessary.
- Robustness and ruggedness are intended to specify a range of operating conditions where the uncertainty budget is valid. For other operating conditions a correction and/or an additional uncertainty component are necessary.
- Linearity is intended to specify that range of analyte content where a linear calibration function applies. Beyond this range a correction and/or an additional uncertainty are necessary.
- Reproducibility aims at establishing a "top-down" estimate of the uncertainty of an analytical method, including interlaboratory bias as an uncertainty component, but excluding method bias.

Discussion and guidance on determination and correction of bias in analytical methods is urgently needed

To date analytical chemists as a rule have been content with stating agreement or disagreement of analytical results, obtained on a certified reference material, with the corresponding certified values. In cases of disagreement, usually no attempt is made to derive a correction. Neither is the uncertainty on the correction taken into account, which is also necessary in cases of agreement, because a correction factor of unity comes also with an uncertainty.

At the workshop, the topic of bias handling was raised on many occasions, indicating an urgent demand for discussion and guidance, for example on

- Practical traceability procedures, i.e. procedures for performing valid comparisons between measured values and reference values and for evaluating the comparison results
- Criteria for when to apply corrections on the basis of bias information
- Procedures for bias correction and for estimating correction uncertainty.

The topic of bias correction will be addressed in the forthcoming revision of the EURACHEM Guide Quantifying Uncertainty in Analytical Measurement.

In practice, rigorous and complete traceability of analytical results to established references will be the exception rather than the rule. Therefore it will be an important task to agree on levels of rigour and completeness of traceability statements required for, and feasible in, specific analytical sectors.

Accred Qual Assur (1998) 3:101–105

S. L. R. Ellison
V. J. Barwick

Estimating measurement uncertainty: reconciliation using a cause and effect approach

Presented at: 2nd EURACHEM Workshop on Measurement Uncertainty in Chemical Analysis, Berlin, 29–30 September 1997

S. L. R. Ellison (✉) · V. J. Barwick
Laboratory of the Government Chemist, Queens Road, Teddington TW11 0LY, UK

Abstract A strategy is presented for applying existing data and planning necessary additional experiments for uncertainty estimation. The strategy has two stages: identifying and structuring the input effects, followed by an explicit reconciliation stage to assess the degree to which information available meets the requirement and thus identify factors requiring further study. A graphical approach to identifying and structuring the input effects on a measurement result is presented. The methodology promotes consistent identification of important effects, and permits effective application of prior data with minimal risk of duplication or omission. The results of applying the methodology are discussed, with particular reference to the use of planned recovery and precision studies.

Key words Measurement uncertainty · Validation · Reconciliation · Cause and effect analysis

Introduction

The approach to the estimation of measurement uncertainty described in the ISO *Guide to the expression of uncertainty in measurement* (GUM) [1] and the EURACHEM interpretation for analytical measurement [2] relies on a quantitative model of the measurement system, typically embodied in a mathematical equation including all relevant factors. The GUM principles differ substantially from the methodology currently used in analytical chemistry for estimating uncertainty [3, 4]. Current practice in establishing confidence and intercomparability relies on the determination of overall method performance parameters, such as linearity, extraction recovery, reproducibility and other precision measures. These are obtained during method development and interlaboratory study [5–7], or by in-house validation protocols, with no formal requirement for a full mathematical model. Whilst there is commonality between the formal processes involved [8], implying that a reconciliation between the two is possible in principle, there are significant difficulties in applying the GUM approach generally in analytical chemistry [4]. In particular, it is common to find that the largest contributions to uncertainty arise from the least predictable effects, such as matrix effects on extraction or response, sampling operations, and interferences. Uncertainties associated with these effects can only be determined by experiment. However, the variation observed includes contributions from some, but not all, other sources of variation, risking "double counting" when other contributions are studied separately. The result, when using this, and other, data to inform GUM-compliant estimates of uncertainty, is substantial difficulty in reconciling the available data with the information required.

In this paper, we describe and illustrate a structured methodology applied in our laboratory to overcome these difficulties, and present results obtained using the methodology. It will be argued that application of the approach can lead to a full reconciliation of validation studies with the GUM approach, and the advantages and disadvantages of the methodology will be considered. Finally, some uncertainty estimates obtained using the methodology are presented, and the relative contributions of different contributions are considered.

Principles of approach

The strategy has two stages:

1. Identifying and structuring the effects on a result. In practice, we effect the necessary structured analysis using a cause and effect diagram (sometimes known as an Ishikawa or "fishbone" diagram) [9].
2. Reconciliation. The reconciliation stage assesses the degree to which information available meets the requirement and thus identifies factors requiring further study.

The approach is intended to generate an estimate of overall uncertainty, not a detailed quantification of all components.

Cause and effect analysis

The principles of constructing a cause and effect diagram are described fully elsewhere [9]. The procedure employed in our laboratory is as follows:

1. Write the complete equation for the result. The parameters in the equation form the main branches of the diagram. (We have found it is almost always necessary to add a main branch representing a nominal correction for overall bias, usually as recovery, and accordingly do so at this stage.)
2. Consider each step of the method and add any further factors to the diagram, working outwards from the main effects. Examples include environmental and matrix effects.
3. For each branch, add contributory factors until effects become sufficiently remote, that is, until effects on the result are negligible.
4. Resolve duplications and re-arrange to clarify contributions and group related causes. We have found it convenient to group precision terms at this stage on a separate precision branch.

Note that the procedure parallels the EURACHEM guide's sequence of preliminary operations very closely; specification of the measurand (step 1), identification of sources of uncertainty (steps 2 and 3) and grouping of related effects where possible (step 4) are explicitly suggested [2].

The final stage of the cause and effect analysis requires further elucidation. Duplications arise naturally in detailing contributions separately for every input parameter. For example, a run-to-run variability element is always present, at least nominally, for any influence factor; these effects contribute to any overall variance observed for the method as a whole and should not be added in separately if already so accounted for. Similarly, it is common to find the same instrument used to weigh materials, leading to over-counting of its calibration uncertainties. These considerations lead to the following additional rules for refinement of the diagram (though they apply equally well to any structured list of effects):

1. Cancelling effects: remove both. For example, in a weight by difference, two weights are determined, both subject to the balance "zero bias". The zero bias will cancel out of the weight by difference, and can be removed from the branches corresponding to the separate weighings.
2. Similar effect, same time: combine into a single input. For example, run-to-run variation on many inputs can be combined into an overall run-to-run precision "branch". Some caution is required; specifically, variability in operations carried out individually for every determination can be combined, whereas variability in operations carried out on complete batches (such as instrument calibration) will only be observable in between-batch measures of precision.
3. Different instances: re-label. It is common to find similarly named effects which actually refer to different instances of similar measurements. These must be clearly distinguished before proceeding.

The procedure is illustrated by reference to a simplified direct density measurement. We take the case of direct determination of the density $d(\mathrm{EtOH})$ of ethanol by weighing a known volume V in a suitable volumetric vessel of tare weight M_{tare} and gross weight including ethanol M_{gross}. The density is calculated from

$$d(\mathrm{EtOH}) = (M_{\mathrm{gross}} - M_{\mathrm{tare}})/V$$

For clarity, only three effects will be considered: equipment calibration, temperature, and the precision of each determination. Figures 1–3 illustrate the process graphically.

A cause and effect diagram consists of a hierarchical structure culminating in a single outcome. For our purpose, this outcome is a particular analytical result ["$d(\mathrm{EtOH})$" in Fig. 1]. The "branches" leading to the outcome are the contributory effects, which include both the results of particular intermediate measurements and other factors, such as environmental or matrix effects. Each branch may in turn have further contributory effects. These "effects" comprise all factors

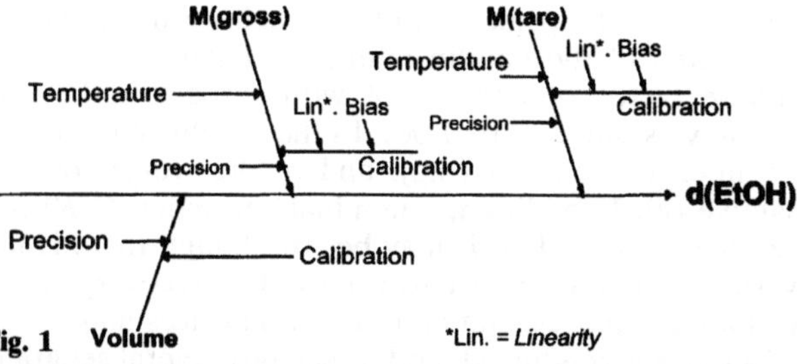

Fig. 1

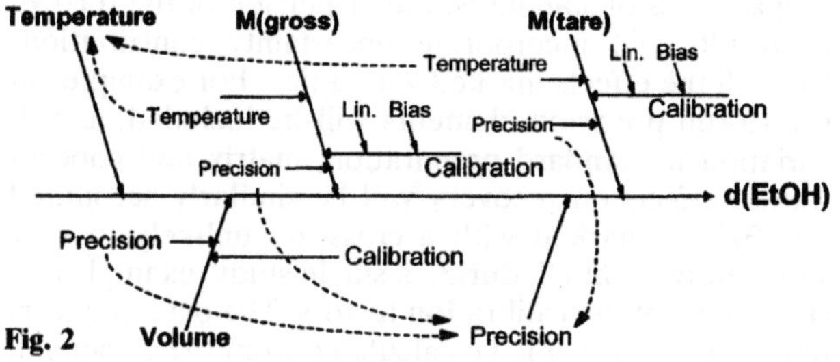

Fig. 2

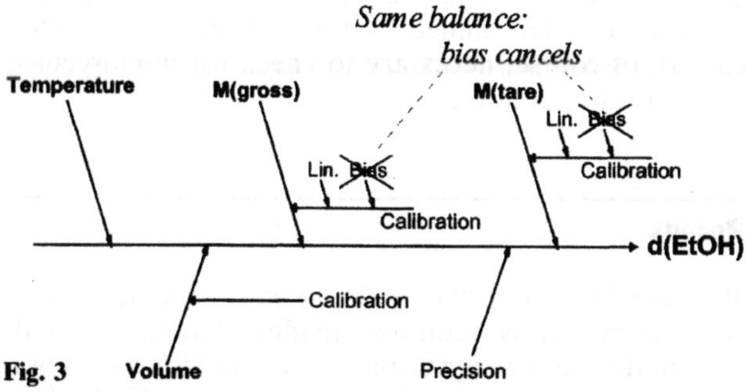

Fig. 3

Figs. 1–3 Stages in refinement of cause and effect diagram. **Fig. 1** Initial diagram. **Fig. 2** Combination of similar effects. **Fig. 3** Cancellation

affecting the result, whether variable or constant; uncertainties in any of these effects will clearly contribute to uncertainty in the result.

Figure 1 shows a possible diagram obtained directly from application of steps 1–3. The main branches are the parameters in the equation, and effects on each are represented by subsidiary branches. Note that there are two "temperature" effects, three "precision" effects and three "calibration" effects. Figure 2 shows precision and temperature effects each grouped together following the second rule (same effect/time); temperature may be treated as a single effect on density, while the individual variations in each determination contribute to variation observed in replication of the entire method. The calibration bias on the two weighings cancels, and can be removed (Fig. 3) following the first refinement rule (cancellation). Finally, the remaining "calibration" branches would need to be distinguished as two (different) contributions owing to possible non-linearity of balance response, together with the calibration uncertainty associated with the volumetric determination.

This form of analysis does not lead to uniquely structured lists. In the present example, temperature may be seen as either a direct effect on the density to be measured, or as an effect on the measured mass of material contained in a density bottle; either could form the initial structure. In practice this does not affect the utility of the method. Provided that all significant effects appear once, somewhere in the list, the overall methodology remains effective.

Once the cause-and-effect analysis is complete, it may be appropriate to return to the original equation for the result and add any new terms (such as temperature) to the equation. However, the reconciliation which follows will often show that additional terms are adequately accounted for; we therefore find it preferable to first conduct the next stage of the analysis.

Reconciliation

Following elucidation of the effects and parameters influencing the results, a review is conducted to determine qualitatively whether a given factor is duly accounted for by either existing data or experiments planned. The fundamental assumption underlying this review is that an effect varied representatively during the course of a series of observations needs no further study. In this context, "representatively" means that the influence parameter has demonstrably taken a distribution of values appropriate to the uncertainty in the parameter in question. For continuous parameters, this may be a permitted range or stated uncertainty; for factors such as sample matrix, this range corresponds to the variety of types permitted or encountered in normal use of the method. The assumption is justified as follows.

The ISO approach calculates a standard uncertainty $u(y)$ in $y(x,x_j\ldots)$ from contributions $u(y_i)=u(x_i)\cdot\partial y/\partial x_i$ (with additional terms if necessary). Each value of $u(x_i)$ characterises a dispersion associated with the value x_i. The sensitivity coefficient $\partial y/\partial x_i$ may be determined by differentiation (analytically or numerically), or by experiment. Consider an increment Δx_i in x_i. This will clearly lead to a change Δy in the result given by

$$\Delta y = y((x_i+\Delta x_i),x_j\ldots) - y(x_i,x_j\ldots) \quad (1)$$

Given the appropriate distribution $f(\Delta x_i)$ of values of Δx_i with dispersion characterised by standard uncertainty $u(x_i)$, the corresponding distribution $g(\Delta y_i)$ of Δy_i will be characterised by $u(y_i)$. This is essentially the basis of the ISO approach [1]. It follows that in order to demonstrate that a particular contribution to overall uncertainty is adequately incorporated into an observed dispersion of results, it is sufficient to demonstrate that the distribution of values taken by the influence parameter in the particular experiment is representative of f(Δx_i). [Strictly, u(x_i) could characterise many possible distributions and not all will yield the same value of u(y_i) for all functions y(x_i,x_j...). It is assumed here that either f(Δx_i) is the particular distribution appropriate to the problem, when g(Δy_i) necessarily generates the correct value of u(y_i), or that y(x_i,x_j...) satisfies the assumptions justifying the first order approximation of Ref. [1], in which case any distribution f(Δx_i) characterised by u(x_i) will generate u(y_i)].

Following these arguments, it is normally straightforward to decide whether a given parameter is sufficiently covered by a given set of data or planned experiment. Where a parameter is already so accounted for, the fact is noted. The parameters which are not accounted for become the subject of further study, either through planned experimentation, or by locating appropriate standing data, such as calibration certificates or manufacturing specifications. The resulting contributions, obtained from a mixture of whole method studies, standing data and any additional studies on single effects, can then be combined according to ISO GUM principles.

An illustrative example of a reconciled cause and effect study is shown in Fig. 4, which shows a partial diagram (excluding long-term precision contributions and secondary effects on recovery) for an internally standardised GC determination of cholesterol in oils and fats. The result, cholesterol concentration C_{ch} in mg/100 g of material, is given by

$$C_{ch} = \frac{A_c \times R_f \times IS}{A_B \times m} \times \frac{1}{R} \times 100, \qquad (2)$$

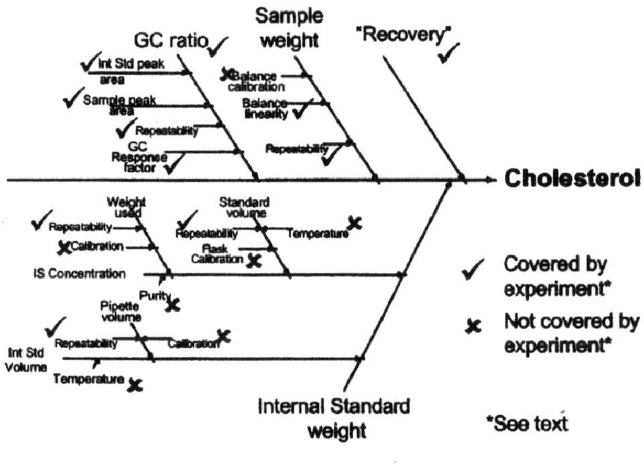

Fig. 4 Partial cause and effect diagram for cholesterol determination. See text for explanation

where A_c is the peak area of the cholesterol, A_B is the peak area of the betulin internal standard, R_f the response factor of cholesterol with respect to betulin (usually assumed to be 1.00), IS the weight of the betulin internal standard (mg), and m the weight of the sample (g). In addition, a nominal correction (1/R) for recovery is included; R may be 1.0, though there is invariably an associated uncertainty. If a recovery study including a representative range of matrices and levels of analyte is conducted, and it includes several separate preparations of standards, the dispersion of the recovery results will incorporate uncertainty contributions from all the effects marked with a tick. For example, all run-to-run precision elements will be included, as will variation in standard preparation; matrix and concentration effects on recovery will be similarly accounted for. Effects marked with a cross are unlikely to vary sufficiently, or at all, during a single study; examples include most of the calibration factors. The overall uncertainty can in principle be calculated from the dispersion of recoveries found in the experiment combined with contributions determined for the remaining terms. Due care is, of course, necessary to check for homoscedasticity before pooling data.

Results

We have found that the methodology is readily applied by analysts. It is intuitive, readily understood and, though different analysts may start with differing views, leads to consistent identification of major effects. It is particularly valuable in identifying factors for variation during validation studies, and for identifying the need for additional studies when whole method performance figures are available. The chief disadvantage is that, in focusing largely on whole method studies, only the overall uncertainty is estimated; individual sources of uncertainty are not necessarily quantified directly (though the methodology is equally applicable to formal parameter-by-parameter studies). However, the structured list of effects provides a valuable aid to planning when such additional information is required for method development. Some results of applying this methodology are summarised in Fig. 5, showing the relative magnitudes of contributions from overall precision and recovery uncertainties u(precision) and u(recovery), before combination. "Other" represents the remaining combined contributions. That is, the pie charts show the relative magnitudes of u(precision), u(recovery) and $\sqrt{\sum u(y_i)^2}$ with u(y_i) excluding u(precision) and u(recovery). It is clear that, as expected, most are dominated by the "whole method" contributions, suggesting that studies of overall method performance, together with specific additional factors,

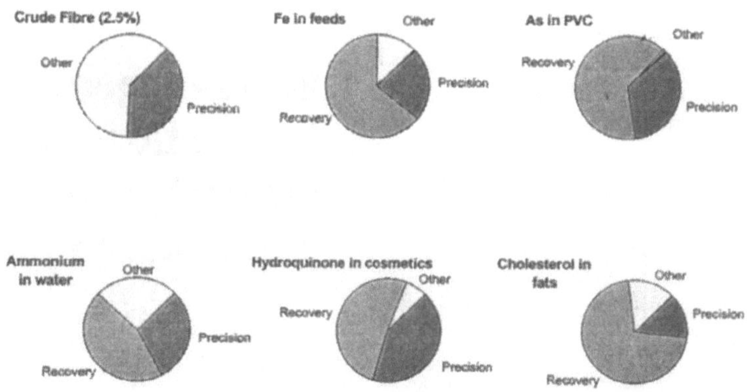

Fig. 5 Contributions to combined standard uncertainty. Charts show the relative sizes of uncertainties associated with overall precision, bias, and other effects (combined). See text for details

should provide adequate estimates of uncertainty for many practical purposes.

Conclusions

We have presented a strategy capable of providing a structured analysis of effects operating on test results and reconciling experimental and other data with the information requirements of the GUM approach. The initial analysis technique is simple, visual, readily understood by analysts and encourages comprehensive identification of major influences on the measurement. The reconciliation approach is justified by comparison with the ISO GUM principles, and it is shown that the two approaches are equivalent given representative experimental studies. The procedure permits effective use of any type of analytical data, provided only that the ranges of influence parameters involved in obtaining the data can be established with reasonable confidence. Use of whole method performance data can obscure the magnitude of individual effects, which may be counter-productive in method optimisation. However, if an overall estimate is all that is required, it is a considerable advantage to avoid laborious study of many effects.

Acknowledgement Production of this paper was supported under contract with the Department of Trade and Industry as part of the National Measurement System Valid Analytical Measurement Programme.

References

1. ISO (1993) Guide to the expression of uncertainty in measurement. ISO, Geneva
2. EURACHEM (1995) Guide: Quantifying uncertainty in analytical measurement. Laboratory of the Government Chemist, London
3. Analytical Methods Committee (1995) Analyst 120:2303
4. Ellison SLR (1997) In: Ciarlini P, Cox MG, Pavese F, Richter D (eds) Advanced mathematical tools in metrology III. World Science, Singapore, pp 56–67
5. Horwitz W (1988) Pure Appl Chem 60:855–864
6. AOAC (1989) Recommendation. J Assoc Off Anal Chem 72:694–704
7. ISO 5725:1994 (1995) Accuracy (trueness and precision) of measurement methods and results. ISO, Geneva
8. Ellison SLR, Williams A, Accred Qual Assur (in press)
9. ISO 9004-4:1993 (1993) Total quality management, part 2. Guidelines for quality improvement. ISO, Geneva

Accred Qual Assur (1998) 3:6–10

Stephen L.R. Ellison
Alex Williams

Measurement uncertainty and its implications for collaborative study method validation and method performance parameters

Abstract ISO principles of measurement uncertainty estimation are compared with protocols for method development and validation by collaborative trial and concomitant "top-down" estimation of uncertainty. It is shown that there is substantial commonality between the two procedures. In particular, both require a careful consideration and study of the main effects on the result. Most of the information required to evaluate measurement uncertainty is therefore gathered during the method development and validation process. However, the information is not generally published in sufficient detail at present; recommendations are accordingly made for future reporting of the data.

S.L.R. Ellison (✉)
Laboratory of the Government Chemist,Queens Road, Teddington, Middlesex, TW11 0LY, UK

A. Williams
19 Hamesmoor Way,Mytchett, Camberley, Surrey, GU16 6JG, UK

Introduction

One of the fundamental principles of valid, cost-effective analytical measurement is that methodology should demonstrably fit its intended purpose [1]. Technical fitness for purpose is usually interpreted in terms of the required "accuracy". To provide reasonable assurance of fitness for purpose, therefore, the analyst needs to demonstrate that the chosen method can be correctly implemented and, before reporting the result, needs to be in a position to evaluate its uncertainty against the confidence required.

Principles for evaluating and reporting measurement uncertainty are set out in the 'Guide to the expression of uncertainty in measurement' (GUM) published by ISO [2]. EURACHEM has also produced a document "Quantification of uncertainty in analytical measurement" [3], which applies the principles in this ISO Guide to analytical measurements. A summary has been published [4]. In implementing these principles, however, it is important to consider whether existing practice in analytical chemistry, based on collaborative trial [5–7], provides the information required. In this paper, we compare existing method validation guidelines with published principles of measurement uncertainty estimation, and consider the extent to which method development and validation studies can provide the data required for uncertainty estimation according to GUM principles.

Measurement uncertainty

There will always be an uncertainty about the correctness of a stated result. Even when all the known or suspected components of error have been evaluated and the appropriate corrections applied, there will be uncertainty on these corrections and there will be an uncertainty arising from random variations in end results.

The formal definition of "Uncertainty of Measurement" given by the GUM is "A parameter, associated with the result of a measurement, that characterises the dispersion of the values that could reasonably be attributed to the measurand. Note (1): The parameter may be, for example, a standard deviation (or a given multiple of it) or the half width of an interval having a stated level of confidence."

For most purposes in analytical chemistry, the "measurand" is the concentration of a particular species. Thus, the uncertainty gives a quantitative indication of the range of the values that could reasonably be attributed to the concentration of the analyte and enables a judgement to be made as to whether the result is fit for its intended purpose.

Uncertainty estimation according to GUM principles is based on the identification and quantification of the effects of influence parameters, and requires an understanding of the measurement process, the factors influencing the result and the uncertainties associated with those factors. These factors include corrections for duly quantified bias. This understanding is developed through experimental and theoretical investigation, while the quantitative estimates of relevant uncertainties are established either by observation or prior information (see below).

Method validation

For most regulatory applications, the method chosen will have been subjected to preliminary method development studies and a collaborative study, both carried out according to standard protocols. This process, and subsequent acceptance, forms the 'validation' of the method. For example, the AOAC/IUPAC protocol [5, 6] provides guidelines for both method development and collaborative study. Typically, method development forms an iterative process of performance evaluation and refinement, using increasingly powerful tests as development progresses, and culminating in collaborative study. On the basis of the results of these studies, standard methods are accepted and put into use by appropriate review or standardisation bodies. Since the studies undertaken form a substantial investigation of the performance of the method with respect to trueness, precision and sensitivity to small changes and influence effects, it is reasonable to expect some commonality with the process of uncertainty estimation.

Comparison of measurement uncertainty and method validation procedures

The evaluation of uncertainty requires a detailed examination of the measurement procedure. The steps involved are shown in Fig. 1. This procedure involves very similar steps to those recommended in the AOAC/IUPAC protocol [5, 6] for method development and validation, shown in Fig. 2. In both cases the same processes are involved: step 1 details the measurement procedure, step 2 identifies the critical parameters that influence the result, step 3 determines, either by experiment or by calculation, the effect of changes in each of these parameters on the final result, and step 4 their combined effect.

The AOAC/IUPAC protocol recommends that steps 2,3 and 4 be carried out within a single laboratory, to optimise the method, before starting the collaborative trial. Tables 1 and 2 give a comparison of this part of the protocol [6] with an extract from corresponding parts of the EURACHEM Guide [3]. The two procedures are very similar. Section 1.3.2 of the method vali-

Fig. 1

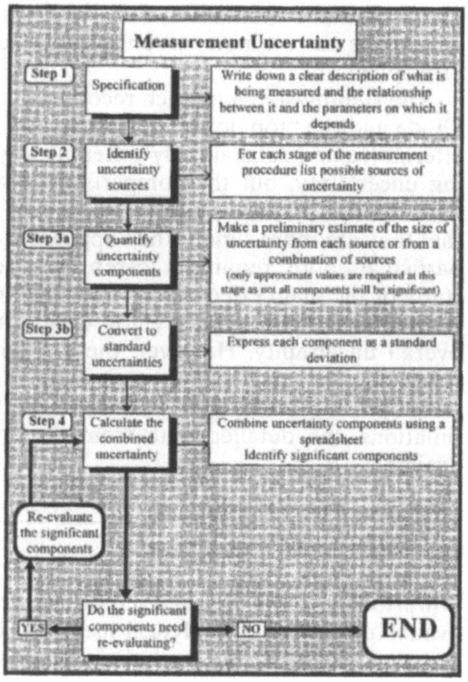

Fig. 2

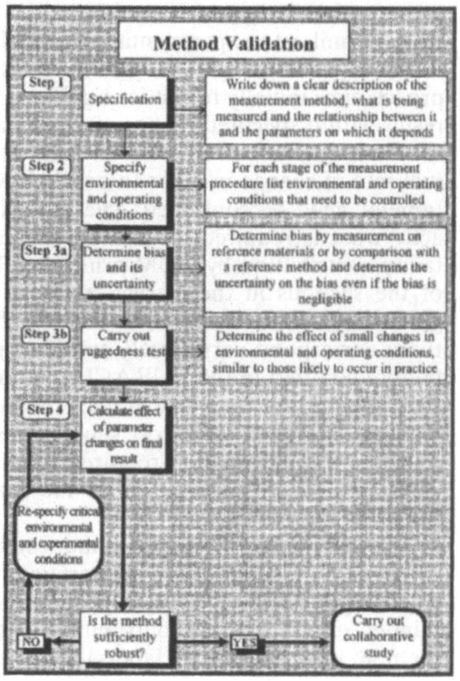

Table 1 Method development and uncertainty estimation

Method validation[1]	Uncertainty estimation
1.3.2 Alternative approaches to optimisation (a) Conduct formal ruggedness testing for identification and control of critical variables. (b) Use Deming simplex optimisation to identify critical steps. (c) Conduct trials by changing one variable at a time.	Having identified the possible sources, the next step is to make an approximate assessment of size of the contribution from each source, expressed as a standard deviation. Each of these separate contributions is called an uncertainty component. Some of these components can be estimated from a series of repeated observations, by calculating the familiar statistically estimated standard deviation, or by means of subsidiary experiments which are carried out to assess the size of the component. For example, the effect of temperature can be investigated by making measurements at different temperatures. This experimental determination is referred to in the ISO Guide as "Type A evaluation".

dation protocol is concerned with the identification of the critical parameters and the quantification of the effect on the final result of variations in these parameters; the experimental procedures (a) and (c) suggested are closely similar to experimental methodology for evaluating the uncertainty. Though the AOAC/IUPAC approach aims initially to test for significance of change of result within specified ranges of input parameters, this should normally be followed by closer study of the actual rate of change in order to decide how closely a parameter need be controlled. The rate of change is exactly what is required to estimate the relevant uncertainty contribution by GUM principles. The remainder of the sections in the extract from the protocol give guidance on the factors that need to be considered; these correspond very closely to the sources of uncertainty identified in the EURACHEM Guide. The data from method development studies required by existing method validation protocols should therefore provide much of the information required to evaluate the uncertainty from consideration of the main factors influencing the result.

The possibility of relying on the results of a collaborative study to quantify the uncertainty has been considered [8], following from a general model of uncertainties arising from contributions associated with method bias, individual laboratory bias, and within- and between-batch variations. Collaborative trial is expected to randomise most of these contributions, with the exception of method bias. The latter would be addressed via combination of the uncertainties associated with a reference material or materials to which results are traceable with the statistical uncertainty associated with any estimation of bias using a finite number of observations. Note that the necessary investigation and reporting of bias and associated statistical uncertainty (i.e. excluding reference material uncertainty), are now recommended in existing collaborative study standards [7]. Where the method bias and its uncertainty are small, the overall uncertainty estimate is expected to be represented by the reproducibility standard deviation. The approach has been referred to as a "top-down" view. The authors concluded that such an approach would be feasible given certain conditions, but noted that demonstrating that the estimate was valid for a particular laboratory required appropriate internal quality control and assurance. Clearly, the controls required would relate particularly to the principal factors affecting the result. In terms of ISO principles, this requirement corresponds to control of the main contributions to uncertainty; in method development and validation terms, the requirement is that factors found to be significant in robustness testing are controlled within limits set, while factors not found individually significant remain within tolerable ranges. In either case, where the control limits on the main contributing factors, together with their influence on the result, are known to an individual laboratory, the laboratory can both check that its performance is represented by that observed in the collaborative trial and straightforwardly provide an estimate of uncertainty following ISO principles.

The step-by-step approach recommended in the ISO Guide and the "top down" approach have been seen as alternative and substantially different ways of evaluating uncertainty, but the comparison between method development protocols and ISO approach above shows that they are more similar than appears at first sight. In particular, both require a careful consideration and study of the main effects on the result to obtain robust results accounting properly for each contribution to overall uncertainty. However, the top down approach relies on that study being carried out during method development; to make use of the data in ISO GUM estimations, the detailed data from the study must be available.

Availability of validation data

Unfortunately, the necessary data are seldom readily available to users of analytical methods. The results of the ruggedness studies and the within-laboratory op-

Table 2 Method performance and measurement uncertainty estimation. Note that the text is paraphrased for brevity and the numbers in parentheses refer to corresponding items in the EURACHEM guide (column 2)

Method validation protocol[1]	EURACHEM guide
1.4 Develop within-laboratory attributes of the optimised method (Some items can be omitted; others can be combined.) 1.41 Determine [instrument] calibration function … to determine useful measurement range of method. (8, 9) 1.4.2 Determine analytical function (response vs concentration in matrix …). (9) 1.4.3 Test for interference (specificity): (a) Test effects of impurities … and other components expected … (5) (b) Test non-specific effects of matrices. (3) (c) Test effects of transformation products … (3) 1.4.4 Conduct bias (systematic error) testing by measuring recoveries … (Not necessary when method itself defines the property or component.) (3, 10, 11) 1.4.5 Develop performance specifications … and suitability tests … to ensure satisfactory performance of critical steps … (8) 1.4.6 Conduct precision testing … [including] … both between-run (between-batch) and within-run (within-batch) variability. (4, 6, 7, 8, 12) 1.4.7 Delineate the range of applicability to the matrices or commodities of interest. (1) 1.4.8 Compare the results of the application of the method with existing tested methods intended for the same purposes, if other methods are available. 1.4.9 If any of the preliminary estimates of the relevant performance of these characteristics are unacceptable, revise the method to improve them, and retest as necessary 1.4.10 Have method tried by analyst not involved in its development. Revise method to handle questions raised and problems encountered.	The evaluation of uncertainty requires a detailed examination of the measurement procedure. The first step is to identify possible sources of uncertainty. Typical sources are: 1. Incomplete definition of the measurand (for example, failing to specify the exact form of the analyte being determined). 2. Sampling – the sample measured may not represent the defined measurand. 3. Incomplete extraction and/or pre-concentration of the measurand, contamination of the measurement sample, interferences and matrix effects. 4. Inadequate knowledge of the effects of environmental conditions on the measurement procedure or imperfect measurement of environmental conditions. 5. Cross-contamination or contamination of reagents or blanks. 6. Personal bias in reading analogue instruments. 7. Uncertainty of weights and volumentric equipment. 8. Instrument resolution or discrimination threshold. 9. Values assigned to measurement standards and reference materials. 10. Values of constants and other parameters obtained from external sources and used in the data reduction algorithm. 11. Approximations and assumptions incorporated in the measurement method and procedure. 12. Variations in repeated observations of the measurand under apparently identical conditions.

[1] Reprint from The Journal of AOAC INTERNATIONAL (1989) 72(4):694–704. Copyright 1989, by AOAC INTERNATIONAL, Inc.

timisation of the method are, perhaps owing to their strong association with the development process rather than end use of the method, rarely published in sufficient detail for them to be utilised in the evaluation of uncertainty. Further, the range of critical parameter values actually used by participants is not available, leading to the possibility that the effect of permitted variations in materials and the critical parameters will not be fully reflected in the reproducibility data. Finally, bias information collected prior to collaborative study has rarely been reported in detail (though overall bias investigation is now included in ISO 5725 [7]), and the full uncertainty on the bias is very rarely evaluated; it is often overlooked that, even when investigation of bias indicates that the bias is not significant, there will be an uncertainty associated with taking the bias to be zero [9], and it remains important to report the uncertainty associated with the reference material value.

Recommendations

The results of the ruggedness testing and bias evaluation should be published in full. This report should identify the critical parameters, including the materials within the scope of the method, and detail the effect of variations in these on the final result. It should also include the values and relevant uncertainties associated with bias estimations, including both statistical and reference material uncertainties. Since it is a requirement of the validation procedure that this information should be available before carrying out the collaborative study, publishing it would add little to the cost of validating the method and would provide valuable information for future users of the method.

In addition, the actual ranges of the critical parameters utilised in the trial should be collated and included in the report so that it is possible to determine their effect on the reproducibility. These parameters will have been recorded by the participating laboratories, who normally provide reports to trial co-ordinators; it should therefore be possible to include them in the final report.

Of course there will frequently be additional sources of uncertainty that have to be examined by individual laboratories, but providing this information from the validation study would considerably reduce the work involved.

Acknowledgements The preparation of this paper was supported under contract with the Department of Trade and Industry as part of the National Measurement System Valid Analytical Measurement (VAM) Programme [10].

References

1. Sargent M (1995) Anal Proc 32:201–202
2. ISO (1993) Guide to the expression of uncertainty in measurement. ISO, Geneva, Switzerland, ISBN 92-67-10188-9
3. EURACHEM (1995) Quantifying uncertainty in analytical measurement. Laboratory of the Government Chemist, London. ISBN 0-948926-08-2
4. Williams A (1991) Accred Qual Assur 1:14-17
5. Horwitz W (1995) Pure Appl Chem 67:331–343
6. AOAC recommendation (1989) J Assoc Off Anal Chem 72:694–704
7. ISO (1994) ISO 5725:1994 Precision of test methods. ISO, Geneva
8. Analytical Methods Committee (1995) Analyst 120:2303–2308
9. Ellison SLR, Williams A (1996) In: Parkany M (ed) The use of recovery factors in trace analysis. Royal Society of Chemistry, London
10. Fleming J (1995) Anal Proc 32:31–32

Accred Qual Assur (1997) 2:180–185

Ilya Kuselman
Avinoam Shenhar

Uncertainty in chemical analysis and validation of the analytical method: acid value determination in oils

Abstract Quantifying uncertainty in chemical analysis, according to EURACHEM document (1995), is based on known relationships between parameters of the analytical procedure and corresponding results of the analysis. This deterministic concept is different from the cybernetic approach to analytical method validation, where the whole analytical procedure is a "black box". In the latter case, analytical results only are the basis for statistical characterization of the method without any direct relationship with intermediate measurement results like weighings, volumes, instrument readings, or other parameters like molecular masses. This difference requires the harmonization of parameters to be validated and to be included in the uncertainty calculation. As an example, results of the uncertainty calculation and validation are discussed for a new method of acid value determination in oils by pH measurement without titration.

Key words Uncertainty of measurements · Analytical method validation · Acid value determination · Oils · pH measurement

I. Kuselman (✉) · A. Shenhar
The National Physical Laboratory of Israel, Danciger A Building, Givat Ram, Jerusalem 91904, Israel
Tel.: +972-2-6536534;
Fax: +972-2-6520797;
e-mail: freddy@vms.huji.ac.il

Introduction

The application to chemistry of the ISO Guide to the Expression of Uncertainty in Measurement [1] was issued by EURACHEM [2] in 1995. The EURACHEM document preserved the basic deterministic concept of the Guide [1] that the relationship between intermediate measurement results or parameters and the value of the measurand can be simple or complex, but that each value is determined by a combination of these parameters only. For example, for the measurement of a concentration $c_e = m_e / V_e$, where m_e is mass and V_e is volume, the intermediate parameters are m_e and V_e. Therefore, for the calculation of the uncertainty in c_e it is enough to know the uncertainties in m_e and V_e. It is recommended to break down more complex relationships into a combination of simpler ones. In general, the uncertainty estimation process according to [2] is simple and includes (1) specification, i.e. a statement of what is being measured, including the relationship between the measurand and the parameters, (2) identification of uncertainty sources for each parameter in this relationship, (3) quantifying uncertainty components associated with each potential source of uncertainty identified, and (4) calculation of total uncertainty as a combination of the quantified uncertainty components [3].

Uncertainty components can be estimated from relevant previous information (for example, the tolerance of volumetric glassware as provided by the manufacturer's catalogue or a calibration certificate), from special experimental work (test of some parameters), or by using the judgement of an analyst based on his experience [2]. Thus, total uncertainty evaluation is possible without any analytical experiment, i.e. from theoretical analysis with "pen and paper" only. Naturally, the result of this evaluation depends on the model of the measurement defined by the specification of the analytical procedure, identified uncertainty sources, and "de-

Table 1 The main AV uncertainty components in the new method [10] of AV determination in oils by pH measurement without titration

Symbol of source	Value	Uncertainty component	
		Standard deviation	Relative standard deviation
C_{st}	0.5 M	0.00022 M	0.00045
m	0.1–40 g	0.000087 g	0.00087–0.0000022
V_{st}	0.05–0.4 mL	0.0005–0.004 mL	0.01
ΔpH	0.25–0.35	0.014	0.056–0.040
M_{KOH}	56.10564	0.0023	0.000041

gree of belief" in the judgement of the analyst and others as in all theoretical results in science.

The principles of the analytical methods validation [4–7], in contrast to those described above, are based on the cybernetic approach, which considers the whole analytical procedure as a "black box". In this case, only the results of analysis can serve as the data for statistical characterization of the analytical method without any direct relationship with intermediate measurement results such as weighings, volumes, instrument readings, or other parameters such as molecular masses. Moreover, from the Horwitz function [8] it follows that the standard deviation of analytical results arising from random errors is practically independent of the specification of the analytical method.

The characteristics of the method used for the validation (validation parameters) such as repeatability, reproducibility and accuracy are certainly correlated with the uncertainty of the analytical results. In particular, the combined uncertainty arising from random effects cannot be less than the repeatability [2].

Obviously, the use of the uncertainty calculation of ref. [2] for the definition of the quality of analytical data should be harmonized with the concepts and practices of the method validation as well as quality control, proficiency testing, and certification of reference materials [9].

In the present paper, the results of the uncertainty calculation and validation are discussed for a new method of acid value determination in oils by pH measurement without titration developed in our laboratory [10].

The calculation of the uncertainty in the method by pH measurement

In this method [10], an oil sample is introduced into the reagent consisting of triethanolamine dissolved in a mixture of water and isopropanol. First, a conditional pH (pH_1) in the reagent-oil system is measured[1] and then a second pH (pH_2) is measured after the addition of standard acid (for example, HCl) to the system. Acid value is calculated according to the following formula:

$$AV = M_{KOH} C_{st} V_{st}/(10^{\Delta pH}-1)m \text{ (mg KOH/g oil)} \quad (1)$$

where M_{KOH} is the molecular mass of KOH (g), C_{st} is the concentration of the standard acid (mol/L), V_{st} is the volume of the standard acid added (mL), $\Delta pH = pH_1 - pH_2$, and m is the mass of the oil sample (g).

The main AV uncertainty components discussed below are presented in Table 1.

Preparation of the standard acid (HCl) solution

A Titrisol (Merck, Germany) solution of HCl containing $m_{HCl} = 18.230$ g HCl is used to prepare $C_{st} = 0.5$ M HCl of volume $V = 1000$ mL. The volumetric flask used for the solution preparation has the volume 1000 ± 0.4 mL at 20 °C (DIN A, Superior, Germany). The appropriate standard deviation of the calibrated volume (a rectangular distribution [1, 2]) is $0.4/\sqrt{3} = 0.23$ mL. Since the difference between the actual temperature and the flask calibration temperature is ~3 °C (with 95% confidence), at volume coefficient of water expansion 2.1×10^{-4}/°C, the possible volume variation is $1000 \times 3 \times 2.1 \times 10^{-4} = 0.63$ mL, and the corresponding standard deviation is $0.63/1.96 = 0.32$ mL. The standard deviation of the flask filling is less than 1/3 of the standard deviations for calibration and temperature variations (mentioned above) and is thus negligible. Combining the two contributions of the uncertainty $u(V)$ we have $u(V)/V = \sqrt{(0.23^2 + 0.32^2)}/1000 = 0.00039$.

The concentration of HCl is $C_{st} = m_{HCl}/M_{HCl} V$, where M_{HCl} is the molecular mass of HCl. The manufacturer of the HCl solution indicates a possible deviation of its titer of 0.02%/°C. Taking a possible temperature variation in the manufacturer's laboratory of ~2 °C (with 95% confidence), the standard uncertainty of m_{HCl} is $u(m_{HCl}) = 18.230 \times 0.02 \times 2/(100 \times 1.96) = 0.004$ g, and $u(m_{HCl})/m_{HCl} = 0.00022$.

The standard uncertainty of the molecular mass of HCl, according to IUPAC atomic masses and rectangular distribution [2], is $u(M_{HCl}) = 0.000043$.

[1] Because the measurements of pH are performed with an aqueous reference electrode calibrated by aqueous buffer solutions, the results of measurements are conditional [10]

Since $u(M_{HCl})/M_{HCl}$ is negligible in comparison with $u(V)/V$ and $u(m_{HCl})/m_{HCl}$, the relative standard uncertainty is $u(C_{st})/C_{st} = \sqrt{(0.00039^2 + 0.00022^2)} = 0.00045$.

Weighing and transfer of an oil sample to the reagent

The final mass of an oil sample is the difference in mass between a beaker with the sample and the empty beaker (after transfer of oil to the reagent). In the range up to 50 g, by analogy with ref. [2], $u(m) = 0.000087$. Since for different AVs the recommended sample mass is from 0.1 to 40 g, $u(m)/m$ values are from 0.00087 to 0.0000022.

Measurement of pH_1

After mixing the system "oil-reagent", pH_1 is measured with a pH meter pHM 95 (Radiometer, Denmark), and the standard uncertainty of pH reading $u(\text{pH}) = 0.01$.

Addition of HCl to the "oil-reagent" system

A recommended standard addition for samples with different AV is 0.05–0.4 mL of 0.5 M HCl; this volume should be negligible in comparison with the volume of the reagent −50 mL. For transfer of the acid to the "oil-reagent" system, a mechanical hand pipette (Gilson, France) was used with a relative standard uncertainty of $u(V_{st})/V_{st} = 0.01$ according to the manufacturer's information.

Measurement of pH_2 and calculation of ΔpH

After mixing the system with HCl added, pH_2 is measured under the same conditions as those for pH_1 (both measurements performed within 2–3 min) with the same uncertainty of pH reading. The expanded uncertainty of the pH measurements can reach [11] 0.05 or even [12] 0.1, but in our case the standard uncertainty of the difference between the two measurements is caused only by repeatability factors. Thus, only the uncertainty of reading is important, and $u(\Delta\text{pH}) = \sqrt{2} \times u(\text{pH}) = 0.014$, which is less than would be expected from [11] and [12].

Calculation of acid value

The acid value calculation is performed using Eq. 1. In this equation, all sources of uncertainty were described by us with the exception of M_{KOH}. The relative standard uncertainty of the molecular mass of KOH, according to IUPAC data and rectangular distribution, is $u(M_{KOH})/M_{KOH} = 0.000041$, which is less than 1/3 of the standard uncertainty of any component in Eq. 1, for example $u(C_{st})/C_{st} = 0.00045$. In their turn, $u(C_{st})/C_{st}$ and $u(m)/m$ are negligible in comparison with the relative standard uncertainty in the standard acid addition $u(V_{st})/V_{st} = 0.01$. The latter is also a negligible component of the uncertainty of AV determination, since pH measurement is the dominant source of $u(\text{AV})/\text{AV}$ (see Table 1). Therefore, after the logarithmic differentiation of Eq. 1 only the following remains valid:

$$u(\text{AV})/\text{AV} = u(10^{\Delta\text{pH}} - 1)/(10^{\Delta\text{pH}} - 1) = 10^{\Delta\text{pH}} \times 2.30 \times u(\Delta\text{pH})/(10^{\Delta\text{pH}} - 1) \quad (2)$$

Therefore

$$u(\text{AV})/\text{AV} = 0.032/(1 - 1/10^{\Delta\text{pH}}) \quad (3)$$

From the relationship of Eq. 3 illustrated in Fig. 1 for the ΔpH range 0.1–1.0, it is clear that $\Delta\text{pH} < 0.2$ leads to an essential increase in the AV uncertainty. At $\Delta\text{pH} > 0.4$, the amount of standard acid added may exceed 3 times the sum of the free fatty acids in the oil sample. This acid addition may cause (1) pH_2 to deviate from the linear range of pH versus AV or (2) a significant change in the concentration of the free form of triethanolamine in the reagent, which is inadmissible [10]. Therefore the recommended ΔpH range is 0.25–0.35, and corresponding values of $u(\text{AV})/\text{AV}$ are 0.07–0.06. The expanded uncertainty $\boldsymbol{U(\text{AV})/\text{AV}} = k\, u(\text{AV})/\text{AV}$ is 0.14–0.12, coverage factor k being 2. This uncertainty is higher than in ref. [13], where some additional simplifications were made.

Note, the interference of atmospheric CO_2 (due to the reaction with triethanolamine) is not taken into account.

Since there are no general criteria for evaluation of expanded uncertainty values, it is worth while to compare the values obtained with corresponding ones for the standard titrimetric method of AV determination in oils [14].

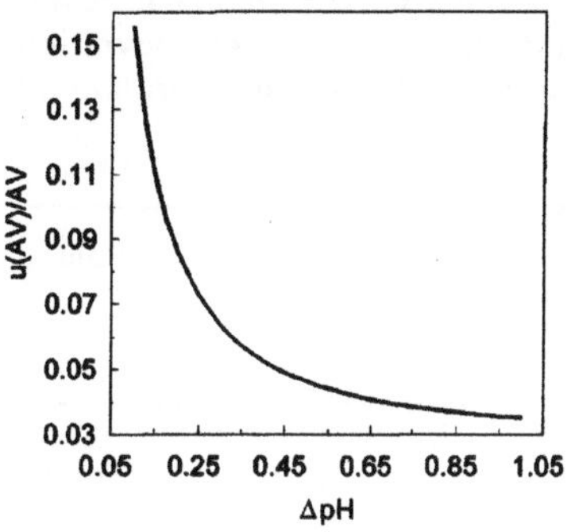

Fig. 1 Dependence of AV uncertainty on ΔpH

The calculation of the uncertainty in the titrimetric method

The standard titrimetric method [14] consists of dissolution of an oil sample in a mixed solvent (diethyl ether + ethanol) and subsequent titration of the free fatty acids contained in the sample against ethanolic potassium hydroxide solution in the presence of phenolphthalein. The acid value is

$$AV_t = M_{KOH} V_{KOH} C_{KOH}/m, \quad (4)$$

where V_{KOH} is the volume (mL) of potassium hydroxide solution used and C_{KOH} is the concentration (mol/L) of this solution. By analogy with the uncertainty calculation for titrimetry in ref. [2], the following enlarged steps of the analysis are examined (the corresponding main uncertainty components are given in Table 2).

Determination of C_{KOH}

The exact concentration of the ethanolic potassium hydroxide solution (0.1 or 0.5 mol/L [14]) is established before its use by titration against the standard HCl solution. Therefore, $C_{KOH} = C_{st} V_{st}^t / V_{KOH}^t$, where V_{st}^t is the volume (mL) of the standard HCl solution used for titration of the volume V_{KOH}^t (mL) of the KOH solution.

As shown above, $u(C_{st})/C_{st} = 0.00045$.

For transfer of an aliquot of the KOH solution to the titration vessel, a glass pipette of volume 5 ± 0.01 mL (Bein Z. M., Israel) is used. Taking a possible temperature variation of ±3 °C with 95% confidence and repeatability of filling the pipette (standard deviation) 0.0033 mL, one can calculate $u(V_{KOH}^t)/V_{KOH}^t = 0.0015$.

The titration is accomplished using a 5-mL microburette graduated in 0.01-mL divisions (Bein Z. M., Israel; calibration accuracy of ±0.01 mL), as the burette recommended in the standard [14], having a capacity of 10 mL and graduated in 0.1-mL divisions, is not suitable for a low AV. The possible temperature variation is the same as that mentioned above, the standard deviation of filling is 0.0033 mL, and the standard deviation of end point detection arising due to the drop size of the burette (0.017 mL) is 0.0098 mL. Thus, the maximum value of $u(V_{st}^t)/V_{st}^t$ may be 0.013 if $C_{KOH} \approx 0.1$ mol/L, and the corresponding $V_{st}^t \approx 1$ mL.

The uncertainties $u(C_{st})/C_{st}$ and $u(V_{KOH}^t)/V_{KOH}^t$ are negligible in comparison to $u(V_{st}^t)/V_{st}^t$; therefore $u(C_{KOH})/C_{KOH} = u(V_{st}^t)/V_{st}^t = 0.013$.

AV determination

The standard [14] recommends the use in Eq. 4 of the rounded KOH molecular mass 56.1 instead of the complete value $M_{KOH} = 56.10564$. Hence, in this case $u(M_{KOH})/M_{KOH} = 0.00564/(\sqrt{3} \times 56.1) = 0.00006$.

For the free fatty acids titration against KOH, we used the 5-mL burette described earlier; therefore $u(V_{KOH})/V_{KOH} = u(V_{st}^t)/V_{st}^t = u(C_{KOH})/C_{KOH} = 0.013$.

The uncertainty of oil sample weighing permitted by the standard [14] can be calculated from the table of accuracy versus mass m. The maximum value of the ratio (accuracy/mass) allowed in [14] for $m = 2.5$ g is $0.01/2.5 = 0.004$. Using rectangular distribution, we have $u(m)/m = 0.004/\sqrt{3} = 0.0023$.

It is clear that the uncertainties of the molecular mass of KOH and of oil sample weighing are negligible here just as in the previous method based on pH measurement. After logarithmic differentiation of Eq. 4 only $u(V_{KOH})/V_{KOH}$ and $u(C_{KOH})/C_{KOH}$ remain valid (see Table 2), and $u(AV_t)/AV_t = 0.018$. The expanded uncertainty is $\mathbf{U(AV_t)/AV_t} = k\, u(AV_t)/AV_t = 0.04$, using coverage factor $k = 2$.

Note, the detection of the end point of the titration is a dominant source of uncertainty. If, for example, the commercial burette used has a drop size of 0.043 mL, the expanded uncertainty will increase to 0.07. Moreover, the color of the oils and the possible change in the indicator behavior near the end point in the oil-solvent mixture (in comparison to water) are not taken into consideration. The same relates also to the influence of atmospheric CO_2 on C_{KOH}.

If we accept that the uncertainty in the standard titrimetric method (0.04) is one third of that in the new method by pH measurement (0.12–0.14), the titrimetric method can be used as a "true" for the new method validation.

Table 2 The main AV uncertainty components in the standard titrimetric method [14] of AV determination in oils

Symbol of source	Value	Uncertainty component	
		Standard deviation	Relative standard deviation
C_{KOH}	0.1 M	0.0013 M	0.013
V_{KOH}	1 mL	0.013 mL	0.013
m	2.5 g	0.0057 g	0.0023
M_{KOH}	56.1	0.0033	0.00006

Validation of the method by pH measurement

For the validation, five kinds of vegetable oils with three different AV values were used (Table 3). Oils with two higher AV values were prepared by adding to the initially purchased oils (with minimal AV) the known amounts of oleic acid. These oils were analyzed by the validated method: four replicates for each sample daily, during a period of five days [15]. Results of the AV determination by the standard titration method (average from ten replicates) were used as "true" or assigned [9] values. Thus, the whole experiment consisted of $5 \times 3 \times 4 \times 5 = 300$ AV determinations by the validated method and $5 \times 3 \times 10 = 150$ determinations by the standard method. Corresponding statistical data are given in Table 3.

Average values of the relative standard deviation S_1 of replicates (within a day – repeatability) and values of the daily relative standard deviation S_2 (within a week – reproducibility or intermediate precision [7]) for all the samples satisfy Horwitz's criterion [4]. The relative bias of the average result by pH measurement for each oil with respect to the corresponding "true" value is less than $t_{0.95}\,S_2$ ($t_{0.95}$ is Student's coefficient at 95% level of confidence), and consequently satisfies Student's criterion.

Table 3 Evaluation of some validation parameters for the new method [10] of AV determination in oils by pH measurement without titration

Oil	"True" AV	S_1	S_2	Bias	$t_{0.95} \cdot S_2$
Sunflower	24.8	0.022	0.016	−0.010	0.045
	1.51	0.025	0.008	−0.021	0.023
	0.055	0.042	0.041	−0.053	0.112
Soya	24.9	0.024	0.014	−0.019	0.039
	1.60	0.025	0.017	−0.032	0.047
	0.107	0.036	0.042	−0.082	0.117
Maize	23.4	0.014	0.012	−0.006	0.034
	1.57	0.016	0.011	−0.016	0.031
	0.096	0.047	0.014	−0.022	0.038
Canola	22.4	0.023	0.012	−0.023	0.033
	1.57	0.029	0.009	−0.010	0.025
	0.063	0.026	0.034	−0.066	0.096
Olive	23.7	0.013	0.003	−0.004	0.009
	6.62	0.019	0.004	−0.008	0.012
	0.579	0.016	0.016	−0.029	0.045

Discussion

Comparing the combined uncertainty arising from random effects [$u(\mathrm{AV})/\mathrm{AV} = 0.06$–$0.07$] with S_1 values, one can see that it is no less real than the repeatability, as is required in ref. [2]. Also, from Table 3 it follows that for all oil samples $S_2 < u(\mathrm{AV})/\mathrm{AV}$, i.e. the combined uncertainty is not less than the reproducibility (intermediate precision) too.

Bias values characterizing the accuracy [4] of the method (its trueness [16]) are less than the expanded uncertainty $\boldsymbol{U}(\mathbf{AV})/\mathbf{AV} = 0.12$–$0.14$ even for minimal AV, where the bias values are naturally the highest.

The uncertainty of the AV determination evaluated from the data collected in Table 3 by the scheme proposed in ref. [9] (as a root of the sum of the variances caused by S_1, S_2 and the bias) is also less than $\boldsymbol{U}(\mathbf{AV})/\mathbf{AV}$, the maximum value obtained being 0.10. This value may be higher when the evaluation is complete, since at present we still have no estimation of interlaboratory deviations in the AV determination, i.e. reproducibility of the method.

As shown above by statistical criteria, the values of the validation parameters can be accepted as satisfactory. Their comparison with the results of the uncertainty calculation according to ref. [2] allows us only to be sure that assumptions made during this calculation (see, for example, the notes) were admissible.

On the other hand, the comparison of the uncertainties of the new method by pH measurement with the standard titrimetric method cleared up the possibility of using the latter as a source of "true" values in the validation process. Moreover, although judgement on the acceptability of the analytical method is based today on the final validation report [7], the uncertainty quantified by ref. [2] may be useful at an earlier stage before experiments are carried out.

For our example, the advantages of the new method based on pH measurement are simplicity, rapidity, low-cost instruments, and suitability for automation [10]. It can be applied for on-line quality control of oils and regulation of the extraction (from oil seeds) and refining processes. Its expanded uncertainty is satisfactory for practical purposes because there are no two species or grades of oil with AV values differing by less than 12–14%.

Thus, the quantification of uncertainty by ref. [2] is a suitable instrument for planning or forecasting method applications. Therefore, relationships between the uncertainty and validation parameters should be analyzed in all possible aspects.

Acknowledgements The authors express their gratitude to Prof. Ya. I. Tur'yan, Prof. E. Schoenberger and Dr. O. Yu. Berezin for helpful discussions.

References

1. ISO (1993) Guide to the expression of uncertainty in measurement, 1st edn. ISBN 92-67-10188-9, Geneva
2. EURACHEM (1995) Quantifying uncertainty in analytical measurement, 1st edn. ISBN 0-948926-08-2, Teddington
3. Williams A (1996) Accred Qual Assur 1:14–17
4. AOAC (1993) AOAC Peer-verified methods program. Manual on policies and procedures, Gaithersburg
5. Accreditation for Chemical Laboratories. Guidance on the interpretation of the EN 45000 series of Standards and ISO/IEC Guide 25 (1993) WELAC Guidance Document No. WGD2/EURACHEM Guidance Document No. 1, 1st edn., Teddington
6. Hokanson GC (1994) Pharmaceut Technol 18:118–130; 92–100
7. Green JM (1996) Anal Chem 68:305 A–309 A
8. Boyer KW, Horwitz W, Albert R (1985) Anal Chem 57:454–459
9. Analytical Methods Committee (1995) Analyst:2303–2308
10. Tur'yan Ya I, Berezin OYu, Kuselman I, Shenhar A (1996) J Amer Oil Chem Soc 73:295–301
11. Danish Standard DS 287 (1978) Vandundersogelse pH (Water Analysis, pH), 2nd edn
12. Jensen H, Nielsen L (1994) Uncertainty of pH Measurements. Report of Danish Institute of Fundamental Metrology DFM-94-R24 on Nordtest-project No. 1194-94, Lyngby, Denmark
13. Tur'yan Ya I, Ruvinsky OE, Sharudina SYa (1991) J Anal Chem (in Russian):917–925
14. International Standard ISO 660 (1983) Animal and Vegetable Fats and Oils – Determination of Acid Value and of Acidity, 1st edn., Switzerland
15. Berezin OYu, Kogan L, Tur'yan Ya I, Kuselman I, Shenhar A (1996) The Proceedings of the Eleventh International Conference of the Israel Society for Quality, Nov. 19–21, Jerusalem, pp 536–538
16. Pocklington WD (1991) In: Rossell JB, Pritchard JLR (eds) Analysis of Oilseeds, Fats and Fatty Foods, Elsevier, London, pp 1–38

Accred Qual Assur (2002) 7:182–188
DOI 10.1007/s00769-002-0447-1

Paolo de Zorzi
Maria Belli
Sabrina Barbizzi
Sandro Menegon
Andrea Deluisa

A practical approach to assessment of sampling uncertainty

Presented at EUROLAB/EURACHEM Workshop "Sampling", 5–6 November 2001, Lisbon, Portugal

P. de Zorzi (✉) · M. Belli · S. Barbizzi
Agenzia Nazionale per la Protezione dell'Ambiente (ANPA) –
Unità Interdipartimentale di Metrologia Ambientale, Via Vitaliano Brancati 48, 00144 Rome, Italy
e-mail: dezorzi@anpa.it
Tel.: +39-06-5007-2086/2952
Fax: +39-06-5007-2313

S. Menegon · A. Deluisa
Ente Regionale per lo Sviluppo Agricolo del Friuli Venezia-Giulia (ERSA),
Via Sabbatini 5,
33050 Pozzuolo del Friuli (UD), Italy

Abstract The paper reports the approach followed in the SOILSAMP project, funded by the National Environmental Protection Agency (ANPA)of Italy. SOILSAMP is aimed at assessing uncertainties associated with soil sampling in agricultural, semi-natural, urban, and industrial environments. The uncertainty assessment is based on a bottom-up approach, according to the Guide to the Expression of Uncertainty in Measurement published by the International Organization for Standardization (ISO). A designated agricultural area, which has been characterized in terms of elemental spatial distribution, will be used in future as a reference site for soil sampling intercomparison exercises.

Keywords Soil sampling · Uncertainty · Reference sampling · Intercomparison · Sampling device

Introduction

Over the past few years, a large effort has been made to improve analytical techniques, laboratory practices and procedures, and reduce sources of uncertainty during laboratory operations. Measurement uncertainty is usually well characterized, understood and controlled by laboratory quality assurance and quality control procedures [1]. However, uncertainty associated with sampling and sample preparation has not yet been fully taken into consideration and the principles to assess uncertainty associated with this phase of environmental monitoring are rarely applied. ISO/IEC 17025 reports that sampling is a factor to be considered as a contributor to the total uncertainty of measurement [2].

Thompson states that: "there is an understandable lack of enthusiasm for rousing the sleeping dogs of sampling when there is a fair chance of being severely bitten" [3]. This statement appears justifiable because there is:

- An attitude of overlooking sampling, which is still circulating the scientific community.
- A lack of specific guidance to quantify uncertainty associated with sampling; measurement uncertainty is generally considered only after the sample has been received in the laboratory for analysis.
- A difficulty in managing and transferring concepts linked to sampling uncertainty to practical problems (i.e. the assessment of the effect of the application of sewage sludge on agricultural land, or the classification of contaminated land).

Nevertheless, a few exercises and studies on soil sampling uncertainty evaluation have recently been carried out. Studies on soil sampling have mainly addressed precision and bias [4–11]. Sampling intercomparison exercises have confirmed the contribution of different sampling protocols and devices [12] to the variability of the final analytical data, However, the best way of evaluating the combined measurement uncertainty, which includes sampling, is still under discussion. In the case of soil, there is no doubt that an assessment of the contribution of sampling operations on the overall measurement uncertainty is necessary to completely understand the meaning of the analytical results [13, 14].

On the basis of these considerations, the National Environmental Protection Agency of Italy (ANPA) has funded a project for the "Assessment of the uncertainty associated with the soil sampling in agricultural, seminatural, urban and contaminated environments (SOILSAMP)". The project covers a three-year period from2001 to2003 and involves collaboration with an Expert Advisory Group (EAG) composed of experts from national and international institutions. The following Institutions are represented in the SOILSAMP EAG:

- National Environmental Protection Agency – ANPA (Italy)
- International Union of Pure and Applied Chemistry – IUPAC (United States)
- International Union of Radioecology – IUR (Belgium)
- Netherlands Energy Research Foundation – ECN (The Netherlands)
- Ente Italiano Nazionale di Unificazione – UNI (Italy)
- Università Cattolica del Sacro Cuore di Piacenza, "Istituto di Chimica Agraria ed Ambientale – ICAA", (Italy)
- Università di Pisa, Area della Ricerca CNR "Istituto di Chimica del Terreno" (Italy)
- Università di Perugia, "Dipartimento di Scienze Agro-ambientali e della Produzione Vegetale – DiSAProV", (Italy)
- University of Barcelona, "Dipartimento Química Analítica" (Spain)
- University of Utrecht, Faculty of Geographical Science, "Utrecht Centre for Environment and Landscape Dynamic" (The Netherlands)
- Regional Environmental Protection Agencies – ARPA within the framework of the Centro Tematico Nazionale – Suoli e Siti Contaminati, CTN-SSC (Italy)
- Ente Regionale per lo Sviluppo Agricolo del Friuli Venezia-Giulia – ERSA (Italy)
- Dr. Herbert Muntau (Germany).

SOILSAMP is aimed at: i) the assessment of uncertainties associated with soil sampling in different environments, based on trace element concentration measurement in soil; ii) the characterization, in terms of trace element spatial variability, of a site to be qualified as a reference site for national and international intercomparison exercises.

This paper reports the methodological approach and a description of the first experimental activities performed in the framework of SOILSAMP, including a few preliminary considerations.

Methodological approach

The evaluation of uncertainty associated with sampling activities is based on a methodological approach including the identification of the different sources of uncertainty attributable to sampling procedures, the characterization of the sampling site (reference sampling) in terms of trace element spatial distribution and the intercomparison exercise.

Identification of uncertainty sources

The combined uncertainty of analytical results $u(\mathrm{r})$ includes uncertainties associated with sampling $u(\mathrm{s})$, sample reduction $u(\mathrm{rd})$ and analysis $u(\mathrm{a})$. In the following equation (Eq. 1) the relationship between the above reported uncertainties (combined standard uncertainty) is given:

$$u(\mathrm{r}) = \sqrt{u(\mathrm{s})^2 + u(\mathrm{rd})^2 + u(\mathrm{a})^2} \qquad (1)$$

The principles of EURACHEM/CITAC Guide [15] indicate several steps to assess the uncertainty associated with an analytical process: a) specification of the measurand, b) identification of the uncertainty sources, c) quantification of the uncertainty components, d) calculation of the combined uncertainty.

The EURACHEM approach requires a clear definition of the measurand and a quantitative expression of the relations existing between the value of the measurand and the parameters affecting its value. The parameters have to be identified; they can be other measurands, quantities not measurable, or constants.

The first phase of the SOILSAMP project has been devoted to the identification of the significant sources of uncertainty linked to soil sampling. To this end, a cause-effects diagram has been used. The diagram (sometimes called fish-bone) easily shows the parameters considered and how they relate to each other. The fish-bone permits visualization of the different sources of uncertainty avoiding over-counting.

Characterization of the sampling sites (reference sampling)

Reference sampling is aimed at the characterization of the sampling site in terms of element spatial distribution:

it allows assessment of the element concentrations at any point of a field with known uncertainty.

In order to be used as a reference sampling site, the site first has to be characterized for long- and short-range spatial variation of trace element concentrations in the soil. The long-range spatial variation is assessed by subdividing the sample site into sub-areas of the same size. The same number of single soil samples is collected from each sub-area. The samples are then pooled to give a composite sample. The comparison of trace element concentrations between the composite soil samples allows evaluation of the long-range spatial variation.

The short-range spatial variation of trace element concentration in the soil is assessed by comparing the analytical results obtained from single soil samples collected from randomly selected sub-areas. The number of sub-areas to be considered for single sampling depends on the expected spatial variability of the trace elements considered.

The selection of the reference site must fulfil some minimal requirements, such as representative size, heterogeneity, easy access, and a suitable trace element gradient within the site.

Intercomparison sampling exercise

The intercomparison exercise is intended to assess the uncertainty component attributable to different sampling devices.

The trace element concentrations in soil samples collected at different point locations differ as a result of spatial variation, effects of soil sampling, sample reduction, and laboratory analyses. The final aim of the project is not the evaluation of spatial variability of trace elements, but the assessment of the contribution of sampling to the uncertainty associated with the analytical data. To this end, the spatial variation must be accounted for, and subsequently eliminate. The *regionalized variable theory* [16, 17] assumes that samples collected at locations close to each other are on average, more similar than samples collected further away from each other. Accordingly, the spatial variation of an attribute is assumed to be the sum of three components: a) a structural component, having a constant mean or trend, b) a spatially correlated component, and c) an uncorrelated random noise. The spatially correlated and noise terms are encapsulated in an *experimental variogram*, plotting the experimental semi-variance as a function of sampling distance. The experimental semi-variance is estimated from the sample data and its value at zero distance is called the *nugget*. Theoretically, the semi-variance should be zero at zero distance, but short-distance variation and other sources of uncertainty make a positive value of the semi-variance.

The *nugget* is an estimation of the spatially uncorrelated noise component mentioned above, including variance due to sampling, measurement and other unexplained spatially uncorrelated sources of variance.

Trace element determination

Trace element measurement in all samples is carried out using instrumental neutron activation analysis (INAA). This technique achieves high precision levels and requires little or no sample processing prior to analysis. This analytical technique also eliminates uncertainty associated with sample processing [18–21]. To rule out variabilities eventually caused by different analytical laboratories, a single laboratory, following a predefined analytical protocol, performs all the analysis.

Field sampling exercise

The SOILSAMP project foresees the evaluation of the sampling uncertainty in four different environments: agricultural, semi-natural, urban and contaminated sites.

The agricultural site (10,000 m^2) has a regular shaped and is characterized by the presence of three sub-areas with different gravel content. These two conditions comply with the pre-requisites of representative size and of a structural heterogeneity of the soil. The site is a research field belonging to a public scientific institution which is easily accessible at any time, and where any accidental or unauthorized use can be prevented (i.e. spreading of unknown substances, transit of vehicles such as tractors). The present and past land use of this site are known. Considering that, generally, agricultural fields are not characterized by high spatial variability of trace elements, a spot-wise addition of fertilizer was performed, to produce a well-marked analyte gradient within the test site. The fertilizer containing 46% P_2O_5 was added manually to two triangular-shaped areas, of about 50 m^2 each. The quantity of fertilizer was sufficient to increase the concentration of phosphate in the first 5 cm of the top soil by about one order of magnitude.

The agricultural test site has been divided into 10× 10 m sub-areas. Figures 1 and 2 report the grid sampling points selected in the sampling exercises.

To assess the long-range spatial variation of trace elements in each sub-area, samples were taken using an Edelman auger (20 cm length, 7 cm diameter) at a 2 m distance from each other, after removing any surface vegetation, resulting in 25 soil samples. These samples were pooled and processed to give one composite sample. The sampling device was cleaned after sampling each sub-area.

To assess the short-range spatial variation of trace elements, on the hypothesis that the spatial variability of trace element is comparable between the different sub-areas, only 2 sub-areas (one where P_2O_5 had been added)

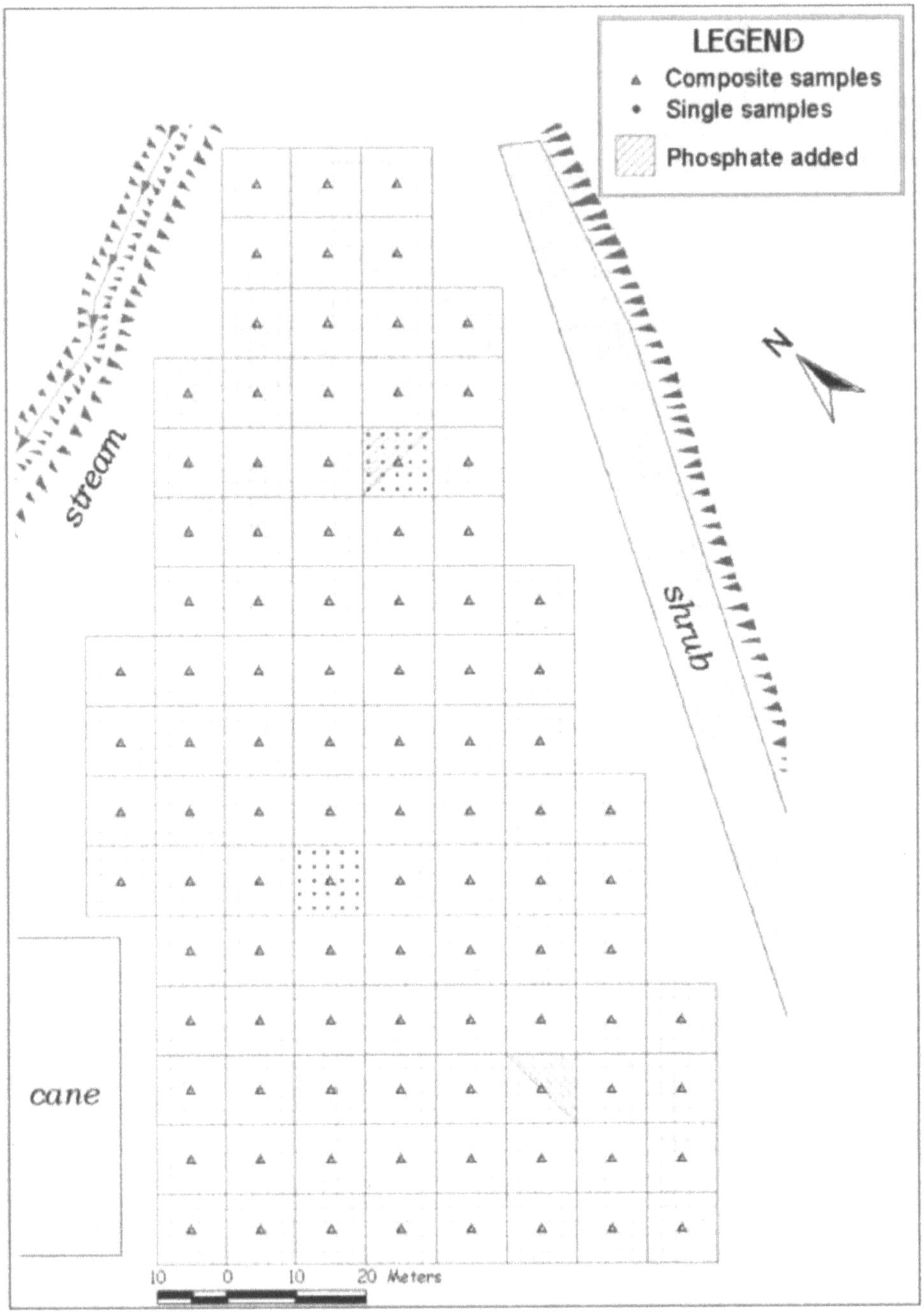

Fig. 1 Scheme of reference sampling. The site (10,000 m^2) was divided into 100 squares of 10×10 m. 100 composite samples, each obtained by pooling 25 increments of the same square, were collected. 50 single samples were also collected (25 samples per 2 squares)

were sampled again. The resulting 25 samples per sub-area will be analyzed separately to explore the within-sub-area variability.

Three different devices, commonly used for sampling in agricultural fields were used for the intercomparison exercise: an Edelman auger (20 cm length, 7 cm diameter), a gauge auger (20 cm length, 3 cm diameter) and a shovel. The devices were cleaned after each sampling.

Sample preparation

The samples were weighted (wet weight) and the data recorded. The samples were then stored in cartons before being dried in a fan oven at 35–40°C for several days (to constant weight). They were then disaggregated with a wooden pestle and passed through an automatic, rotating stainless steel sieve (2 mm mesh). The fraction above

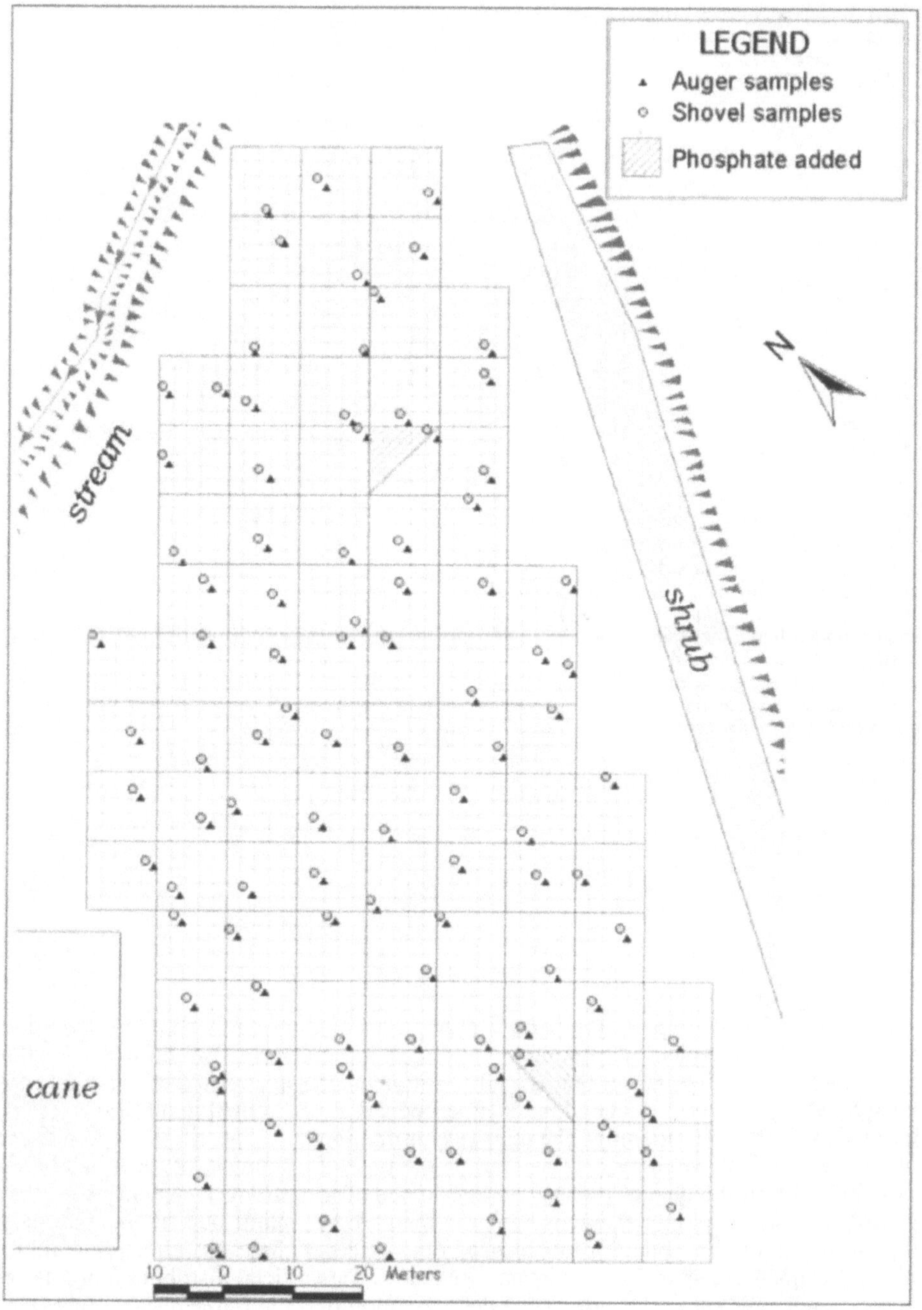

Fig. 2 Representation of the intercomparison exercise comparing different sampling devices (Edelman auger and shovel)

2 mm was removed and was not considered in the analytical phases.

The samples sieved at 2 mm were reduced in order to obtain the laboratory samples. The reduction phase was carried out to obtain samples representative of the soil collected but having a reduced size to make them more manageable in the laboratory. The reduction phase began with coning and quartering of the sample sieved at 2 mm and ended with the reduction by a riffle divider.

Preliminary considerations

By applying the EURACHEM principles a general cause-effect diagram (fish-bone) of the sampling phase

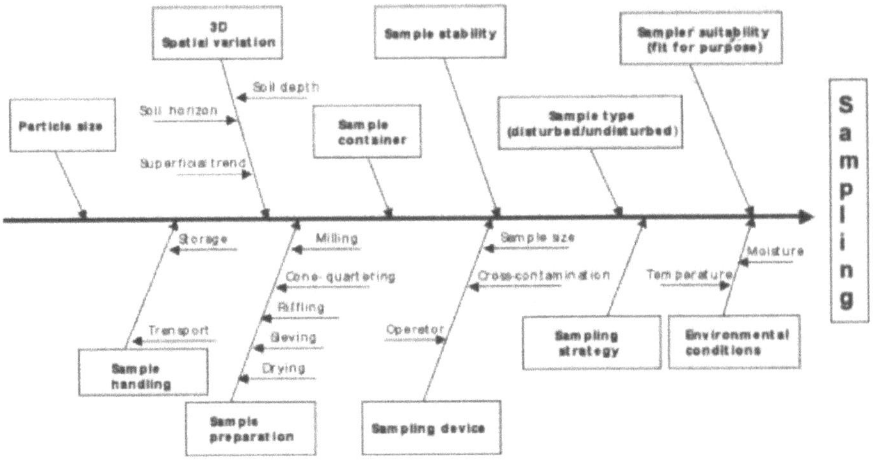

Fig. 3 Preliminary cause-effect diagram for the sampling phase based on the EURACHEM/CITAC approach

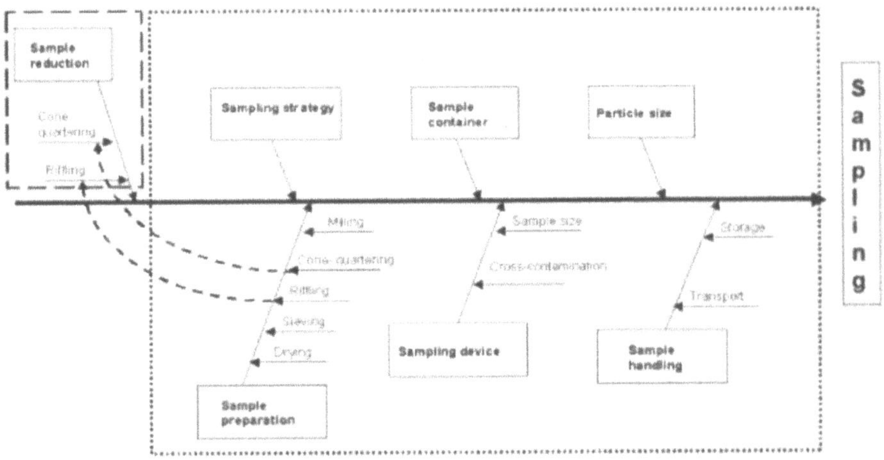

Fig. 4 Aggregated cause-effect diagram for the sampling phase in the agricultural SOILSAMP experimental design. The dashed-line (sample reduction) and the dotted-line (other sampling uncertainty sources) represent the sampling uncertainty components that can be quantified as separate blocks

can be established. Figure 3 reports all the potential sources of uncertainty in soil sampling.

It is necessary to point out that not all the sources of uncertainty have to be considered in all experimental activities involving soil sampling. The relative contribution of the sources of uncertainty is dependent on the type of sampling and on the type of the analyte considered. The contribution of sampling strategy is higher in the case of the evaluation of "hot spots" or in the case of the assessment of elements distribution in contaminated sites. This aspect is not relevant in the case of the determination of the mean value of an analyte in an agricultural field. Sample type (disturbed/undisturbed) and 3D-spatial variation give an important contribution in the determination of vertical distribution of an analyte along the soil profile. Sample stability and sample handling have a high relative contribution for volatile elements determination. Environmental conditions, like moisture content of the soil and temperature can influence the depth of soil sampled (in wet conditions, the layer sampled with an auger or corer samplers can increase due to compression of the soil). The influence of this source is higher in unmanaged soil (semi-natural ecosystems) than in managed soil (agricultural field).

The above reported considerations indicate that the sources of uncertainty associated with sampling are strongly dependent on the ecosystem investigated, sampling objectives and analytes studied. Figure 4 reports the cause-effect diagram for the assessment of the superficial distribution of trace elements in agricultural soil. In the frame of SOILSAMP project, the influence of the operator is ruled out by selecting only one operator for sampling. Environmental conditions are ruled out as well, because all sampling activity is carried out with similar temperature and moisture content in soil.

Another aspect that has to be considered is that in some cases it is extremely difficult to quantify each single uncertainty independently. In this case is more useful to select the uncertainty sources that can be evaluated as a "block". In the agricultural SOILSAMP experimental design, some aggregated uncertainties were defined as reported in Fig. 4. The influence of particle size, sampling device, sampling strategy, sample handling, sample container, and part of sample preparation are included in the first block, while the critical phases of cone quartering and riffling (reduction of the sample) in sample preparation are considered separately.

Each step of the sample reduction phase has its own quantifiable uncertainty and it is possible to quantify the uncertainty of the sample reduction as a "block". The determination of this uncertainty will be quantified experimentally as standard deviation after several repetition of the reduction phase in three different samples. To quantify the contribution of the uncertainty linked with the reduction phase, ^{137}Cs and uranium series radionuclides activity concentrations will be determined. These radionuclides have been selected as appropriate elements, because it is possible to carry out their determinations both on the entire sample before reduction and on the different fractions resulting from reduction, without any treatment before radionuclide analysis. In addition, ^{137}Cs and uranium series radionuclides show similar environmental behavior to many others trace elements in soil. ^{137}Cs and uranium series radionuclides activity concentrations will be determined in at least 10 replicates by gamma-spectrometry.

Acknowledgements The authors would like to thank all the participants of the SOILSAMP external advisory group for their scientific contribution during the development of the activities. A special thanks to Dr. Luisa Stellato, consultant ANPA, for the assessing the georef sampling points. Moreover, we are grateful to Valter Coletti, ERSA – Ente Regionale per lo Sviluppo Agricolo del Friuli Venezia-Giulia, for support and technical assistance during the field activity.

References

1. ISO (1993) Guide to the expression of uncertainty in measurement. International Organization for Standardization (ISO), Geneva
2. ISO/IEC 17025:1999 (1999) General requirements for the competence of testing and calibration laboratories. International Organization for Standardization (ISO), Geneva
3. Thompson M (1999) J Environ Monit 1: 19–21
4. ISO 3534–1 (1993) Statistics, vocabulary and symbols – Part 1. Probability and general statistical terms. International Organization for Standardization, Geneva
5. Thompson M, Ramsey MH (1995) Analyst 120: 261–270
6. Ramsey MH, Argyraki A, Thompson M (1995) Analyst 120: 1353–1356
7. Ramsey MH, Argyraky A (1997) Sci Total Environ 198: 243–257
8. Ramsey MH (1997) Analyst 122: 1255–1260
9. Ramsey MH (1998) J Anal At Spectrom 13: 97–104
10. Ramsey MH, Squire S, Gardner MJ (1999) Analyst 124: 1701–1706
11. Squire S, Ramsey MH, Gardner MJ (2000) Analyst 125: 139–145
12. Belli M, de Zorzi P, Menegon S, Sansone U (2000) In the Proceedings of XXXI National Congress of Radioprotection, 20–22 September 2000. Ancona (Italy), pp. 97–105
13. Ramsey MH, Thompson M, Hale M (1992) J Geochem Explor 44: 23–36
14. Muntau H, Rehnert A, Desaules A, Wagner G, Theocharopoulos S, Quevauviller P (2001) Sci Total Environ 264: 27–49
15. EURACHEM-CITAC Guide (2000) Quantifying uncertainty in analytical measurement. 2nd edn. EURACHEM
16. Isaaks EH, Srivastava RM (1989) An introduction to applied geostatistics. Oxford University Press, Oxford, UK
17. Mulla DJ, McBratney AB (2000) Soil spatial variability. In: Sumner ME (ed) Handbook of soil science. CRC Press, Boca Raton, FL
18. Smodiš B (1992) Vestn Slov Kem Drus 39(4): 503–519
19. Smodiš B, Jacimovic R, Jovanovic S, Stegnar P (1990) Biol Trace Elem Res 26: 43–51
20. Smodiš B, Jacimovic R, Medin G, Jovanovic S (1993) J Radioanal Nucl Chem Artic 169(1): 177–185
21. Svetina M, Smodiš B, Jeran Z, Jacimovic R (1996) J Radioanal Nucl Chem Artic 204: 45–55

Accred Qual Assur (2002) 7:106–110
DOI 10.1007/s00769-001-0420-4

Zhengzhi Hu
Li Liu

Quality assurance for the analytical data of micro elements in food

Abstract The micro element content of food is an important quality index due to the action of these elements on human health. In this article, we discuss how to ensure the reliability of analytical data on micro elements in order to truly represent the condition of food. Sampling, treatment of the analytical sample, selection of the analytical method, standard solution, and certified reference material, blank test, calibration of the instrument and equipment, application of the quality control chart, assessment of the final analytical result, and quality assurance system are briefly described.

Keywords Quality assurance · Analytical data · Micro elements · Food testing

Z. Hu, L. Liu (✉)
Chinese National Center for Food Quality Supervision & Testing
32 Xiaoyun Road, Chaoyang District, Beijing 100027, China
Tel.: +86-10-64645551
Fax:+86-10-64625604

Introduction

Micro elements may be divided into three classes according to their function and action on human health: essential element, non-essential element, and harmful element.

Although the essential element is a necessary mineral nutrient for the human body, it becomes harmful when amounts exceeds a definite limit. For example, the physiological requirement of selenium is 40 μg day^{-1} and poisoning results if the Se intake exceeds 800 μg day^{-1}. Some elements, such as Pb and Hg, are harmful even very small quantities due to accumulation in the human body. The effect of some elements change depending on the oxidation state or form present, for example, chromium (III) is an essential element while chromium (VI) is a carcinogen. The toxicity of organomercury compounds is different from inorganic mercury; methyl mercury is the most toxic of these due to its ease of absorption which is highest in digestive tract of human body [1].

Food containing many kinds of micro element is a primary source of mineral nutrients for the human body, however contaminated food is often a source of harmful elements. The micro elements contained in food are an important quality index due to their effect on human health. For example, the tolerance limits (mg kg^{-1}) of cadmium in food are less then 0.2 in rice, 0.1 in wheat flour, fish, and meat, and 0.05 in egg and vegetables; the tolerance limits of mercury in food are less than 0.01 in potatoes and milk, 0.02 in grain, 0.05 in meat and egg, and 0.3 in fish (0.2 for methyl mercury) which are provided in food standards and must be controlled rigorously [2].

Product quality control depends on analytical data and we must therefore ensure the accuracy and reliability of such data in order to represent its true condition. This condition is essential for the sample determination result to ensure the control of product quality, prevent product contamination, and protection of human health.

In this paper, we discuss how to ensure the quality of the analytical data of micro elements in food against problems which can introduce error.

Sampling

Our laboratory performed the following two parts of the task: (i) the mandatory inspection (including quality supervisory inspection, productive license inspection, and products attested inspection that are assigned by government) and (ii) the entrust inspection (including common sample analysis and arbitrate analysis).

Samples used for the mandatory inspection were randomly sampled from the qualifying products within their guarantee dates at the products' factory storehouse or market goods cabinet. Samples used for the entrust inspection were delivered to our laboratory in person by the sampling person. A sufficient amount of the sample was selected by a suitable method according to the provision in the relative standard.

Only the determination value of each ingredient in the sample was used to determine whether the product was up to standard or not; we did not deduce whether this product set was up to standard or not.

Treatment of the analytical sample

An appropriate sample treatment method was selected before analysis according to the rules of relative standard in order to reduce the sampling error and misrepresentation.

Common liquids, powders, and small pellet foods were homogenized by simply shaking, however, for solid food, especially non-uniform solid food, it was first broken into pieces and then mixed to achieve sufficient homogeneity. Canned foods were poured into a blender and thoroughly mixed since the element content is quite different at the can center and at surfaces in contact with the can wall.

Most of the foods examined were multi-component organisms. For the accurate determination, the micro elements must be free from organic matter before analysis. The decomposition procedure of organisms and extractive procedures for inorganic elements were generally applied. However, for all procedures applied, the effective steps necessary to avoid contamination or loss of the elements determined during the sample treatment were always taken.

1. Steps to avoid contamination:
   (a) Clear the air in laboratory by filtration to avoid analytical sample contamination by elements in the floating dust
   (b) Soak the glass vessel in dilute acid and then wash it with deionized water to avoid contamination by elements adsorbed on the vessel walls
2. Steps to avoid loss:
   (a) Keep the ashing temperature controlled and below 500 °C to avoid the loss of volatile elements (e.g., Cd and Pb) when the dry ashing method is used during sample treatment. Ashing aids used in dry ashing, may promote the decomposition of the organic matter, ash solubilization, and also help avoid the loss of the elements determined because of the solute produced. For example, $Mg(NO_3)_2$ has been applied as an ashing aid to avoid the loss of the Se, As, Cd, and Pb in fish, milk, and fruit juice samples, when these samples have been treated with the high temperature ashing method. In the determination of volatile halogens, NaOH has been used to fix fluorine, and $Mg(NO_3)_2$ has been used as an ashing aid; the alkaline dry ashing method may be used for the destruction of organic matter and the liberation of fluorine without any loss. $ZnSO_4$ has been used as an ashing aid in the iodine determination; the sample may then be ashed at 550–600 °C without any iodine loss.
   (b) When the wet-digestive method is used for food sample treatment, the appropriate digestant must be selected for the sample and the element being determined. For example, the oxidizing HNO_3-H_2SO_4 must be used as the digestant for arsenic determination in food containing high amounts of salt in order to ensure that the arsenic present is all arsenic (V), otherwise, the arsenic (III) may be lost as the volatile $AsCl_3$ (b.p.= 130 °C). When HNO_3-H_2SO_4 is used in canned food digestion, the acid-insoluble *meta*-stannic acid is produced and adsorbed on the inner wall of the Kjeldahl flask, so that the tin is then lost. At this time, 4 mol l^{-1} NaOH solution must be added and then gently heated with swirling until *meta*-stannic acid is fully dissolved in the sample solution. However, if H_2SO_4-H_2O_2 used in this digestion, the aforementioned trouble may be avoided. H_2SO_4 should not be used in digestions for the determination of trace lead in samples containing large amounts of calcium because insoluble $CaSO_4$ produced at the end of the digestion causes the adsorption-loss of lead.
   (c) The original oxidation state (OS) of the elements must be retained when the different OSs of the elements have been determined individually. For example, in the simultaneous determination of total chromium and chromium (VI) in food by atomic absorption spectrophotometry(AAS), the sample must be treated with 10% aqueous tetramethyl ammonium hydroxide in an ultrasonic water bath at 60±2°C until all solid matter is dissolved; all the chromium ions are extracted into the alkaline solution without any change in OS. For the determination of total iron and iron (II) in

infant food by AAS, the sample is treated with an acid-extraction procedure using hydrochloric acid and ultrasonic vibration under nitrogen flow; all iron is extracted into the acid solution without any change in OS.

(d) When the hydride generation method is applied to the determination of total arsenic in food, all of the organic and inorganic arsenic-containing components in sample are completely converted into arsenic (V) by digestion with HNO_3-H_2SO_4; the reduction of As^{5+} to arsine is very slow by borohydride and the As^{5+} must therefore be completely pre-reduced to As^{3+} by potassium iodide-ascorbic acid (KI-VC). However, when graphite furnace AAS is used for the total arsenic determination, the arsenic-containing components in the sample solution must be completely oxidized to As^{5+}, since the atomization temperature of As^{3+} is very different from As^{5+}. When a sodium diethyldithiocarbamate- methyl isobutyl ketone (DDTC-MIBK) system is applied to the determination of chromium concentration, the Cr^{3+} must be completely oxidized to Cr^{6+} to ensure an accurate analytical result because Cr^{3+} is not readily chelated by DDTC.

Selection of the analytical method

Many analytical methods can be applied to the determination of micro elements in food. Suitable analytical methods should have the following features:

1. The uncertainty of the methods should be minimized (good precision). In general, the relative standard deviation of method should be lower than ±5%.
2. The sensitivity and detection limits of the methods should meet the needs of the standard (high sensitivity and low detection limit). In general, the detection limit of the method should lower than the permitted content in the sample (provided in the product standard) by at least one order of magnitude.
3. A fair agreement between the true content and the expected content observed by the method is sufficient (good accuracy).
4. The method used for investigation is different from the reference method and should have a better precision than the method generally used for the determination of the parameters.

In general, we selected the suitable analytical method according to the element, content, and matrix in the determined sample in order to ensure the accuracy and reliability of the analytical result.

Our principle for method selection is that the standard method should always be used where possible. Under the normal conditions, the existing national standard, international standard, and trade standard have been selected for the mandatory inspection. The national standard is preferential for the entrust inspection and arbitrate inspection. If it is not available the normal standard and then the trade standard or contract standard will be used. When no standard method was available for a certain sample, the reliable method published or developed by us was selected, however, these methods must be passed through assessment. Furthermore, it should be emphasized that we must carefully consider the following factors when the standard method is applied to certain element determinations in practical samples, because the standard method has been worked out for many kinds of food:

1. We must consider how to remove the various interferences for the determination of trace elements in food by AAS, according to the elements determined and its content:
   (a) Prepare the standard solution in the same composition as the sample or apply a standard addition method in order to remove physical interference from the differences in viscosity, surface tension, and vapor pressure.
   (b) Suitable chelated-extraction or ion-exchange methods should be applied to collect the determined element in order to separate off the high amounts of interfering inorganic salts or extract out the micro elements.
   (c) Make use of the characteristic gaseous state of the hydride (at normal atmospheric temperature and pressure); it can therefore be decomposed at lower temperatures. As, Sn, Bi, Pb, Se, Sb, Te, and Ge may therefore be readily separated from their mother solutions at normal temperature and pressure.
   (d) When the alkaline metal and part of the alkaline-earth metals present have been determined, a readily ionizable element (another alkaline metal) must be added in an analytical solution in order to increase the free electronic concentration in the flame, therefore effectively controlling or removing the effect of the ionization interference.
   (e) The chemical interference can be removed using the temperature effect, gaseous state of the flame, addition of the release agent, protective agent, flux agent, and organic solvent etc, or by pre-seperating off the interference matters.
   (f) The molecular absorption interference can be removed by the adjustment of the zero point, deduct with continuative light source and the Zeeman effect.
   (g) When the graphite furnace-AAS is used for the determination of trace elements in food, the matrix effect is more serious. A matrix improver must be used for the removal of the matrix inter-

ference. For example, phosphoric acid must be added to the sample solution for the determination of trace lead; the ash temperature may be increased to 900–1000 °C and the matrix interference for the lead determination is therefore removed. In the determination of arsenic, $Mg(NO_3)_2$ and $Ni(NO_3)_2$ are added as a matrix improver increasing the ash temperature to 1100 °C; thus the interference of the anion and cation that coexisted in the sample solution is removed allowing the detected of concentrations as low as 6 ng g^{-1}.

Standard solution and certified reference materials

1. Standard solutions

A standard solution of very reliable quality is a necessity for quantitative analysis. In our laboratory, the certified reference reagents used as stock solutions for the micro element analysis were prepared by American Fisher Scientific Company or the China National Research Center for CRM.

Each working standard solution was prepared by diluting the stock solution with deionized water or dilute acid using a calibrated burette and volumetric flask to a known concentration before use. The container for the storage of the standard solution were soaked with acid and cleaned thoroughly with deionized water prior to use in order to avoid contamination. The standard working solutions for super micro element analysis were stored in Teflon containers to avoid the adsorption loss and dissolving element contamination from the container. The storage conditions of the standard working solutions were vigorously maintained. For example, the standard tin working solution was prepared before use with dilute acid (0.1 mol l^{-1} HCL, HNO_3, or H_2SO_4) and may be stored for several months in glass, polypropylene, polyvinyl chloride, polycarbonate, and Teflon containers. If prepared with water, obvious losses occurred when the standard working solution was stored for as little as one day.

In our laboratory, the standard solutions were prepared by two people, calibrated with each other, and then checked against newly purchased standards in order to eliminate the risk of error.

2. Certified reference material

The certified reference material (CRM) is used as quality assurance samples for the assessment of the analytical methods and results. The CRM used for the elements' analysis in our laboratory were prepared by the American National Bureau of Standards, MBH Analytical Limited, and the China National Research Center for CRM. For example, oyster tissue (SRM 1566), bovine liver (SRM 1577a; 308F185), wheat flour (SRM 1567; GBW08503), rice flour (SRM 1568), milk powder (SRM 1549; 304 F063; GBW08509), cabbage (GBW08504), mussel (GBW08571), prawn (GBW08572) and pork (GBW08552).

Blank test

The blank test is a scale for the inspection of reagents and methods used in analysis to detect whether or not they correspond to the requirements of trace analysis. Blank test must be carried out for each set of the sample and treated with high-purity reagent and water passed through re-distillation, ion-exchange, or sub-boiling distillation apparatus in order to reduce the reagent blank value to a sufficiently low level. For example, the nitric acid solution containing large amounts of chromium should not be used in the sample digestion for chromium determination in food; the H_2SO_4-H_2O_2 digestive method is however suitable. The high pressure digestion (performed in a sealed container) is applied to the analysis of food because only small amounts of reagent are used in the sample digestion, thus its blank value is lower than the other method. If a microwave heater was used in this method, the effect may be even better.

Calibration of the instrument and equipment

All of the instruments and equipment in our laboratory, including atomic absorption spectrophotometer, UV-spectrophotometer, balance, thermometer, pressure gauge, vacuum meter, and high capacity glass container were calibrated at regular intervals and operative inspections were preformed every day before use to ensure a good operative state, accuracy, and reliability.

The calibration of the instruments are carried out every year by the legal measurement department (China National Institute of Metrology, China National Research Center for CRMs) according to the national rules for calibration. Its verified value may be traced to the national measurement standards. Operative inspections were carried out by analytical personnel prior to each use.

For instruments without calibration rules, calibration was carried out by comparison with a similar instrument so as to make its verified value comparable.

Any erroneous instrument or equipment was not used further and the data for samples recently analyzed were re-inspected.

Our laboratory had one set of small capacity glass containers which had been passed for calibration by the legal measurement department. Other glass containers were self-calibrated by personnel who had passed the special training at the legal measurement department, using above the verified container as the measurement standard in order to ensure its quantitative value could be traced back to the national measurement standard.

Application of the quality control chart [3]

The quality control chart for the single measurement of micro element determination in food has been applied in our laboratory. The analytical results can therefore be directly expressed in order to discover and correct problems during the inspection. The determination data are therefore accurate.

Preparation of the quality control chart: a suitable reference material was selected as a quality control sample. The certified value (A) of the reference material was taken for the center line and certified values plus, minus two times the standard deviation ($A \pm 2\sigma$) of the analytical method for the upper and lower control limits, respectively; take the determined values and data for the vertical and horizontal coordinates, respectively, draw out the quality control chart of the single measurement for every element.

These quality control charts were always available in our laboratory, so that determination result of the quality control sample could be pin-pointed on its graph in order to control the analytical quality at any time.

The analytical quality control test for the assessment of trueness of the analytical result was carried on each of the sample determinations; at least one CRM was analyzed with each set of the practice sample. As the determination result of the quality control sample fell within the upper and lower control limits, it showed that the determination process of this sample set was situated within the statistical control state; these results were therefore effective. If the determination result of the CRM fell outside the control limits, it showed that the determination process was out of control. These results within the period from this time to former time were ineffective and the reason for the deviation was investigated and corrected.

Assessment of the final analytical result [4]

The important analytical result (including the arbitration analytical result, specific sample analytical result, and determination by non-standard methods) obtained in our laboratory were verified by following techniques:

1. The CRM with a similar matrix to the sample was selected and analyzed using the same procedure together with the sample. If the analytical result of the CRM fell within the range of the certified value, it showed that the sample analytical result was accurate and reliable.
2. When there no suitable CRM could be used, the sample was determinate by classical methods or other methods based on a different principle. Statistical analysis was then carried on their means obtained by the above two methods. If there was no obvious difference between them, we considered the sample analytical result to be accurate and effective.
3. The analytical result of the unknown sample is also corrected by the "recovery" of the certified value in the analysis of the CRM.
4. We joined the comparison test with national or international laboratories at irregular intervals and obtained good results. This indicated that our analytical results were accurate and reliable. For example, in 1998, we joined the proficiency test between 108 international laboratories in the Asia Pacific region for the determination of As, Cd, Pb, Hg, and Zn content in a canned fish sample labeled APLAC T009. The between-laboratory Z-score and the within-laboratory Z-score was –0.67~0.98 and –0.24~0.73, respectively. Our achievement is elegant.

Quality Assurance System [5]

Our laboratory is a qualified laboratory through the examination and accreditation by the China National Accreditation Committee for Laboratory (CNACL). We have set up a quality assurance system and sufficiently ensured the accuracy and reliability of the inspection data in six fields (including environmental condition, instrument and equipment, personnel quality, inspective process, quality appeal treatment, and accident treatment). It has been continuously revised and progressively standardized through the management review (including internal quality audits and review) each year.

Our laboratory is also subject to one re-examination every five years and one selective examination every year by the CNACL.

In our laboratory, the CRM is used as a blind sample in order to allow examination of personnel. Our personnel are all sufficiently competent and qualified.

References

1. Hu Zhengzhi (1996) China Encyclopedia of Chemical Industry. Chemical Industry Publisher, Beijing, 10:57–138
2. China National Standard, GB2762–94 and GB4810–94
3. Pan Xiurong (1993) Introduction to quality assurance of analysis and inspection. Scientific and Technical Information Network for Standard Material, Beijing, pp 1–63
4. Wang Shuchun (1991) Mathematical statistics and quality control for food analysis. China People's Hygienic Publisher, pp 24–49
5. Quality Manual (1998) China National Center of Food Quality Supervision and Testing

Accred Qual Assur (1998) 3:227–230

Manfred Golze

Customers' needs in relation to uncertainty and uncertainty budgets

Presented at: EUROLAB Workshop on Confidence in Testing – Customers Needs, Copenhagen, 11 September 1997

M. Golze (✉)
Federal Institute for Materials Research and Testing (BAM),
Unter den Eichen 87,
D-12205 Berlin, Germany
Tel.: +49-30-8104 1943
Fax: +49-30-8104 3717
e-mail: manfred.golze@bam.de

Abstract The general requirement of Quality Management standards to include in test reports a statement of the uncertainty of the results reflects the fact that a test result is rather useless without a knowledge of its accuracy. After an outline of the basic concepts of uncertainty, the need for uncertainty statements is illustrated for different ranges of applications.

Key words Uncertainty · Uncertainty statement

Introduction

The European standard EN 45001 [1] and its international counterpart ISO/IEC Guide 25 [2] both require that a laboratory include in its calibration or test reports "a statement of the estimated uncertainty of the calibration or test result (where relevant)" [2]. This requirement reflects the fact that a measurement or test result is rather useless without a knowledge of its accuracy. Therefore the standards authorities are clearly taking a position which represents the interests of customers in their dealings with laboratories.

In this paper, I mainly want to illustrate this statement on the basis of examples, but it is first necessary to outline, at least roughly, the basic concepts of uncertainty.

Uncertainty of measurements and tests – basic concepts

The true value of an unambiguously defined measurand (i.e. the specific quantity subject to measurement) would be obtained by a perfect measurement. But, in a real measurement, there always exists an error, i.e. an unknown difference between the measurement result and the true value. Such an error consists of two kinds of elements: random and systematic components.

The random components arise from unpredictable or stochastic variations of influence quantities. They cause the variations in repeated measurements and often lead to the well-known normal distribution of measurement results. If no systematic component is present one would obtain the true value as the mean of an infinite number of replicates.

A systematic component means that the centre of the distribution of measurement results is shifted away from the true value because of, e.g., an incorrect cali-

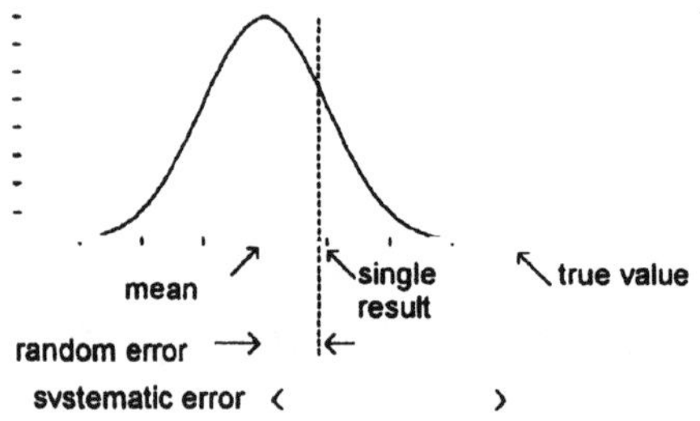

Fig. 1 Distribution of results with random and systematic error components

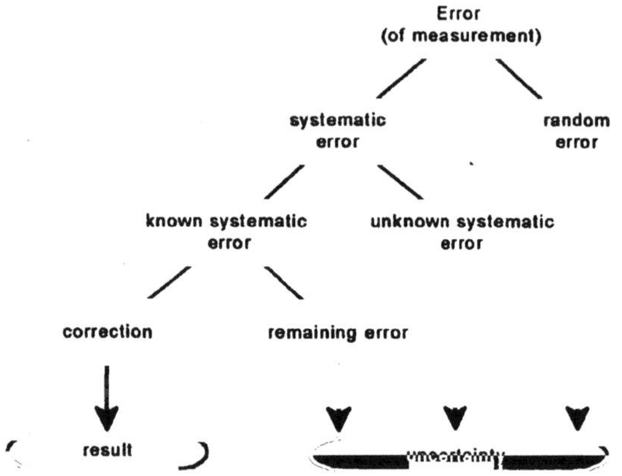

Fig. 2 The different error components and their influence on a measurement or test result and its uncertainty (according to [4])

bration (see Fig. 1), i.e. the results are biased. To some extent, systematic effects are known and the results can therefore be adequately corrected.

But, even after correction for known systematic effects, a measurement result is still only an estimate of the true value and should be accompanied by a statement of its uncertainty. In [3], uncertainty (of measurement) is defined as:

"A parameter, associated with the result of a measurement, that characterizes the dispersion of the values that could reasonably be attributed to the measurand."

Thus, uncertainty arises from random effects and unknown systematic error components but also from imperfect correction of known systematic errors, as shown in Fig. 2 [4].

According to the Guide to the expression of uncertainty in measurement (GUM) [3], the measurement uncertainty should be estimated by making up an uncertainty budget which takes all relevant components into account. While in the field of metrology the necessary tools for uncertainty evaluations are well established, further development is required in the field of testing, particularly in the case of qualitative results.

The need for uncertainty statements

It becomes clear from the above that an uncertainty statement is only an estimate and is less accurate than the result itself. Nevertheless, knowledge of the uncertainty is essential for the assessment of the reliability and conclusiveness of the result, i.e. its quality.

Comparison of different results

Uncertainties are the criteria for the decision whether two test results obtained on identical items are compatible or different. Often such decisions have to be taken in customer/supplier relations and can influence essentially the price of a commodity. In a routine case there is no need to estimate the uncertainty of each individual analysis, but both parties can agree on a particular analytical procedure and on an uncertainty which can be attributed to it.

An example is the delivery of ferromolybdenum, a binary alloy with a high molybdenum content, which is used as an additive for the production of molybdenum-containing steels. Because the content of molybdenum is essential for the price of a batch, usually both buyer and supplier perform an analysis according to an agreed gravimetric procedure. By agreement, each party can ask for an arbitration analysis if the difference exceeds a previously fixed value. For example, if the result of the supplier's laboratory is $w_{Mo} = 72.58$ wt%, that of the buyer's laboratory is $w_{Mo} = 72.06$ wt%, and the maximum difference accepted by both is 0.3 wt%, our institute could be asked to perform an independent analysis. Our finding might be $w_{Mo} = 72.16$ wt%, which would then agree reasonably well with that of the buyer.

Control of tolerances or detection of deviations

In industrial quality assurance it is a common task to assess whether manufactured parts comply with specified tolerances. Equally it might be necessary to detect a specific deviation with high reliability. In both cases one needs test methods with an adequate capability. For instance in the former case the so-called "golden rule of metrology" requires that the uncertainty u of the test method be less than 1/10 of the tolerance interval T [4]: $u \leq T/10$. As far as this rule applies, one can neglect the uncertainty of an individual measurement because the spread in the final results mainly reflects the variability of the manufactured parts.

Often geometrical tolerances of manufactured parts are controlled by use of calliper gauges. According to a German standard [5], the highest permissible error of such a calliper gauge should be $u = 0.03$ mm. Thus the measurable tolerance using this gauge should be $T \geq 0.3$ mm. If the tolerance is below this value one should use a more accurate device or measuring procedure.

Compliance with limiting values

The important task of control of whether products or samples comply with limiting values defined by regulations or laws for reasons of health, safety or environmental protection is related to the control of tolerances. It is only mentioned here for the sake of completeness and is dealt with in detail in [6]. Again the uncertainty has to be taken into account when making decisions.

Laboratory quality control

Reference materials are widely used in chemical analysis to establish traceability and to control analytical procedures e.g. by use of control charts. For this purpose two uncertainties should be known:

- The uncertainty of the certified reference value
- The uncertainty of the analytical procedure to be checked.

Usually the latter is dominant.

As an example (Fig. 3), a control chart is shown set up by a BAM laboratory for the analysis of aluminium in steel by spark emission spectroscopy. The reference material used is the EURONORM-CRM No. 194-1

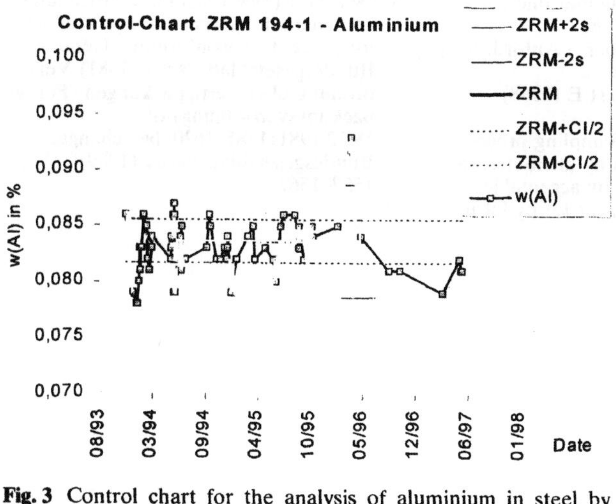

Fig. 3 Control chart for the analysis of aluminium in steel by spark emission spectroscopy. The EURONORM-CRM No. 194-1 is used as reference standard

and the inner limits depicted in Fig. 3 give the confidence interval of the certified value. The total interval reflects the uncertainty of the analytical procedure estimated as $\pm 2\,s$ (s = standard deviation). As can be seen, the procedure was out of control in early 1996 and had to be readjusted.

Uncertainty of tests caused by sampling

Often the uncertainty associated with a test is not mainly caused by the measurement process itself but by the sampling procedure performed beforehand. Therefore it is important to include this component into the uncertainty budget, which otherwise would be misleading.

In connection with sampling, two question arise:

- The representativeness of the sample
- The deduction of a result for the whole batch based on the sample result.

The first question causes severe problems e.g. in the field of environmental analysis, but cannot be treated here.

The second problem which is of importance e.g. in the fields of industrial quality control and market surveillance, can be treated appropriately by statistical means [7, 8]. An application is the market surveillance with regard to the so-called e-mark. This mark on prepacked food and consumer goods is intended to assure the consumer that the actual contents of the prepack conform with the nominal contents within certain limits. For example, a German regulation [9] applied to a lot of 10 000 packages of butter with a nominal weight of 250 g each stipulates the sampling instruction 125–7/8, which means that a sample of 125 packages is randomly taken and each pack has to be weighed. If x gives the number of items in the sample, with m_i < 241 g (the minimum permissible weight), then the criteria for acceptance or rejection of the whole batch are $x \leq 7$ and $x \geq 8$, respectively. The probability of acceptance P_a as a function of the quality level of the lot can be derived from the so-called operational characteristic of this specific sampling instruction and is given in Table 1 for some selected values. It can be seen that a batch containing 8% packages with a weight less than 241 g will

Table 1 Probability of acceptance P_a as function of percentage of defective items in the lot (selected values). p, percentage of defective items; P_a, probability of acceptance of the lot

p [%]	P_a [%]
2	99.62
4	87.09
8	20.90
12	1.33

still be accepted with a probability of approx. 20%. This example may demonstrate the limited resolving power of such a sampling instruction, and should be kept in mind.

In the case of the e-mark, according to a second requirement, a batch has also to be rejected if the mean weight of the sample items is less than the nominal weight with a significance level of 99%.

Concluding remarks

It is the aim of this paper to demonstrate that uncertainty statements are essential for the users of measurement and test results when they assess these results and have to take decisions based on them. However, for this purpose it is often sufficient to know the generic uncertainty of the type of test performed instead of the uncertainty of the particular result. But we are faced with some problems.

Confusion of the customers, hesitation of the laboratories

Laboratory practitioners know from their contacts with the customers that often the latter are not familiar with the concepts of uncertainty and are rather confused when they are confronted with uncertainty statements. On the other hand many laboratories fear that a comprehensive and honest statement of uncertainty might affect their reputation and competitiveness.

Concerted actions of the testing community, accreditation and standardization bodies

These problems cannot be solved by individual laboratories. Instead, the testing community as a whole is asked to co-operate with accreditation and standardization bodies aiming at:

- Evaluation of the (generic) uncertainty of measurement and test procedures
- Inclusion of uncertainty characteristics (e.g. repeatability, reproducibility) in testing standards
- Education of the customers.

Provision of the necessary funds

It is the task of national and European authorities to provide the necessary funds for these concerted actions

- Because by this means the testing infrastructure can be improved
- Because authorities are also customers of testing laboratories and important administrative and political decisions are based on their results.

Organizations like EURACHEM, EUROLAB and NORDTEST can help to initiate and stimulate these co-operative processes.

Acknowledgements The author would like to thank colleagues from EUROLAB, NORDTEST and BAM for fruitful discussions. In particular the contributions of Nazmir Presser, Rolf Oberhauser, Siegfried Noack and Thomas Goedecke are gratefully acknowledged.

References

1. EN 45001 (1989) General criteria for the operation of testing laboratories, Brussels.
2. ISO/IEC Guide 25 (1990) General requirements for the competence of calibration and testing laboratories, 3rd edn. Geneva
3. BIPM, IEC, IFCC, ISO, IUPAC, IUPAP, OIML (1993) Guide to the expression of uncertainty in measurement, 1st edn
4. Hernla M (1996) Qualität und Zuverlässigkeit 41:1156–1162
5. DIN 862, Meßschieber – Anforderungen, Prüfungen, 1988
6. Christensen JM, Holst E (1998) Accred Qual Assur
7. ISO 2859-1 (1989) Sampling procedures for inspection by attributes; sampling plans indexed by acceptable quality level (AQL) for lot-by-lot inspection. Geneva
8. ISO 3951 (1989) Sampling procedures and charts for inspection by variables for percent nonconforming. Geneva
9. Bundesgesetzblatt Part I (1981) Verordnung über Fertigpackungen (Fertigpackungsverordnung) of 18.12.1981:1585–1620, last change: Bundesgesetzblatt Part I (1989): 1557–1567

Accred Qual Assur (1998) 3:237–241

Rouvim Kadis

Evaluating uncertainty in analytical measurements: the pursuit of correctness

Presented at: 2 nd EURACHEM Workshop on Measurement Uncertainty in Chemical Analysis, Berlin, 29–30 September 1997

R. L. Kadis (✉)
D. I. Mendeleyev Institute for Metrology (VNIIM), 19 Moskovsky pr., 198005 St. Petersburg, Russia
Fax: +7-812 327-97-76
e-mail: hal@onti.vniim.spb.su

Abstract Simple in principle, the evaluation of uncertainty, especially in chemical analysis, is not a routine task and needs great care to be correct. This can be seen, particularly, from an examination of the EURACHEM Guide, *Quantifying Uncertainty in Analytical Measurement* (1995), which is the most important document on the subject. The examination reveals, in the author's opinion, a shortage of correctness in some principal details of the uncertainty estimation process as presented in worked examples in the Guide, and the author has therefore formulated some "in pursuit of correctness" rules for estimating uncertainty. The rules and respective comments are concerned with the following items: (1) choosing an appropriate distribution function in type B evaluation of uncertainty, (2) the necessity for consideration of separate contributions to the combined uncertainty, and (3) taking account of actual influence factors in the uncertainty estimation process. Furthermore, the problem of estimation of conditional versus overall uncertainty is touched upon in connection with comparative trials where only internal consistency of results is required.

Key words Chemical analysis · Measurement uncertainty · Estimation process

Introduction

The term "measurement uncertainty", in common use for the characterization of physical measurements, has so far been a little difficult to adapt to the requirements of chemical analysis. However, introducing the concept into this field of (amount-of-substance, analytical, chemical) measurements is quite natural. A number of papers [1–4] published on the topic have considered the important issues of using and evaluating uncertainty in the context of analytical data quality. This paper, however, focuses on the things that determine the quality of the uncertainty estimates themselves.

The current state of implementation of the measurement uncertainty concept in analytical practice may be compared with the design of a unique building at the stage where a consensus is reached on the key issues, but there remain some details which cannot be considered as fixed, as they require further correction. The same can be said about the procedures in evaluating uncertainties as they are presented in the EURACHEM Guide on uncertainty in analytical measurement [5]. The document is undoubtedly the most important contribution to the development of the concept as applied to analytical chemistry problems. However, some of the practical directions in the uncertainty estimation process which are indicated in worked examples

[5, Appendix A] do not seem entirely correct and call for comment. This is due to some oversimplified or incorrect insights into the subject, which may result in unrealistic estimation. Regardless how much the "error" in an estimate of the uncertainty may be, these issues are of principal importance in view of the educational significance of the Guide. Therefore, these "trifles" are worth drawing attention to in order to formulate some (obvious enough) rules, so that uncertainty estimates can be as correct as possible. These "in pursuit of correctness" rules are given below, with reference to the respective examples of the Guide.

Three "in pursuit of correctness" rules in estimating uncertainty

Rule 1: The choice of an appropriate distribution function (in type B evaluation of uncertainty) should be made on the basis of all the available information on the quantity at issue.

An estimate of standard uncertainty is often made from bounds a_- and a_+ within which values of the quantity in question X are expected to lie. [The range a_- to a_+ is commonly symmetric with respect to the best estimate of X and has half-width $a=(a_+-a_-)/2$.] It is a fairly frequent task in the practice of measurement data evaluation, as assigning maximum bounds (based on objective knowledge or personal judgment) is often the only thing to do. One may simply divide the value of a above by an appropriate conversion factor depending on what kind of probability distribution is assumed. It is essential here that all the relevant information available be used. The three typical cases of what may be known are as follows (see Fig. 1):

1. A (statistically) estimated confidence interval having a stated confidence level
2. An expected value and assigned maximum bounds about it
3. Assigned maximum bounds only.

Unless otherwise stated, it is quite natural to assume in case 1 that a normal (Gauss) distribution was used to calculate the interval and recover the standard uncertainty by using a suitable quantile of the distribution. (The quantile is taken equal to 2.0 for 95% confidence level.) In contrast, extremely little information about the quantity in question is available in case 3, and all one can do is to model it by symmetric uniform (rectangular) distribution. Then, the expected value of the quantity is the midpoint of the range and the conversion factor is equal to $\sqrt{3}$. Case 2 occurs where additional information such as an expected value allows us to regard values of the quantity near this value as being more likely than values near the bounds. This situation differs from that in case 3. It is because of this that item F.2.3.3 of the ISO Guide to the Expression of Uncertainty in Measurement [6] recommends in such instances the adoption of a triangular distribution as a compromise between the two extremes, normal and rectangular distributions. The conversion factor is equal to $\sqrt{6}$ in this case, and the standard uncertainty obtained proves to be about 30% smaller than that obtained using the rectangular distribution model. It can be said that increasing uncertainty in going from case 2 to case 3 is in a sense a "payment" for our ignorance about the distribution of possible values of the quantity between the bounds.

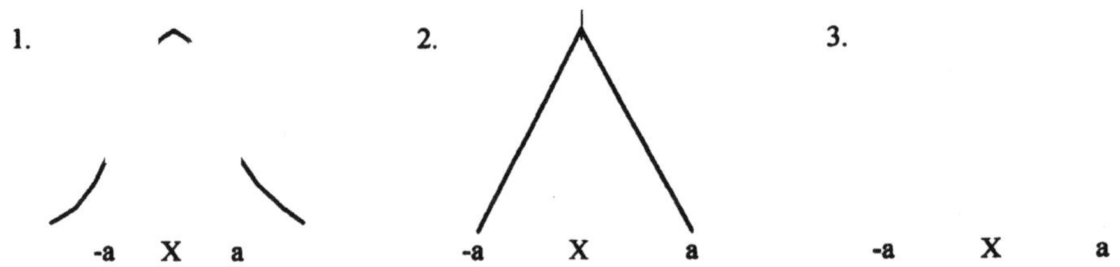

P=0,95 $u(X) = a/2,0$ | $u(X) = a/\sqrt{6}$ | $u(X) = a/\sqrt{3}$

Examples

Confidence interval for a weighing result:

$(m \pm 0,1)$ mg $(P = 0,95)$
$u(m) = 0,1/2,0 = 0,05$ mg
(Example 1, Step 1 [5])

Nominal values and specification limits for volumetric glassware:

for a 250-ml standard flask
$V = (250,00 \pm 0,15)$ ml
$u(V) = 0,15/\sqrt{6} = 0,061$ ml
Example 1, Step 2 [5] gives:
$u(V) = 0,15/\sqrt{3} = 0,087$ ml

The purity of a material as being "not less than p (%) level":

$100 - p = 2a$
$u(p) = (100 - p)/2\sqrt{3}$

Fig. 1 Type B evaluation of standard uncertainty for a quantity X given as: (1) an estimated confidence interval, (2) an expected value and assigned maximum bounds about it, and (3) assigned maximum bounds only

It should be recognized that all the instances of using rectangular distributions in the EURACHEM Guide examples fall in fact under case 2, not case 3. Such are, in particular, the evaluations of the uncertainty concerned with volumetric glassware: a nominal capacity is simply an expected value. Thus, all the standard uncertainties calculated by applying the conversion factor of $\sqrt{3}$ appear to be overestimated. Although this approach based on Bayes "principle of equal ignorance" is common practice in estimating uncertainty in metrology, we cannot regard it as correct in all the cases of specifying measurement errors in the form of maximum limits. Though simple and universal, this scheme comes into conflict with common sense. Rectangular distribution is only to be assumed when nothing but the limits for possible values of the quantity are available. For example, the uncertainty associated with purity of a material and expressed as being "not less than the p (%) level" might be one such case insofar as an expected (or nominal) value of purity is unknown here. Other examples of the application of model distributions are shown in Fig. 1.

Rule 2: It is necessary to consider uncertainty components as making independent contributions to the combined uncertainty as far as possible.

There are, however, situations in the worked examples where one of the components combined encompasses in fact another resulting in "double counting" and hence in redundancy in the uncertainty estimation. Evaluation of an uncertainty in volumetric measurements is a characteristic example of this.

So, three contributions to the uncertainty are considered to be essential here:

A. Specification limits for the glassware of a given type
B. Repeatability of filling the article to the mark
C. Ambient temperature effects.

Standard volumetric glassware with specified capacity tolerances is used everywhere. It is important, however, that the tolerance, i.e. the limit of volumetric error, is a single and sufficient error characteristic for an article of volumetric ware of a given type for each usage under standard conditions (see the ISO standards [7, 8]). In other words, the difference between the actual capacity and the nominal capacity is to lie within the limits in each case of filling the article (for instance, the volumetric flask) to the mark. To consider variations in filling as a separate uncertainty component is therefore superfluous where the tolerances are used. Of course, if data on repeatability of filling with water are available, they may be accounted for. But, as a result of the repeatability experiments, one has actually the article individually calibrated and one can replace the nominal capacity by the estimated value of it, immediately eliminating the need for the application of specified tolerances. Only the two contributions to the uncertainty remain in such a case, and a substantially reduced uncertainty value is achieved in the final analysis. The cases considered are schematically depicted in Fig. 2. [It is necessary to note that an additional contribution to the uncertainty may arise due to a substantial difference between the properties (such as viscosity, surface tension, and so on) of a liquid to be measured and those of water, for instance, in the case of nonaqueous solutions. These effects are taken into account by means of individual calibration with an appropriate calibration liquid.]

Let us examine another situation, "Determination of organophosphorus pesticides in bread" (Example 3 of the Guide). This is a multistage procedure consisting of several sequential steps, beginning with homogenization and ending with a GC determination. The combined correction F_c and the combined uncertainty u_c for the procedure as a whole are derived from individual values of the correction factor F_i and the uncertainty u_i each relating to a stage i as follows:

$$\text{Combined correction } F_c = \prod_{i=1}^{n} F_i$$

$$\text{Combined uncertainty } u_c = \sqrt{\sum_{i=1}^{n} u_i^2}$$

(The values may be known from a recovery experiment; if R_j is the recovery, then $F_j = 1/R_j$). The protocol shows all these calculations. However, this would only be true if the components in the above formulae were strongly independent. In fact, this is not so. For instance, the correction factor F_3 experimentally obtained at the extraction stage includes systematic effects of all the subsequent stages, as the uncertainty u_3 incorporates all the following uncertainties. The same applies to the values of F_5 and u_5 for the stage of concentration of the washed extract. So, we have in fact summary estimates instead of individual ones. To get the individual

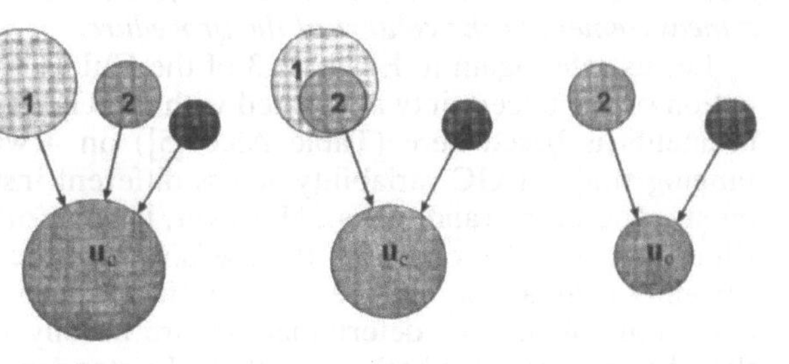

Fig. 2 Evaluation of uncertainty in volumetric measurements. Contributions to the combined uncertainty u_c: (1) specification limits for the glassware, (2) variation in filling to the mark, and (3) ambient temperature effects

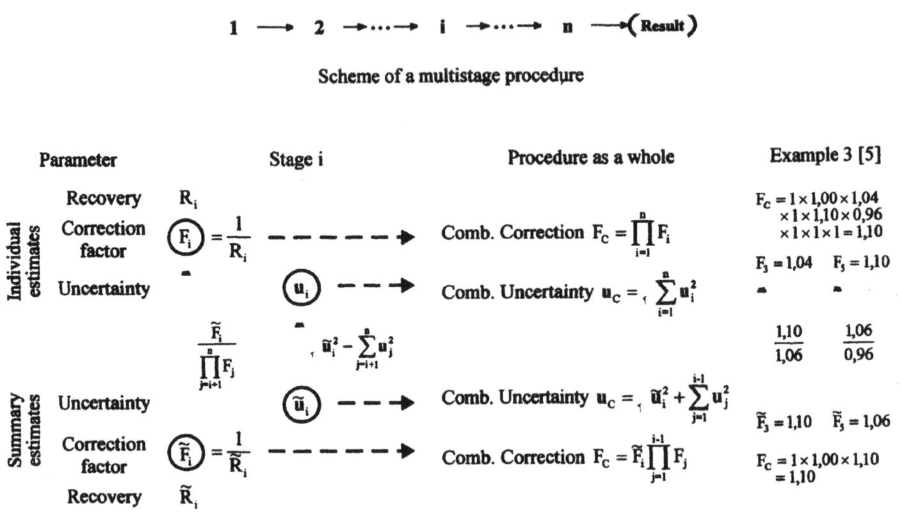

Fig. 3 Evaluation of combined correction factor and combined uncertainty in a multistage procedure. Components relating to a stage i (shown in *circles*) are either individual estimates F_i, u_i (*upper part of the table*) or summary estimates $\tilde{F}_i$, u_i (*lower part of the table*)

correction factor F_i for such a stage one must divide the summary value $\tilde{F}_i$ by the correction factor(s) relating to all the following stages (the available data in Example 3 permit us to do this), and one can then calculate the combined factor for the procedure.

It is also possible to do this without finding the individual correction factors in the calculation of the combined correction. It is sufficient to stop at the first summary factor $\tilde{F}_i$ (such as F_3 in Example 3) in the product of the factors. This leads the multistage procedure to be broken up into the elements that make independent contributions to the combined value. Figure 3 demonstrates the two possible ways of calculating: getting individual estimates from summary estimates is depicted by upright arrows, and obtaining the combined correction as a product is depicted by horizontal arrows. (The appropriate procedures in calculating the combined uncertainty are also included in the table.) The right-hand column of the table shows the relevant calculations associated with Example 3.

Rule 3: When estimating uncertainty only those influence factors are to be considered that really affect the result of a measurement in the context of the procedure.

Let us refer again to Example 3 of the Guide. Evaluation of the uncertainty associated with the GC measurements is based here (Table A3.6 [5]) on a wide-ranging study of GC variability across different instruments, operators, (and times). However, in spite of the wide variation of the factors, the usefulness of the uncertainty estimate so obtained is doubtful. Indeed, the conditions of the GC determinations are usually such that the responses for both a sample and a standard are registered with the same instrument, by the same analyst, and over a short period of time. This is why these influence factors largely cancel in the result of analysis. Therefore, accounting for these factors (with separate estimation of variabilities for GC determination and calibration stages) seems to be based on a misunderstanding in the context of the procedure.

Evaluation of an uncertainty associated with weighing in the same example should also be mentioned. The two contributions to the uncertainty taken into account in this case are: a standard deviation for "repeatability experiments" (0.03 g) and a standard deviation of the mean of the long-term data (0.008 g). The two components are combined, giving the value 0.031 g.

Note first of all that the use of the standard deviation of the mean as a measure of long term variability is not correct in estimating the uncertainty sought for. If the data available have covered not 11 months, as in the example, but say 22 months, the long-term contribution would be $\sqrt{2}$ times smaller following this way of thinking. The available monthly check weights data (Table A3.1(3) [5]) give a standard deviation of 0.026 g, and this value alone characterizes the long-term variability of a single weighing. At the same time, Table A3.1(1) [5] shows repeat weighings / replicate readings results that may be very useful for detailed examination of the precision of weighing. Application of one-way analysis of variance according to a standard scheme [9] allows us to estimate separate components of total weighing variation, inasmuch as the standard deviation of 0.03 g mentioned above was evidently obtained by treating the data as one large sample, without a separation.

So, a replicate readings standard deviation is found to be 0.020 g (with the number of degrees of freedom f being equal to 36). One can easily prove by means of the F test that the long term standard deviation of 0.026 g ($f = 10$) does not differ significantly from this estimate, even at the 10% level. This means here that time is not a factor at all, so that a contribution due to long-term variability need not be considered. The analysis further leads to a repeat weighings standard deviation of roughly 0.05 g ($f = 11$) for a single weighing. This estimate is significantly greater (F test) than 0.020 g, the standard deviation for replicate readings, and hence there is some kind of additional source of variation in weighing in the laboratory. Whatever the source may be, it is the value 0.05 g which should be taken as an actual contribution of weighing to the combined standard uncertainty required.

And one more remark on the subject

One of the most important points relating to the evaluation of uncertainty is that all relevant error sources should be taken into account, with the corresponding contributions being combined. The estimate so obtained quantifies the *overall* uncertainty inherent in the analytical procedure at issue. Apart from the meaningful reporting and interpretion of an analytical result, the overall uncertainties are applicable, for instance, for quality control purposes when reference materials are used.

There are, however, many cases in analytical practice in which the overall uncertainty estimates seem inappropriate to be handled and thus unnecessary. Suppose, for instance, one has to compare two articles of ceramic ware with respect to cadmium release according to BS 6478 (see Example 2 of the Guide). The experiment is carried out in such a way that the tested vessels filled with the same leaching solution are allowed to stand during the same period of time at the same temperature (both measured with reasonable accuracy), and the two extract solutions obtained are analyzed by AAS using the same bracketing reference solutions. The question is whether the two samples differ from each other with respect to the test or not, or, in terms of statistics, whether the difference between the two measurement results is significant against the background of their own variabilities.

Appropriate conformity criteria based on Bayesian theory as well as those of the usual statistics can be applied to the problem of comparison of two measurement results, taking into account their uncertainties [10], regardless of the fact that the over-all uncertainties, as calculated in Example 2, would be unsuitable to solve the problem correctly with respect to the comparative experiment. They are "excessive" for this. Clearly, a number of error components caused in particular by deviations of actual experimental conditions from nominal ones are the same for the two results to be compared, and the corresponding contributions vanish in the uncertainty budget for the difference. Thus, if only an internal consistency of results, not absolute trueness, is of interest, the influence quantities which are not variable in the scope of such a comparative trial may be disregarded, with the overall uncertainty being reduced to that suited to the particular conditions and referred to as a *conditional* uncertainty.

The possibility of such an approach, albeit with respect to the "top-down" method of dealing with uncertainty, was noticed in [1]. The term "conditional uncertainty" or a similar one is likely to gain currency in analytical data treatment, since a considerable part of everyday tasks in analytical laboratories only requires such an internal consistency of results. It should not be regarded as a "loophole" in order to reduce an uncertainty that may otherwise be too large. It is to be reckoned rather as an instance of applying the fitness-for-purpose principle. The notion of fitness for purpose is apparently quite applicable to uncertainty estimates as well as to data produced by the measurement process itself.

In conclusion, it would be relevant to cite a very true and profound passage (item 3.4.6) from the ISO Guide [6], which was fully carried over to the EURACHEM document (item 5.4.16): "The evaluation of uncertainty is neither a routine task nor a purely mathematical one; it depends on detailed knowledge of the nature of the measurand and of the measurement method and procedure used. The quality and utility of the uncertainty quoted for the result of a measurement therefore ultimately depends on the understanding, critical analysis, and integrity of those who contribute to the assignment of its value."

References

1. Analytical Methods Committee, RSC (1995) Analyst 120:2303–2308
2. Cortez L (1995) Microchim Acta 119:323–328
3. Williams A (1996) Accred Qual Assur 1:14–17
4. Wegscheider W, Zeiler H-J, Heindl R, Mosser J (1997) Annal Chim 87:273–283
5. EURACHEM (1995) Quantifying Uncertainty in Analytical Measurement
6. ISO, IEC, OIML, BIPM (1992) Guide to the Expression of Uncertainty in Measurement, 1st edn. ISO (The 1993 edition in the name of the seven organizations including IFCC, IUPAC, IUPAP is also available)
7. ISO 384 (1979) Laboratory glassware. Principles of design and construction of volumetric glassware
8. ISO 4787 (1984) Laboratory glassware. Volumetric glassware. Methods for use and testing of capacity
9. Doerffel K (1990) Statistik in der analytischen Chemie, 5th edn, chap 8. Deutscher Verlag für Grundstoffindustrie, Leipzig
10. Weise K, Wöger W (1994) Meas Sci Technol 5:879–882

Accred Qual Assur (1998) 3:14–19

Angel Ríos
Miguel Valcárcel

A view of uncertainty at the bench analytical level

A. Ríos (✉) · M. Valcárcel
Department of Analytical Chemistry,
Faculty of Sciences,
University of Córdoba,
E-14004 Córdoba, Spain

Abstract The problem with which analytical laboratories are confronted, after *traceability* of their results has been demonstrated, is correctly estimating their *uncertainty* – to which traceability is also to some extent subject. While the general principles for calculating the uncertainty of physical measurements are applicable to chemical metrology, some refinements are needed, especially careful selection and planning the level at which uncertainty will be estimated by each laboratory in accordance with its capacity and required demands. Depending on the particular decision to be made, the mechanism to be used to estimate the uncertainty varies markedly; also, the rigour of the estimation increases with increasing stringency of the demands. This paper describes the primary sources of uncertainty in chemical metrology and discusses different approaches to its estimation in relation to the type of analytical laboratory concerned. The view presented tries to be close to the bench analytical level, in order to be practical and flexible for laboratories, although it could sometimes be considered slightly heterodox.

Key words Uncertainty · Quality Assurance · Chemical measurements · Metrology

Introduction

Traceability and uncertainty have become two major paradigms for quality systems in testing laboratories. While traceability must be demonstrated (it cannot be represented by a mathematical figure), uncertainty has to be calculated. The ease with which these demands can be met varies with the type of laboratory and where the metrological principles are applied (physical, chemical or biological field). Broadly speaking, metrology in the physical field poses no serious problems at present as regards traceability of its measurements; rather, its chief concern is to decrease the uncertainty of measurements. By contrast, traceability is the principal concern of biological metrology owing to a shortage of reference materials; uncertainty at the biological laboratory is typically very high and still regarded as secondary, as acknowledged by the Accreditation Guide for Microbiological Laboratories, which is accepted by EAL (European cooperation for Accreditation of Laboratories). Metrology in chemical measurements lies between these two, though possibly closer to physical metrology. While a number of reference materials and certified reference materials are available, they are clearly inadequate to meet the needs and frequently entail using an alternative route to traceability (e.g. comparisons with validated methods or interlaboratory exercises). Also, the preliminary operations of the analytical process occasionally escape the control required to assure traceability. However difficult, traceability can be demonstrated, supported and documented in most instances. One must concede, however, that calculating the uncertainty of the results obtained by an analytical chemical laboratory is currently an issue of great concern for laboratories and also, occasionally, a source of controversy between auditors themselves. It is clear that the direct use of metrological principles in analytical laboratories as they are used in the physical field is pretty useless and produces a strong aversion on the part of the laboratories, which see the process as artificial and far from reality.

This paper is aimed at clarifying the way analytical laboratories should approach the problem and should adopt a solution consistent with their role and competence. An assumption is made that there are several levels at which uncertainty can be estimated and which make up global uncertainty (a more rigorous and valuable concept). While every laboratory should aim to estimate this last value, it would be foolish to ignore the fact that most analytical control laboratories – those accredited included – estimate other types of uncertainty that are numerically more accessible; in so doing, they restrict the diversity and intrinsic heterogeneity of the samples they receive.

Features of metrology in chemical measurements

It is important to examine in some detail the essential features of chemical metrology and its differences from physical metrology, where traceability is more immediate. These features can help one better understand the difficulties involved in estimating uncertainty in chemical metrology. The essential feature of *physical metrology* is its direct association with *transfer standards*, which are available for every quantity measured. Also, the measured quantity is always independent of the sample or object tested (examined). Thus, a length measurement is always referred to the metre, for which a modern definition [1] containing no "artefacts" has been given – the meter also exists as a physical entity in the form of various transfer standards. Whether it is used to measure the length of a table, the height of a tree or the distance between two objects, the property actually measured is independent of the nature of the sample or the body of the experimental object. In this type of metrology, *uncertainty* is a factor (feature) introduced by the measuring equipment rather than by the sample itself. Hence the chief concern of physical metrologists is to establish the uncertainty of measuring equipment, which is associated with the result for the sample or object examined (whether as such or as a combination of several uncertainties).

The picture in *chemical metrology* is rather different. This type of metrology is chiefly associated with *analytical chemical standards* [2]; also, the essence is not only *what* is measured (i.e. the analyte or measurand) but also *where* it is measured (i.e. samples and their matrices). Obviously, determinations of iron in rocks and human blood are not the same. The "tools" to be used in each case vary, i.e. the analytical process that follows sample collection differs (viz. samples are treated differently and subjected to measuring methods and techniques that are dictated by their analytical properties). The analytical problem addressed, which demands a solution, is obviously not the same either [3]. As a result, in chemical metrology the *sample* (as the physical materialisation of the analytical problem) is the decisive factor. Whatever chemical measurement is to be made will be dictated by the type of sample; also, even if the measurand is the same, the standard to be used and the way measurements are to be made (viz. the analytical technique of choice) can vary markedly. The result of a determination will be subject to a global uncertainty arising from three distinct but closely related agents of the analytical process, namely: (a) the measuring instrument, (b) the analytical method (sample treatment included) and (c) the sampling and sub-sampling procedure. One other distinct feature of chemical metrology is qualitative analysis, which also requires appropriate standards and is absent from physical metrology.

Sources of uncertainty in chemical metrology

The three characteristic steps mentioned above comprise various sub-steps or basic activities of analytical work that are the actual sources of uncertainty in the measurement process. A brief description of each follows.

Uncertainty in the measuring instrument

The sources of uncertainty in the measuring instrument are easy to identify and diminish by improving the equipment itself or using an alternative, more precise analytical method. Analytical instruments require *maintenance*, *calibration* (calibration of the equipment) and *standardisation* (analytical calibration) of their response. Maintenance has virtually no effect on uncertainty in the absence of underlying malfunctioning. On the other hand, calibration and, especially, standardisation, can be a source of high uncertainty. Calibration is in fact a typical activity of physical metrology and is intended to ensure that the instrument will be in perfect condition to make the measurements for which it has been designed and constructed. It is done by using transfer standards and produces an uncertainty inherent in the measured quantity (i.e. intrinsic to the equipment) and independent of the measured parameter. The calibration of an analytical balance is a good example; however, any other laboratory instrument is tied to this activity (e.g. the use of holmium or didymium filters to calibrate wavelengths in UV-visible spectrophotometers).

The standardisation of an instrument's response – commonly referred to as "calibration", as in "calibration curve", despite the fact that calibration is a different activity – is a purely analytical activity typical of chemical metrology. It affects analytical instruments only and defines their response to the measurand(s) to be measured. Standardisation is crucial to ensuring traceability in the results subsequently obtained. One must make several decisions and take several steps to reach this goal, namely: (a) select an appropriate standard, (b) choose a suitable standardisation method, (c) derive a mathematical relation between the analytical signal and concentration, and (d) validate the model established. There are some manuals and literature references of help in this context, particularly those with a chemometric slant [4, 5]. It is worth emphasising the significance of validating the experimental model in terms of quality. Validation of a model involves experimentally confirming that it is a correct simplification of the series of experimental points it contains in such a way that it can accurately predict future unknown values (in the samples to be analysed). Univariate linear calibration is usually done by using least-squares regression, which involves checking fulfilment of various statistical hypotheses -alternatively, residual analysis can be used to check for homoscedasticity. However, one must also determine how closely the model fits experimental points by using analysis of variance (ANOVA) in order to confirm whether a different, more precise type of fitting is needed. The validation process also involves determining the confidence region for the model, its sensitivity and its valid lower limit (represented by the limit of determination).

Validating the standardisation model allows one to ensure traceability in the results produced by inverse interpolation in the analysis of samples; however, the model will obviously be subject to an uncertainty u_3 that will be a function of the standard error, $S_{y/x}$; this latter characterises the standard deviation associated with the mathematical definition for the regression line. This is a standard uncertainty essentially subject to the model's random errors, which arise from variations in signal measurements (the independent variable, y). Because the uncertainties in x_i values are small relative to the previous ones, they can usually be neglected. In any case, calculating this type of uncertainty poses no special problem.

Uncertainty in the analytical method

Analytical methods are currently divided according to traceability into *primary methods* (formerly designated "absolute" and "stoichiometric" methods) and *secondary methods*, which involve a longer traceability chain and are commonly referred to as "relative" or "comparative" methods. One prominent part of relative analytical methods is the standardisation of the instrument's response, described in the previous section. However, a relative method also comprises other steps that are collectively designated the *preliminary operations* of the analytical process. In fact, these operations significantly complicate chemical metrology as they are varied and difficult to control and reproduce in a systematic manner [6]. They are thus the source of major errors not only of the random but also of the systematic type that have a decisive influence on uncertainty.

Method validation is thus a central activity in laboratory quality systems in as much as it assesses adherence of the laboratory to its quality policy. The validation process is closely related to representativeness of the results [7], which depends on the analytical objectives and types of sample. Table 1 shows the basic landmarks of the process. No doubt, demonstrating that the results obtained are accurate is essential proof and an unavoidable requisite. In addition to meeting other objectives, including compatibility with the sample matrix and adequate robustness for use in routine work, one must estimate the degree of uncertainty associated with the results produced by a given method.

The primary source of uncertainty lies in the preliminary operations required to treat samples. Such operations as digestion/disaggregation of solid samples or extraction and clean-up processes (fairly frequent) intro-

Table 1 The process involved in validating an analytical method

1. Checking fitness to the analytical problem:
   - Choosing a suitable method (to be subsequently confirmed or rejected by validation)
2. Preliminary study:
   - Clear, detailed description
   - Checking fitness to the analytical goal via
     - Comparability with the sample matrix (applicability to real samples)
     - Limit of detection
     - Determination range
     - Selectivity
   - Robustness tests
3. Experimentally demonstrating that the system is "under statistical control" (by means of control graphs):
   - The means of measured values should remain constant over long periods (at high and low analyte concentrations)
   - The precision should be adequate and constant
4. Demonstrating accuracy:
   - Recovery tests
   - Comparison with an independent, previously validated method
   - Comparison with CRMs
   - Interlaboratory studies
5. Compiling the SOP after the method has been validated

duce significant "uncontrolled" sources of error that contribute to uncertainty. Type A evaluation (viz. uncertainty that can be experimentally evaluated from the statistical distribution of the results from a series of measurements) and type B evaluation (uncertainty evaluated from assumed probability distributions based on experience or other information) uncertainties can be used to estimated the global uncertainty (as the ISO recognises) of the analytical method concerned, denoted by u_2. Recovery tests must be conducted very carefully if their uncertainty is to be correctly estimated. Thus, the analyte must be added in the same chemical form as it is likely to be present in the samples; also, the spiked sample must be thoroughly homogenised and additions must include variable amounts of analyte. Finally, the recoveries must be evaluated in statistical terms (usually by regression analysis).

The possibility of evaluating uncertainty under a type B approach in this analytical step is a distinct feature of chemical metrology that entails obtaining additional information (mainly about the preliminary operations involved in the method used) frequently obtained outside the laboratory (from the analytical literature or other laboratories). This type of evaluation is closely related to the variety of samples where the analyte can occur or with the fact that sometimes it is not a single analyte but a group of ill-defined individual analytes that are to be determined (e.g. bitter compounds in beer or the hydrocarbon index in waters).

Uncertainty in sampling and sub-sampling

It is widely admitted that sampling poses special problems and influences the representativeness of the results. The portions extracted from a sample for analysis should contain essentially the same information as the population or system studied as a whole. Unsurprisingly, this activity has been the subject of abundant literature [8–10]. The sampling strategy can be suited to the analytical problem addressed by using four different types of approach, viz. intuitive or judgmental, random, systematic and protocol-based. This is therefore the first decision with which one is confronted. The *sampling manual* describes in detail the conditions, equipment and procedures used in this step. In such a widely variable activity, it is utterly important to develop clear, well-documented protocols in order to release operators from the need to improvise or undertake responsibilities beyond their qualification. Even if these cautions are exercised, there remain the enormous variability of samples and their also variable representativeness of the problem addressed.

Equally important is sub-sampling, which involves withdrawing aliquots from previously collected samples for subjection to the analytical process at the laboratory. This step is also documented in quality schedules; however, the process is complicated by heterogeneity in most samples – particularly solid samples – and introduces appreciable variability between sub-samples that ultimately leads to significant differences between the results obtained for the same sample. No doubt, sampling and sub-sampling will demand greater interest in the future; such standards as EN 45000 and similar ones should provide a more systematic and extensive description of the minimum requirements in this respect, engage laboratories in these activities and encourage the release of sampling guides for specific fields. This will decisively increase the quality of results and connect them to the real world – from which samples ultimately come – rather than only to the portion that reaches the laboratory. Simultaneously, the bodies and institutions concerned with or responsible for quality on a national or international scale should promote efforts in the direction pointed out by Thompson and Ramsey [11] in order to develop reference sampling targets (RST), analogue sampling of an RM or CRM, collaborative trials in sampling, and the possibility to organise future proficiency testing in sampling.

The uncertainty produced by these steps, denoted by u_1, is undoubtedly very high and exceeds that resulting from the previous two steps. Also, its estimation is rather complex in most cases: it entails using vast amounts of information about the samples analysed and their origin to ensure a correct assessment. Obviously, as the tools noted in the previous section become available, this task will be easier and more reliable. *Heterogeneity* within and between samples is the origin of this problem and one more feature that clearly distinguishes chemical metrology from physical metrology. It confirms that the primary target of chemical metrology is the sample rather than the equipment used in the analytical process, which is also important but only secondarily. There is thus a highly significant underlying problem awaiting solution in order to assure quality of chemical measurements: sampling and sub-sampling. Under the influence of our fellow engineers and physicists, who deserve due credit for establishing metrological principles and starting and systematising *Quality Assurance* systems, we have placed too much emphasis on measuring equipment to the detriment of our true goal and primary source of variability: samples.

Estimating uncertainty

The "Guide to the Expression of Uncertainty in Measurement", published jointly by the ISO and other bodies [12], sets general rules for assessing and expressing uncertainty, and applying them to chemical metrology. Uncertainty is assigned various sources including the following: an incorrect definition of the measurand, sampling, incomplete extraction or preconcentration of the measurand, matrix and interfering effects, carry-over during sampling or sample preparation, unknown effects of the environmental conditions on the sample, instrument bias, tolerances of weights and volumetric material, reagent purity, values assigned to standards and reference materials, calibration, etc. A document released by EURACHEM [14] refines these principles as applied to chemical measurements and establishes the steps to estimating uncertainties; it shows how to express them properly and – especially useful – provides examples of variable complexity. Although the document can be seen as very academic and not very close to the bench analytical level, it confirms the significance of the early steps of the analytical process (sampling, sub-sampling and sample treatment) in the overall process with a view to estimating uncertainty, its high contribution to global uncertainty and the difficulty involved in its evaluation. It is not an exaggeration to state that the mere reading of these examples can "frighten" laboratories, most of which are bound to feel unable to calculate their uncertainty in the proposed ways. Where does the problem lie? Is this way of managing things too demanding? The answer, as almost always, is to rationalise computations by simplifying or discarding those steps whose uncertainty is known to be small relative to other key steps with a decisive influence on global uncertainty. While this can facilitate estimating uncertainty, it does not solve the actual problem.

As is typical of quality systems, the goals of an analytical laboratory are established in its *Quality Policy*. Because estimating uncertainty and demonstrating traceability are two paradigms for these systems, as noted above, it seems obvious that, depending on the competence of the laboratory concerned, it should answer this preliminary question: What type of uncertainty is to be calculated? In fact, *total uncertainty* is the combined contribution of the three above-described sources of uncertainty (denoted by u_1, u_2 and u_3) and the most real and demanding of all. As a rule, $u_1 \gg u_2 > u_3$, but we know that u_1 is very difficult and laborious to estimate reliably. It is seemingly clear that these laboratories do not feel compelled to estimate u_1, so they oversimplify the problem; there is the question, however, as to how useful can the information they provide be. Even if the laboratory is not concerned with sampling, if the samples it receives are variegated (e.g. those to be used in determining pesticide residues in fruits of different kinds) and essentially heterogeneous, it will be in the position depicted graphically in Fig. 1.

Figure 1 can be viewed as a general case that only excludes the variability of sampling carried out outside the laboratory. It represents the general origin of the uncertainties that effect the results delivered by the laboratory. The horizontal direction of the figure represents the specific analytical process that produces the results. If a *certified reference material* (CRM) or, failing this, a control sample suited to the type of analysis to be performed is available, one can not only check whether the results are traceable, but also estimate the uncertainty of the analytical process (U_{process}) by using an appropriate number of replicates (under reproducibility conditions). This is essentially an uncertainty estimated as type A, that is obtained with the same material (a CRM). The vertical direction of Fig. 1 represents the different real samples that reach the laboratory in various degrees of heterogeneity. The total uncertainty (U_{total}) will be a combination of U_{process} and the uncertainties introduced by the variability of the samples processed (U_{sample}), which sometime are estimated as type B (not experimentally). Obviously $U_{\text{sample}} > U_{\text{process}}$, since, by definition, the CRM used to calculate U_{process} must be a homogeneous material – and must exhibit some difference from the various types of samples to be analysed. Therefore, these laboratories must estimate u_1 from U_{sample} (one of the principal sources of uncertainty). Estimating total uncertainty from a CRM only is inadequate, especially in control laboratories that analyse no replicates, since each result they produce is subject to the *uncertainty associated with the process*. Strictly speaking, another additional uncertainty term should be taken into account to calculate U_{total}. This is the uncertainty certified for the CRM (again classified as type B by the ISO). Note that the contributions of the uncertainty calculated as U^2_{total} are: the uncertainty of the CRM obtained after the certification process (a wide variety of analytical methods having been used and combined); the uncertainty of the analytical process under the "ideal" (use of the CRM) situa-

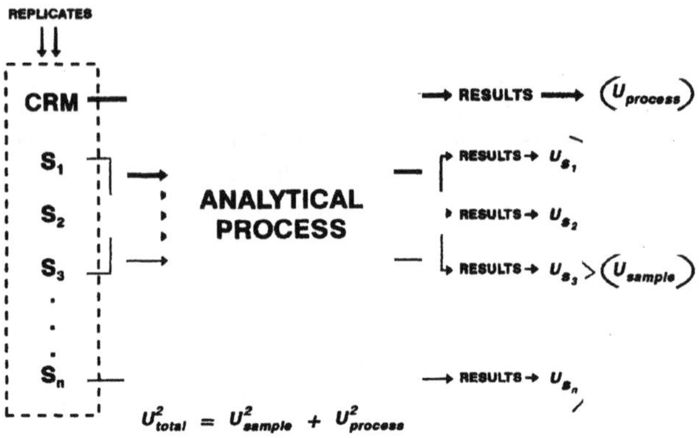

Fig. 1 General process for calculating uncertainties in chemical metrology. The total variance (U^2_{total}) is the summation of the variances resulting from sample diversity and heterogeneity (U^2_{sample}) and that stemming from the analytical process (U^2_{process}). CRM certified reference material, S_i sample i, U uncertainty

tion used for its validation (by comparison this is the analogous variation that introduces the tolerance stated for the internal volume of a volumetric flask); and, finally, the uncertainty introduced by the variety of samples analysed (real samples, basically heterogeneous). Thus, U_{total} should represent the maximum uncertainty that a particular laboratory could have in its reported results.

Estimating uncertainty under these circumstances is especially complex. The experience of the laboratory concerned (or others) in the analytical process involved and the type of sample processed can be of great assistance as they may allow one to use existing data or plan experiments for different samples or parts of samples in order to derive information on the degree of variability resulting from sample heterogeneity and sample types. Interlaboratory exercises are one other valuable source of information for assigning uncertainties due to sample variability and heterogeneity. Finally, scanning the specialised literature – an essential task for laboratories wishing to sustain their competence – may also provide estimations of variability which, duly justified, can be used to estimate uncertainties in specific cases.

Conclusions

The inherent problems and peculiarities of chemical metrology entail introducing rational adaptations of the general principles established for physical measurements in estimating uncertainties. As W. Horwitz and R. Albert have recently said, "the calculation of uncertainty as recommended for physical measurements cannot be transferred readily to chemical measurements" [14], because both testing fields have entirely different error patterns and bias is difficult to identify and eradicate in chemical systems. In chemical metrology, the sample (the physical entity of the analytical problem to be solved) is the primary target. Also, the greatest source of variability is the sample (its nature, its heterogeneity, the matrix that accommodates it, the forms under which the analytes are present, etc.). Therefore, an analytical result can hardly be representative of the object from which information is to be derived unless it is accompanied by the uncertainty introduced at this early stage of the analytical process. The principal virtue of the metrological concept of uncertainty is that it explicitly encompasses aspects related to the representativeness of results that have traditionally been disregarded in much analytical work. On the other hand, the orthodox step-by-step procedure to estimate the uncertainty in chemical measurements can be alternatively replaced by a overall procedure, as it has been presented in this paper, which is simpler, more rational and closer to the bench level. Under this approach, laboratories see as more practical and realistic the way to calculate their uncertainties by themselves.

Editors note

The above paper reflects the authors' point of view as well as their understanding of uncertainty. Readers are invited to comment and report points of view of other practitioners.

References

1. XVII General Conference on Weights and Measurements (1983)
2. Valcárcel M, Ríos A (1995) Analyst 120:2291–2297
3. Valcárcel M, Ríos A (1997) Trends Anal Chem 16:385–393
4. Massart DL, Bandeginste BGM, Deming SN, Nichotte Y, Kaufman L (1988) Chemometrics: a textbook. Elsevier, Amsterdam, pp 75–92
5. Miller JC, Miller JN (1993) Statistics for analytical chemistry (3rd edn), chap 5. Ellis Horwood, New York
6. Valcárcel M, Luque de Castro MD, Tena MT (1993) Anal Proc 30:276–280
7. Ríos A, Valcárcel M (1994) Analyst 119:109–112
8. Crosby T, Patel Y (1995) General principles of good sampling practice. Royal Society of Chemistry, London
9. Smith R, James GV (1981) The sampling of bulk materials. Royal Society of Chemistry, London
10. Gy PM (1995) Trends Anal Chem 14:67–76
11. Thompson M, Pansey MH (1995) Analyst 120:261–270
12. Guide to the Expression of Uncertainty in Measurements (1995) ISO, Geneva, Switzerland
13. Quantifiying Uncertainty in Analytical Measurement, version 6 (1995) EURACHEM
14. Horwitz W, Albert R (1997) Analyst 122:615–617

Accred Qual Assur (1998) 3:117–121

Michael Thompson

Uncertainty of sampling in chemical analysis

Presented at: 2nd EURACHEM Workshop on Measurement Uncertainty in Chemical Analysis, Berlin, 29–30 September 1997

M. Thompson (✉)
Department of Chemistry, Birkbeck College, Gordon House, 29 Gordon Square, London WC1H 0PP, UK
Tel.: +44-171-3807469;
Fax: +44-171-3807464;
email: m.thompson@chem.bbk.ac.uk

Abstract Uncertainty of sampling is the contribution from sampling errors to the combined uncertainty associated with an analytical measurement when the measurand is the concentration of the analyte in the 'target', the total bulk of material that the sample is meant to represent. Of the errors considered to contribute to uncertainty, random errors of sampling, characterised by precision, are much more accessible to investigation than those due to bias. Where an approximation to random sampling can be achieved, realistic precisions can normally be estimated. In some instances reproducibility precision is significantly greater than repeatability precision, and the contribution of between-sampler variations to sampling uncertainty must be acknowledged. However, the collaborative trial of a sampling method is an expensive and difficult exercise to execute. A system of internal quality control for routine sampling can be introduced. Fitness for purpose has been defined in terms of the required combined uncertainty of sampling and analysis.

Key words Uncertainty · Sampling · Fitness for purpose · Collaborative trial · Internal quality control

Definitions of terms in this paper

Uncertainty of sampling contribution to the combined uncertainty of an analytical measurement that results from the production of the laboratory sample from the sampling target.
Sampling target mass of material that the laboratory sample is designed to represent.
Fitness for purpose property of the result of a measurement when its associated uncertainty minimises a cost function comprising all terms that are functions of the uncertainty.

Introduction

In analytical science, measurements are not usually made on the whole amount of the material of interest (here called the 'target'), but on a much smaller amount, the sample, which is selected from the target in some manner. As a consequence, metrologists in chemistry have hitherto concentrated on the analytical measurement process in isolation. As that process involves estimating the concentration of an analyte in the laboratory sample, the uncertainty of measurement for the analyst refers to that specific measurand. For the end-user of the data, however, the measurand of interest is the concentration of the analyte in the target. Hence the uncertainty relevant to the end-user should include the uncertainty contribution that is introduced by preparing the laboratory sample from the target.

Samples taken from the same target will usually vary in composition, from each other and from the average composition of the target, partly because of the heterogeneity of the target but also because of shortcomings in the sampling procedure such as contamination, loss of analyte or use of an incorrect sampling procedure [1]. Hence, to obtain the uncertainty (u_t) associated with the target measurand, the uncertainty of the 'pure analytical measurement' must be augmented by a contribution from the sampling, so that we have:

$$u_t = \sqrt{u_a^2 + u_s^2}$$

where u_a is the standard uncertainty of 'pure measurement' and u_s is the standard uncertainty resulting from errors in sampling. It is stressed here that sampling uncertainty characterises only those errors made during the process of producing the laboratory sample. Errors introduced during the selection and weighing of a test portion from the laboratory sample are subsumed into the measurement uncertainty.

In most sectors of analytical science sampling procedures regarded as 'best practice' or 'fit-for-purpose' have been developed. Usually, however, we have very little information on the performance of such procedures, because the validation of sampling is far less developed than that of analysis. In such cases the uncertainty of sampling usually needs to be estimated ab initio. As such estimation can present considerable practical difficulties, it is currently attempted in only a limited number of sectors of analytical practice.

Sampling errors can be quantified only after analysis of the samples, so the results of ordinary measurements carry both sampling and analytical errors. As a consequence, results used to estimate sampling uncertainty must usually be obtained from designed experiments (with replication and randomisation) for interpretation by anova (analysis of variance) methods.

Sampling precision and sampling bias

The estimation of sampling precision in the manner to be described is analogous to the estimation of precision in the validation of an analytical method. Hence we can think in terms of characterising a particular sampling strategy or protocol for use with a particular type and size of target.

If a number of samples are extracted from a target by repeated application of a sampling protocol, the variation in true concentration among the samples is characterised for a particular analyte by the sampling variance σ_s^2. A single analysis on a sample will have a variation characterised by $\sqrt{\sigma_s^2 + \sigma_a^2}$, where σ_a^2 is the analytical variance. An experimental design for estimating σ_s^2 and σ_a^2 is shown in Fig. 1. A reasonably reliable estimate of σ_s^2 can be made by anova if $\sigma_s > 3\sigma_a$ (i.e., the analytical precision must be somewhat better than the sampling precision) and if there are a sufficient number of replicate samples (i.e., more than ten). If the mean squares between and within samples from the anova are designated *MSB* and *MSW* respectively, the estimate $\hat{\sigma}_s^2$ of the sampling variance is given by

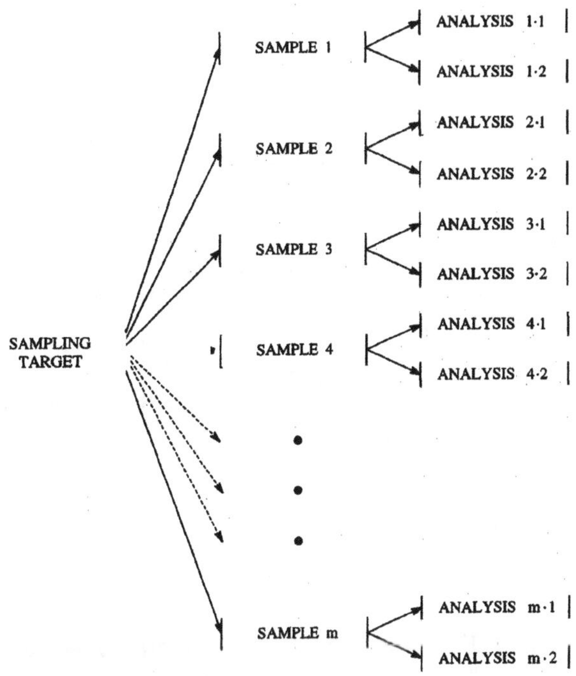

Fig. 1 Design for replicate sampling and analysis of a single sampling target, for the estimation of sampling and analytical precisions

$$\hat{\sigma}_s^2 = (MSB - MSW)/n$$

where $n = 2$ for duplicate analyses.

Another design for estimating σ_s^2 is shown in Fig. 2. In this design a number of distinct targets similar in composition are each sampled in duplicate and each sample analysed in duplicate. This design gives a more rugged estimate of sampling precision than the example discussed above, as it avoids reliance on a single target that might turn out to be atypical of the material as a whole. In this design the sampling precision is 'averaged' over a number of targets.

For the result of such an experiment to be useful, the replicated samples must be taken independently and at random. Obviously a single sampling protocol will be used, but its implementation must be randomised. For example, if the protocol specifies that the sam-

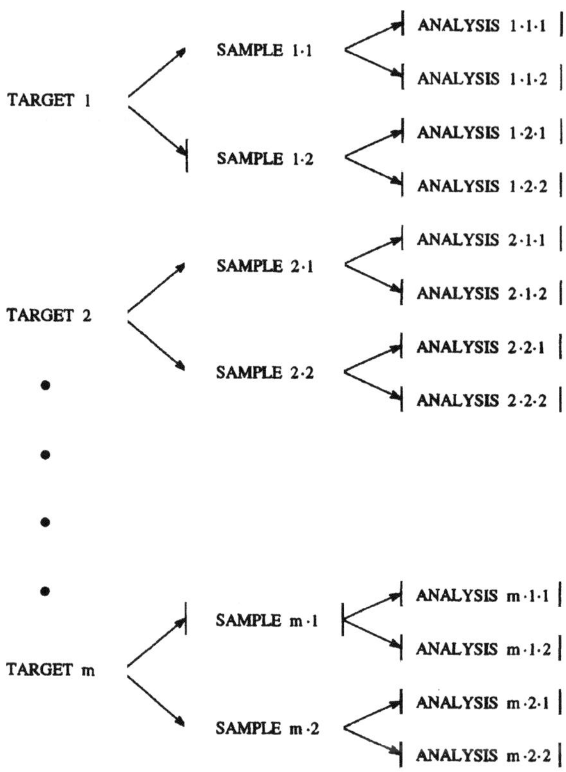

Fig. 2 Design for duplicate sampling and analysis of a number *m* of similar sampling targets, for the estimation of sampling and analytical precisions

ple should be made up of a number of increments taken from the target in a specific two-dimensional pattern, the replicate samples should be obtained by relocating the origin and orientation of the pattern at random for each sample. Any deviation from randomness would tend to give rise to an underestimate of sampling precision. In practice the extraction of a random sample from particular targets may be difficult or impracticable. Samplers must do their best under prevailing circumstances.

In all of these experiments, the set of samples collected should be analysed in a random order under repeatability conditions. This strategy avoids confusing analytical problems like drifts with genuine differences between the samples.

Bias in sampling is a difficult topic. Some experts on sampling argue that sampling bias is not a meaningful concept: sampling is either 'correct' or 'incorrect' [2]. Certainly sampling bias may well be difficult to detect or estimate, because we need an alternative sampling technique or protocol, regarded as a reference point, with which to compare our method under test, before we can claim that bias is present. However, it is easy to see how bias could arise in sampling methodology. For example, the samples could be consistently contaminated by the sampling tools such as containers or grinding devices (example – rock chips from a borehole contaminated by chromium from the drill bit). Alternatively, some of the analyte could be consistently lost from the samples because of inappropriate handing (example – loss of elemental mercury from a rock sample during grinding). A sample can be biased if the sampler misunderstands the protocol (example – sampler collects a-horizon soil contaminated with b-horizon soil). Finally a sample can be biased if the sampler is selective (instead of random) in the selection of increments that form the aggregate sample (example – always picking up large pieces of the target materials rather than small pieces).

It is often possible to avoid these biases once they are known. All too often it is very difficult to establish the existence of such a bias in sampling for want of (a) a reference sampling method for comparison, or (b) the will to carry out the comparison, or because sampling precision is too large to allow the existing bias to be demonstrated at a significant level. In such cases it is usual to regard the sampling method as 'empirical' (by analogy with the empirical analytical method, where the result is dependent on the analytical method). An empirical sampling method would have zero bias by definition. In the absence of any readily available information on sampling bias it is best in most circumstances to calculate combined uncertainty estimates only from precision contributions.

Internal quality control in sampling

Given estimates of σ_s and σ_a for a particular type of material and a fixed sampling protocol we can consider the application of an internal quality control method to ensure that the measurement system, including sampling, stays in statistical control. Let us assume that some or all of the successive routine sampling targets are sampled in duplicate and each of the two samples is analysed once, to give two values (x_1, x_2) for each target. For monitoring this data, a control chart, based on $N(0,\sigma^2=2(\sigma_s^2+\sigma_a^2))$, could be constructed for the difference $d=x_1-x_2$ between the two values. As usual the warning limits should be at $\pm 2\sigma$ and the action limits at $\pm 3\sigma$. That would be useful as a control on sampling as long as σ_s was the dominant precision term. An out-of-control situation could arise if, for example, the target were more heterogeneous than usual, or if an inadequate number of increments were combined to form the aggregate sample. In contrast to analytical quality control, there exists an extra possibility, namely that the two samples could be too similar. That could arise if the two samples were not independent, for instance, in an extreme case, if both laboratory samples were splits from a single aggregate sample. A possible approach to monitoring too great a similarity would be to set up additional control lines at (say) $\pm 0.5\sigma$. Four successive points within these inner lines would occur only rarely under statistical control and suggest problems with sampling.

The collaborative trial in sampling

Until recently sampling precision has been treated implicitly as if it were independent of conditions under which sampling is executed, so that repeatability conditions should suffice for its estimation. However, it is well recognised that in analytical measurement precisions measured under reproducibility conditions (one method, one material, different laboratories) are greater than repeatability precisions (one analyst, one material, one method, one instrument, short time period) by a factor of about two [3], i.e., $\sigma_R/\sigma_r \approx 2$. Consequently it is worth enquiring whether, for sampling, using the same protocol and test material, reproducibility precisions are greater than repeatability precisions to any important degree. Such a finding would have considerable implications for the correct estimation of sampling uncertainty.

The established method of considering reproducibility precisions in analysis is the collaborative trial (method performance study) [4]. Applied to sampling, the collaborative trial would require a number of samplers each to take independent duplicate samples from a target, at random, using a fixed sampling protocol [5]. If all of the samples were then analysed in duplicate, together under randomised repeatability conditions, then hierarchical anova could be used to decide whether either within-sampler or between-sampler precision had reached statistically significant levels. If that were so, the sampling precision under repeatability and reproducibility conditions could be estimated from the mean squares.

Very little experimentation has been conducted along those lines so far. The findings are suggestive, but insufficient work has been done for general conclusions. Studies on sampling contaminated land [5] showed contrasting results for different analytes. At the sites investigated the major contaminant (lead) was spatially distributed in a very heterogeneous manner. As a result, the within-sampler precision was so large (RSD $\approx$30%) that no significant between-sampler variation could be detected for this element. In contrast, elements present at near the background levels (i.e., present because of natural processes rather than contamination and therefore not wildly heterogeneous), significant levels of between-sampler precision were found. Values of the ratio σ_R/σ_r found for sampling were ≈ 2, close to the value found in analytical collaborative trials. This finding suggests that in some instances reproducibility precisions might be most appropriate for estimating sampling uncertainty. At the other extreme, if the target comprises material that is nearly homogeneous in the analyte, it may be impossible to find significant between-sampler or within-sampler variation, because they are both small relative to the analytical variation. This situation has been found in a collaborative trial in sampling wheat (unpublished data).

The lack of information in this area is hardly surprising. It is sometimes an unpopular activity, often technically difficult, and always expensive to organise a collaborative exercise in sampling. The samplers all have to travel to the sampling target(s) and, as they must work independently, they must visit the target in succession. This might delay the shipment of a large amount of a valuable commodity. Often a commodity (especially a packaged material) would be spoilt to an unacceptable degree by multiple sampling. However, there is no doubt that an optimised sampling/analytical system can maximise profits for a manufacturer (see below), and there are industrial instances known to the author where proper attention to sampling errors have resulted in a substantial net gain.

Uncertainty and fitness for purpose

Given that errors are introduced into measurements by both sampling and analysis, we need to consider two questions relating to fitness for purpose, namely: (a) given limited resources, how can we divide them optimally between expenditure on sampling and on analysis; and (b) how can we decide whether a combined uncertainty is adequate for the end-user needs (i.e., is the end-user able to make valid decisions given the uncertainty of the measurements)?

A simple answer to the former question is to consider the relative contributions of sampling and analysis to the combined uncertainty. If sampling uncertainty is the dominant term there is no point in utilising an expensive highly accurate analytical method, because the combined uncertainty will not be usefully improved. For example if $u_a < 0.2u_s$, then $u_t < \sqrt{(u_s^2 + (0.2u_s)^2)} = 1.02u_s$, so that $u_t \approx u_s$ regardless of how small u_a becomes. At the other 'extreme', if $u_a > 0.5u_s$, then u_t is dilated to a level substantially greater than u_s. Therefore u_a should best fall within the approximate range $\{0.2u_s - 0.5u_s\}$. The same argument, applied to a dominant analytical uncertainty, produces the corresponding result. These considerations, although informative in themselves, pay no attention to the relative costs of sampling and analysis as functions of the precision obtained.

To obtain a more realistic picture, including an elementary consideration of costs, we need to examine the apportionment of a fixed financial resource between sampling and analysis [6]. We first consider the cost A of procuring a sample with unit sampling uncertainty. We can achieve an uncertainty of 1/2 by collecting and thoroughly mixing four independent samples, collected using the same protocol, at a cost of 4A. The cost L_s of achieving any uncertainty will therefore generally be

$$L_s = A/u_s^2.$$

The same type of consideration would apply to analysis. We could assume with reasonable confidence that the cost of analysis L_a would be given by

$$L_a = B/u_a^2.$$

Both costs escalate steeply with requirements for decreasing uncertainty. To apportion the costs we need to minimise the total cost $L_t = L_s + L_a$ of the measurement operation for a particular combined uncertainty $u_t = \sqrt{u_a^2 + u_s^2}$. It can be shown [6] that the minimum is defined by

$$u_s^2 = u_t^2 \left(\frac{\sqrt{A}}{\sqrt{A}+\sqrt{B}}\right), \quad u_a^2 = u_t^2 \left(\frac{\sqrt{B}}{\sqrt{A}+\sqrt{B}}\right),$$

which gives

$$\frac{u_s^2}{u_a^2} = \sqrt{\frac{A}{B}}$$

at a total cost of

$$L_t = \frac{(\sqrt{A}+\sqrt{B})^2}{u_t^2} = \frac{D}{u_t^2}.$$

We now consider the end-users cost function L_e. The exact form of this function would depend on particular circumstances, but would be almost invariably an increasing function of the combined uncertainty (i.e., the bigger the uncertainty in the measurement, the greater the likelihood of an error of judgement based on the measurement). For the sake of a simple example, we take a linear function such as

$$L_e = Q + Ru_t$$

where Q and R are constant costs. We are now in a position to define fitness for purpose in an operational manner: it is the combined uncertainty that minimises the total cost L, which is given by

$$\frac{dL}{du_t} = 0$$

where

$$L = L_e + L_t = Q + Ru_t + D/u_t^2.$$

Concrete applications of this idea have yet to be published, and it is likely that the necessary information would be difficult to obtain in many practical circumstances. However, it provides a useful conceptual framework for defining the relationship between uncertainties of sampling and analysis and fitness for purpose.

Conclusions

Uncertainty of sampling is a topic that has to date received scant attention by metrologists or analytical chemists. The difficulties of studying sampling errors are great, and the cost of such studies may be substantial. Where it has been studied, sampling uncertainty has often been found to be of considerable magnitude, sometimes much greater than the pure measurement uncertainty. Few analytical chemists or end-users of data are aware that only fit-for-purpose sampling uncertainty combined with appropriate analysis will maximise their cost-effectiveness. Analytical chemists should be willing to confront the subject: there is no doubt that it needs thorough investigation, with the appropriate investment of money.

References

1. Thompson M, Ramsey MH (1995) Analyst 120:261–270
2. Gy PM (1992) Sampling of heterogeneous and dynamic materials. Elsevier, Amsterdam
3. Boyer KW, Horwitz W, Albert R (1985) Anal Chem 57:454–459
4. Horwitz W (1995) Pure Appl Chem 67:331–343
5. Ramsey MH, Argyraki A, Thompson M (1995) Analyst 120:2309–2312
6. Thompson M, Fearn T (1996) Analyst 121:275–278

Accred Qual Assur (2002) 7:274–280
DOI 10.1007/s00769-002-0489-4

Michael H. Ramsey

Appropriate rather than representative sampling, based on acceptable levels of uncertainty

Received: 28 December 2001
Accepted: 25 April 2002

Presented at EUROLAB/EURACHEM Workshop "Sampling", 5–6 November 2001, Lisbon, Portugal

M.H. Ramsey (✉)
Centre for Environmental Research, School of Chemistry, Physics and Environmental Science, University of Sussex, Falmer, Brighton BN1 9QJ, UK
e-mail: m.h.ramsey@sussex.ac.uk
Tel.: +44-1273-678085
Fax:+44-1273-677196

Abstract Appropriate sampling, that includes the estimation of measurement uncertainty, is proposed in preference to representative sampling without estimation of overall measurement quality. To fulfil this purpose the uncertainty estimate must include contribution from all sources, including the primary sampling, sample preparation and chemical analysis. It must also include contributions from systematic errors, such as sampling bias, rather than from random errors alone. Case studies are used to illustrate the feasibility of this approach and to show its advantages for improved reliability of interpretation of the measurements. Measurements with a high level of uncertainty (e.g. 50%) can be shown to be fit for some specified purposes using this approach. Once reliable estimates of the uncertainty are available, then a probabilistic interpretation of results can be made. This allows financial aspects to be considered in deciding upon what constitutes an acceptable level of uncertainty. In many practical situations "representative" sampling is never fully achieved. This approach recognises this and instead, provides reliable estimates of the uncertainty around the concentration values that imperfect appropriate sampling causes.

Keywords Representative sampling · Uncertainty of measurement

Introduction

The traditional approach to sampling is to select a "correct" sampling protocol that is assumed to give a representative sample, and to eliminate sampling bias by definition [1]. This approach has the advantage of simplicity, but can lead to unsuspected errors in estimates of concentration, caused by sources such as variations in the practical application of the protocol.

This situation in sampling has a useful analogy in the chemical analysis of test materials. There was a time when analytical chemists strove to find a "correct" analytical method, which would determine the true value of an analyte concentration. If all laboratories used this single "correct" method, it was argued, then all measurements would be true and comparable. It has now been realised by the analytical community that achieving this ideal is impossible. All analytical measurements are only estimates of concentration and are never exactly equal to the true value of concentration. The emphasis in analytical chemistry has therefore changed to estimating the uncertainty of a concentration measurement, as well as its value. This uncertainty of measurement can be defined informally as "the interval around the result of the measurement that contains the true value with high probability" [2]. With this information, all estimates of concentration can be compared, and the measurements can be interpreted with a known level of uncertainty.

This approach can usefully be extended to the primary sampling procedure. Instead of assuming that a "correct" sampling protocol will produce a representative sample, it is possible to select an appropriate sampling protocol that will give measurements with an acceptable level of uncertainty. Representivity is still an objective, but there is an explicit admission that it is never achieved perfectly, and the range within which the true value lies is stated together with the measured value of concentration.

There is one crucial concept that must be accepted in order to make this approach applicable; the action of primary sampling must be considered as the first step in the making of a measurement of concentration. In this way the uncertainty of the measurement includes the contribution from all of the sources [3]. These include the primary sampling, the physical preparation, the laboratory sub-sampling, the chemical preparation and the chemical analysis of the sample. The word "measurement" refers to the final estimate of analyte concentration, but the phrase "measurement process" in this paper is used to denote all of these processes collectively. Estimates of measurement uncertainty that omit some of these sources, particularly the sampling, will inevitably be too small and therefore unrealistic.

A second important step is to include systematic errors (unknown or uncorrected) into the estimates of uncertainty. If the uncertainty interval is to include the true value, then its calculation cannot be restricted to just the random errors that are unrelated to the true value. In primary sampling, the random errors are traditionally well characterised and used, for example, to judge the amount of sample that is required to achieve a specified sampling precision. It is however the systematic errors (e.g. sampling bias) that can often cause unsuspected errors, and therefore both types of error need to be estimated.

If the actual uncertainty is estimated for every measurement, rather than relying on an assumption of correctness or representivity of the sampling, then the reliability of measurements (and the sampling component) will improve. Moreover, once the assumption of perfectly representative samples is set aside, it is possible to decide how close to the true value the measurements are required to be, for any particular application. There are cases where relatively high levels of uncertainty (e.g. 80%) can be shown to be appropriate for some purposes (i.e. the measurements are "fit-for-purpose") [3]. The practical limitation on the number and quality of measurements made is frequently financial, and this "appropriate sampling" approach also allows a optimal balance to be made between the quality of the measurements (from sampling and analysis), and the cost of both the measurements and the consequences of undetected measurement errors [4].

The traditional approach to quality sampling is usually linked to an equally traditional but separate approach to quality in chemical analysis. In traditional analytical quality control (AQC), various AQC materials are analysed in the same batch as the samples. If the measurements made on these materials fall within predetermined limits, then the measurements made on the samples are reported to a customer as single concentration values (e.g. in μg of analyte per g of sample). The customer who interprets these measurements often assumes that they are "correct" and does not use, or have access to, any of the information gained from the AQC materials. In the alternative "appropriate" approach, the laboratory reports an uncertainty value with each concentration value. This estimate of uncertainty is not just that arising from the chemical analysis (currently being reported in some labs), but also includes components from the primary sampling and physical preparation. This uncertainty value is estimated using information derived in part from the AQC materials, and gives the customer access to this information in a form that is useful.

This paper aims to give an overview of the research that underpins this approach to "appropriate sampling", with references given to sources of more detailed information. It will cover definitions of uncertainty, methods to estimate uncertainty that include all sources, acceptable levels of uncertainty (fitness-for-purpose), implications of uncertainty for interpretation of measurements, and conclusions, with some examples from case studies to help clarify the explanations.

Definitions of uncertainty

The formal definition of uncertainty of measurement is "A parameter associated with the result of a measurement, that characterises the dispersion of the values that could reasonably be attributed to the measurand" [5]. The meaning of this definition is not entirely clear, and depends heavily on the definition of the word "measurand", which is formally "the particular quantity subject to measurement" [6]. The previously quoted informal definition of uncertainty as "the interval around the result of the measurement that contains the true value with high probability", clearly interprets measurand as being the "true value" of the analyte concentration, and not just as the "analyte concentration" as has been implied by some sources. This interpretation has the important implication that uncorrected systematic errors should be included within the estimates of uncertainty. It differentiates uncertainty from the traditional "error bars" often quoted for analytical measurements, which are invariably based upon random errors alone, with no reference to the "true value" of the analyte concentration.

The estimation of uncertainty for analytical measurements has now been widely advocated, at least in Europe [7]. However, these estimates specifically exclude the contribution to uncertainty arising from the primary sampling, and often from the physical preparation of the sample (e.g. drying, grinding, splitting). Several studies have shown that these are often the largest sources of uncertainty in measurements [3, 8–10].

Methods to estimate uncertainty that include all sources

Methods for the estimation of uncertainty from analytical methods are well developed. They often use "bottom up" method which sum all of the individual components in the uncertainty budget [7], but they can also use information from "top down" approaches that use estimates of the total uncertainty, from inter-laboratory trial for example, without necessarily subdividing it. The step of primary sampling is traditionally excluded from these estimates of measurement uncertainty, but recent attempts have been described to apply bottom up methods to this aspect [10].

New methods have been devised to estimate the uncertainty of measurements, which is caused by all of the sources, including procedures used for primary sampling. Uncertainty of measurement (u_c) can be estimated by summing contributions from the four types of error in the methods of measurement. These include two random components (sampling precision and analytical precision) and two systematic components (sampling bias and analytical bias). The expanded uncertainty (U) is estimated in this case by the use of a coverage factor of two, to give approximately 95% confidence for a normal distribution.

Well established methods are available to estimate three of these four components (Table 1). Analytical precision can effectively be estimated most cost-effectively using duplicate chemical analyses. Sampling precision can be estimated similarly by taking duplicated samples at points separated in space (or time) by a distance reflecting the possible ambiguity in the sampling protocol [11]. Analytical bias can be estimated using certified reference materials that have a chemical composition that is well matched to the samples.

The estimation of sampling bias is potentially much more problematic, but two methods have been described. One method requires the use of a Reference Sampling Target (RST), which is the sampling equivalent of a reference material for the estimation of bias. The RST can either be created synthetically to have a known concentration of analyte [12], or it can be a routine sampling target selected for the purpose [13]. The accepted or certified value of concentration (and its uncertainty) can either be taken from the known concentration of analyte added, for the synthetic RST [12], or established by the consensus of an inter-organisational sampling trial (IOST) [13]. The accepted value can also include a specification of the spatial distribution of the analyte, and its uncertainty [14].

The second method of estimating the contribution of sampling bias to the uncertainty of the measurement, is to apply more than one sampling protocol to a sampling target, ideally with more than one sampler (i.e. the person who takes the sample). One extreme example of this approach is the inter-organisational sampling trial in which eight or more samplers take samples from the same sampling target for the same specified purpose. If all participants use the same protocol it constitutes a Collaborative Trial in Sampling (CTS) [15], but if they all select their own protocols, based on their professional judgement, it constitutes a Sampling Proficiency Test (SPT) [13]. The variability of the estimates of analyte concentration between the participants can then be used to estimate the uncertainty of the measurement procedure as a whole, as applied to a particular site. Any bias caused by the sampling of any participant then becomes part of the random error across the whole sampling trial, and hence is automatically included in the uncertainty [3].

There are therefore four methods that can be identified for the estimation of the overall uncertainty of measurement (Table 2). None of these methods use the traditional "bottom-up" approach of adding all of the separate components of uncertainty together [5]. They rely on a fundamentally "top-down" approach that aims to get the most reliable estimate of the uncertainty overall, without necessarily identifying the contributions from all of the possible sources [16, 7].

These four methods estimate the uncertainty with increasing rigour, but at increasing cost. Method 1 is the least expensive, but it does not include an estimate of sampling bias, although this can be added by the independent use of an RST. Separation of the main sources of uncertainty requires the use of analysis of variance, usually of the robust type, to allow for non-normal frequency distributions. Detailed description of these methods is given elsewhere [3]. These methods have the advantage of estimating the actual uncertainty for a particular investigation, and should in that way be more realistic than estimates made by bottom-up methods. These later methods will need to use generalised values for the component variances, and cannot easily reflect the special contribution to variance at each site, such as those made by the local levels of analyte heterogeneity. Top-down methods are therefore particularly appropriate for estimation of uncertainty from sampling especially for variable matrices such as those found in environmental materials. Further research will be needed to investigate the relative merits of top-down and bottom-up methods for this purpose.

Table 1 The four types of errors in methods that contribute to the uncertainty of measurements, and examples of how they might be estimated

Error type → Process ↓	Random (*precision*) Estimate using:	Systematic (*bias*) Estimate using:
Analysis	Duplicate analyses	Certified reference materials
Sampling	Duplicate samples	RST, IOST

RST, reference sampling target; IOST, inter-organisational sampling trial

Table 2 Four methods for estimating uncertainty in measurements (including that from sampling)

No	Method	Samplers	Protocols	Uncertainty components estimated			
				Analytical precision	Analytical bias	Sampling precision	Sampling bias
1	Duplicates+CRMs	Single	Single	Y	Y	Y	No
2	Protocols+CRMs	Single	Multiple	Y	Y	Between protocols	
3	CTS+CRMs	Multiple	Single	Y	Y	Between samplers	
4	SPT (+CRMs optional)	Multiple	Multiple	Y	Y	Between protocols +between samplers	

CTS, collaborative trial in sampling; SPT, sampling proficiency test; CRM, certified reference material

Applications of uncertainty estimation that includes contribution from sampling

Estimation of uncertainty in the measurement of lead in top soils has been reported using duplicate samples [3] (i.e. Method 1 in Table 2). The 1.8 hectares site in Derbyshire UK, was contaminated by a lead smelter operating in the 14–16th century, and demonstrates the general principle.

A regular grid of 40 sampling points at 20 m spacing was applied initially with single samples. This grid was repeated using 5-fold composite samples (i.e. 5 increments taken within 1 m^2 around each sampling point) in order to investigate how composite sampling would affect the measurement uncertainty. The duplicate samples were taken at around 20% of the sampling points (Table 3), at a distance of 2 m away from the original sample point. This distance represents the spatial uncertainty caused by the method of surveying employed (i.e. measuring tape) on this undulating site. Duplicated analytical measurements were then taken on both of the sample duplicates in a balanced design, and the three components of the variance separated using robust analysis of variance (ANOVA) [17], according to the model:

$$s^2_{total} = s^2_{geochem} + s^2_{samp} + s^2_{anal} \quad (1)$$

The three components of the total variance are the analytical variance (s^2_{anal}), the sampling variance (s^2_{samp}) and the geochemical variance ($s^2_{geochem}$). The measurement variance can be considered as the sum of the sampling and analytical variance:

$$s^2_{meas} = s^2_{samp} + s^2_{anal} \quad (2)$$

The expanded uncertainty (U) can be estimated as $2s_{meas}$, using a coverage factor of 2 for 95% confidence.

The use of 5-fold composite samples reduces the overall expanded measurement uncertainty (U) by a factor of two (3742 to 1881 μg g^{-1}). This is not significantly different from the reduction of 2.2 (i.e. $\sqrt{5}$), predicted by the theory that sample mass is inversely proportional to sample variance [1]. In this case the 5-fold increase in the sample mass would be predicted to reduce the variance by a factor of 5, and hence the uncertainty by the square root of 5.

Acceptable levels of uncertainty (fitness-for-purpose)

Once estimates of the uncertainty of measurements are known, it is possible to judge whether that level of uncertainty is acceptable for a particular stated purpose. This can be used to judge whether measurements (rather than all of the measurement procedures themselves) are "fit-for-purpose" (FFP), where fitness for purpose is defined as "The property of data produced by a measurement process that enables a user of the data to make technically correct decisions for a stated purpose" [18].

Three basic types of FFP criteria have been suggested, in somewhat different contexts. The first, and widely accepted criterion, is based on the relative precision of measurement method, usually specified within an Analytical Quality Control Scheme. A typically criterion is that the relative analytical precision should be better than 10% (at 95% confidence). AQC is normally used to check the measurement process and to check that this process step is in statistical control and comparable with the performance pertaining at the time of validation. This target performance is often set however, so as to enable users of the data to make technically correct decisions. It can therefore be considered as crude type of FFP criterion. The main problem with this approach is that this criterion is set by the laboratory, often without reference to the specific purpose for which the customer will use the data. It could be, for example, that a precision of 30% would be quite good enough for some of the user's purposes.

The second FFP Criterion that has been suggested, is that the uncertainty of the measurement (including that from the sampling) should not contribute more than 20% of the total variance for the analyte across all of the samples in a particular survey [11]. The relative contributions to the total variance can be usefully represented using pie charts (Fig. 1). In the case study in Derbyshire, it

Table 3 Estimates of uncertainty made at a site in Derbyshire using Method 1 (Table 2). The use of composite samples reduces the expanded measurement uncertainty (U) by nearly a factor of 2

Sampling design	Points duplicated	Mean Pb ($\bar{x}$) robust (μg g^{-1})	s_{total} robust (μg g^{-1})	s_{meas} robust (μg g^{-1})	s^2_{meas}/ s^2_{total} (%)	U=2s_{meas} (μg g^{-1})	U=200s_{meas}/$\bar{x}$ (%)
Regular grid, single sample	7	7516	8185	1871	5.2	3742	49.8
Regular grid, comp. samples	9	6093	5600	940.5	2.7	1881	30.9

Despite the high level of relative uncertainty using single samples (U%=50%), the variance caused by the measurements only contributes 5.2% to the overall total variance of Pb measurements across the site (s^2_{meas}/s^2_{total}%). This is well within the second FFP criterion of 20% (Fig. 1)

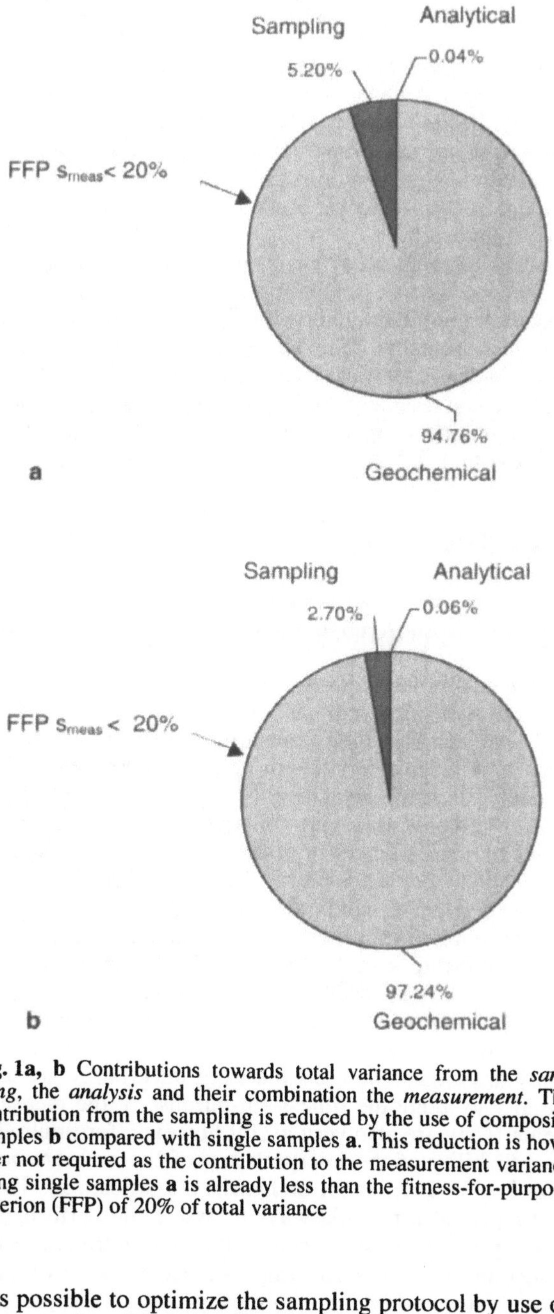

Fig. 1a, b Contributions towards total variance from the *sampling*, the *analysis* and their combination the *measurement*. The contribution from the sampling is reduced by the use of composite samples **b** compared with single samples **a**. This reduction is however not required as the contribution to the measurement variance using single samples **a** is already less than the fitness-for-purpose criterion (FFP) of 20% of total variance

was possible to optimize the sampling protocol by use of this criterion. The use of 5-fold composite samples in place of single non-composite samples was shown to reduce the variance contributed by the sampling protocol by a factor of approximately $\sqrt{2}$. However, this improvement was shown to be unnecessary as the variance contributed by the measurements overall (5.2%) was already well below the limit of 20% of the total variance, even when using the non-composted samples [3]. This judgement that the measurements were fit-for-purpose was

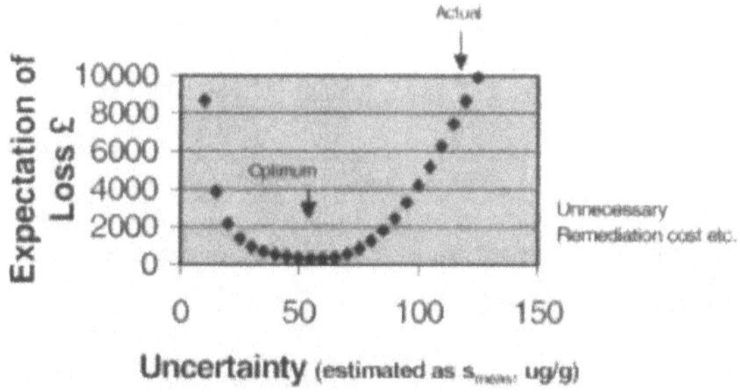

Fig. 2 Optimisation of measurement uncertainty against cost, demonstrated for a metal contaminated site in West London. It shows the economic loss function with a clear minimum cost at the optimal uncertainty (estimated by s_{meas}), which is 30× lower than the expectation of loss at the actual uncertainty of the measurements

reached despite the fact that the value of the measurement uncertainty was 50% (relative to concentration value, at 95% confidence), using the protocol with single samples. This level of uncertainty is much higher than is often considered acceptable. These samples are not "representative" in the usually accepted sense of the word, but they have been shown to be "appropriate", in providing measurements that are fit-for-purpose.

The third FFP Criterion that has been suggested incorporates a balance between uncertainty and financial loss [4]. The optimal value of uncertainty is identified as that which incurs the minimum value for the expectation of loss. This loss includes both the cost of making the measurement and also the loss that may arise due to incorrect decisions made on the basis of the uncertain measurements. In an initial application of these ideas to a site investigation in West London [19], the actual measurement uncertainty of 110 $\mu g\ g^{-1}$ resulted in an expectation of financial lost of around £9000 per sampling point (Fig. 2). When this uncertainty was optimised to a value of 55 $\mu g\ g^{-1}$ the expectation of loss was reduced by a factor of thirty to around £300. A second stage of the optimisation allocates expenditure in an optimal way between the sampling and the chemical analysis, based on their respective contributions to the overall uncertainty. A recent application of this approach to the sampling of food has shown that a substantial reallocation of expenditure to the sampling process (+300%) gave a substantial reduction in the overall uncertainty (–31%) and a consequent saving of £428 per batch [20].

Implications of uncertainty for interpretation of measurements

One of the main advantages of reporting realistic estimates of uncertainty together with measurements of concentration, is that end-users of the analyses can consider

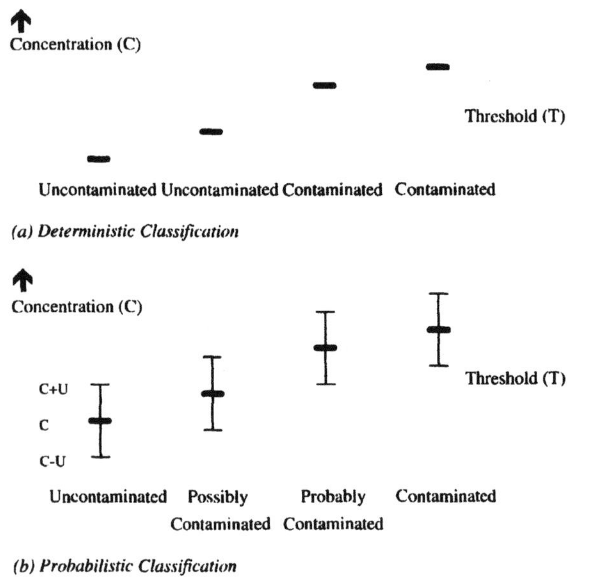

Fig. 3a, b Comparison between **a** deterministic, and **b** probabilistic classification of contaminated land, to show the effect of using estimates of uncertainty to improve the interpretation of measurements (derived from general case described previously [7])

the implications of the uncertainty [7]. One example that illustrates this is in the classification of contaminated land [3]. The traditional deterministic approach is to compare the measured concentration values with an appropriate regulatory threshold value (Fig. 3a). Any sampling point that has a reported concentration value below the threshold is classified as uncontaminated, and those above as contaminated. This approach ignores the presence of uncertainty, but once it is known a probabilistic approach can be taken (Fig. 3b). If the measured concentration is below the threshold but the uncertainty interval extends above it, the point is classified as "possibly contaminated" rather than "uncontaminated". Similarly if the measured concentration is above the threshold but the uncertainty interval extends below it, the point is classified as "probably contaminated" rather than "contaminated".

In one application of this probabilistic approach to a disused landfill site in West London, the effect of the uncertainty was to totally change the interpretation of the extent of the lead contamination at the site [3]. From the 100 samples taken from the site in a regular grid pattern, the deterministic interpretation showed only eight samples, scattered across the site, to be over the appropriate threshold value (500 μg Pb g^{-1} soil). However, the uncertainty of the measurements at the site was estimated as 83.6%, using Method 1 (Table 2). This large value is due primarily to the high degree of small-scale heterogeneity of lead distribution at the site. After taking this into account 91% of the sampling point were classified as either possibly, probably and definitely contaminated (at 95% confidence). Only 9% of the site was classified as uncontaminated, which is only marginally greater than the proportion of that which would be expected by chance at this confidence interval. The interpretation of the site has changed therefore, from "basically uncontaminated with a few patches of contamination" using the deterministic method, to "probably all contaminated" with the probabilistic approach.

More sophisticated interpretations of contaminated land use a risk assessment approach, but this can also be made more reliable by allowing for the uncertainty in the raw measurements. The uncertainties in the measurements, both in the investigation and in the construction of the risk assessment model, can be propagated through the calculation to give more realistic uncertainty values for the calculated exposure or risk.

Conclusions

Appropriate sampling is potentially much more reliable and more cost-effective than representative sampling. The assumption that a "correct" protocol will give representative samples, does not give either the rationale for estimation of sampling bias, or the flexibility to vary the quality of the sampling depending on the proposed objective. An "appropriate" protocol can be selected to give an acceptable level of uncertainty that can allow for consideration of financial constraints (e.g. the potential financial consequences of errors, or logistical constraints such as local conditions, time limitations). All "appropriate" protocols do require the estimation of uncertainty to be incorporated into their design. The simplest method, using duplicate samples at a small proportion of sampling points, is adequate for most purposes. More elaborate methods to estimate uncertainty will only be required where the consequences of unsuspected uncertainty are large. These estimates of uncertainty are used initially to judge the fitness-for-purpose of the measurements. They are also very useful however, for improving the reliability of the interpretation of the measurements (e.g. in risk assessment or hazard classification). These techniques also provide ways to assess the performance of sampling protocols (using collaborative trials in sampling) and to assess and improve performance of samplers (using sampling proficiency tests). Sampling is never perfect, it is better therefore to measure the uncertainty that an "appropriate" sampling protocol generates, rather than to assume the perfect application of a "correct" protocol.

References

1. Gy P (1979) Sampling of particulate materials – theory and practice. Elsevier, Amsterdam
2. Thompson M (1995) Analyst 120:117N
3. Ramsey MH, Argyraki A (1997) Science of the Total Environment 198:243
4. Thompson M, Fearn T (1996) Analyst 121:275
5. ISO (1993) Guide to the expression of uncertainty in measurement. ISO, Geneva
6. ISO (1995) VIM: 1995 Vocabulary of metrology. Part 1. Basic and general terms (international). International Organisation for Standardisation (Geneva) [and published by the British Standards Institution (London) as PD6461:Part 1:1995, 59 pp]
7. CITAC (2000) Quantifying uncertainty in analytical measurement. Eurochem
8. Rios A, Valcarcel M (1998) Accred Qual Assur 3:14
9. BCR Report (1998), EUR 18405 EN Metrology in chemistry and biology: a practical approach. Office for Official Publications of the European Communities, Luxembourg
10. de Zorzi P, Belli M, Barbizzi S, Menegon S, Deliusa A (2002) Accred Qual Assur this issue
11. Ramsey MH, Thompson M, Hale M (1992). Journal of Geochemical Exploration 44:23
12. Ramsey MH, Squire S, Gardner MJ (1999). Analyst 124:1701
13. Argyraki A, Ramsey MH, Thompson M (1995) Analyst 120:2799
14. Squire S, Ramsey MH, Gardner MJ, Lister D (2000) Analyst 125:2026
15. Ramsey MH, Argyraki A, Thompson M (1995) Analyst 120:2309
16. Analytical Methods Committee (1995) Anayst 120:2303
17. Analytical Methods Committee (1989) Analyst 114:1699
18. Thompson M, Ramsey MH (1995) Analyst 120:261
19. Hulls J (1998) Optimising sampling and analytical strategies for assessing contaminated land. MSc thesis, Imperial College, London
20. Ramsey MH, Lyn JA, Wood R (2001) Analyst 126: 1777

Accred Qual Assur (2001) 6:368–371

John R. Cowles
Simon Daily
Stephen L.R. Ellison
William A. Hardcastle
Carole Williams

Experimental sensitivity analysis applied to sample preparation uncertainties: are ruggedness tests enough for measurement uncertainty estimates?

Abstract It has been suggested that typical ruggedness tests might lead directly to uncertainty estimates. This assertion is tested using simple experimental studies of uncertainties associated with sample grinding and oven-drying operations. The results are used to predict the outcome of typical ruggedness tests on the same systems. It is concluded that uncertainty estimation from ruggedness tests is appropriate only where a strong effect can be observed. Since current practice in ruggedness testing is predisposed to confirming insignificance, typical ruggedness tests are not likely to lead to reliable uncertainty estimates; instead, lack of statistical significance in ruggedness tests is better interpreted as reason to leave an effect out of the uncertainty budget. Only where the ruggedness study is modified in order to achieve statistically significant change is it useful for uncertainty estimation.

Keywords Measurement uncertainty · Ruggedness tests · Sample pre-treatment · Moisture · Grinding

J.R. Cowles · S. Daily · S.L.R. Ellison (✉)
W.A. Hardcastle · C. Williams
LGC (Teddington) Ltd., Teddington, England TW11 0LY, UK
e-mail: slre@lgc.co.uk
Tel.: +44–181–943 7000
Fax: +44–181–943 2767

Introduction

There is a general trend in modern analytical science towards providing quantitative estimates of the reliability of measurements. Analysts are increasingly coming under pressure from accreditation bodies to present such information in the form of a statement of measurement uncertainty conforming to International Organisation for Standardization (ISO) recommendations [1, 2].

The effects of pre-treatment operations have been identified by EURACHEM as important considerations in measurement uncertainty estimation [2]. The general approach to experimental measurement uncertainty estimation is to vary experimental conditions and estimate the sensitivity of the result to the conditions [3–5]. This is sometimes referred to as 'sensitivity analysis'. Its close relationship to ruggedness testing as described, for example, by Youden and Steiner [6] has suggested the possibility of using ruggedness tests as the basis for measurement uncertainty estimates [2–5]. Since ruggedness and related tests are likely to be the most detailed experimental information to hand in most chemical testing laboratories, due to the practical problems of resourcing larger studies, their practical utility and interpretation for uncertainty estimation is an important issue.

In this paper, two methods of pre-treatment, viz. oven drying, and grinding, are examined using a simple sensitivity analysis approach to estimating uncertainty contributions in sample pre-treatment. The substantial literature on the effects of varying drying conditions commonly shows that quantitation of the analyte is sensitive to the drying temperature employed and to the nature of both the analyte and its matrix; temperature uncertainties associated with drying accordingly provide an example of uncertainty estimation for a readily detectable effect. The literature on sample grinding and milling shows more variable results. Some authors have found no effect on the final result, others reported effects of varying magnitude. The milling study was therefore expected to test the utility of sensitivity analysis for a variable or ill-defined effect.

It is important to bear in mind that under normal circumstances, ruggedness tests are typically designed as screening experiments; restricted to very short checks, using a nominal level of a method control parameter (such as operating temperature or grinding time) and one or at most two alternate levels corresponding to the expected (or permitted) variation in the parameter. The studies here are significantly larger, in order to provide more information against which to assess the likely efficacy of routine ruggedness studies in uncertainty estimation.

Experimental

For the determination of moisture, two different feeds were analysed in triplicate using the oven-loss method specified in appropriate United Kingdom legislation [7]. A drying time of 3 h was used for all test portions. For the study of temperature dependence, three 5 g portions of each sample were dried separately at each temperature (that is, in separate drying runs), with the oven held within ±1°C of the target temperature for the whole of the heating period. The weight loss for each 5°C increment in temperature from 85°C to 115°C was determined for each sample.

In the grinding/milling experiments, a legislative pretreatment (grinding) method [7] was used. The method is intended to reduce materials to <1 mm particle size in three or fewer grinding cycles, which allows variation upward from an experimentally determined minimum cycle time. Approximately 1200 g of one feed was thoroughly mixed in a tumbler mixer overnight and divided into five equal parts, which were ground using cycle times of 8, 10, 12, 15 and 20 s, respectively. Ground samples were analysed for dimetridazole by extraction with dichloromethane and clean-up using a Sep-Pak silica cartridge (Waters Corporation), followed by reverse phase high performance liquid chromatograph (HPLC) with UV detection. The solvent employed was acetonitrile with ammonium acetate buffer.

Results and discussion

Case study 1 – Uncertainties from oven temperatures in moisture determination

Moisture content is usually determined using a calculation of the form:

$$l = \frac{m_w - m_d}{m_w}$$

where l is the fractional loss in mass (usually expressed as a percentage), m_w the mass before drying and m_d the mass after drying.

The variability of the method is controlled by restricting the drying temperature to a narrow range, typically ±1°C. This figure represents an uncertainty in temperature.

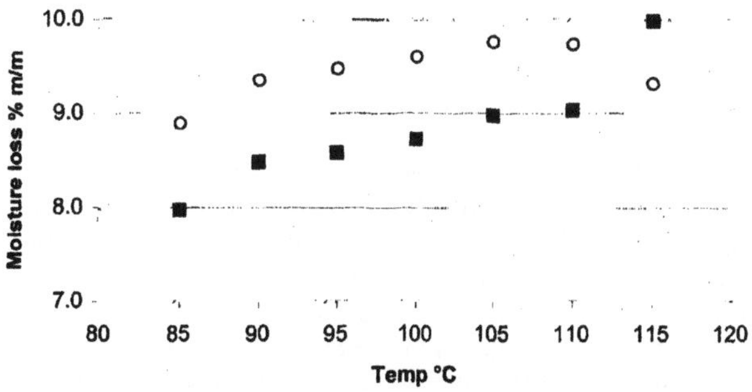

Fig. 1 Mean weight loss vs drying temperature. Weight loss with temperature for two samples (■ and ○ respectively)

ture. Results obtained by methods of this type are accordingly subject to an uncertainty component due to the allowable variation in temperature. In this study, a series of experiments was carried out on two samples of pelleted animal feed. The drying temperature was varied over a range around the target temperature of 100°C and the effect on the weight loss noted. The range chosen, 85 C to 115 C, is substantially larger than the permissible range, to permit both an investigation of the linearity or otherwise of the effect and a sound sensitivity analysis.

The results, which are typical for moisture determination, are shown in Fig. 1. Both curves are approximately linear in the range 90–105°C, but depart from linearity at the extremes. At 85°C, the mean weight loss is lower than might be expected from the trend at the higher temperatures. At higher extremes, factors such as progressive oxidation or thermal degradation lead to different directions of departure from linearity.

To estimate the uncertainty associated with weight loss using the *Guide to the Expression of Uncertainty in Measurement* (GUM) principles [1], the gradient $\partial l/\partial T$ at the nominal temperature (100°C) is multiplied by the temperature uncertainty u(T). Here, linear regression applied to the linear temperature range in each case produces a gradient $\partial l/\partial T$=0.030%m/m C^{-1} for sample 1 and $\partial l/\partial T$=0.019%m/m°C^{-1} for sample 2. The temperature uncertainty u(T) is 0.577°C, estimated from the permitted variation of ±1°C taken as the limits of a rectangular distribution. Calibration uncertainties in the thermometer used are under 0.1°C, so can be neglected by comparison. This gives $u(l)$=0.577×0.030=0.017%m/m for sample 1 and $u(l)$=0.577×0.019=0.010%m/m for sample 2. Comparing this with a repeatability estimate of 0.05%m/m obtained from the replicate data by analysis of variance (ANOVA), it is clear that the uncertainty estimates associated with temperature are just on the margin of practical significance compared to the repeatability estimate (assuming an uncertainty is practically significant if greater than a fifth of the largest component – here, the precision is the largest component so far found). Under these circumstances, therefore, there is some evidence that the uncertainty arising from a permitted temperature variation of 1°C could be practically significant compared to the repeatability precision. In the context of normal use, of course, these are both small uncertainties; for most practical purposes, therefore, the temperature-related uncertainty can be considered sufficiently small. Further, note that the study is carried out under the most precise conditions available; the uncertainties found will almost certainly prove negligible compared, for example, to between-run variation.

Returning to the principal aim of the study, it is useful to consider whether a typical ruggedness test, operating at one or both extremes of the permitted range, would have given comparable results in the present case. Given the largest calculated sensitivity of 0.030%m/m °C^{-1}, the expected variations in l across a 1 or 2°C range are 0.030%m/m and 0.060%m/m, respectively. Clearly, neither would reliably lead to statistically significant effects with the present repeatability precision unless a prohibitively large number of replicates were undertaken in the ruggedness test; uncertainty estimates would accordingly be extremely variable. For small to modest effects, then, uncertainty estimation from sensitivity experiments requires substantially wider variation than the 'expected range'. Further, if the 'expected range' is based on control limits intended to render an effect insignificant – as in most standard methods – it is generally to be expected that the change in influence quantity will not provide useful uncertainty estimates.

Case study 2 – Milling and particle size effects on HPLC determination of dimetridazole

Another common requirement is for the determination of analytes in samples of agglomerated materials such as soils or animal feeds. A common preparative method is to grind or mill the material so that it passes through a sieve of specified aperture. The grinding time in such cases is not generally specified, leaving open the possibility of variation in grinding time and particle size. Both constitute potential sources of uncertainty. Longer grinding times may affect the results through greater production of fines (that is, a particle size effect) or by thermal degradation or loss of the analyte. Grinding/milling therefore constitutes a possible source of uncertainty. In this study, a series of experiments with different grinding times was carried out on a sample of medicated pelleted animal feed. A rough particle size distribution was also estimated using different sieve sizes. The feed contained approximately 100 mg kg^{-1} of dimetridazole (a coccidiostat) and the experiments were designed to assess the effect of grinding on the subsequent determination of this compound.

A plot of the variation of observed concentration of dimetridazole with grinding time is shown at Fig. 2. Although there appears to be a decrease in the concentration at longer grinding times, the plot does not suggest strong linearity, and ANOVA does not show this effect to be statistically significant (P = 0.45, CL = 95%). It remains possible to provide an estimate of uncertainty using linear regression and the first-order GUM expression [1]; in this case, we obtain a linear regression result of y=-0.45t+112.41, where y is analyte concentration in mg kg^{-1}, and t the cycle time (in seconds). For an uncertainty u(t) of 1 s in grinding time (based on the practical difficulties of controlling grinding time more closely), the uncertainty in analyte concentration would be ±0.45 mg kg^{-1}. In comparison, the precision of the analytical method for dimetridazole, at the analyte level encountered for the whole sample (ca. 105 mg kg^{-1}), was 3.9 mg kg^{-1}. Reassuringly, this very rough uncertainty estimate confirms the insignificance implied by the ANOVA result.

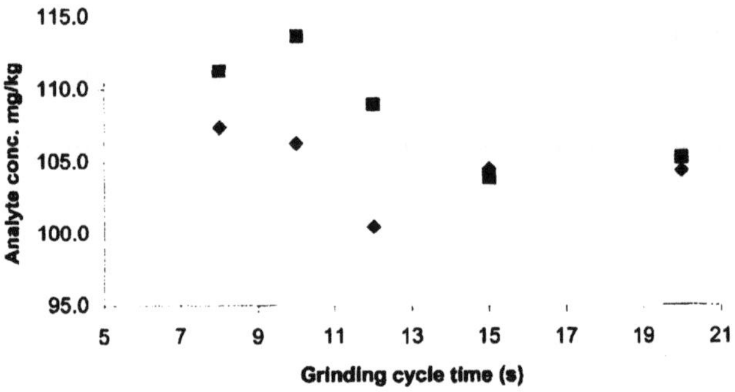

Fig. 2 Effect of grinding time on analyte concentration. Duplicate analyses of samples ground for different grinding cycle times (■ and ◆ are replicates 1 and 2, respectively)

Considering the possible outcome of a ruggedness test aimed at establishing the effect of grinding time across a range of, for example, 3–5 s about the nominal time of 10 s (much smaller variation would be impractical), again we find that such a test would almost certainly fail to find a significant effect. In this instance, however, it would have been entirely correct to ignore the effect in comparison with observed precision.

Conclusions

The experiments were designed with the aim of using sensitivity values, generated by linear regression, to obtain estimates of the contribution to overall measurement uncertainty from two methods of sample pre-treatment. By comparison, the minimal range typical of current practice in ruggedness testing would not be expected to give useful or reliable uncertainty estimates in either case. This is consistent with recent studies on derivatisation effects [8] which show that as modelling coefficients (such as gradients) become statistically insignificant, uncertainty estimates become progressively more unreliable and can be misleadingly large, even though the average remains negligible. However, the lack of statistical significance predicted for such minimal studies was generally consistent with the finding that the uncertainties were small compared to repeatability precision.

This has important implications for the use of data from simple ruggedness studies in uncertainty estimation. Given a typical ruggedness study, properly designed to confirm insignificance of an expected range for an influence quantity, the sensitivity estimates will generally be too unreliable for useful uncertainty estimation even though the study is a valid check on the potential influence quantities' effects. It follows that typical ruggedness studies are not generally appropriate sources of data for reliable uncertainty estimation. However, where no significant effect is found in such a study, sensitivity coefficients obtained from a study redesigned to assure a significant change (for example by increasing the influence quantity range well beyond that expected) will generally only confirm practical insignificance of the effect. It is clearly more sensible to use a recorded lack of statistical significance in typical ruggedness tests as justification for omitting an effect from the measurement model and associated uncertainty budget (which is, in fact, the traditional statistical view), than to attempt to construct an unreliable estimate for a practically insignificant effect.

Acknowledgements Production of this paper was supported under contract with the Department of Trade and Industry as part of the National Measurement System Valid Analytical Measurement Programme.

References

1. ISO- GUM (1993) Guide to the expression of uncertainty in measurement. ISO, Geneva, Switzerland; ISBN 92–67–10188–9
2. EURACHEM (1995) Quantifying uncertainty in analytical measurement. EURACHEM, London; ISBN 0–948926–08–2. Second edition now available at: http://www.vtt.fi/ket/eurachem/publications.htm
3. Ellison SLR, Williams A (1998) Accred Qual Assur 3: 6–10
4. Barwick VJ, Ellison SLR (2000) Accred Qual Assur 5: 47–53
5. Barwick VJ, Ellison SLR, Rafferty MJQ, Gill RS (2000) Accred Qual Assur 5: 104–113
6. Youden WJ, Steiner EH (1975) Statistical manual of the AOAC. Association of Official Analytical Chemists International (AOAC), Arlington, Va., USA
7. The Feeding Stuffs (Sampling and Analysis) Regulations (1999). Statutory Instrument No 1633. Her Majesty's Stationary Office (HMSO), London
8. Ellison SLR, Burns M, Holcombe DG (2001) Analyst 126: 199–210

Accred Qual Assur (1998) 3:462–467

Adriaan M.H. van der Veen
A.J.M. Broos
Anton Alink

Relationship between performance characteristics obtained from an interlaboratory study programme and combined measurement uncertainty: a case study

A.M.H. van der Veen (✉)
A.J.M. Broos
A. Alink
Nederlands Meetinstituut,
P.O. Box 654,
2600 AR Delft,
The Netherlands
e-mail: AvanderVeen@NMi.nl
Tel.: +31-15-2691733
Fax: +31-15-2612971

Abstract In the interlaboratory study programme "ILS Coal Characterisation", eight interlaboratory studies were organised based on the ISO standards for coal analysis. The use of blind samples in each round allows comparability of measurement results between rounds to be assessed. Based on the results, it could be demonstrated that the vast majority of the measurement results of the laboratories were traceable to results obtained in previous rounds of this programme. The hypothesis has been formulated that the combined standard uncertainty obtained from an interlaboratory study is equal to the reproducibility standard deviation. Whether the reproducibility can be used as the basis for the certification depends on whether the interlaboratory study includes all effects to be taken into account for establishing an uncertainty statement.

Key words Interlaboratory study · Traceability · Comparability · Reference material

Introduction

Over a period of 3 years, an interlaboratory study programme was organised that aimed to supply the coal community with a range of suitable reference materials that can be used as measurement standards in a wide variety of experiments and common analyses. The programme was entitled "ILS Coal Characterisation" and consisted of eight interlaboratory studies. Apart from the objective of supplying measurement standards, it was also aimed to investigate several metrological aspects, such as establishing traceability of measurement results throughout the programme. Summarising, the interlaboratory study programme aimed to [1]:

1. Establish a series of well-characterised coal samples (reference materials) in order to support coal research
2. Investigate the statistical parameters of coal analysis
3. Study the influencing factors such as sample preparation, subsampling and statistics on measurement results.

The interlaboratory studies were evaluated and reported during the programme [2–9]. After completion of the programme, a report was published that covers the evaluation of the interlaboratory study programme as a whole [1]. This second evaluation cycle aimed to establish links between interlaboratory studies and thus establish performance characteristics for selected methods common to coal analyses throughout the programme. These performance characteristics are believed to be better established than those from a single interlaboratory study obtained with the procedure given in ISO 5725:1994, parts 1 and 2 [10, 11].

The main question remaining after the validation of the data from the complete programme is how the values obtained for repeatability standard deviation and reproducibility standard deviation compare to the (combined) measurement uncertainty at that level. In metrology, the VIM [12] and GUM [13] are the basis for evaluating uncertainty in measurement. However, the implementation of the principles of the GUM is far from straightforward for matrix materials, especially when parameters are defined by the measurement process. As a result, an interpretation of uncertainty analysis of this kind of parameter/matrix combination is required that explains the experimental results and is in agreement with the basic principles of the GUM [13]. A problem that was already addressed in a paper by Van der Veen and Alink [14] is that it is impossible to quantify several sources of uncertainty when dealing with matrix materials.

Set-up of the programme

The objectives of the programme have been described in the previous section. With respect to the acceptance of laboratories as participants, no specific requirements were set other than that these laboratories should be involved in the analysis of coal in support of trade. This requirement implies that the laboratories are involved in one-to-one comparisons between coal buyer and coal seller. The implementation of a quality assurance system (QAS) was, however, a requirement. An accreditation was not asked for.

For the characterisation of coal there are two series of written standards: ASTM and ISO. In this interlaboratory study programme the ISO standards were requested. The laboratories were allowed to use their own methods if the results of these methods are comparable to those obtained with the ISO method. It was the responsibility of the laboratory to verify whether its method is comparable to the ISO method.

Several methods, such as the determination of the ash content, define the parameter: ash is the result of a chemical conversion of coal. Its formation (and as a result, the ash content) highly depends on the conditions under which the coal is combusted. This fact has some consequences. The first consequence is that it is generally not possible to determine the parameter with independent methods. As a result, the best realisation of the parameter depends on how closely the written standard is followed by the participants and how much freedom is still left in the measurement method. Traceability of the parameter is limited to the measurement results being traceable to the written standard.

The evaluation protocol was merely based on a combination of ISO 5725-1:1994 [10], ISO 5725-2:1994 [11], ISO Guide 43[1] [15], and ISO Guide 35:1989 [16].

[1] During the interlaboratory study programme the most recent draft of this ISO Guide was used

Outliers and/or stragglers [11] were identified by computing a Z-score [15] based on the mean and the standard deviation of the laboratory averages. The criterion was that this "Z" should not exceed a value of 2. The criterion was developed based on requirements set by all parties involved in coal production, trade, and consumption. For all samples involved in the programme, the performance characteristics were computed after removal of stragglers and outliers. A database of values of grand mean, repeatability standard deviation, and reproducibility standard deviation resulted.

In each interlaboratory study, a blind sample was used, except in round I [2]. The link between the results of this interlaboratory study and the results of the other rounds in the programme was established by using the materials of ILS Coal Characterisation I in later rounds [1]. Figure 1 shows the principle of establishing these traceability links in an interlaboratory study programme. The use of a blind sample enables the evaluation of whether the results of a round are comparable to those of other interlaboratory studies due to the fact that each laboratory was requested to perform the measurement of all samples of the suite as independent measurements under repeatability conditions. This way of implementing comparability also enables the assessment of traceability of measurement results to the written standard [12]. All laboratories involved use internationally accepted certified reference materials as a part of their QAS. So, from that point a traceability link is established between these reference materials and the results of the interlaboratory study programme.

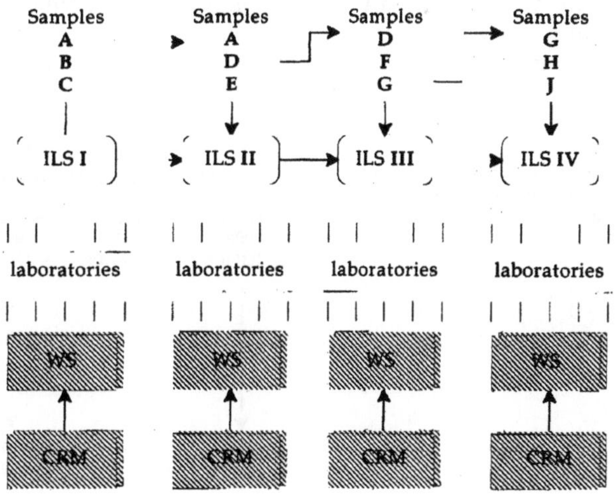

Fig. 1 Framework for assessing traceability of results in the interlaboratory study programme

Results

Based on the evaluation of the results of the blind samples, it may be concluded that, generally speaking, there is a good agreement between results obtained in two interlaboratory studies on a single batch of samples [1, 3–9]. A good agreement was also obtained when comparing the results of two different batches of samples, even if one batch was prepared to analysis grade (<200 μm top size), whereas the other was prepared to 3 or 10 mm top size.

These results make it possible to establish fairly homogeneous performance characteristics for the coal analyses involved in the programme [1]. They also lead to a hypothesis that needs further investigation: the values for the repeatability and reproducibility standard deviations are estimators for the expectation value of the combined measurement uncertainty.

This hypothesis is supported by several facts:

1. If in two independent interlaboratory studies the same values for the grand mean, the repeatability, and the reproducibility are obtained, then it may be assumed that these values are characteristic of the variability of the data that can be obtained in such an exercise.
2. Linking the results, grouped per parameter and per interlaboratory study, leads to performance characteristics that show a regular behaviour that can be described by a mathematical model.
3. An attempt to determine the combined measurement uncertainty of each step separately would lead to a serious overestimation of the combined measurement uncertainty of the complete measurement process.

The third fact needs some explanation. As indicated, in some interlaboratory studies different batches of the same coal were used. These batches differed not only in preparation route, but they also differed in grain size. Usually, sample preparation and extra subsampling steps were required to obtain suitable material for performing the measurements. The results of rounds III and IV show that no significant difference is found between the samples of analysis grade and of samples that needed further treatment.

Figure 2 provides an outline of the set-up in rounds I and III on the Göttelborn coal. In ILS Coal Characterisation III, samples of 10 mm top size and about 1 kg were distributed, whereas in round I samples of analysis grade were distributed. In Fig. 1 the ash analysis (ISO 1171) has been taken as an example, but any other coal parameter is determined in the same manner. The laboratories are denoted by A..D in ILS I, and A'..D' in ILS III, denoting that there is no relationship between A and A' etc. The measurement chains in round III are more different than those in ILS Coal Characterisation I, which might lead to greater differences in measurement uncertainty.

It was expected that including aspects of sample preparation and subsampling (other than subsampling of a sample of analysis grade) would result in an increase in both the repeatability and reproducibility standard deviations. As already stated, this was not the case. This implies that the combined measurement uncertainty for both groups of measurement chains is expected to show no significant difference either.

This observation seems to clearly contradict the commonly accepted "rule" that the preparation and subsampling steps contribute considerably to the combined measurement uncertainty. When assessing the quality of coal sampling, subsampling, and sample preparation, usually the ash content is used [17, 18].

Fig. 2 Outline of measurement chains within laboratories for the determination of ash

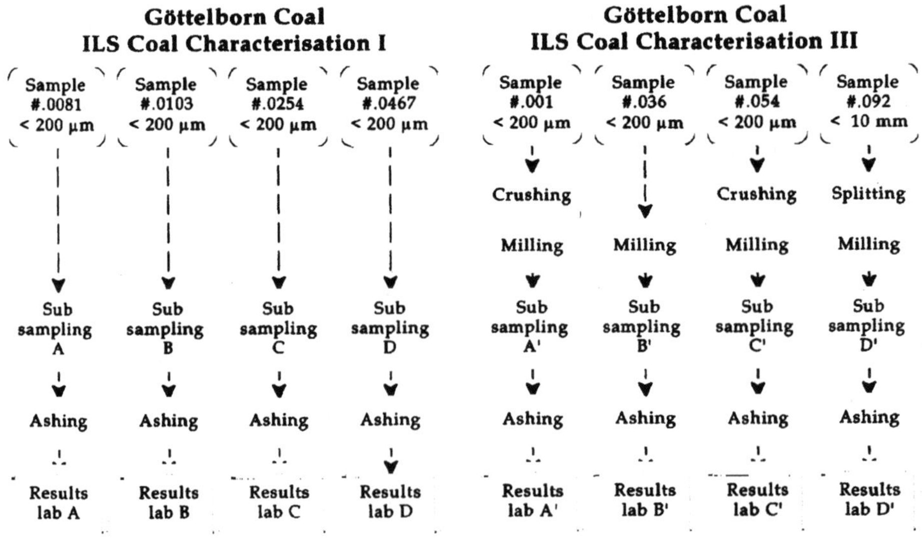

Uncertainty analysis

Starting with the working hypothesis, an uncertainty analysis can be carried out. The basic problem with the approach of the "Guide to the expression of uncertainty in measurement" (GUM) is that many sources of uncertainty are difficult, not to say impossible, to quantify in evaluating measurement results from matrix materials. There must also be a relationship between the performance characteristics obtained from the interlaboratory study programme and the combined standard uncertainty that can be obtained by applying the GUM directly.

In an interlaboratory study programme the measurement chain starts with the arrival of one or more samples to be analysed. These samples may or may not undergo further treatment prior to the measurement. The principles of this part of the measurement chain, as well as the role that can be played by (certified) reference materials, have been the subject of a previous paper [19]. In the interlaboratory study programme the role of the (C)RM has been taken over by the blind sample. The structure of the interlaboratory study programme met the requirements set by ISO Guide 35 [16], so that in principle the blind samples would be suitable to serve as (at least) reference materials.

The basic expression for the combined measurement uncertainty after splitting up the measurement chain reads as follows

$$u^2_{\text{combined}} = u^2_{\text{crushing}} + u^2_{\text{subdividing}} + u^2_{\text{measurement}} \qquad (1)$$

where it should be realised that each of the terms on the right-hand side of this equation consists of one or more contributions from various sources. This approach is in agreement with the rule in statistics that variances of parts of a process can be added in order to obtain the total variance of the whole process [20–23].

The dominant factor in terms of uncertainty budgets is the heterogeneity of the material. Unfortunately, the heterogeneity of the material affects all terms on the right-hand side of Eq. 1, and as a result these contributions are heavily correlated. A crushing step for instance increases the heterogeneity on the level of particles, but decreases heterogeneity on the level of, say, a few grams of powder. So, from the point of view of evaluating uncertainty, it is no use to make an attempt to quantify the contribution of heterogeneity. A better approach is to select the opposite way, starting with investigating a measurement chain and – by experiment – breaking it up into smaller parts. The measurement will, however, always be part of the uncertainty evaluation, as was shown in a previous paper [14].

During the processing of the material in a measurement chain, the critical property (say, for instance, ash content) undergoes (from a statistical perspective) changes. In principle, there are two changes: the expectation value μ changes, or the variance σ^2 changes. Both changes can happen simultaneously. Procedures such as milling may well change the probability density function of the content on the level of particles. Changes in this distribution function are sometimes wanted (reduction of the combined measurement uncertainty by increasing the total number of particles (crushing/milling), sometimes unwanted. An example of the latter is the loss of volatiles and moisture during milling of coal.

The evaluation model to be developed should therefore (1) avoid "double counting" of sources of uncertainty and (2) comply with the additivity rule of uncertainties as expressed in Eq. 1. This has been done by working with a reference term, i.e. the uncertainty is expressed in terms of the uncertainty of the measurement, followed by several correction terms that may account for extra budgets due to other steps in the measurement chain.

The statistical fundamentals read as follows. Let the random variable X denote the content of the critical component (in this example: ash content). The expectation of X is given by

$$E(X) = \mu \qquad (2)$$

If, during the process represented by the measurement chain, changes in this expectation take place, then it is said that the method is biased. For the determination of the ash content, it is very hard to find out whether the measurement method is biased, as the parameter is determined by the method. Ash is as such not present in coal; the precursor of ash is the mineral matter in coal. During combustion, this mineral matter is converted into ash, a chemical process. The ash composition is a function of the temperature at which the ash formation takes place. As a result, the ash content (expressed as weight-% of the coal on a dry basis) is also a function of the temperature.

The requested method (ISO 1171) requires a constant temperature of 815 °C. Insufficient control of this temperature, or a deviation of the sample temperature in the oven may lead to a change in expectation value, and thus in a bias of the measurement method. A convenient way of expressing the expectation value of the critical content could be

$$\mu = E(X) + E(B_{\text{crushing}}) + E(B_{\text{subsampling}}) + E(B_{\text{measurement}}) \qquad (3)$$

where B denotes the bias of the given step. Equation 3 expresses the expectation value in a sum of random variables, where the first term denotes the expectation value of the method, followed by several correction terms. In the ideal case each of these terms has the expectation value 0 ($\Rightarrow$ expected bias = 0). Now a match must be found between (3) and (1). The expression for the variance of μ reads as

$$\sigma^2 = \mathrm{Var}(X) + \mathrm{Var}(B_{\mathrm{crushing}}) + \mathrm{Var}(B_{\mathrm{subsampling}}) + \mathrm{Var}(B_{\mathrm{measurement}}) \quad (4)$$

where Var() denotes variance. Equation 4 provides a mathematical model for expressing the variance of the measurement in terms of the variance of the measurand X and the variances of the bias terms. This equation is only valid if the biases of the steps involved in the measurement chain are not correlated. Otherwise, covariance terms should be introduced [13, 23] in Eq. 4. It is unlikely that the bias terms are truly uncorrelated, as a bias is defined by a systematic difference between the expectation value of the measurand before and after a specific treatment. However, it is always possible to modify Eq. 4 in such a way that it can be expressed as a set of independent variables.

However, even without assuming that the bias terms are uncorrelated, Eq. 4 provides an explanation for the insignificant difference in the performance characteristics in the interlaboratory studies as shown in Fig. 1. As all laboratories maintain the operational conditions in steps such as crushing, milling, subdividing and analysis by means of a QAS, this will eventually result in comparable measurement results. Maintenance of the operation conditions during sample preparation and analysis steps will also minimise the variance of the biases associated with these steps in the measurement chain. That is, the terms $\mathrm{Var}(B_i)$ in Eq. 4 are minimised by a detailed description of the processing of the sampled material. If the QAS is successful, it may be expected that the contributions of the $\mathrm{Var}(B_i)$ terms in Eq. 4 will be small in comparison with the value of the term $\mathrm{Var}(X)$.

There are still two terms to be interpreted: $E(X)$ and $\mathrm{Var}(X)$. It is well known that the concept of a true value is not very useful. The value $E(X)$ is the expectation of the measurand, given the matrix and given the measurement method. The same holds for $\mathrm{Var}(X)$. Both parameters account for the performance characteristics of the test method involved. So, the fundament of the model given in Eq. 3 complies with the GUM. Moreover, it does not contain parameters that are inaccessible in practice. The concept of expectations also complies with the GUM, as it forms the basis for statistics. The concept of uncertainty is very closely related to the standard deviation [13].

Similarly, the QAS will also aim to reduce $E(B_{\mathrm{crushing}})$ and $E(B_{\mathrm{subdividing}})$ (and other bias terms) to values close to zero. Likewise, the values for $\mathrm{Var}(B_{\mathrm{crushing}})$ and $\mathrm{Var}(B_{\mathrm{subdividing}})$ will be minimised by maintaining the procedure as well as is feasible. In a separate paper [14] the determination of these terms has been discussed in more detail.

Finally, the combined measurement uncertainty of the measurement chain can also be calculated from the experimental biases and variances by the procedures as given in the GUM. Thus, with each of the terms known in Eqs. 3 and 4, the combined measurement uncertainty can be calculated. Whether the resulting combined standard uncertainty equals the reproducibility standard deviation depends on the answer to the question whether all sources of uncertainty have been included in the interlaboratory study. If this is the case, both processes should lead to the same result. If not, the reproducibility standard deviation will be lower than the combined standard uncertainty that would be obtained by identifying and quantifying all sources of uncertainty.

Conclusions

The concepts of the GUM are also valid for solid-state materials. With a careful interpretation of the statistical concepts of the standard for the organisation of interlaboratory studies, ISO 5725:1994 [10, 11, 24–26] can be brought into agreement with the concepts of the GUM.

A measurement chain is best evaluated when taking the consensus/certified value of a reference material as a reference term and expressing all other terms in the chain in the form of bias terms. The sum of the reference term and the bias terms defines the expectation value of the critical content typical for the laboratory, which cannot be better than the reference term (critical content and stated uncertainty of the reference material). The expectation value of the critical content is independent of a possible correlation of the bias terms.

The expression of the measurement chain in a reference term in combination with several bias terms enables the evaluation of the effectiveness of the reduction of the bias (and its variance).

An exact evaluation of the combined measurement uncertainty requires (at least) knowledge about the relationship between the critical content X and any of the bias terms B. The functional relationships between the bias terms may be left out in a first approximation, as it may be expected that successful implementation of a QAS will minimise the variance of these terms, and as a result will lead to a very low covariance value between any of the bias terms when compared to the variance of the critical content, $\mathrm{Var}(X)$.

The evaluation of the functional relationship between X and B is problematic for matrix materials (i.e. coal) due to "matrix effects", but can be well established for synthetic, more homogeneous systems.

Acknowledgements The European Coal and Steel Community (ECSC) is acknowledged for its financial support of this work done under contract number ECSC 7220/EC-036, "Preparation and characterisation of coal samples and maceral concentrates for studies on gasification and combustion reactivity of coals in combined cycle processes". The participants in the interlaboratory study programme are thanked for their work and their expression of interest during the project.

References

1. Veen AMH van der, Broos AJM (1996) Preparation and characterisation of coal samples and maceral concentrates for studies on gasification and combustion reactivity of coals in combined cycle processes. Draft final report, ECSC 7220/EC-036, Eygelshoven
2. Veen AMH van der (1994) ILS Coal Characterisation I, Evaluation report. NMi Van Swinden Laboratorium B.V., Eygelshoven
3. Veen AMH van der, Broos AJM (1994) ILS Coal Characterisation II, Evaluation report. NMi Van Swinden Laboratorium B.V., Eygelshoven
4. Veen AMH van der, Broos AJM (1995) ILS Coal Characterisation III, Evaluation report. NMi Van Swinden Laboratorium B.V., Eygelshoven
5. Veen AMH van der, Broos AJM (1995) ILS Coal Characterisation IV, Evaluation report. NMi Van Swinden Laboratorium B.V., Eygelshoven
6. Veen AMH van der, Broos AJM (1995) ILS Coal Characterisation V, Evaluation report. NMi Van Swinden Laboratorium B.V., Eygelshoven
7. Veen AMH van der, Broos AJM (1996) ILS Coal Characterisation VI, Evaluation report. NMi Van Swinden Laboratorium B.V., Eygelshoven
8. Veen AMH van der, Broos AJM (1996) ILS Coal Characterisation VII, Evaluation report. NMi Van Swinden Laboratorium B.V., Eygelshoven
9. Veen AMH van der, Broos AJM (1996) ILS Coal Characterisation VIII, Evaluation report. NMi Van Swinden Laboratorium B.V., Eygelshoven
10. International Organization for Standardization (1994) ISO 5725-1:1994 Accuracy (trueness and precision) of measurement methods and results, part 1. General principles and definition. Statistical methods for quality control, vol 2, pp 9–29
11. International Organization for Standardization (1994) ISO 5725-2:1994 Accuracy (trueness and precision) of measurement methods and results, part 2. Basic method for the determination of repeatability and reproducibility of a standard measurement method. Statistical methods for quality control, vol 2, pp 30–74
12. BIPM, IEC, IFCC, ISO, IUPAC, IUPAP, OIML (1993) International vocabulary of basic and general terms in metrology, 2nd edn. ISO, Geneva
13. BIPM, IEC, IFCC, ISO, IUPAC, IUPAP, OIML (1993) Guide to the expression of uncertainty in measurement, 1st edn. ISO, Geneva
14. Veen AMH van der, Alink A (1998). Accred Qual Assur 3:20–26
15. International Organization for Standardization (1996) ISO/IEC Guide 43-1:voting draft. Proficiency testing by interlaboratory comparisons, part 1. Development and operation of proficiency testing schemes
16. International Organization for Standardization (1989) ISO Guide 35:1989 – Certification of reference materials – general and statistical principles, 2nd edn. ISO, Geneva
17. International Organization for Standardization (1975) ISO 1988 Hard coal – sampling. ISO, Geneva
18. International Organization for Standardization (1997) ISO 1171 Solid mineral fuels – determination of ash. ISO, Geneva
19. Veen AMH van der, Alink A, Verkuil D, Lecq B van der (1996). Accred Qual Assur 1:207–212, 250
20. International Organization for Standardization (1994) ISO 3534-1:1993 Statistics – vocabulary and symbols, part 1. Probability and general statistical terms. Statistical methods for quality control, vol 1, pp 9–57
21. International Organization for Standardization (1994) ISO 3534-2:1993 Statistics – vocabulary and symbols, part 2. Statistical quality control. Statistical methods for quality control, vol 1, pp 58–92
22. International Organization for Standardization (1994) ISO 3534-3:1993 Statistics – vocabulary and symbols, part 3. Design of experiments. Statistical methods for quality control, vol 1, pp 93–134
23. DeGroot MH (1989) Probability and statistics, 2nd edn. Addison-Wesley
24. International Organization for Standardization (1994) ISO 5725-3:1994 Accuracy (trueness and precision) of measurement methods and results, part 3. Intermediate measures of the precision of a standard measurement method. Statistical methods for quality control, pp 75–104
25. International Organization for Standardization (1994) ISO 5725-4:1994 Accuracy (trueness and precision) of measurement methods and results, part 4. Basic methods for the determination of the trueness of a standard measurement method. Statistical methods for quality control, pp 105–130
26. International Organization for Standardization (1994) ISO 5725-6:1994 Accuracy (trueness and precision) of measurement methods and results, part 6. Use in practice of accuracy values. Statistical methods for quality control, pp 131–176

Accred Qual Assur (2000) 5:47–53

Vicki J. Barwick
Stephen L.R. Ellison

The evaluation of measurement uncertainty from method validation studies

Part 1: Description of a laboratory protocol

V.J. Barwick (✉) · S.L.R. Ellison
Laboratory of the Government Chemist, Queens Road, Teddington, Middlesex, TW11 0LY, UK
e-mail: vjb@lgc.co.uk
Tel.: +44-20-89437421
Fax: +44-20-89432767

Abstract A protocol has been developed illustrating the link between validation experiments, such as precision, trueness and ruggedness testing, and measurement uncertainty evaluation. By planning validation experiments with uncertainty estimation in mind, uncertainty budgets can be obtained from validation data with little additional effort. The main stages in the uncertainty estimation process are described, and the use of trueness and ruggedness studies is discussed in detail. The practical application of the protocol will be illustrated in Part 2, with reference to a method for the determination of three markers (CI solvent red 24, quinizarin and CI solvent yellow 124) in fuel oil samples.

Key words Measurement uncertainty · Method validation · Precision · Trueness · Ruggedness testing

Introduction

In recent years, the subject of the evaluation of measurement uncertainty in analytical chemistry has generated a significant level of interest and discussion. It is generally acknowledged that the fitness for purpose of an analytical result cannot be assessed without some estimate of the measurement uncertainty to compare with the confidence required. The *Guide to the Expression of Uncertainty in Measurement* (GUM) published by ISO [1] establishes general rules for evaluating and expressing uncertainty for a wide range of measurements. The guide was interpreted for analytical chemistry by EURACHEM in 1995 [2]. The approach described in the GUM requires the identification of all possible sources of uncertainty associated with the procedure; the estimation of their magnitude from either experimental or published data; and the combination of these individual uncertainties to give standard and expanded uncertainties for the procedure as a whole. Some applications of this approach to analytical chemistry have been published [3, 4]. However, the GUM principles are significantly different from the methods currently used in analytical chemistry for estimating uncertainty [5–8] which generally make use of "whole method" performance parameters, such as precision and recovery, obtained during in-house method validation studies or during method development and collaborative study [9–11]. We have previously described a strategy for reconciling the information requirements of formal (i.e. GUM) measurement uncertainty principles with the data generated from method validation studies [12–14]. The approach involves a detailed analysis of the factors influencing the result using cause and effect analysis [15]. This results in a structured list of the possible sources of uncertainty associated with the method. The list is then simplified and reconciled with existing experimental and other data. We now report the application of this approach in the form of a protocol for the estimation of measurement uncertainty from validation studies [16]. This paper outlines the key stages in the protocol and discusses the use of data from trueness and ruggedness studies in detail. The practical application of the protocol will be described in Part 2 with reference to a high performance liquid chromatography (HPLC) procedure for the determination of markers in road fuel [17].

Principles of approach

The stages in the uncertainty estimation process are illustrated in Fig. 1. An outline of the procedure discussed in the protocol is presented in Fig. 2. The first stage of the procedure is the identification of sources of uncertainty for the method. Once the sources of uncertainty have been identified they require evaluation. The main tools for doing this are precision, trueness (or bias) and ruggedness studies. The aim is to account for as many sources of uncertainty as possible during the precision and trueness studies. Any remaining sources of uncertainty are then evaluated either from existing data (e.g. calibration certificates, published data, previous studies, etc.) or via ruggedness studies. Note that it may not be necessary to evaluate every source of uncertainty in detail, if the analyst has evidence to suggest that some are insignificant. Indeed, the EURACHEM Guide states that unless there are a large number of them, uncertainty components which are less than one-third of the largest need not be evaluated in detail. Finally, the individual uncertainty components for the method are combined to give standard and expanded uncertainties for the method as a whole. The use of data from trueness and ruggedness studies in uncertainty estimation is discussed in more detail below.

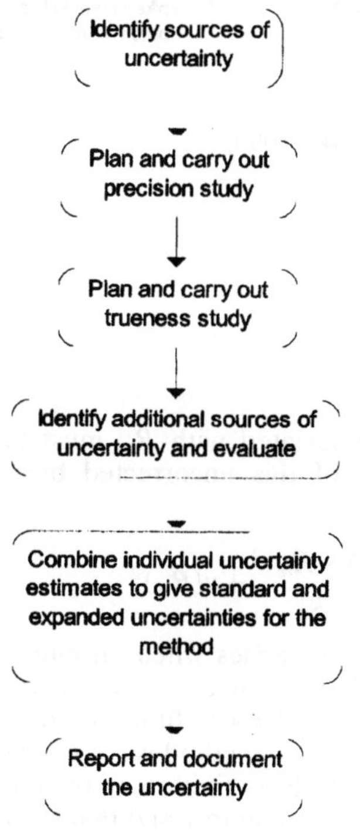

Fig. 1 Flow chart summarising the uncertainty estimation process

Trueness studies

In developing the protocol, the trueness of a method was considered in terms of recovery, i.e. the ratio of the observed value to the expected value. The evaluation of uncertainties associated with recovery is discussed in detail elsewhere [18, 19]. In general, the recovery, R, for a particular sample is considered as comprising three components:

- $\bar{R}_m$ is an estimate of the mean method recovery obtained from, for example, the analysis of a CRM or a spiked sample. The uncertainty in $\bar{R}_m$ is composed of the uncertainty in the reference value (e.g. the uncertainty in the certified value of a reference material) and the uncertainty in the observed value (e.g. the standard deviation of the mean of replicate analyses).
- R_s is a correction factor to take account of differences in the recovery for a particular sample compared to the recovery observed for the material used to estimate $\bar{R}_m$.
- R_{rep} is a correction factor to take account of the fact that a spiked sample may behave differently to a real sample with incurred analyte.

These three elements are combined multiplicatively to give an estimate of the recovery for a particular sample, R, and its uncertainty, $u(R)$:

$$R=\bar{R}_m\times R_s\times R_{rep}, \tag{1}$$

$$u(R)=R\times\sqrt{\left(\frac{u(\bar{R}_m)}{\bar{R}_m}\right)^2+\left(\frac{u(R_s)}{R_s}\right)^2+\left(\frac{u(R_{rep})}{R_{rep}}\right)^2}. \tag{2}$$

$\bar{R}_m$ and $u(\bar{R}_m)$ are calculated using Eq. (3) and Eq. (4):

$$\bar{R}_m=\frac{\bar{C}_{obs}}{C_{RM}}, \tag{3}$$

$$u(\bar{R}_m)=\bar{R}_m\times\sqrt{\left(\frac{s_{obs}}{\bar{C}_{obs}}\right)^2+\left(\frac{u(C_{RM})}{C_{RM}}\right)^2}, \tag{4}$$

where $\bar{C}_{obs}$ is the mean of the replicate analyses of the reference material (e.g. CRM or spiked sample), s_{obs} is the standard deviation of the mean of the results, C_{RM} is the concentration of the reference material and $u(C_{RM})$ is the standard uncertainty in the concentration of the reference material. To determine the contribu-

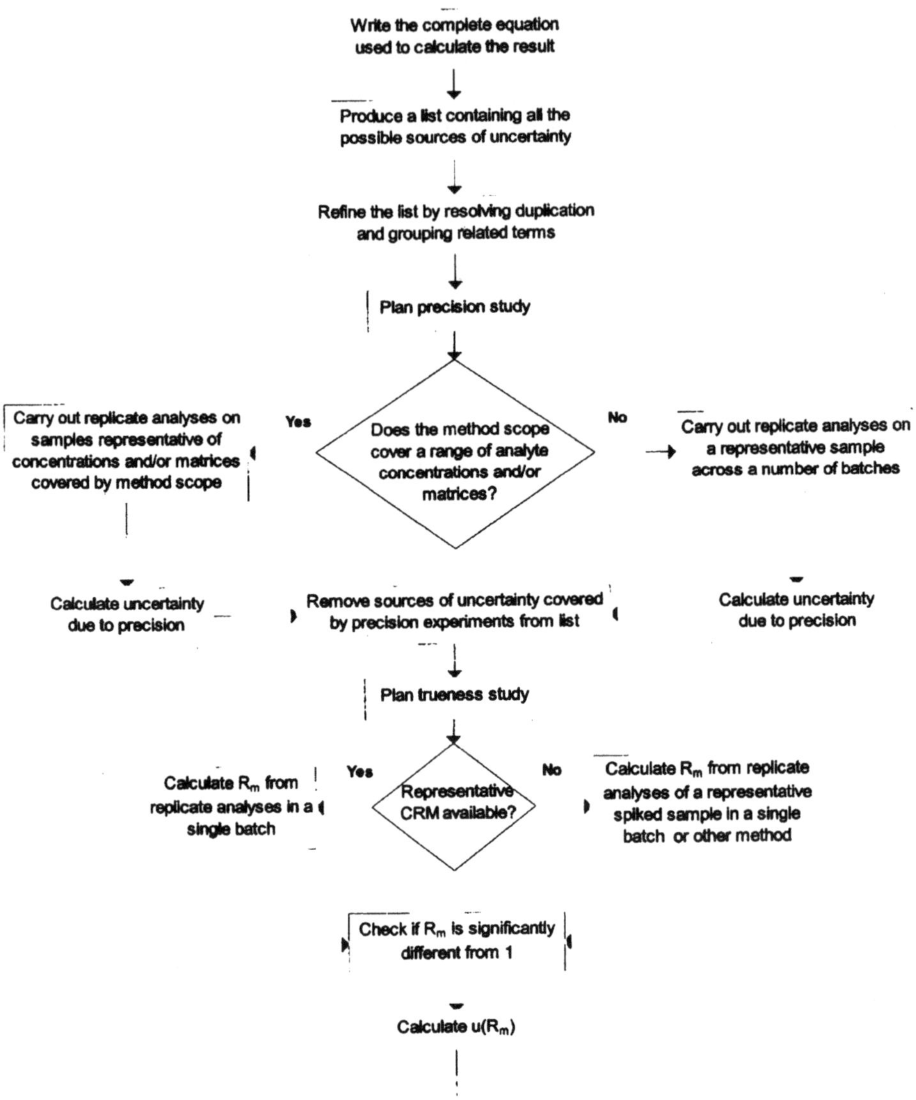

Fig. 2 Flow chart illustrating the stages in the method validation/measurement uncertainty protocol

tion of $\bar{R}_m$ to the combined uncertainty for the method as a whole, the estimate is compared with 1, using an equation of the form:

$$t = \frac{|1-\bar{R}_m|}{u(\bar{R}_m)} \,. \tag{5}$$

To determine whether $\bar{R}_m$ is significantly different from 1, the calculated value of t is compared with the coverage factor, $k=2$, which will be used to calculate the expanded uncertainty [19]. A t value greater than 2 suggests that $\bar{R}_m$ is significantly different from 1. However, if in the normal application of the method, no correction is made to take account of the fact that the method recovery is significantly different from 1, the uncertainty associated with $\bar{R}_m$ must be increased to take account of this uncorrected bias. The relevant equation is:

$$u(\bar{R}_m)' = \sqrt{\left(\frac{1-\bar{R}_m}{k}\right)^2 + u(\bar{R}_m)^2} \,. \tag{6}$$

A special case arises when an empirical method is being studied. In such cases, the method defines the measurand (e.g. dietary fibre, extractable cadmium from ceramics). The method is considered to define the true value and is, by definition, unbiased. The presumption is that $\bar{R}_m$ is equal to 1 and that the only uncertainty is that associated with the laboratory's particular application of the method. In some cases, a reference ma-

Fig. 2 Continued

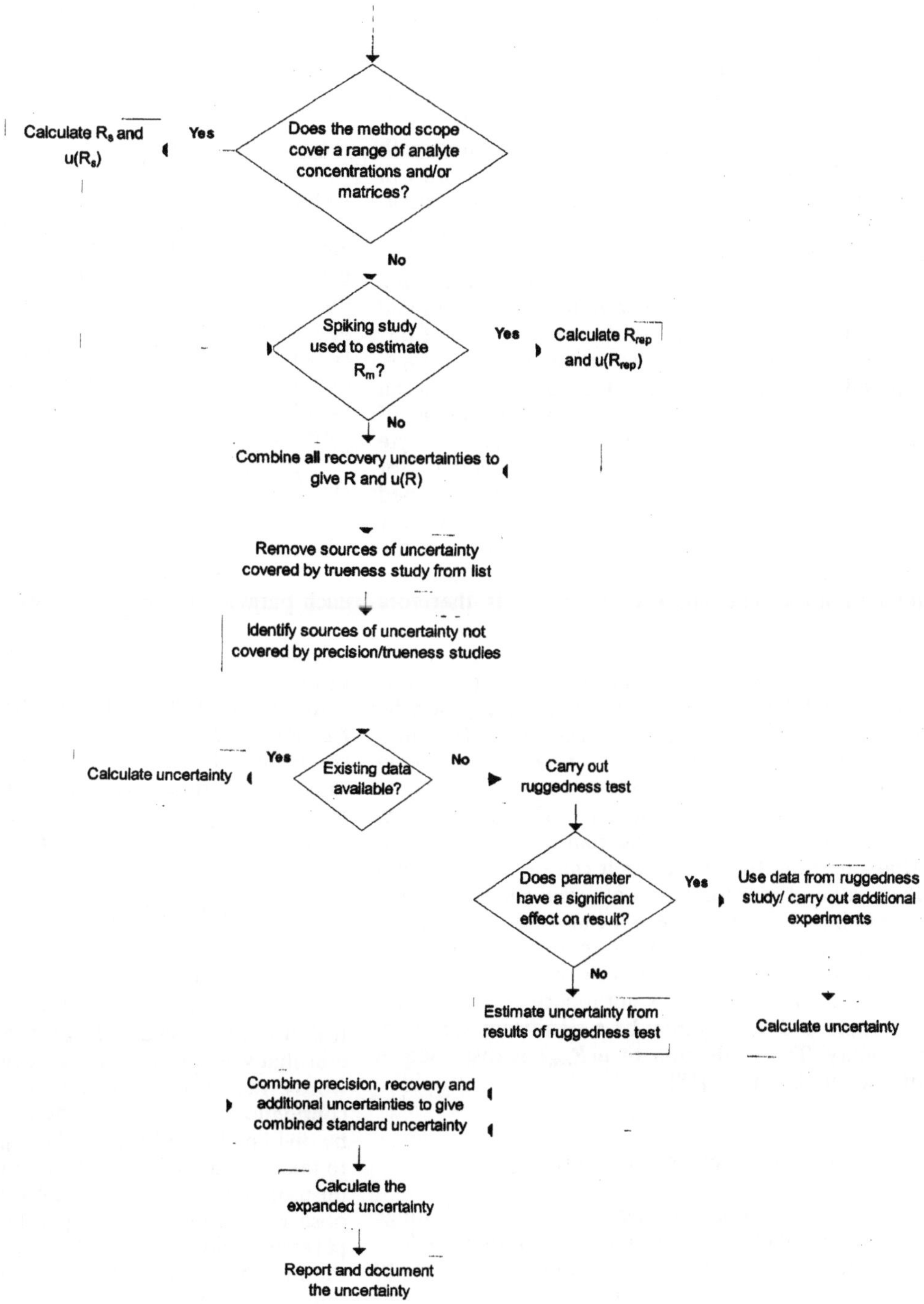

terial certified for use with the method may be available. Where this is so, a bias study can be carried out and the results treated as discussed above. If there is no relevant reference material, it is not possible to estimate the uncertainty associated with the laboratory bias. There will still be uncertainties associated with bias, but they will be associated with possible bias in the temperatures, masses, etc. used to define the method. In such cases it will normally be necessary to consider these individually.

Where the method scope covers a range of sample matrices and/or analyte concentrations, an additional uncertainty term R_s is required to take account of differences in the recovery of a particular sample type, compared to the material used to estimate $\bar{R}_m$. This can be evaluated by analysing a representative range of spiked samples, covering typical matrices and analyte concentrations, in replicate. The mean recovery for each sample type is calculated. R_s is normally assumed to be equal to 1. However, there will be an uncertainty associated with this assumption, which appears in the spread of mean recoveries observed for the different spiked samples. The uncertainty, $u(R_s)$, is therefore taken as the standard deviation of the mean recoveries for each sample type.

When a spiked sample, rather than a matrix reference material, has been used to estimate $\bar{R}_m$ it may be necessary to consider R_{rep} and its uncertainty. In general, R_{rep} is assumed to equal 1, indicating that the recovery observed for a spiked sample is truly representative of that for the incurred analyte. The uncertainty, $u(R_{rep})$, is a measure of the uncertainty associated with that assumption. In some cases it can be argued that a spike is a good representation of a real sample, for example in liquid samples where the analyte is simply dissolved in the matrix; $u(R_{rep})$ can therefore be assumed to be small. In other cases there may be reason to believe that a spiked sample is not a perfect model for a test sample and $u(R_{rep})$ may be a significant source of uncertainty. The evaluation of $u(R_{rep})$ is discussed in more detail elsewhere [18].

Evaluation of other sources of uncertainty

An uncertainty evaluation must consider the full range of variability likely to be encountered during application of the method. This includes parameters relating to the sample (analyte concentration, sample matrix) as well as experimental parameters associated with the method (e.g. temperature, extraction time, equipment settings, etc.). Sources of uncertainty not adequately covered by the precision and trueness studies require separate evaluation. There are three main sources of information: calibration certificates and manufacturers' specifications, data published in the literature and specially designed experimental studies. One efficient method of experimental study is ruggedness testing, discussed below.

Ruggedness studies

Ruggedness tests are a useful way of investigating simultaneously the effect of several experimental parameters on method performance. The experiments are based on the ruggedness testing procedure described in the *Statistical Manual of the AOAC* [20]. Such experiments result in an observed difference, D_{x_i}, for each parameter studied which represents the change in result due to varying that parameter. The parameters are tested for significance using a Student's *t*-test of the form [21]:

$$t=\frac{\sqrt{n}\times D_{x_i}}{\sqrt{2}\times s}, \tag{7}$$

where s is the estimate of the method precision, n is the number of experiments carried out at each level for each parameter ($n=4$ for a seven-parameter Plackett-Burman experimental design), and D_{x_i} is the difference calculated for parameter x_i. The values of t calculated using Eq. (7) are compared with the appropriate critical values of t at 95% confidence. Note that the degrees of freedom for t_{crit} relate to the degrees of freedom for the precision estimate used in the calculation of t. For parameters identified as having no significant effect on the method performance, the uncertainty in the final result y due to parameter x_i, $u(y(x_i))$, is calculated using Eq. (8):

$$u(y(x_i))=\frac{\sqrt{2}\times t_{\mathrm{crit}}\times s}{\sqrt{n}\times 1.96}\times\frac{\delta_{\mathrm{real}}}{\delta_{\mathrm{test}}}, \tag{8}$$

where δ_{real} is the change in the parameter which would be expected when the method is operating under control in routine use and δ_{test} is the change in the parameter that was specified in the ruggedness study. In other words, the uncertainty estimate is based on the 95% confidence interval, converted to a standard deviation by dividing by 1.96 [1, 2]. The $\delta_{\mathrm{real}}/\delta_{\mathrm{test}}$ term is required to take account of the fact that the change in a parameter used in the ruggedness test may be greater than that observed during normal operation of the method. For parameters identified as having a significant effect on the method performance, a first estimate of the uncertainty can be calculated as follows:

$$u(y(x_i))=u(x_i)\times c_i, \tag{9}$$

$$c_i=\frac{\textit{Observed change in result}}{\textit{Change in parameter}}, \tag{10}$$

where $u(x_i)$ is the uncertainty in the parameter and c_i is the sensitivity coefficient.

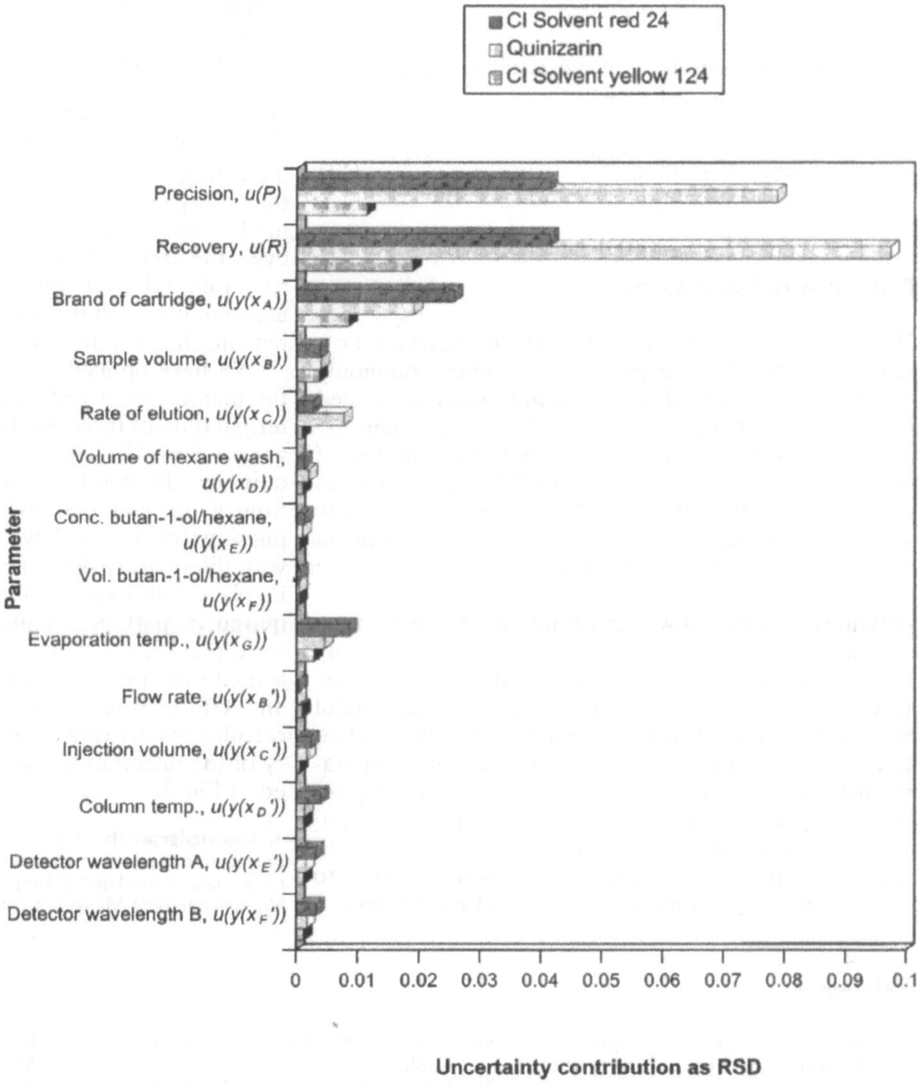

Fig. 3 Contributions to the measurement uncertainty for the determination of CI solvent red 24, quinizarin and CI solvent yellow 124 in fuel oil

The estimates obtained by applying Eqs. 8–10 are intended to give a first estimate of the measurement uncertainty associated with a particular parameter. If such estimates of the uncertainty are found to be a significant contribution to the overall uncertainty for the method, further study of the effect of the parameters is advised, to establish the true relationship between changes in the parameter and the result of the method. However, if the uncertainties are found to be small compared to other uncertainty components (i.e. the uncertainties associated with precision and trueness) then no further study is required.

Calculation of combined measurement uncertainty for the method

The individual sources of uncertainty, evaluated through the precision, trueness, ruggedness and other studies are combined to give an estimate of the standard uncertainty for the method as a whole. Uncertainty contributions identified as being proportional to analyte concentration are combined using Eq. (11):

$$\frac{u(y)}{y} = \sqrt{\left(\frac{u(p)}{p}\right)^2 + \left(\frac{u(q)}{q}\right)^2 + \left(\frac{u(r)}{r}\right)^2 + \ldots}, \qquad (11)$$

where the result y is affected by parameters p, q, r.... which each have uncertainties $u(p)$, $u(q)$, $u(r)$.... Uncertainty contributions identified as being independent of analyte concentration are combined using Eq. (12):

$$u(y)' = \sqrt{u(p)^2 + u(q)^2 + u(r)^2 + \ldots} \quad (12)$$

The combined uncertainty in the result at a concentration y' is calculated as follows:

$$u(y') = \sqrt{(u(y)')^2 + \left(y' \times \frac{u(y)}{y}\right)^2} \quad (13)$$

Discussion and conclusions

We have developed a protocol which describes how data generated from experimental studies commonly undertaken for method validation purposes can be used in measurement uncertainty evaluation. The main experimental studies required are for the evaluation of precision and trueness. These should be planned so as to cover as many of the possible sources of uncertainty identified for the method as possible. Any remaining sources are considered separately. If there is evidence to suggest that they will be small compared to the uncertainties associated with precision and trueness, then no further study is required. However for uncertainty components where no prior information is available, further experimental study will be required. One useful approach is ruggedness testing which allows the evaluation of a number of sources of uncertainty simultaneously. It should be noted that ruggedness testing really only gives a first estimate of uncertainty contributions. Further study is recommended to refine the estimates for any sources of uncertainty which appear to be a significant contribution to the total uncertainty.

The main disadvantage of this approach is that it may not readily reveal the main sources of uncertainty for a particular method. In previous studies we have typically found the uncertainty budget to be dominated by the precision and trueness terms [14]. In such cases, if the combined uncertainty for the method is too large, indicating that the method requires improvement, further study may be required to identify the stages in the method which contribute most to the uncertainty. However, the approach detailed here will allow the analyst to obtain, relatively quickly, a sound estimate of measurement uncertainty, with minimum experimental work beyond that required for method validation.

We have applied this protocol to the evaluation of the measurement uncertainty for a method for the determination of three markers (CI solvent red 24, CI solvent yellow 124 and quinizarin (1,4-dihydroxyanthraquinone)) in road fuel. The method requires the extraction of the markers from the sample matrix by solid phase extraction, followed by quantification by HPLC with diode array detection. The uncertainty evaluation involved four experimental studies which were also required as part of the method validation. The studies were precision, trueness (evaluated via the analysis of spiked samples) and ruggedness tests of the extraction and HPLC stages. The experiments and uncertainty calculations are described in detail in Part 2. A summary of the uncertainty budget for the method is presented in Fig. 3.

Acknowledgements The work described in this paper was supported under contract with the Department of Trade and Industry as part of the United Kingdom National Measurement System Valid Analytical Measurement (VAM) Programme.

References

1. ISO (1993) Guide to the expression of uncertainty in measurement. ISO, Geneva
2. EURACHEM (1995) Quantifying uncertainty in analytical measurement. Laboratory of the Government Chemist (LGC), London
3. Pueyo M, Obiols J, Vilalta E (1996) Anal Commun 33:205–208
4. Williams A (1993) Anal Proc 30:248–250
5. Analytical Methods Committee (1995) Analyst 120:2303–2308
6. Ellison SLR (1997) In: Ciarlini P, Cox MG, Pavese F, Tichter D (eds) Advanced mathematical tools in metrology III. World Scientific, Singapore
7. Ellison SLR, Williams A (1998) Accred Qual Assur 3:6–10
8. Ríos A, Valcárcel M (1998) Accred Qual Assur 3:14–29
9. IUPAC (1988) Pure Appl Chem 60:885
10. AOAC (1989) J Assoc Off Anal Chem 72:694–704
11. ISO 5725:1994 (1994) Accuracy (trueness and precision) of measurement methods and results. ISO, Geneva
12. Ellison SLR, Barwick VJ (1998) Accred Qual Assur 3:101–105
13. Ellison SLR, Barwick VJ (1998) Analyst 123:1387–1392
14. Barwick VJ, Ellison SLR (1998) Anal Comm 35:377–383
15. ISO 9004-4:1993 (1993) Total quality management Part 2: Guidelines for quality improvement. ISO, Geneva
16. Barwick VJ, Ellison SLR (1999) Protocol for uncertainty evaluation from validation data. VAM Technical Report No. LGC/VAM/1998/088, available on LGC website at www.lgc.co.uk
17. Barwick VJ, Ellison SLR, Rafferty MJQ, Gill RS (1999) Accred Qual Assur
18. Barwick VJ, Ellison SLR (1999) Analyst 124:981–990
19. Ellison SLR, Williams A (1996) In: Parkany M (ed) The use of recovery factors in trace analysis. Royal Society of Chemistry, Cambridge
20. Youden WJ, Steiner EH (1975) Statistical manual of the association of official analytical chemists. Association of Official Analytical Chemists (AOAC), Arlington, Va.
21. Vander Heyden Y, Luypaert K, Hartmann C, Massart DL, Hoogmartens J, De Beer J (1995) Anal Chim Acta 312:245–262

Accred Qual Assur (2000) 5:104–113

Vicki J. Barwick
Stephen L.R. Ellison
Mark J.Q. Rafferty
Rattanjit S. Gill

The evaluation of measurement uncertainty from method validation studies

Part 2: The practical application of a laboratory protocol

V.J. Barwick (✉) · S.L.R. Ellison
M.J.Q. Rafferty · R.S. Gill
Laboratory of the Government Chemist,
Queens Road, Teddington, Middlesex,
TW11 0LY, UK
e-mail: vjb@lgc.co.uk,
Tel.: +44-20-8943 7421,
Fax: +44-20-8943 2767

Abstract A protocol has been developed illustrating the link between validation experiments and measurement uncertainty evaluation. The application of the protocol is illustrated with reference to a method for the determination of three markers (CI solvent red 24, quinizarin and CI solvent yellow 124) in fuel oil samples. The method requires the extraction of the markers from the sample matrix by solid phase extraction followed by quantification by high performance liquid chromatography (HPLC) with diode array detection. The uncertainties for the determination of the markers were evaluated using data from precision and trueness studies using representative sample matrices spiked at a range of concentrations, and from ruggedness studies of the extraction and HPLC stages.

Key words Measurement uncertainty · Method validation · Precision · Trueness · Ruggedness · High performance liquid chromatography

Introduction

In Part 1 [1] we described a protocol for the evaluation of measurement uncertainty from validation studies such as precision, trueness and ruggedness testing. In this paper we illustrate the application of the protocol to a method developed for the determination of the dyes CI solvent red 24 and CI solvent yellow 124, and the chemical marker quinizarin (1,4-dihydroxyanthraquinone) in road fuel. The analysis of road fuel samples suspected of containing rebated kerosene or rebated gas oil is required as the use of rebated fuels as road fuels or extenders to road fuels is illegal. To prevent illegal use of rebated fuels, HM Customs and Excise require them to be marked. This is achieved by adding solvent red 24, solvent yellow 124 and quinizarin to the fuel. A method for the quantitation of the markers was developed in this laboratory [2]. Over a period of time the method had been adapted to improve its performance and now required re-validation and an uncertainty estimate. This paper describes the experiments undertaken and shows how the data were used in the calculation of the measurement uncertainty.

Experimental

Outline of procedure for the determination of CI solvent red 24, CI solvent yellow 124 and quinizarin in fuel samples

Extraction procedure

The sample (10 ml) was transferred by automatic pipette to a solid phase extraction cartridge containing 500 mg silica. The cartridge was drained under vacuum until the silica bed appeared dry. The cartridge was then washed under vacuum with 10 ml hexane to remove residual oil. The markers were eluted from the cartridge under gravity with 10 ml butan-1-ol in hexane (10% v/v). The eluent was collected in a glass specimen vial and evaporated to dryness by heating to 50 °C under an air stream. The residue was dissolved in acetonitrile (2.5 ml) and the resulting solution placed in an ultrasonic bath for 5 min. The solution was then passed through a 0.45 μm filter prior to analysis by high performance liquid chromatography (HPLC).

HPLC conditions

The samples (50 μl) were analysed on a Hewlett Packard 1050 DAD system upgraded with a 1090 DAD optical bench. The column was a Luna 5 μm phenyl-hexyl, 250 mm × 4.6 mm maintained at 30 °C. The flow rate was 1 ml min^{-1} using a gradient elution of acetonitrile and water as follows:

Time (min)	0	3	4	5	9	10	20	21	23
Water	40	40	30	10	10	2	2	40	40
% Acetonitrile	60	60	70	90	90	98	98	60	60

Calibration was by means of a single standard in acetonitrile containing CI solvent red 24 and CI solvent yellow 124 at a concentration of approximately 20 mg l^{-1} and quinizarin at concentration of approximately 10 mg l^{-1}. CI solvent red 24 and quinizarin were quantified using data (peak areas) recorded on detector channel B (500 nm), whilst CI solvent yellow 124 was quantified using data recorded on detector channel A (475 nm). The concentration of the analyte, C in mg l^{-1}, was calculated using Eq. (1):

$$C=\frac{A_S \times V_F \times C_{STD}}{A_{STD} \times V_S}, \quad (1)$$

where A_S is the peak area recorded for the sample solution, A_{STD} is the peak area recorded for the standard solution, V_F is the final volume of the sample solution (ml), V_S is the volume of the sample taken for analysis (ml) and C_{STD} is the concentration of the standard solution (mg l^{-1}).

Experiments planned for validation and uncertainty estimation

A cause and effect diagram [3–5] illustrating the main parameters controlling the result of the analysis is presented in Fig. 1. Note that uncertainties associated with sampling are outside the scope of this study, as the uncertainty was required for the sample as received in the laboratory. The uncertainty contribution from sub-sampling the laboratory sample is represented by the "inhomogeneity" branch in Fig. 1. Initially, two sets of experiments were planned – a precision study and a trueness study. These were planned so as to cover as many sources of uncertainty as possible. Parameters not adequately covered by these experiments (i.e. not varied representatively) were evaluated separately using ruggedness tests or existing published data. Whilst these studies are required for the method validation process, it should be noted that they do not form a complete validation study [6].

Fig. 1 Cause and effect diagram illustrating sources of uncertainty for the method for the determination of markers in fuel oil

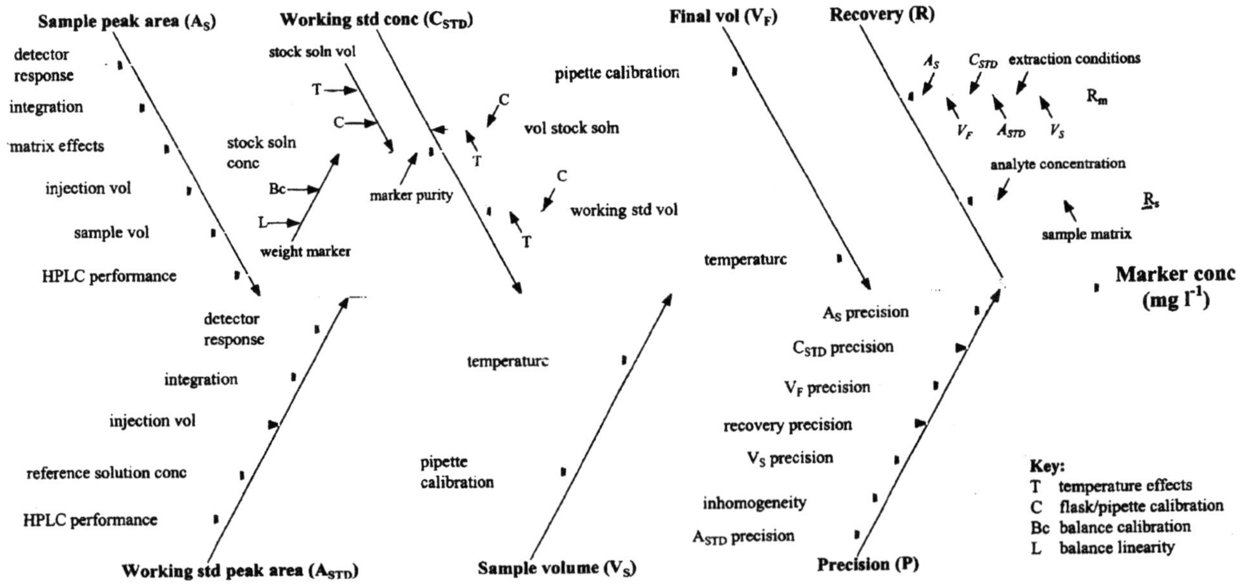

Precision experiments

Samples of 3 unmarked fuel oils (A–C) were fortified with CI solvent red 24 at concentrations of 0.041, 1.02, 2.03, 3.05 and 4.06 mg l^{-1}; quinizarin at concentrations of 0.040, 0.498, 0.996, 1.49 and 1.99 mg l^{-1}; and CI solvent yellow 124 at concentrations of 0.040, 1.20, 2.40, 3.99 and 4.99 mg l^{-1} to give a total of 15 fortified samples. Oil B was a diesel oil, representing a sample of typical viscosity. Oil A was a kerosene and oil C was a lubricating oil. These oils are respectively less viscous and more viscous than oil B.

Initially, 12 sub-samples of oil B with a concentration of 2.03 mg l^{-1} CI solvent red 24, 0.996 mg l^{-1} quinizarin and 2.40 mg l^{-1} CI solvent yellow 124 were analysed. The extraction stage was carried out in two batches of six on consecutive days. The markers in all 12 sub-samples were quantified in a single HPLC run, with the order of the analysis randomised. This study was followed by the analysis, in duplicate, of all 15 samples. The sample extracts were analysed in three separate HPLC runs such that the duplicates for each sample were in different runs. For each HPLC run a new standard and a fresh batch of mobile phase was prepared.

In addition, the results obtained from the replicate analysis of a sample of BP diesel, prepared for the trueness study (see below), were used in the estimate of uncertainty associated with method precision.

Trueness experiments

No suitable CRM was available for the evaluation of recovery. The study therefore employed representative samples of fuel oil spiked with the markers at the required concentrations. To obtain an estimate of $\bar{R}_m$ and its uncertainty, a 2-l sample of unmarked BP diesel was spiked with standards in toluene containing CI solvent red 24, quinizarin and CI solvent yellow 124 at concentrations of 0.996 mg ml^{-1}, 1.02 mg ml^{-1} and 1.97 mg ml^{-1}, respectively, to give concentrations in the diesel of 4.06 mg l^{-1}, 1.99 mg l^{-1} and 4.99 mg l^{-1}, respectively. A series of 48 aliquots of this sample were analysed in 3 batches of 16. The estimate of R_s and its uncertainty, $u(R_s)$, was calculated from these results plus the results from the analysis of the samples used in the precision study.

Evaluation of other sources of uncertainty: Ruggedness test

The effects of parameters associated with the extraction/clean-up stages and the HPLC quantification stage were studied in separate experiments. The parameters studied for the extraction/clean-up stage of the method and the levels chosen are shown in Table 1a. The ruggedness test was applied to the matrix B (diesel oil) sample containing 2.03 mg l^{-1} CI solvent red 24, 0.996 mg l^{-1} quinizarin and 2.40 mg l^{-1} CI solvent yellow 124 used in the precision study. The eight experiments were carried out over a short period of time and the resulting sample extracts were analysed in a single HPLC run. The HPLC parameters investigated and the levels chosen are given in Table 1b. For this set of experiments a single extract of the matrix B (diesel oil) sample, obtained under normal method conditions, was used. The extract and a standard were run under each set of conditions required by the ruggedness test. The effect of variations in the parameters was monitored by calculating the concentration of the markers observed under each set of parameters, using the appropriate standard.

Results and uncertainty calculations

Precision study

The results from the precision studies are summarised in Table 2. Estimates for the standard deviation for a single result were obtained from the results of the duplicate analyses of the 15 samples, by taking the standard deviation of the differences between the pairs and dividing by $\sqrt{2}$. Estimates of the relative standard deviations were obtained by treating the normalised differences in the same way [7]. The results from the analysis of the BP diesel sample represented three batches of 16 replicate analyses. An estimate of the total precision (i.e. within and between batch variation) was obtained via ANOVA [8]. The precision estimates cover different sources of variability in the method. The estimates obtained from the duplicate samples and the BP oil sample cover batch to batch variability in the extraction and HPLC stages of the method (including the preparation of new standards and mobile phase). The estimate obtained from matrix B does not cover batch to batch variability in the HPLC procedure as all the replicates were analysed in a single HPLC run. The precision studies also cover the uncertainty associated with sample inhomogeneity as they involved the analysis of a number of sub-samples taken from the bulk.

CI solvent red 24

No significant difference was observed (*F*-tests, 95% confidence) between the three estimates obtained for the relative standard deviation (0.0323, 0.0289 and 0.0414). However, the test was borderline and across the range studied (0.04 mg l^{-1} to 4 mg l^{-1}) the method

Table 1 Results from the ruggedness testing of the procedure for the determination of CI solvent red 24, quinizarin and CI solvent yellow 124 in fuel oil. **a** Ruggedness testing of the extraction/clean-up procedure

Parameter	Values		$\delta_{real}/\delta_{test}$	$u(x_i)$	CI solvent red 24			Quinizarin			CI solvent yellow 124		
					D_{x_i} (mg l^{-1})	c_i	$u(y(x_i))$ (mg l^{-1})	D_{x_i} (mg l^{-1})	c_i	$u(y(x_i))$ (mg l^{-1})	D_{x_i} (mg l^{-1})	c_i	$u(y(x_i))$ (mg l^{-1})
Brand of silica cartridges	A Varian	a Waters	1/1[a]		0.00750*	–	0.0493	−0.00750*	–	0.0174	−0.00250*	–	0.0199
Sample volume	B 10 ml	b 12 ml	–	0.04 ml	−0.353	0.176	0.00705	−0.180	0.090	0.00360	−0.423	0.212	0.00845
Rate of elution of oil with hexane	C vacuum	c gravity	1/10[a]	–	0.0275*	–	0.00493	0.070	–	0.0070[a]	−0.020*	–	0.00199
Volume of hexane wash	D 12 ml	d 8 ml	–	0.04 ml	0.213	0.0531	0.00213	0.176	0.0444	0.00177	0.225	0.0563	0.00225
Concentration of butan-1-ol/hexane	E 12%	e 8%	0.2% (v/v)/ 4% (v/v)		−0.0425*	–	0.00247	0.0175*	–	0.000868	−0.010*	–	0.0010
Volume 10% of butan-1-ol/hexane	F 12 ml	f 8 ml	0.08 ml/ 4 ml	0.04 ml	0.0625*	–	0.000986	−0.0050*	–	0.000347	0.080	0.020	0.00080
Evaporation temperature	G 50 °C	g 80 °C	10 °C/30 °C	2.89 °C	−0.0275*	–	0.0164	−0.0425	0.00142	0.00409	0.00750*	–	0.00663

[a] See text for explanation
*No significant effect at the 95% confidence level

Table 1b Ruggedness testing of the HPLC procedure

Parameter	Values		$\delta_{real}/\delta_{test}$	$u(x_i)$	CI solvent red 24			Quinizarin			CI solvent yellow 124		
					D_{x_i} (mg l^{-1})	c_i	$u(y(x_i))$ (mg l^{-1})	D_{x_i} (mg l^{-1})	c_i	$u(y(x_i))$ (mg l^{-1})	D_{x_i} (mg l^{-1})	c_i	$u(y(x_i))$ (mg l^{-1})
Type of acetonitrile in mobile phase	A′ Far-UV grade	a′ HPLC grade	–	a	−0.0748	a	a	−0.101	a	a	−0.0228*	a	a
Flow rate	B′ 0.8 ml min^{-1}	b′ 1.2 ml min^{-1}	–	0.00173 ml min^{-1}	−0.124	0.309	0.000535	0.0283	0.0707	0.000122	−0.0465	0.116	0.00020
Injection volume	C′ 40 µl	c′ 60 µl	–	0.75 µl	0.115	0.00576	0.00432	−0.0406	0.00203	0.00152	0.0284	0.00142	0.00107
Column temperature	D′ 25 °C	d′ 35 °C	2 °C/10 °C	1 °C	−0.130	0.0130	0.00752	0.0201	0.00201	0.00116	−0.0233*	–	0.00282
Detector wavelength (A)	E′ 465 nm	e′ 485 nm	4 nm/20 nm	1.15 nm	0.104	0.00520	0.00598	0.0239	0.00120	0.00138	−0.0161*	–	0.00282
Degassing of mobile phase	F′ Degassed	f′ Not degassed	–	a	0.108	a	a	0.0641	a	a	0.00907*	a	a
Detector wavelength (B)	G′ 490 nm	g′ 510 nm	4 nm/ 20 nm	1.15 nm	0.105	0.00525	0.00604	−0.0112*	–	0.00154	0.0198*	–	0.00282

[a] See text for explanation
*No significant effect at the 95% confidence level

Table 2 Summary of data used in the estimation of $u(P)$

Analyte/Matrix	n	Mean (mg l^{-1})	Standard deviation (mg l^{-1})	Relative standard deviation
CI solvent red 24				
Matrix B	12	1.92	0.0621	0.0323
BP diesel	48[a]	3.88	0.112	0.0289
Matrices A–C	15[b]	–	0.0376	0.0414
Quinizarin				
Matrix B	11	0.913	0.0216	0.0236
BP diesel	48[a]	1.89	0.0256	0.0136
Matrices A–C	15[b]	–	0.0470	0.0788
CI solvent yellow 124				
Matrix B	12	2.35	0.0251	0.0107
BP diesel	48[a]	4.99	0.0618	0.0124
Matrices A–C	15[b]	-	0.0247	0.0464

[a] Standard deviation and relative standard deviation estimated from ANOVA of 3 sets of 16 replicates (see text)
[b] Standard deviation and relative standard deviation estimated from duplicate results (15 sets) for a range of concentrations and matrices (see text)

precision was approximately proportional to analyte concentration. It was decided to use the estimate of 0.0414 as the uncertainty associated with precision, $u(P)$, to avoid underestimating the precision for any given sample. This estimate was obtained from the analysis of different matrices and concentrations and is therefore likely to be more representative of the precision across the method scope.

Quinizarin

The estimates of the standard deviation and relative standard deviation were not comparable. In particular, the estimates obtained from the duplicate results were significantly different from the other estimates (*F*-tests, 95% confidence). There were no obvious patterns in the data so no particular matrix and/or concentration could be identified as being the cause of the variability. There was therefore no justification for removing any data and restricting the coverage of the uncertainty estimate, as in the case of CI solvent yellow 124 (see below). The results of the precision studies indicate that the method is more variable across different matrices and analyte concentrations for quinizarin than for the other markers. The uncertainty associated with the precision was taken as the estimate of the relative standard deviation obtained from the duplicate results, 0.0788. This estimate should ensure that the uncertainty is not underestimated for any given matrix or concentration (although it may result in an overestimate in some cases).

CI solvent yellow 124

There was no significant difference between the estimates of the relative standard deviation obtained for samples at concentrations of 2.4 mg l^{-1} and 4.99 mg l^{-1}. However, the estimate obtained from the duplicate analyses was significantly greater than the other estimates. Inspection of that data revealed that the normalised differences observed for the samples at a concentration of 0.04 mg l^{-1} were substantially larger than those observed at the other concentrations. Removing these data points gave a revised estimate of the relative standard deviation of 0.00903. This was in agreement with the other estimates obtained (*F*-tests, 95% confidence). The three estimates were therefore pooled to give a single estimate of the relative standard deviation of 0.0114. At present, the uncertainty estimate cannot be applied to samples with concentrations below 1.2 mg l^{-1}. Further study would be required to investigate in more detail the precision at these low levels.

Trueness study

Evaluation of $\bar{R}_m$ and $u(\bar{R}_m)$

The results are summarised in Table 3. In each case $\bar{R}_m$ was calculated using Eq. (2):

$$\bar{R}_m = \frac{\bar{C}_{obs}}{C_{RM}}, \tag{2}$$

where $\bar{C}_{obs}$ is the mean of the replicate analyses of the spiked sample and C_{RM} is the concentration of the

Table 3 Results from the replicate analysis of a diesel oil spiked with CI solvent red 24, quinizarin and CI solvent yellow 124

Analyte	Target concentration, C_{spike} (mg l^{-1})	Mean, $\bar{C}_{obs}$ (mg l^{-1})	Standard deviation of the mean, s_{obs} (mg l^{-1})[a]
CI solvent red 24	4.06	3.88	0.0360
Quinizarin	1.99	1.89	0.00370
CI solvent yellow 124	4.99	4.99	0.0167

[a] Estimated from ANOVA of 3 groups of 16 replicates according to ISO 5725:1994 [9]

spiked sample. The uncertainty, $u(\bar{R}_m)$, was calculated using Eq. (3):

$$u(\bar{R}_m)=\bar{R}_m\times\sqrt{\left(\frac{s_{obs}}{\bar{C}_{obs}}\right)^2+\left(\frac{u(C_{RM})}{C_{RM}}\right)^2}, \quad (3)$$

where $u(C_{RM})$ is the standard uncertainty in its concentration of the spiked sample. The standard deviation of the mean of the results, s_{obs}, was estimated from ANOVA of the data according to Part 4 of ISO 5725:1994 [9].

Using information on the purity of the material used to prepare the spiked sample, and the accuracy and precision of the volumetric glassware and analytical balance used, the uncertainty in the concentration of CI solvent red 24 in the sample, $u(C_{RM})$, was estimated as 0.05 mg l^{-1}.[1] The uncertainties associated with the concentration of quinizarin and CI solvent yellow 124 were estimated as 0.025 mg l^{-1} and 0.062 mg l^{-1}, respectively. The relevant values are:

CI solvent red 24:	$\bar{R}_m=0.957$	$u(\bar{R}_m)=0.0148$
Quinizarin:	$\bar{R}_m=0.949$	$u(\bar{R}_m)=0.0121$
CI solvent yellow 124:	$\bar{R}_m=1.00$	$u(\bar{R}_m)=0.0129$

Applying Eq. (4):

$$t=\frac{|1-\bar{R}_m|}{u(\bar{R}_m)} \quad (4)$$

indicated that the estimates of $\bar{R}_m$ obtained for CI solvent red 24 and quinizarin were significantly different from 1.0 ($t>2$) [7, 10]. During routine use of the method, the results reported for test samples will not be corrected for incomplete recovery of the analyte. Equation (5) was therefore used to calculate an increased uncertainty for $\bar{R}_m$ to take account of the uncorrected bias:

$$u(\bar{R}_m)'=\sqrt{\left(\frac{1-\bar{R}_m}{k}\right)^2+u(\bar{R}_m)^2}, \quad (5)$$

$u(\bar{R}_m)'$ was calculated as 0.0262 for CI solvent red 24 and 0.0283 for quinizarin. The significance test for CI solvent yellow 124 indicated that $\bar{R}_m$ was not significantly different from 1.0. The uncertainty associated with $\bar{R}_m$ is the value of $u(\bar{R}_m)$ calculated above (i.e. 0.0129).

$u(R_s)$ is the standard deviation of the mean recoveries obtained for the samples analysed in the precision studies and the BP diesel sample used in the study of $\bar{R}_m$. This gave estimates of $u(R_s)$ of 0.0322 for CI solvent red 24, 0.0932 for quinizarin and 0.0138 for CI solvent yellow 124. The estimate for CI solvent yellow 124 only includes concentrations above 1.2 mg l^{-1}, for the reason discussed in the section on precision.

[1] Detailed information on the estimation of uncertainties of this type is given in Ref. [7].

Calculation of R and u(R)

The recovery, R, for a particular test sample and the corresponding uncertainty, $u(R)$, is calculated using Eqs. (6) and (7):

$$R=\bar{R}_m\times R_s\times R_{rep}, \quad (6)$$

$$u(R)=R\times\sqrt{\left(\frac{u(\bar{R}_m)}{\bar{R}_m}\right)^2+\left(\frac{u(R_s)}{R_s}\right)^2+\left(\frac{u(R_{rep})}{R_{rep}}\right)^2}. \quad (7)$$

In this study a spiked sample can be considered a reasonable representation of test samples of marked fuel oils. There is therefore no need to correct the estimates of $\bar{R}_m$ and $u(\bar{R}_m)$ by including the R_{rep} and $u(R_{rep})$ terms. Both $\bar{R}_m$ and R_s are assumed to be equal to 1. R is therefore also equal to 1. Combining the estimates of $u(\bar{R}_m)$ and $u(R_s)$, the uncertainty $u(R)$ was calculated as 0.0415 for CI solvent red 24, 0.0974 for quinizarin and 0.0187 for CI solvent yellow 124.

Ruggedness test of extraction/clean-up procedure

The results from the ruggedness study of the extraction/clean-up procedure are presented in Table 1a. The precision of the method for the analysis of the sample used in the ruggedness study had been estimated previously as 0.0621 mg l^{-1} ($\nu=11$) for CI solvent red 24, 0.0216 mg l^{-1} ($\nu=10$) for quinizarin and 0.0251 mg l^{-1} ($\nu=11$) for CI solvent yellow 124. Parameters were tested for significance using Eq. (8):

$$t=\frac{\sqrt{n}\times D_{x_i}}{\sqrt{2}\times s}, \quad (8)$$

where s is the estimate of the method precision, n is the number of experiments carried out at each level for each parameter ($n=4$ for a seven-parameter Plackett-Burman experimental design), and D_{x_i} is the difference calculated for parameter x_i [1, 11]. The degrees of freedom for t_{crit} relate to the degrees of freedom for the precision estimate used in the calculation of t.

The parameters identified as having no significant effect on method performance, at the 95% confidence level are highlighted in Table 1a. For these parameters the uncertainty in the final result was calculated using Eq. (9):

$$u(y(x_i))=\frac{\sqrt{2}\times t_{crit}\times s}{\sqrt{n}\times 1.96}\times\frac{\delta_{real}}{\delta_{test}}, \quad (9)$$

where δ_{real} is the change in the parameter which would be expected when the method is operating under control in routine use and δ_{test} is the change in the parameter that was specified in the ruggedness study. The estimates of δ_{real} are given in Table 1a. For parameter A, brand of silica cartridge, the conditions of the test (i.e. changing between two brands of cartridge) were considered representative of normal operation of the method. δ_{real} is therefore equal to δ_{test}. The effect of the rate of elution of oil by hexane from the cartridge was investigated by comparing the elution under a vacuum and with elution under gravity. In routine analyses, the oil will be eluted under vacuum. Variations in the vacuum applied from one extraction to another will affect the rate of elution of the oil and the amount of oil eluted. However, the effect of variations in the vacuum will be small compared to the effect of having no vacuum present. It can therefore be assumed that variations in the observed concentration of the markers, due to variability in the vacuum, will be small compared to the differences observed in the ruggedness test. As a first estimate, the effect of variation in the vacuum during routine application of the method was estimated as one-tenth of that observed during the ruggedness study. This indicated that the parameter was not a significant contribution to the overall uncertainty for CI solvent red 24 and CI solvent yellow 124, so no further study was required. The estimates of δ_{real} for the concentration and volume of butan-1-ol in hexane used to elute the column were based on the manufacturers' specifications and typical precision data for the volumetric flasks and pipettes used to prepare and deliver the solution.

For the parameters identified as having a significant effect on method performance, the uncertainty was calculated using Eqs. (10) and (11):

$$u(y(x_i)) = u(x_i) \times c_i, \tag{10}$$

$$c_i = \frac{\textit{Observed change in result}}{\textit{Change in parameter}} \tag{11}$$

The estimates of the uncertainty in each parameter, $u(x_i)$, are given in Table 1a. The uncertainties associated with the sample volume, volume of hexane wash and volume of the 10% butan-1-ol/hexane solution were again based on the manufacturers' specifications and typical precision data for the volumetric flasks and pipettes used to prepare and deliver the solutions. The uncertainty in the evaporation temperature was based on the assumption that the temperature could be controlled to ±5 °C. This was taken as a rectangular distribution and converted to a standard uncertainty by dividing by $\sqrt{3}$ [7]. As discussed previously, the effect on the final result of variations in the vacuum when eluting the oil from the cartridge with hexane was estimated as one-tenth that observed in the ruggedness test.

The effects of all the parameters were considered to be proportional to the analyte concentration. The uncertainties were therefore converted to relative standard deviations by dividing by the mean of the results obtained from previous analyses of the sample under normal method conditions (see results for Matrix B in Table 2).

Ruggedness test of the HPLC procedure

The results from the ruggedness study of the HPLC procedure, and the values of δ_{real} and $u(x_i)$ used in the uncertainty calculations, are presented in Table 1b. Replicate analyses of a standard solution of the three markers gave the following estimates of the precision of the HPLC system at the concentration of the sample used in the study: CI solvent red 24, $s = 0.0363$ mg l^{-1} ($n = 69$); quinizarin, $s = 0.0107$ mg l^{-1} ($n = 69$); CI solvent yellow 124, $s = 0.0196$ mg l^{-1} ($n = 69$). Parameters were tested for significance, at 95% confidence, using Eq. (8). The uncertainties for parameters identified as having no significant effect on the method performance were calculated using Eq. (9). Based on information from manufacturers' specifications for HPLC systems, the uncertainty associated with the column temperature was estimated as ±1 °C, giving an estimate of δ_{real} of 2 °C. Again, based on manufacturers' specifications for DAD detectors, the uncertainty associated with the detector wavelengths was estimated as ±2 nm, giving a δ_{real} value of 4 nm.

The uncertainties due to significant parameters were estimated using Eqs. (10) and (11). Information in the literature suggests that a typical variation in flow rate is ±0.3% [12]. The uncertainty in the flow rate was therefore estimated as 0.00173 ml min^{-1} assuming a rectangular distribution. Data in the literature gave 1.5% as a typical coefficient of variation for the volume delivered by an autosampler [13]. The uncertainty associated with the injection volume of 50 µl was therefore estimated as 0.75 µl.

Two remaining parameters merit further discussion; the type of acetonitrile used in the mobile phase and whether or not the mobile phase was degassed. The method was developed using HPLC grade acetonitrile. The ruggedness test indicated that changing to far-UV grade results in a lower recovery for all three analytes. The method protocol should therefore specify that for routine use, HPLC grade acetonitrile must be used. The ruggedness test also indicated that not degassing the mobile phase causes a reduction in recovery. The method was developed using degassed mobile phase, and the method protocol will specify that this must be the case during future use of the method. As these two parameters are being controlled in the method protocol, uncertainty terms have not been included.

The effects of all the parameters were considered to be proportional to the analyte concentration. The uncertainties were therefore converted to relative standard deviations by dividing by the mean of results obtained from previous analyses of the sample under normal method conditions (see results for Matrix B in Table 2).

Other sources of uncertainty

The precision and trueness studies were designed to cover as many of the sources of uncertainty as possible (see Fig. 1), for example, by analysing different sample matrices and concentration levels, and by preparing new standards and HPLC mobile phase for each batch of analyses. Parameters which were not adequately varied during these experiments, such as the extraction and HPLC conditions, were investigated in the ruggedness tests. There are however, a small number of parameters which were not covered by the above experiments. These generally related to the calibration of pipettes and balances used in the preparation of the standards and samples. For example, during this study the same pipettes were used in the preparation of all the working standards. Although the precision associated with the operation of the pipette is included in the overall precision estimate, the effect of the accuracy of the pipettes has not been included in the uncertainty budget so far. A pipette used to prepare the standard may typically deliver 0.03 ml above its nominal value. In the future a different pipette, or the same pipette

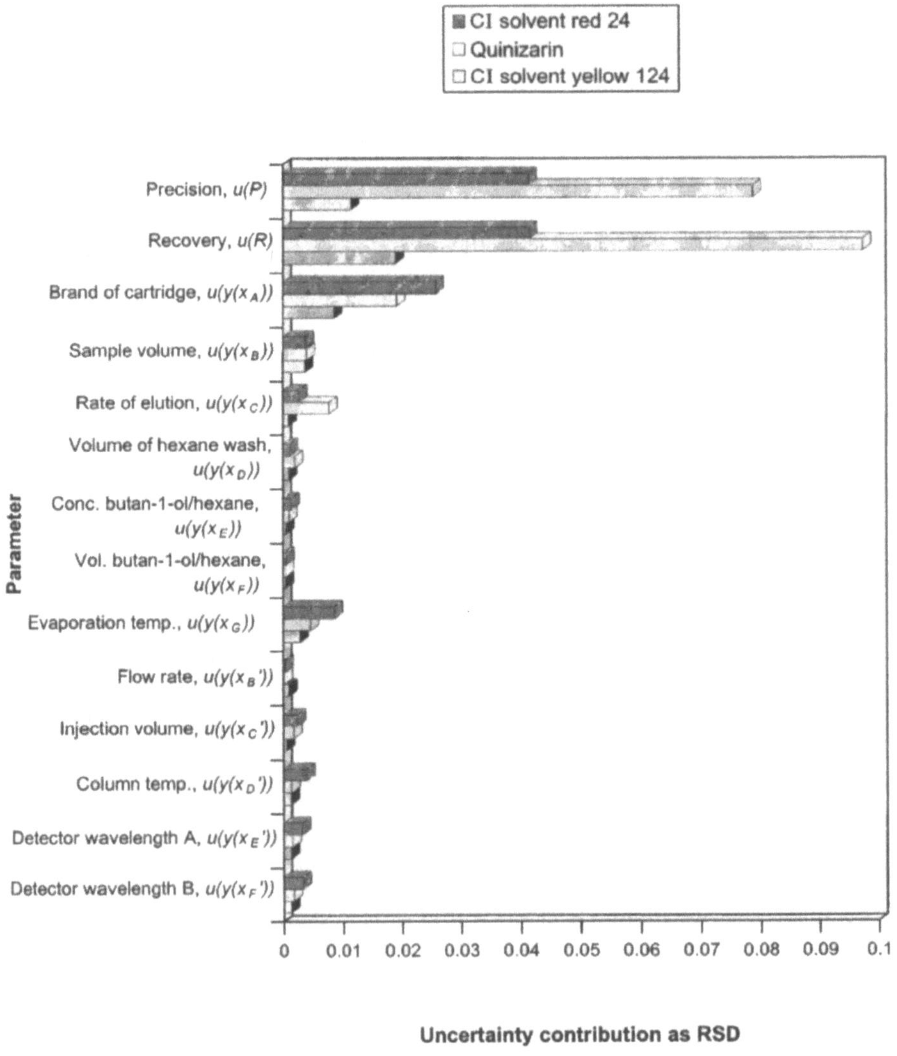

Fig. 2 Contributions as relative standard deviations (RSDs) to the measurement uncertainty for the determination of CI solvent red 24, quinizarin and CI solvent yellow 124 in fuel oil

after re-calibration, may deliver 0.02 ml below the nominal value. Since this possible variation is not already included in the uncertainty budget it should be considered separately. However, previous experience [14] has shown us that uncertainties associated with the calibration of volumetric glassware and analytical balances are generally small compared to other sources of uncertainty such as overall precision and recovery. Additional uncertainty estimates for these parameters have not therefore been included in the uncertainty budgets.

Calculation of measurement uncertainty

The contributions to the uncertainty budget for each of the analytes are illustrated in Fig. 2. In all cases the sources of uncertainty were considered to be proportional to analyte concentration. Using Eq. (12):

$$\frac{u(y)}{y}=\sqrt{\left(\frac{u(p)}{p}\right)^2+\left(\frac{u(q)}{q}\right)^2+\left(\frac{u(r)}{r}\right)^2+\ldots} \qquad (12)$$

the uncertainty in the final result, $u(y)$, was calculated as 0.065 for CI solvent red 24, 0.13 for quinizarin and 0.024 for CI solvent yellow 124, all expressed as relative standard deviations. The expanded uncertainties, calculated using a coverage factor of $k=2$ which gives a confidence level of approximately 95%, are 0.13, 0.26 and 0.048 for CI solvent red 24, quinizarin and CI solvent yellow 124, respectively.

Discussion

In the case of CI solvent red 24 and CI solvent yellow 124, the significant contributions to the uncertainty budget arose from overall precision and recovery, and the brand of the solid phase extraction cartridge used. If a reduction in the overall uncertainty of the method was required, useful approaches would be to specify a particular brand of cartridge in the method protocol, or to adopt matrix specific recovery corrections for test samples.

The combined uncertainty for quinizarin, which is significantly greater than that calculated for the other markers, is dominated by the precision and recovery terms. The results of the precision study indicated variable method performance across different matrices and analyte concentrations. The uncertainty, $u(R_s)$, associated with the variation in recovery from sample to sample was the major contribution to the recovery uncertainty, $u(R)$. This was due to the fact that the recoveries obtained for matrix B were generally higher than those obtained for matrices A and C. However, in this study, a single uncertainty estimate for all the matrices and analyte concentrations studied was required. It was therefore necessary to use "worst case" estimates of the uncertainties for precision and recovery to adequately cover all sample types. If this estimate was found to be unsatisfactory for future applications of the method, separate budgets could be calculated for individual matrices and concentration ranges.

Conclusions

We have developed a protocol which describes how data generated from experimental studies commonly undertaken for method validation purposes can be used in measurement uncertainty evaluation. This paper has illustrated the application of the protocol. In the example described, the uncertainty estimate for three analytes in different oil matrices was evaluated from three experimental studies, namely precision, recovery and ruggedness. These studies were required as part of the method validation, but planning the studies with uncertainty evaluation in mind allowed an uncertainty estimate to be calculated with little extra effort. A number of areas were identified where additional experimental work may be required to refine the estimates. However the necessary data could be generated by carrying out additional analyses alongside routine test samples. Again this would minimise the amount of laboratory effort required.

For methods which are already in routine use there may be historical validation data available which could be used, in the same way as illustrated here, to generate an uncertainty estimate. If no such data are available, the case study gives an indication on the type of experimental studies required. Again, with careful planning, it is often possible to undertake the studies alongside routine test samples.

Acknowledgments The work described in this paper was supported under contract with the Department of Trade and Industry as part of the United Kingdom National Measurement System Valid Analytical Measurement (VAM) Programme.

References

1. Barwick VJ, Ellison SLR (1999) Accred Qual Assur (in press)
2. May EM, Hunt DC, Holcombe DG (1986) Analyst 111:993–995
3. Ellison SLR, Barwick VJ (1998) Accred Qual Assur 3:101–105
4. Ellison SLR, Barwick VJ (1998) Analyst 123:1387–1392
5. ISO 9004-4:1993 (1993) Total quality management Part 2. Guidelines for quality improvement. ISO, Geneva, Switzerland
6. EURACHEM (1998) The fitness for purpose of analytical methods, a laboratory guide to method validation and related topics. Laboratory of the Government Chemist, London
7. EURACHEM (1995) Quantifying uncertainty in analytical measurement. Laboratory of the Government Chemist, London
8. Farrant TJ (1997) Practical statistics for the analytical scientist: a bench guide. Royal Society of Chemistry, Cambridge
9. ISO 5725:1994 (1994) Accuracy (trueness and precision) of measurement methods and results. ISO, Geneva, Switzerland
10. Ellison SLR, Williams A (1996) In: Parkanay M (ed) The use of recovery factors on trace analysis. Royal Society of Chemistry, Cambridge
11. Youden WJ, Steiner EH (1975) Statistical manual of the association of official analytical chemists. Association of Official Analytical Chemists, Arlington, Va
12. Brown PR, Hartwick RA (eds) (1989) High performance liquid chromatography. Wiley, New York
13. Dolan JW (1997) LC-GC International 10:418–422
14. Barwick VJ, Ellison SLR (1998) Anal Comm 35:377–383

Accred Qual Assur (1998) 3:412–415

Stephan Küppers

Is the estimation of measurement uncertainty a viable alternative to validation?

Presented at: Analytica Conference 98, Symposium 2: "Uncertainty budgets in chemical measurements", Munich, 21–24 April 1998

S. Küppers (✉)
Schering AG,
In-Process-Control,
Müllerstrasse 170–178,
D-13342 Berlin, Germany
Tel.: +49-30-468-1-7819
Fax: +49-30-468-9-7819
e-mail: stephan.kueppers@schering.de

Abstract Two examples of the use of measurement uncertainty in a development environment are presented and compared to the use of validation. It is concluded that measurement uncertainty is a good alternative to validation for chemical processes in the development stage. Some advantages of measurement uncertainty are described. The major advantages are that the estimations of measurement uncertainty are very efficient, and can be performed before analysis of the samples. The results of measurement uncertainty influence the type of analysis employed in the development process, and the measurement design can be adjusted to the need of the process.

Key words Chemical analysis · Development process · Measurement uncertainty · Validation · Practical examples

Introduction

The "know how" in quality assessment has grown over the last few years. Quality management systems are improving and validation procedures are being optimized. But the problem that validation procedures are time consuming still remains. Often validations are performed too late. Especially in a development environment where the advance from one development step to the next is based on a decision resulting from the previous step. Therefore, valid analytical data is needed directly after each method development.

The estimation of measurement uncertainty is an alternative to validation if a quantitative result is used for the assessment of the development process. In this case only one parameter of the validation is needed. The precision of the analysis is sometimes taken as a quality figure instead of a validation. But precision cannot replace validation because in this case the environment, i.e. the sample preparation is neglected and precision is only a representative of one contribution to the uncertainty of measurement: in high performance liquid chromatography (HPLC) usually the uncertainty contribution of the sampler.

To illustrate the point, two examples taken from a process research environment in a pharmaceutical company are presented. It is demonstrated that measurement uncertainty has practical advantages compared to validation.

Examples of process development

The first example presents a typical process from our research environment, where a chemical synthesis is transferred from development to production. In this situation validation of the chemical process is performed throughout, including variations of process parameters for intermediates that are not isolated. For these intermediates a pre-selected analytical method is normally

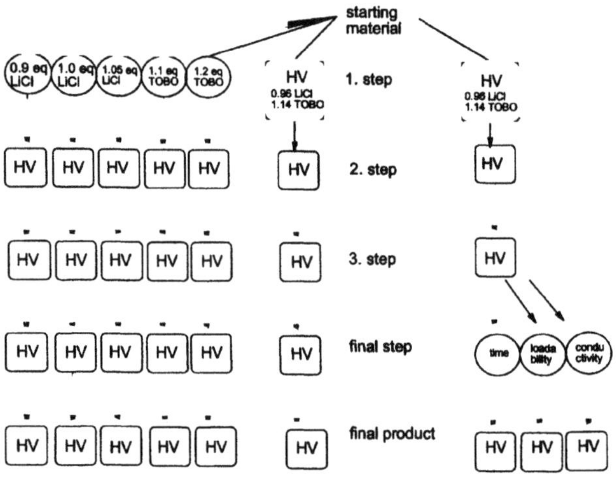

Fig. 1 The synthesis and variations performed in a process research validation shown schematically. HV stands for performed according to the manufacturing formula. The example shown illustrates the first step of the chemical synthesis. *Circles* show the variations of the two parameters tested in the validation (LiCl and TOBO are internal names for chemicals used in the synthesis, eq stands for equivalent)

available, usually a method that has been employed for some time. Because no validation report is requested by regulatory authorities, no formal validation is performed.

An example of a typical validation scheme for a chemical synthesis in process research is given in Fig. 1. In this example the process research chemist sets up a number of experiments. The most important question presented by the chemist to the analytical department is: Is the variability of the analytical process small compared to the variation performed in the process validation [1, 2].

The original manufacturing formula (HV) and five variations are performed in the first step of the synthesis. Six samples are analysed. The results of these six analyses are used to assess the validation of this process step. In this case validation of the analytical method is a prerequisite for any decision that is made about the validity of the process. This information is needed before the process research chemist can start variations of the process otherwise it is possible that the data received cannot be assessed. The difficulty of assessing the data of the process validation results from the fact that the data is influenced by the analytical method and the uncertainty of the chemical process. If the uncertainty of the analytical method is larger or in the same range as the variations of the chemical process, assessment of the data is not possible.

To assess uncertainty, a type B estimation of uncertainty is performed. After analysis of the samples the type B estimation can be verified by a type A estimation analogous to the concept presented by Henrion et al. [3]. In our laboratory, a type B estimation can be performed with confidence because we have performed this type of analysis (an HPLC method with external standard calibration) about 50 times before. A control sample is included in every analysis and the results are plotted on a control chart. The control chart, representing the total measurement uncertainty for the analytical method, is then reviewed for the estimation of uncertainty. The standard deviation for the assay of a control sample for 50 analyses was 1.51%. An example of the way in which a control sample can be used for measurement uncertainty is presented in detail in [1]. The typical uncertainty components for this type of analysis are (Table 1):

— inhomogeneity of the sample and sampling
— weighing of reference materials (about 50 mg)
— weighing of samples (about 50 mg)
— uncertainty of the instrumentation (sampler uncertainty)
— evaluation uncertainty (uncertainty caused by integration)
— uncertainty of reference materials.

The case presented here is simple because all six samples can be analysed as one set of samples in one analytical run. For a single run the estimation of uncertainty can be reduced to the calibration of the method with an in-house reference material and the uncertainty of the analysis itself. In this example inhomogeneity of the sample, the evaluation uncertainty and the uncertainty of the reference material (i.e. inhomogeneity) as given in Table 1 can be neglected because the results of only one analysis are compared. The weighing uncertainties can be neglected because they are small compared to the sampler uncertainty.

If the measurement uncertainty is estimated using the parameters in Table 1 but for a longer time period

Table 1 First step in the estimation of the measurement uncertainty of high performance liquid chromatography (HPLC) analysis as a method used for the assessment of a chemical process

Uncertainty component	Min	Max
Inhomogeneity of the sample and inhomogeneity of sampling	<0.2%	0.5%
Weighing of the reference materials or the sample (about 50 mg)	<0.1%	0.1%
Uncertainty of the instrumentation (sampler uncertainty) and reference material(s) depending on the number of injections	0.5%	1.5%
Evaluation uncertainty (uncertainty caused by integration)	<0.1%	0.1%
Uncertainty of the reference material(s)	–	–

Table 2 Results of the estimation of the measurement uncertainty for HPLC analysis

	Option 1[a]	Option 2[b]
Calibration	1%	1%
Analysis of samples	1.5%	1%
Total uncertainty (received from uncertainty propagation)	1.8%	1.4%

[a] Option 1: two weights of the calibration standard with six injections for each of them, two weights of the sample with two injections for each weight
[b] Option 2: two weights of the sample with four injections for each weight

Table 3 Comparison of the results of the estimation and the analysis (type B estimation compared to type A estimation)

	Estimation	Found (one example)
Calibration	1%	0.8%
Analysis of samples	1%	0.8%
Total uncertainty (by uncertainty propagation)	1.4%	1.14%

as for example covered in the control chart, the inhomogeneity of the sample has to be included in the estimation. Using the mean values between the min and max column in Table 1 the uncertainty estimated by uncertainty propagation is 1.46%, which is close to the value found in the control chart. In the case presented here the estimation of measurement uncertainty can be performed in only two steps as shown in Table 2. Calibration and analysis of samples represent the major uncertainties and combined they provide the complete uncertainty of our experiment; in this case the uncertainty of the HPLC sampler. The influence of the HPLC sampler is known, therefore, there are two options to perform the analysis.

Together with the customer it was decided to perform the analysis according to option 2. The samples were analysed in one analytical run. The result is shown in Table 3. The result of the estimation compares well with the "found" results. The found uncertainty is smaller than the estimation. Assessment of the results of the validation of the manufacturing formula becomes easier from the customers point of view because the customer is able to deceide if a variation of his result is related to his process or to the uncertainty of the analytical method. Additionally, the influence of the individual contributions to uncertainty becomes smaller because of the uncertainty propagation. Therefore, the difference between the estimated and found uncertainty becomes smaller with an increasing number of parameters that influence uncertainty.

The estimation of uncertainty replaces a full validation of the analytical method. It generates the necessary information at the right time. The statistical information received from the analysis can be used for the interpretation of the data and finally the analysis is designed to the customers needs. In this case measurement uncertainty is a good alternative to validation.

The second example illustrates the determination of water content which is an important characteristic for chemical substances and is needed in many chemical reactions. It is usually determined by Karl-Fischers (KF) titration. The water content determined in our laboratory ranges from <0.1% to about 30%. It is widely known that KF water titration's may be influenced by the sample and, depending on the range, some other parameters may significantly affect the uncertainty.

Because the concentration of a reagent is often determined on the basis of the water content of the reaction mixture, uncertainty information for the water determination is needed. The problem in a development environment is that various synthesis routes are tested. If salts are used in a chemical reaction it is usual that chemists test different counter ions for the optimization of the synthesis. However, different salts are rarely tested in the analytical department. One of the problems of the variation of counter ions is that the hygroscopicity of the salts is often different.

Two independent steps have to be followed:

1. Substance dependent influences have to observed. In most cases the chemical structure of the component is known and therefore serious mistakes can be avoided. Titration software and various KF reagents have to be available and standard operation procedures have to be established.
2. The individual uncertainty has to be considered. Therefore a preliminary specification has to be set and a type B estimation of uncertainty can be used to show if there is any problem arising from the data and the specification limit [4].

The first step is a general task using the appropriate equipment and has been established in our laboratory. However, the second part needs to be discussed in detail.

Suppose there is a chemical reaction with reagent A where:

$$A + B \rightarrow C. \qquad (1)$$

The alternative reaction for A may also be with water to D.

$$A + H_2O \rightarrow D. \qquad (2)$$

The reaction with water is often faster than with B. Because of the low molecular weight of water 0.5% (w/w) of water in B may be 10 mol %. Therefore an excess of at least 10 mol % of A might be needed to complete the reaction. The water content is determined in the

Table 4 Estimation of the measurement uncertainty for the titration of water (example performed manually)

Results from the titrations (% (w/w) from the weight of the sample)	0.20%; 0.23%	0.215% +0.0150%
Minimum uncertainty of 1.1%	1.1%	+0.00236%
Hygroscopicity (estimated on the basis of experience with amin compounds)	5%	+0.00621%
Uncertainty of the titer	1%	+0.00124%
Reaction of the sample with the solvent	5%	+0.00621%
Result reported including uncertainty		0.25%

analytical department. For example the value determined by KF titration is 0.22% (w/w) from two measurements with 0.2% (w/w) and 0.23% (w/w) as the individual values. The question that has to be asked is: Is 0.22% of the weight always smaller than 0.25% (w/w)? What we need is the measurement uncertainty added to the "found" value. If this value is smaller than the limit (0.25%) the pre-calculated amount of the reagents can be used.

The model set up consists of two terms:

$$Y = X^* (1 + U)$$

where X is the mean value of the measurements plus the standard uncertainty. U is the sum of the various influence parameters on measurement uncertainty:
- hygroscopicity
- uncertainty of the titer
- reaction of the KF solvent with the sample
- a minimum uncertainty constant of 1.1% (taken from the control chart of the standard reference material) covering balance uncertainties, the influence of water in the atmosphere and the instrument uncertainty with the detection of the end of titration.

The factors for the three contributions mentioned above are estimated on the basis of experience. The calculation is performed using a computer program [5]. This makes the decision easy and fast. In this case the type A estimation on the basis of the results and the type B estimation of the influence factors are combined. An example is given in Table 4 in a compressed form.

The alternative would be an experimental validation. In this case the uncertainty estimation has proven to be a very useful alternative to validation, although, on the basis of experience, the estimate of hygroscopicity is difficult and may lead to incorrect values.

Conclusions

Method validation is a process used to confirm that an analytical procedure employed for a specific test is suitable for the intended use. The examples above show that the estimation of measurement uncertainty is a viable alternative to validation. The estimation of measurement uncertainty can be used to confirm that an analytical procedure is suitable for the intended use. If the estimation of measurement uncertainty is used together with validations both the uncertainty estimation and the validation have their own place in a developmental environment. The major advantages of measurement uncertainty are that it is fast and efficient. Normally, if the analytical method is understood by the laboratory very similar results are found for the estimation of uncertainty and for the classical variation of critical parameters, namely, validation. The decision on how to perform a validation should be made on a case to case basis depending on experience.

Acknowledgements Fruitful scientific discussions with Dr. P. Blaszkiewicz, Schering AG, Berlin and Dr. W. Hässelbarth, BAM, Berlin are gratefully acknowledged.

References

1. Küppers S (1997) Accred Qual Assur 2:30–35
2. Küppers S (1997) Accred Qual Assur 2:338–341
3. Henrion A, Dube G, Richter W (1997) Fres J Anal Chem 358:506–508
4. Renger B (1997) Pharm Tech Europe 9:36–44
5. Evaluation of uncertainty (1997) R. Metrodata GmbH, Grenzach-Whylen, Germany

Accred Qual Assur (1998) 3:155–160

Ricardo J. N. Bettencourt da Silva
M. Filomena G. F. C. Camões
João Seabra e Barros

Validation of the uncertainty evaluation for the determination of metals in solid samples by atomic spectrometry

Presented at: 2nd EURACHEM Workshop on Measurement Uncertainty in Chemical Analysis, Berlin, 29–30 September 1997

R. J. N. Bettencourt da Silva (✉)
M. F. G. F. C. Camões
CECUL, Faculdade de Ciências da Universidade de Lisboa, P-1700 Lisbon, Portugal

J. Seabra e Barros
Instituto Nacional de Engenharia e Tecnologia Industrial, Estrada do Paço do Lumiar, P-1699 Lisbon Codex, Portugal

Abstract Every analytical result should be expressed with some indication of its quality. The uncertainty as defined by Eurachem ("parameter associated with the result of a measurement that characterises the dispersion of the values that could reasonably be attributed to the, ..., quantity subjected to measurement") is a good tool to accomplish this goal in quantitative analysis. Eurachem has produced a guide to the estimation of the uncertainty attached to an analytical result. Indeed, the estimation of the total uncertainty by using uncertainty propagation laws is components-dependent. The estimation of some of those components is based on subjective criteria. The identification of the uncertainty sources and of their importance, for the same method, can vary from analyst to analyst. It is important to develop tools which will support each choice and approximation. In this work, the comparison of an estimated uncertainty with an experimentally assessed one, through a variance test, is performed. This approach is applied to the determination by atomic absorption of manganese in digested samples of lettuce leaves. The total uncertainty estimation is calculated assuming 100% digestion efficiency with negligible uncertainty. This assumption was tested.

Key words Uncertainty · Validation · Quality control · Solid samples · Atomic spectrometry

Introduction

The presentation of an analytical result must be accompanied by some indication of the data quality. This information is essential for the interpretation of the analytical result. The comparison of two results cannot be performed without knowledge of their quality. Eurachem [1] defined uncertainty as the "parameter associated with the result of a measurement that characterises the dispersion of the values that could reasonably be attributed to the, ..., quantity subjected to measurement" and presented it as a tool to describe that quality.

The Eurachem guide for "Quantifying uncertainty in analytical measurement", which is based on the application of the ISO guide [2] to the chemical problem, was observed. ISO aims at the estimation of uncertainty in the most exact possible manner, in order to avoid excess of confidence in overestimated results. The application of these guides turns out to be a powerful tool. The exact estimation of uncertainties is important for the detection of small trends in analytical data. The time and effort used in such estimations can avoid many further doubts concerning observation of legal limits and protects the user of the analytical data from financial losses. The use of uncertainty instead of less informative percentage criteria brings considerable benefits to the daily quality control.

Despite the analyst's experience, some analytical steps like sampling and recovery are of particularly difficult estimation. Mechanisms should be developed to support certain choices or approximations. The comparison of an estimated uncertainty with the experimentally assessed one can be of help.

In this work the Eurachem guide [1] was used for the estimation of uncertainties involved in the determination by electrothermic atomic absorption spectrometry (EAAS) of manganese in digested lettuce leaves. The total uncertainty estimation was calculated assuming a 100% digestion efficiency with negligible uncertainty. The experimental precision was compared with an estimated one for the purpose of validation of the proposed method of evaluation. After this validation the uncertainty estimation was used in an accuracy test and in routine analysis with the support of a spreadsheet programme.

The uncertainty estimation process

The uncertainty estimation can be divided into four steps [1]: (1) specification, (2) identification of uncertainty sources, (3) quantification of uncertainty components, and (4) total uncertainty estimation.

Specification

A dry-base content determination method is proposed, the sample moisture determination being done in parallel. Figure 1 represents the different steps of the analysis. The analytical procedure was developed for laboratory samples. Sampling uncertainties were not considered.

The dry-base correction factor, $f_{\mathrm{corr.}}$, is calculated from the weights of the vial (z), vial plus non-dried sample (x) and vial plus dry sample (y)

$$f_{\mathrm{corr.}} = 1 - \frac{x-y}{x-z} \qquad (1)$$

The sample metal content, M, is obtained from the interpolated concentration in the calibration curve, C_{inter}, the mass of the diluted digested sample, a, and the dilution factor, $f_{\mathrm{dil.}}$, (digested sample volume times dilution ratio).

$$M = \frac{C_{\mathrm{Inter.}} \times f_{\mathrm{dil.}}}{a} \qquad (2)$$

The dry-base content, D, is obtained by application of the correction factor, $f_{\mathrm{corr.}}$, to the metal content, M.

$$D = f_{\mathrm{corr.}} M \qquad (3)$$

Identification of uncertainty sources

The uncertainty associated with the determination of $f_{\mathrm{corr.}}$ is estimated from the combination of the three involved weighing steps, Fig. 1a.

The uncertainty associated with the sample metal content is estimated from the weighing, dilution and interpolation sources (Fig. 1b). The model used for the calculation of the contribution from the interpolation source assumes negligible standards preparation uncertainty when compared with the instrumental random oscillation [5, 10].

Quantification of the uncertainty components

The quantification of the uncertainty is divided into equally treated operations:

Gravimetric operations

The weighing operations are present in the dry-base correction factor (three) and in the sample metal content (one). Two contributions for the associated uncertainty, $\sigma_{\mathrm{Weighing}}$, were studied:

1. Uncertainty associated with the repeatability of the weighing operations, $\sigma^{\mathrm{Balance}}_{\mathrm{Repeat.}}$, is obtained directly from the standard deviation of successive weighing operations. The corresponding degrees of freedom are the number of replicates minus 1.

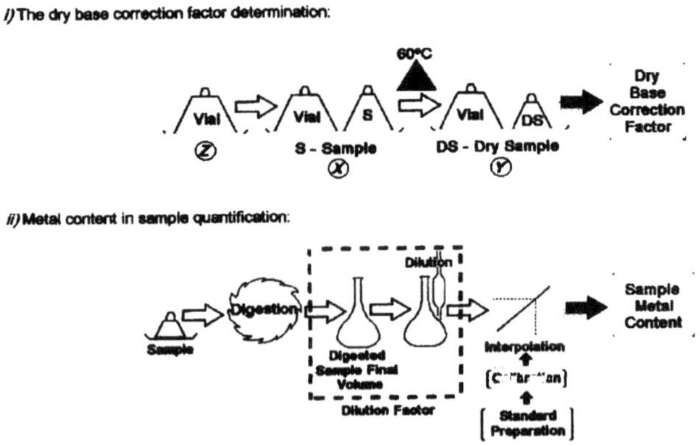

Fig. 1 Proposed method for dry-base metal content determination in lettuce leaves

2. Uncertainty associated with the balance calibration, $\sigma_{\text{Calib.}}^{\text{Balance}}$, defined by

$$\sigma_{\text{Calib.}}^{\text{Balance}} = \frac{2 \times \textit{Tolerance}}{\sqrt{12}} \tag{4}$$

where the *Tolerance* is obtained from the balance calibration certificate.

The Eurachem guide suggests that when the uncertainty components are described by a confidence interval, $\alpha \pm \beta$, without information on degrees of freedom, the associated uncertainty is $2\beta/\sqrt{12}$, which represents the uncertainty of a 2β amplitude rectangular distribution. These uncertainties are designated type B. The number of degrees of freedom associated with the $\sigma_{\text{Calib.}}^{\text{Balance}}$ type B estimation, $\nu_{\text{Calib.}}^{\text{Balance}}$, is approximately [1] equal to

$$\nu_{\text{Calib.}}^{\text{Balance}} \cong \frac{1}{2}\left[\frac{\sigma_{\text{Calib.}}^{\text{Balance}}}{m_{\text{Calib.}}^{\text{Balance}}}\right]^{-2} \tag{5}$$

were $m_{\text{Calib.}}^{\text{Balance}}$ is the mass associated with the balance calibration tolerance.

The two uncertainties are then combined

$$\sigma_{\text{Weiging}} = \sqrt{(\sigma_{\text{Calib.}}^{\text{Balance}})^2 + (\sigma_{\text{Repeat.}}^{\text{Balance}})^2} \tag{6}$$

The corresponding degrees of freedom are calculated by the Welch-Satterwaite equation [1–3]. When the pairs (uncertainty, degrees of freedom)

$(\sigma_a, \nu_a); (\sigma_b, \nu_b); (\sigma_c, \nu_c); (\sigma_d, \nu_d); \ldots$

for the quantities $a, b, c, d, \ldots$, in a function $\nabla = f(a, b, c, d, \ldots)$ are taken into account, then the effective number of degrees of freedom associated with ∇, ν_∇, is

$$\nu_\nabla = \frac{\sigma_\nabla^4}{\left(\frac{\partial f}{\partial a}\right)^4 \frac{\sigma_a^4}{\nu_a} + \left(\frac{\partial f}{\partial b}\right)^4 \frac{\sigma_b^4}{\nu_b} + \left(\frac{\partial f}{\partial c}\right)^4 \frac{\sigma_c^4}{\nu_c} + \left(\frac{\partial f}{\partial d}\right)^4 \frac{\sigma_d^4}{\nu_d} + \ldots} \tag{7}$$

The calculation of the uncertainty and of the degrees of freedom associated with the sample weight is by the direct application of Eqs. 4–7. The calculations of the dry-base correction factor are more elaborate.

3. Uncertainty associated with the dry base factor

The dry base correction factor is a function of three weighing operations (Eq. 1). To estimate the uncertainty, $\sigma_{f\text{corr.}}$, associated with the $f_{\text{corr.}}$, the general equation (Eq. 8) was used [1]

$$\sigma_{f_{\text{corr.}}} = \sqrt{\left(\frac{\partial f_{\text{corr.}}}{\partial x}\right)^2 \sigma_x^2 + \left(\frac{\partial f_{\text{corr.}}}{\partial y}\right)^2 \sigma_y^2 + \left(\frac{\partial f_{\text{corr.}}}{\partial z}\right)^2 \sigma_z^2} \tag{8}$$

It is therefore

$$\sigma_{f_{\text{corr.}}} = \sqrt{\left(\frac{y-z}{(x-z)^2}\right)^2 \sigma_x^2 + \left(-\frac{1}{(x-z)}\right)^2 \sigma_y^2 + \left(\frac{x-y}{(x-z)^2}\right)^2 \sigma_z^2} \tag{9}$$

The values of σ_x, σ_y and σ_z are then calculated as described in the section "Gravimetric operations" above. The number of degrees of freedom is calculated by the Welch-Satterwaite equation (Eq. 7). The application of a spreadsheet program available in the literature simplifies this task [3]. However, the classical approach is more flexible for different experimental configurations or for one or more dilution steps, and is also easily automated.

Volumetric operations

The uncertainties associated with to the volumetric operations were calculated from the combination of two [(1) and (2) below] or three [(1), (2) and (3) below] components:

1. Uncertainty associated with volume calibrations, $\sigma_{\text{Calib.}}^{\text{Vol.}}$

$$\sigma_{\text{Calib.}}^{\text{Vol.}} = \frac{2 \times \textit{Tolerance}}{\sqrt{12}} \tag{10}$$

where the information on this tolerance is normally available with the instrument in the form: volumetric instrument volume ± tolerance. This type B uncertainty estimation has the same treatment as the one reported in Eq. 5 for the degrees of freedom

$$\nu_{\text{Calib.}}^{\text{Vol.}} \cong \frac{1}{2}\left[\frac{\sigma_{\text{Calib.}}^{\text{Vol.}}}{V}\right]^{-2} \tag{11}$$

where $\nu_{\text{Calib.}}^{\text{Vol.}}$ is the number of degrees of freedom associated with $\sigma_{\text{Calib.}}^{\text{Vol.}}$ for a certain volume V.

2. Uncertainty associated with volume repeatability tests, $\sigma_{\text{Repeat.}}^{\text{Vol.}}$

The $\sigma_{\text{Repeat.}}^{\text{Vol.}}$ and the corresponding degrees of freedom, $\nu_{\text{Repeat.}}^{\text{Vol.}}$, are also extracted directly from the repeatability tests. Such tests consist of successive weighings of water volumes measured by the instrument. The observed standard deviation is a function of the analyst's expertise.

3. Uncertainty associated with the use of volumetric equipment at a temperature different from that of calibration, $\sigma_{\text{Temp.}}^{\text{Vol.}}$

This third component corrects for errors associated with the use of 20 °C calibrated material in 20 ± 3 °C solutions. When two consecutive volumetric operations are performed at the same temperature, as is the case in dilution stages, they become self-corrected for this effect.

The glass instrument expansion coefficient is much smaller than that of the solution. For this reason we have only calculated the latter. For a temperature oscillation of $\Delta T = \pm 3K$ with a 95% significance level and for a volumetric expansion coefficient of pure water of $2.1 \times 10^{-4}\,°C^{-1}$ (our solutions can be treated as pure water because of their low concentrations), the 95% volume confidence interval becomes $V \pm V \times 3 \times 2.1 \times 10^{-4}$. Dividing the expanded uncertainty by the Student t value, $t(\infty,95\%)=1.96$, we obtain the temperature effect component uncertainty

$$\sigma^{Vol.}_{Temp.} = \frac{V \times 3 \times 2.1 \times 10^{-4}}{1.96} \quad (12)$$

The number of degrees of freedom due to the temperature effect can also be estimated as for $\nu^{Vol.}_{Calib.}$ (Eq. 11), substituting $\sigma^{Vol.}_{Calib.}$ by $\sigma^{Vol.}_{Temp.}$

These components are then combined to calculate the volume uncertainty, $\sigma_{Vol.}$

$$\sigma_{Vol.} = \sqrt{(\sigma^{Vol.}_{Calib.})^2 + (\sigma^{Vol.}_{Repeat.})^2 + (\sigma^{Vol.}_{Temp.})^2} \quad (13)$$

The number of degrees of freedom associated with $\sigma_{Vol.}$ can also be calculated by the Welch-Satterwaite equation.

4. Uncertainty associated with the dilution factor

Our analytical method has three volumetric steps that can be combined as a dilution factor, $f_{dil.}$, whose uncertainty, $\sigma_{f_{dil.}}$, can easily be estimated by:

$$\frac{\sigma_{f_{dil.}}}{f_{dil.}} = \sqrt{\left(\frac{\sigma^{DSV}_{Vol.}}{V^{DSV}}\right)^2 + \left(\frac{\sigma^{P}_{Vol.}}{V^{P}}\right)^2 + \left(\frac{\sigma^{V}_{Vol.}}{V^{V}}\right)^2} \quad (14)$$

were the DSV, P and V stand respectively for digested solution volume, dilution operation pipette and dilution operation vial; $\sigma_{Vol.}$ and V represent respectively each corresponding volumetric uncertainty and volume. As in the other cases, the degrees of freedom were calculated by the Welch-Satterthwaite equation.

Sample signal interpolation from a calibration curve

The mathematical model used to describe our calibration curve was validated by the Penninckx et al. [4] method. At this stage we proved the good fitting properties of the unweighted linear model to our calibration curve. With this treatment we aimed not only at the accuracy but also at the estimation of more realistic sample signal interpolation uncertainties. These uncertainties were obtained by the application of an ISO international standard [5].

The instrument was calibrated with four standards (0–2–4–6 µg/L for Mn) with three measurement replicates each [4]. Samples and control standard (4 µg/L for Mn) were also measured three times. The control standard was analysed for calibration curve quality control (see "Quality control").

Total uncertainty estimation

The total uncertainty estimation, σ_T, is a function of the dry-base correction factor uncertainty, $\sigma_{f_{corr.}}$, of the uncertainty associated to the analysis sample weighing operation, $\sigma^{Sample}_{Weighing}$, of the dilution factor, $\sigma_{f_{dil.}}$, and of the instrumental calibration interpolated uncertainty, $\sigma_{C_{inter.}}$. These four quantities combine their uncertainties in the equation

$$\frac{\sigma_T}{D} = \sqrt{\left(\frac{\sigma^{Sample}_{Weighing}}{a}\right)^2 + \left(\frac{\sigma_{f_{corr.}}}{f_{corr.}}\right)^2 + \left(\frac{\sigma_{f_{dil.}}}{f_{dil.}}\right)^2 + \left(\frac{\sigma_{C_{Inter.}}}{C_{Inter.}}\right)^2} \quad (15)$$

where D represents the dry-base sample metal content and a has the same meaning as in Eq. 2. The other quantities have already been described.

The expanded uncertainty can then be estimated after the calculation of the effective number of degrees of freedom, df (Eq. 7). Therefore the coverage factor used was the Student t defined for that number and a 95% significance level (t(df,95%)). The estimated confidence interval is defined by

$$D \pm \sigma_T \cdot t(\text{df},95\%) \quad (16)$$

Quality control

Ideally, the readings of the instruments for each sample and for each standard should be random [6–7]. Normally, the instrument software separates the calibration from the sample reading. Although this allows an immediate calculation, it can produce gross errors if the operator does not verify the drift of the instrument response. For this reason, the calibration curves should be tested from time to time by reading a well-known control standard. This standard can also be prepared from another mother solution in respect to the calibration standards, for stability and preparation checking.

Normally, the laboratories use fixed and inflexible criteria for this control. They define a limit to the percentage difference between the expected and the obtained value, and in low precision techniques they are obliged to increase this value. Assuming the uncertainty associated with the control standard preparation to be negligible when compared to the instrumental uncertainty, the case-to-case interpolation uncertainties can be used as a fit for each case. If the observed confidence interval includes the expected value, there is reason to think that the system is not under control. The instrumental deviation from control can be used as a guide for instrumental checking or as a warning of the inadequacy of the chosen mathematical model for the calibration.

Validation of the uncertainty estimation

Method validation is the process of demonstrating the ability of a method to produce reliable results [8]. An analytical result should be expresed along with a confidence interval and a confidence level. The confidence interval can be described by a mean value and a interval width. Therefore the validation depends on the reliability of the confidence interval width estimation. The accuracy test can be performed exactly only, after that step.

The statistical equivalence between the estimated and the observed values can be used to confirm that quality. The F-test [10] is a good tool for comparing (non-expanded) uncertainties.

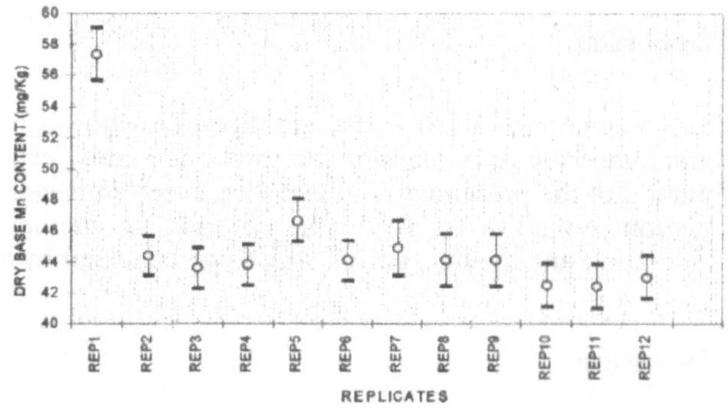

Fig. 2 Repeatability test. The confidence intervals are represented by the average value plus the estimated expanded uncertainty for a 95% confidence level

Application of uncertainty validation schemes

The proposed uncertainty validation method was applied to the dry-base determination of manganese in digested lettuce leaves by electrothermic atomic absorption spectrometry. The proposed quality control scheme was also applied.

The 200-mg samples were digested with nitric acid in a microwave-irradiated closed system [11]. The instrumental determination was performed in a GBC atomic spectrometer with D_2 lamp background correction. A Pd/Mg mixture [12] was used as chemical modifier. The dry-base correction factor was calculated by a parallel assay. The samples were dried in an oven at 60 °C under atmospheric pressure, and the CRM (certified reference material – NIST 1570a) was treated as specified by NIST [13].

Repeatability test

The estimated uncertainties were compared with the experimental ones by an F-test for the 95% confidence level [10]. Figure 2 represents the obtained experimental values associated with the estimated expanded uncertainty (95% confidence level). The coverage factor used was 1.96 for the average effective number of degrees of freedom, df, of 57500. The Eurachem [1] proposal of a coverage factor of 2 is adequate for this case.

The replicates 1 and 5 (REP1, REP5) are consecutive single outliers (Grubbs test) for a 95% confidence level [9]. Therefore, they have not been used for the experimental uncertainty calculation. The two uncertainties are statistically equivalent for the test used (experimental uncertainty: 0.82 mg/Kg for 9 df; estimated uncertainty: 0.73 mg/Kg for 57500 df) at the 95% confidence level.

Accuracy test

The accuracy test was performed with spinach leaves (NIST 1570a) because of their claimed similarity with lettuce leaves in terms of proteins, carbohydrates, fibre and inorganic matter content [14]. The validated uncertainty estimation was used for the comparison of obtained values with certified ones (Fig. 3).

The loss of precision in EAAS with the time of use of furnace is taken into account in the case-to-case interpolation uncertainty calculation. The accuracy is retained with a larger confidence interval.

The overlapping of these intervals indicates that there is no reason to think that our method lacks accuracy. Our analytical method can be considered to perform as badly or as well as the NIST methods. This assumption seems sufficiently valid to consider the method validated.

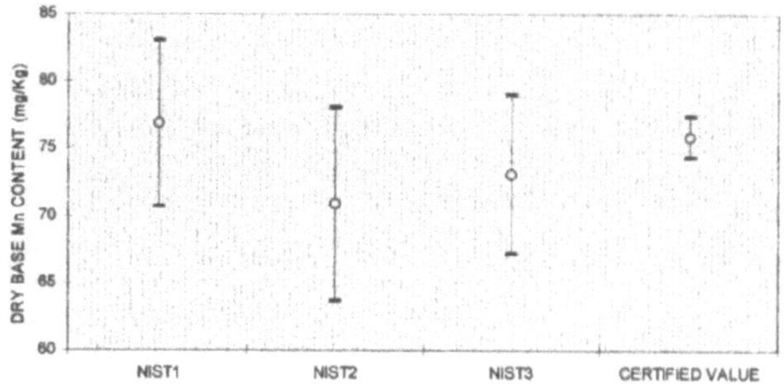

Fig. 3 Accuracy test over spinach leaves NIST CRM. The obtained values (NIST 1, 2 and 3) were associated with a 95% confidence level expanded uncertainty. The certified value is also presented for the same confidence level

Conclusions

The assumption of 100% efficient digestion with negligible uncertainty is valid for the total uncertainty estimation of the presented example. This uncertainty estimation proved to be a valuable criterion for method validation and quality control, which can be tested by a very simple procedure. The easy routine use of an exact treatment in a spreadsheet program can be useful for the more demanding situations. Nevertheless, further approximations can be easily tested.

Acknowledgements Thanks are due to JNICT for financial support to CONTROLAB LDA, for the instrumental facilities that made this work possible, as well as to the teams of INETI and CONTROLAB LDA, for their support.

References

1. Eurachem (1995) Quantifying uncertainty in analytical measurement, version 6
2. ISO (1993) Guide to the expression of uncertainty in measurement, Switzerland
3. Kargten J (1994) Analyse 119:2161–2165
4. Penninckx W, Hartmann C, Massart DL, Smeyers-Verbeke J (1996) J Anal At Spectrom 11:237–246
5. ISO International Standard 8466-1 (1990) Water quality – calibration and evaluation of analytical methods and estimation of performance characteristics – Part 1: Statistical evaluation of performance characteristics, Geneva
6. Analytical Methods Committee (1994) Analyse 119:2363–2366
7. Staats G (1995) Fresenius J Anal Chem 352:413–419
8. Taylor JK (1983) Anal Chem 55:600A–608A
9. Grubbs FE, Beck G (1972) Technometrics 14:847–854
10. Miller JC, Miller JN (1988) Statistics for analytical chemistry (2nd edn). Wiley, UK
11. Deaker M, Maher W (1995) J Anal At Spectrom 10:423–431
12. Soares ME, Bastos ML, Carvalho F, Ferreira M (1995) At Spectrosc 4:149–153
13. NIST (1994) Certificate of Analysis SRM1570a Trace elements in spinach leaves
14. Penninckx W, Smeyers-Verbeke J, Vankeerberghen P, Massart DL (1996) Anal Chem 68:481–489

Accred Qual Assur (2000) 5:495–498

Hans Malissa
Wolfgang Riepe

Statistical evaluation of uncertainty for rapid tests with discrete readings – examination of wastes and soils

H. Malissa · W. Riepe (✉)
University of Salzburg,
Institute of Chemistry and Biochemistry,
Hellbrunnerstrasse 34, 5020 Salzburg,
Austria
e-mail: wriepe@natur.sbg.ac.at

Abstract In the course of the colorimetric determination of analytes using a procedure with discrete readings the measurement uncertainly cannot be calculated in the normally practiced manner. The basic principle of the analytical method used is a stepwise and non-equidistant reading. Based on the fact that half a step can be estimated, a calculation of the measurement uncertainty for the 95% confidence level is possible; this is needed to allow a reliable decision of whether a critical value is exceeded or not.

Keywords Colorimetric · Rapid tests · Discrete reading · Statistical evaluation

Introduction

Analytical rapid tests attain increasing importance for the prompt and reliable characterization of soil and waste samples. Advantages are simple and convenient performance, almost immediate delivery of results, and low costs [1]. Thus it is possible to characterize and classify samples [2] directly on-site in the field.

Various situations can be realized which demand an instantaneous result rather than an accurate one because immediate decisions need to be taken, e.g., a rapid survey of contaminations caused by an accident, instant classification of wastes upon delivery at the disposal area, or preliminary analytical assessment of a suspected landfill site. However, in order to evaluate the analytical results obtained with rapid tests, typical quality criteria like measurement uncertainty, limit of detection, and limit of quantitation must be known or must be calculated, respectively, as is required for any other analytical method.

Any of these situations – we may call it the "site" to be assessed – renders one or several groups of samples with similar composition, which we designate as "sample families". For example, a batch of galvanic sludge is a site from which a sample family of *m* samples is taken, which are each analyzed with *n* parallel determinations. In contrast, the samples taken from a site of an accident can be divided into two sample families, one comprising samples from the obvious contaminated area and the other from the surroundings. However, it should be noted that in waste and soil analysis the classification of samples into sample families is very often a matter of discretion.

Principle of discrete readings

For a number of elements (Table 1), commercial colorimetric test-sets for water analysis [3] were modified and extended to be applicable for measurements in soil and waste eluates [4]. In all these tests the analytical results were obtained by visual comparison of the sample color with that of a series of permanent color standards specifically designed for that particular test and mounted on a disc comparator. Each color standard represents a defined concentration. A typical test set for one element contains five to ten concentration steps. These concentration intervals are normally not equidistant (Table 1). In practice it will be found that the color intensity of the unknown lies between two successive color standards. An interpretation may be difficult in these cases but it is reasonable that the analyst will be able to read with some certainty a value halfway between two color

Table 1 Working ranges of the test sets

color standard #.	Cu concentr. mg/L	Ni concentr. mg/L	Zn concentr. mg/L	Cr concentr. mg/L	As concentr. mg/L	Cd concentr. mg/L	Pb concentr. mg/L
1	0	0	0	0	0	0	0
2	0.3	0.5	0.1	0.1	0.1	0.01	0.3
3	0.6	1.0	0.2	0.2	0.5	0.03	0.6
4	1.0	1.5	0.3	0.35	1.0	0.05	1.0
5	1.5	2.0	0.4	0.6	1.7	0.07	
6	2.0	3.0	0.5	1.0	3.0	0.1	
7	3.0	4.0	0.7	1.8		0.3	
8	5.0	6.0	1.0	3.0		0.5	
9	7.0	8.0	2.0	6.0		1.0	
10	10.0	10.0	5.0	10.0			

standards (half-value reading). The exact procedure for the rapid tests is described in [5].

Requirements for rapid tests

Typically, the aim of any analytical characterization of wastes and contaminated soils is to answer the question whether a specific critical value is exceeded or not. This may have several consequences: for instance a waste is classified as hazardous waste and must be deposed of accordingly, or a contaminated area must be cleaned up by decontamination procedures.

A critical value (*cv*) is definitely exceeded if the analytical result ($\bar{c}$) and its confidence interval ($\Delta\bar{c}$) is greater than that critical value.

The total error of an analytical result is given by three terms: Uncertainty of the sampling process, of sample preparation, and analytical determination. Uncertainty must be expressed in terms of the variance, which is equal to the square of the standard deviation s for the propagation.

$$s^2_{total} = s^2_{samp.} + s^2_{prep.} + s^2_{anal.}$$

If we take a numerical example from the domain of inhomogeneous samples like solid wastes and soils the sampling error is predominant by far: we assume the relative standard deviation due to the sampling error as being 50%, that due to the sample preparation 10%, and that due to the analytical determination 5%. This gives:

$$s_{total} = \sqrt{50^2 + 10^2 + 5^2} = \sqrt{2625} = 51\%$$

It is obvious from this very simple calculation that in practice the total error is in fact determined by the sampling process and only insignificantly by the analytical operation. Only if the analytical error is in the same order as the sampling error will its contribution to total error be remarkable. If we examine very heterogeneous materials like soils and wastes the sampling error is predominant. This fact has consequences for the analytical determination: It is not at all necessary to employ a very precise analytical method which is in general costly, complex, and time-consuming but rather a rapid method which is less precise but can be performed readily. These rapid tests allow us to generate a greater number of analytical data points in obviously less time, which provides for a more representative declaration of any inhomogeneous material.

Statistical requirements

Some important parameters and relations shall be defined:

Critical value (cv)

The decision whether a measured analytical result, given as a mean concentration $\bar{c}$ is definitely (95% significance in our study) below the critical value, can be expressed as: $\bar{c} + \Delta\bar{c} < cv$ where $\bar{c}$ = mean value and $\Delta\bar{c}$ = confidence interval.

Confidence interval

The confidence interval ($\Delta\bar{c}$) is given by

$$\Delta\bar{c} = \frac{t(P,f) \times s}{\sqrt{n}} \quad (1)$$

where n = number of parallel measurements obtained from one sample, t = Student's factor, s = standard deviation of $\bar{c}$, and P = level of significance.

The term parallel measurements comprises the entire analytical procedure including sampling and sample preparation.

Degree of freedom

The number of degrees of freedom (n–1) for one sample denotes the number of control measurements performed on that sample, which are supposed to confirm the first result. If measurements are performed on a number of similar samples (sample family), this number m is accounted for the calculation of f. Introducing j as the sample number within a sample family ranging between 1 and m and on the condition that the number of parallel determinations n_j is equal for each sample, the number of degrees of freedom is:

$$f = m \times (n_j - 1)$$

where m = number of samples in a sample family, f = number of degrees of freedom, and n_j = number of parallel measurements obtained from sample j.

Range

The spread of a smaller number of results of a measurement series (<10) is characterized preferably by the range, which is the difference between maximum and minimum values [6]:

$$R = c_{\max} - c_{\min}$$

The condition stated above allows us to define an average range that can be defined for m samples of a sample family:

$$\bar{R} = \frac{\sum_{j=1}^{m} R_j}{m}$$

Estimated value for the standard deviation s calculated from the range R

An approximation of the standard deviation can be calculated using factors $d(n_j)$ tabulated as a function of m and n_j [7].

$$s = \frac{\bar{R}}{d(n_j)}$$

In the next step this is introduced into Eq. 1 for the confidence interval:

$$\Delta\bar{c} = \frac{t(P;f) \times \bar{R}}{d(n_j) \times \sqrt{n}} \quad (2)$$

Then, all variables of Eq. 2, which are not dependent on the quantity to be measured $\bar{c}$ itself, are combined into a factor F:

$$F = \frac{t(P;f)}{d(n_j) \times \sqrt{n}} \quad (3)$$

Fig. 1 shows the results of a calculation of F (Eq. 3) versus the number of samples in a sample family. $n = 2$ or 4 means duplicate or quadruplicate determinations, respectively, from each sample in that family as well as from any forthcoming single sample, which can be assigned to that family.

On the other hand the influence of n is considerable. However, in the practice of waste and soil analysis hardly more than duplicate measurements will be made. Therefore a factor F=2 will be quite adequate for an established method.

Confidence interval of the mean for procedures with discrete readings

At this point the problem of discrete readings is introduced: in colorimetric procedures using color standards on a comparator wheel, the operator selects that specific standard window which fits the photometric density of the sample as closely as possible. Thus, it will occur in many cases that both readings of a duplicate measurement are identical although the concentrations are different. No variation of measured values, and hence

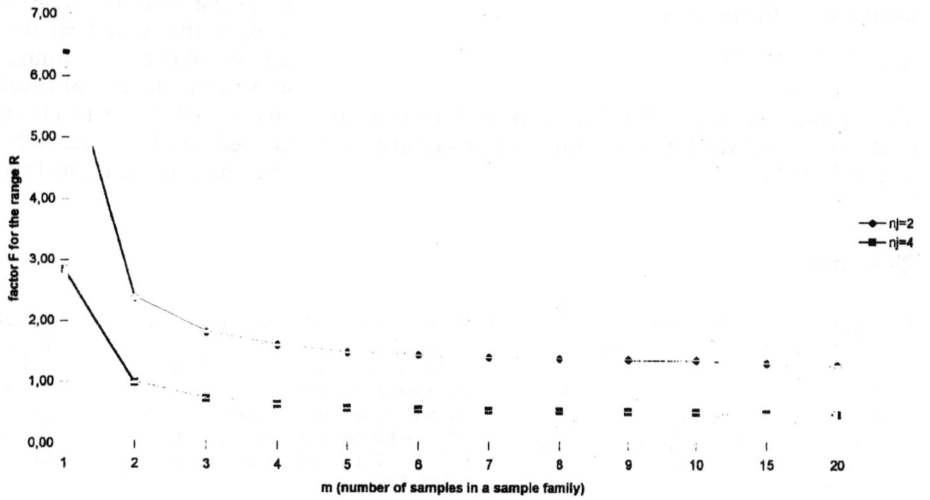

Fig. 1 Dependence of the factor F for multiplying the range of number of samples and parallel determinations

no range R, is observed in these cases because the color standard sequence is too coarse. The question is, how can we still estimate a realistic figure for the confidence interval even in these situations.

An example may illustrate the relevance: two samples ($m=2$) are taken from a batch of galvanic sludge (site) and both are measured in duplicate ($n=2$). The graduation of the standard window may lead to four identical readings. In this case $\bar{R}$, and hence the confidence interval would be zero, although it is well known that a certain degree of inhomogeneity simply exists. Therefore we must derive a figure from the measurement process which is, in fact, the range that could be just recognized and introduce it instead of $\bar{R}$.

It can be recognized from Table 1 that the levels Δc in the concentration scales are not equidistant. When taking the actual readings it will be not be easy for the operator to assign intermediate concentrations by interpolation. However, it can be reasonably expected that he is able to set an imagined limit of photometric density, which will be situated halfway between two concentration levels and may be able to allocate the actual sample density to the higher or lower level, respectively.

An upper and lower limit for each concentration reading c_i may be defined:

Upper limit: $\frac{c_{(i+1)}+c_i}{2}$,

Lower limit: $\frac{c_i+c_{(i-1)}}{2}$,

where c_i = concentration reading for window i, c_{i+1} = concentration reading for window $i+1$, and c_{i-1} = concentration reading for window i-1. All color densities within the interval between these limits are assigned to one and the same concentration c_i. This is equivalent to the statement that the distance between both limits can be regarded as a range R^l of a method, for which a range cannot be derived from the measurement values themselves:

$$R^l = \frac{c_{(i+1)}-c_{(i-1)}}{2}$$

This is introduced into Eq. 2 instead of $\bar{R}$ to give the confidence interval for a colorimetric procedure with discrete readings:

$$\Delta\bar{c} = \frac{t(P;f)}{d(n_j)\times\sqrt{n}} \times \frac{c_{(i+1)}-c_{(i-1)}}{2} \qquad (4)$$

With this expression a confidence interval for a sample family is obtained, which is dependent on the number of samples and determinations used to characterize the object as well as on the level of significance selected. Only those critical values that are different from the mean concentration by more than that confidence interval can be recognized as unambiguously different. It must be emphasized that this is a conversion of the uncertainty of the reading into a concentration uncertainty and does not necessarily reflect a real inhomogeneity.

Conclusions

When doing parallel analysis the first term in Eq. 4 is about 2 (for P=95% and three and more samples) as can be seen from Fig. 1.

Therefrom follows the confidence interval for the mean value:

$$\Delta\bar{c} \approx 2 \times \frac{c_{(i+1)}-c_{(i-1)}}{2}$$

$$\Delta\bar{c} \approx c_{(i+1)}-c_{(i-1)}$$

This means that the difference between the adjacent higher and lower values of the reading step added to the measured concentration $\bar{c}$ is the test figure, which must be compared with the critical value.

The critical value is not exceeded (with 95% level of significance) when the following condition is satisfied:

$$\bar{c}+(c_{(i+1)}-c_{(i-1)}) < \text{critical value}$$

This easy procedure allows us to use rapid tests with a chosen level of significance as a valuable analytical tool. If the analytical uncertainty is estimated as described above, it is found that $\Delta\bar{c}$ has about the same size as the analytical result $\bar{c}$ itself. A reliable decision (95% confidence level) that a critical value is not exceeded can be made only if the result $\bar{c}$ is not greater than half of the critical value itself.

References

1. Unger-Heumann M (1996) Strategy of Analytical Test Kits. Fresenius J Anal Chem 354:803–806
2. Valcárcel M, Cárdenas S, Gallego M (1999) Sample Screening Systems in Analytical Chemistry. Trends in Analytical Chemistry, vol 18, no 11, pp 685–694
3. Merck (1974) Untersuchungen von Wasser Darmstadt. 9. Auflage
4. Götzl A, Malissa H, Riepe W (1997) Analytische Schnellerkennungsmethoden: Bewertung abzulagender Abfälle und Kontrolle von Deponien. UWSF – Z Umweltchem Ökotox 9:245–248
5. Merck (1994) Applications
6. Doerffel K (1990) Statistik in der analytischen Chemie. Dt Verlag der Grundstoffind, Leipzig, p 28
7. Doerffel K (1990) Statistik in der analytischen Chemie. Dt Verlag der Grundstoffind, Leipzig, p 82

Accred Qual Assur (1998) 3:122–126

Gino Stringari
Ivo Pancheri
Frank Möller
Osvaldo Failla

Influence of two grinding methods on the uncertainty of determinations of heavy metals in atomic absorption spectrometry electrothermal atomisation of plant samples

Presented at: 2nd EURACHEM Workshop on Measurement Uncertainty in Chemical Analysis, Berlin, 29–30 September 1997

G. Stringari (✉) · I. Pancheri
Agrarian Institute, I-38010 San Michele all'Adige, Trento, Italy
Tel.: +39-461-615258;
Fax: +39-461-650872;
e-mail: Gino.Stringari@ismaa.it

F. Möller · O. Failla
Faculty of Agriculture, University of Milan, Milan, Italy

Abstract Chemical analyses of trace elements are affected by relatively high analytical errors due to the different steps of the laboratory procedures: samples grinding, mineralisation and instrumental measurements. In the present communication, the influence of the grinding phase on the global uncertainty of Pb, Cd, Ni and Cr determinations in plant samples by the classical method of atomic absorption spectrometry/electrothermal atomisation (AAS-ETA) after dry ashing is quantified. Two grinding machines, a planetary mill with balls and jars of agate versus a stainless steel grinder were compared by analysing leaf samples of cucumber, strawberry, kiwivines, apple trees and grapevines from agricultural experimental plots under controlled conditions. Variance components due to the difference between grinding methods and experimental plots were estimated. Further, the simultaneous effects of the grinding methods on all considered metals have been evaluated by analysis of variance. With the stainless steel grinder, on average, higher levels of the considered heavy metals were obtained (up to 67% of the mean values). On average, the increments were similar for metals contained in steel (Ni and Cr) and those not contained (Pb and Cd). The true causes of these differences need further investigation to determine whether the higher metal detection is due to possible contamination, to a different grinding quality or to other reasons. Finally, the grinding methods did not seem to affect the combined uncertainty of the analyses.

Key words Atomic absorption spectrometry/electrothermal atomisation · Grinding machines · Trace elements · Uncertainty · Variance components

Introduction

Chemical analyses of trace elements still present many problems of uncertainty despite the progress in analytical techniques and instrumental performance. The analyses of trace elements in plant tissues are no exception, although the difficulties associated with these matrices are generally lower than those associated with other organic or mineral matrices.

Several interlaboratory studies [1] indicate that in spite of the standardisation of the procedures [2] data dispersion is considerable and difficult to explain. The same confidence intervals reported for the analysis of the *certified reference materials* highlight that the uncertainty associated with trace elements is generally higher than those associated with macro- and microelements [3].

The possible causes of variability are present in all the analytical steps, which, in atomic absorption spectrometry, can be narrowed to the following three: sample preparation, mineralisation and instrumental measurement. In the recent past, to achieve satisfactory precision or reproducibility, the errors due to the instrumental techniques and/or the matrix mineralisation were investigated [4, 5, 6]. The quality control of these two steps is indeed simplified by the availability of reference materials.

Our attention was centred on sample grinding, an important step in sample preparation upon which sample homogeneity and possible contamination depend.

The object of this communication is to present the results obtained comparing the effects of two grinding devices on analyses for Pb, Cd, Ni and Cr determined in routine procedures by atomic absorption spectrometry with a graphite furnace.

Materials and methods

Chemical analysis

Two grinding machines, a planetary ball mill (*pbm*) (PM 4000-Retsch) with grinding jars and balls of agate versus a rotor-speed mill (Pulverisette 14-Fritsch) stainless steel grinder (*ssg*), were compared, by analysing leaf samples of cucumber (*Cucumis sativus* L.), strawberry (*Fragaria x anassa* Duch.), kiwivines (*Actinidia deliciosa* Liang et Ferg.), apple trees (*Malus pumila* Mill.) and grapevines (*Vitis vinifera* L.) from agricultural experimental plots under controlled conditions consisting in mulching treatments (field treatments) with composts of different origins.

Leaf samples were dried in a dust-free forced draft oven at 70 °C overnight, then coarsely ground by hand, and two 25-g portions were taken for each grinding method (Table 1). Sample mineralisation was carried out according to the procedure recommended by the CII (Comité Inter-Instituts d'étude des techniques analytiques) [3] in a platinum capsule. The ash was treated with HNO_3.

The instrumental measurement was performed on a Varian spectrometer equipped with a graphite tube atomiser and programmable autosampler (Spectra AA-400 Zeeman) with the parameters reported in Table 2.

Statistical analysis

Quality control

An experiment was performed a_x times with two determinations each time, i.e. a total number of $2a_x = N_x$ determinations. Following Stringari et al. [7], the ANOVA SS due to the time, SSA_x, and error, SSE_x, the mean square errors MSA_x have been determined, followed by the test ratio:

$$F = MSA_x/MSE_x$$

the variance components:

$$s^2_{e_x} = MSE_x \textit{ and } s^2_{a_x} = (MSA_x - s^2_{e_x})/N_x$$

and the reproducibility variance which is the square uncertainty for the measurand:

$$u^2(x) = s^2_{R_x} = \begin{cases} s^2_{e_x} + s^2_{a_x}, & \text{for heterogeneous means, i.e. } F \text{ significant} \\ s^2_{e_x}, & \text{for homogenous means.} \end{cases}$$

The control limits for the mean values are given by

$$M_x \pm k s^2_{R_x} \sqrt{1/2 - 1/(2N_x)}$$

Table 1 Sample number and grinding parameters with planetary ball mill

Species	Number of samples	Tissue	No. of balls	Grinding time (min.)	Speed (rpm)
Cucumber	25	Leaf	10	10	300
Strawberry	25	Leaf	12	40	300
Kiwivines	10	Leaf	12	30	300
Apple trees	24	Leaf	11	20	300
Grapevines	32	Leaf	12	40	300

Table 2 Main parameters of the instrumental measurement performed on a Varian spectrometer equipped with a graphite tube atomiser and programmable autosampler (Spectra AA-400 Zeeman)

	PB	Cd	Ni	Cr
Tube		Positions graphites tubes without platform		
Wavelength (nm)	283.3	228.8	232.0	357.9
Ashing (°C)	850	475	975	1100
Modifier of matrix		$NH_4H_2PO_4$ + $Mg(NO_3)_2$	Salts of Pd	
Atomisation (°C)		2100	2500	2550
Measurement mode		Peak height		Peak area
Calibration mode		Standard additions		Calibration curve

with $k=2$, or 2.6, depending on the significance level (5% or 1%).

The upper control limit for the ranges $|x_{i2}-x_{i1}| = 1/2$ standard deviation is $k s_{e_x}$.

Mean values or ranges which lie outside the above control limits are discarded.

Analysis of variance

Two different approaches to analysing the data have been followed, both based on the same linear model

$$y_{ij}=\mu+\alpha_i+\varepsilon_{ij}; \qquad (1)$$

$$i=1, \ldots, a; \quad j=1, \ldots, n_i \quad \Sigma n_i=n$$

The testing of hypotheses and the estimation of variance components are based on the usual analysis of variance.

Source	SQ, SS	*df*	MS	*F*	*E*
Total	SQT	n			
$\varepsilon_{ij} \mid \mu, \alpha_i$	SSE = SQT-SQA	$n-a$	MSE		σ_e^2
$\alpha_i \mid \mu,$	SSA = SQA − SQM	$a-1$	MSA	$F(\alpha)$	σ_a^2
μ	SQM	1	MSM	$F(\mu)$	σ_y^2

where: $SQT=\Sigma\Sigma y_{ij}^2$; $SQA=\Sigma n_i \bar{y}_i^2$; $SQM=n\bar{y}^2$

Analysis based on the single griding method

First the two grinding methods are considered separately putting:

$$y_{ij}=\begin{cases} x_{ij1} \text{ for the griding method } pbm \\ x_{ij2} \text{ for the grinding method } ssg \end{cases} \quad i=1,\ldots,a; j=1,\ldots,n_i \Sigma n_i=n$$

The parameters of the model (Eq. 1) are

μ: the overall mean of the two grinding methods
α_i: the a main effects due to treatments.

The hypothesis to test is

$$H_0: \alpha_i=0, \quad \forall i \qquad (2)$$

The estimates of the above parameters with the "usual" restrictions, are

$$m=\bar{y} \begin{cases} \text{the general mean } m \text{ for } \mu \\ \text{the difference between treatment and} \\ \text{general mean } a_i-m_i-m \text{ for } a_i \end{cases}$$

The hypothesis testing proceeds along the lines of the flow diagram represented in Fig. 1. If the hypothesis of Eq. 2 is refused, the variance components σ_e^2 and σ_a^2 are estimated by

$$s_e^2=\text{MSE}; \; s_a^2=(\text{MSA}-s_e^2)/n_a, \text{ with } n_a=(n-n_m)/(a\text{-}1)$$

leading to a combined uncertainty of $u(y)=\sqrt{s_a^2+s_e^2}$.

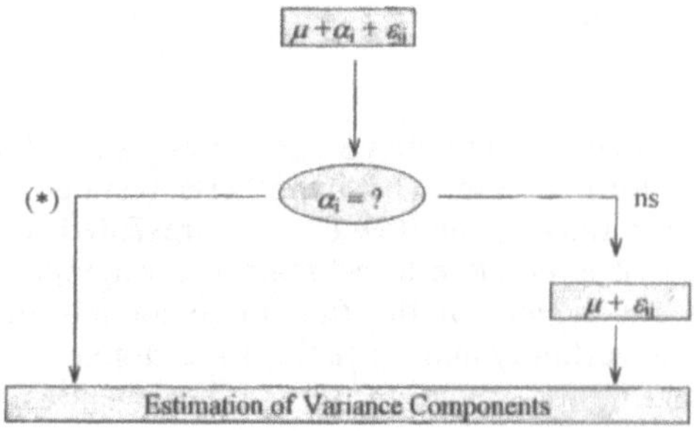

Fig. 1 Statistical analysis based on the single grinding method. Flow diagram for hypothesis testing. (*): significant; ns: not significant

Alternatively the model reduces to $y_{ij}=\mu+\varepsilon_{ij}$, with estimation of the error variance by

$$s_e^2=(\text{SSE}+\text{SSA})/(n-1)$$

and an uncertainty $u(y)=s_e$.

Analysis based on the differences of the grinding methods

The second approach of the analysis is based on the difference and on the relative differences

$$y_{ij}=\begin{cases} x_{ij1}-x_{ij2} \\ 2(x_{ij1}-x_{ij2})/(x_{ij1}+x_{ij2}) \end{cases} \quad i=1, \ldots, a; j=1, \ldots, n_i; \Sigma n_i=n$$

The parameters of the linear model (Eq. 1) have now the following meaning:

μ, the overall mean difference between the two griding methods;
α_i, the a differences due to the treatments.

The variance components are determined according to the acceptance or refusal of the following hypotheses

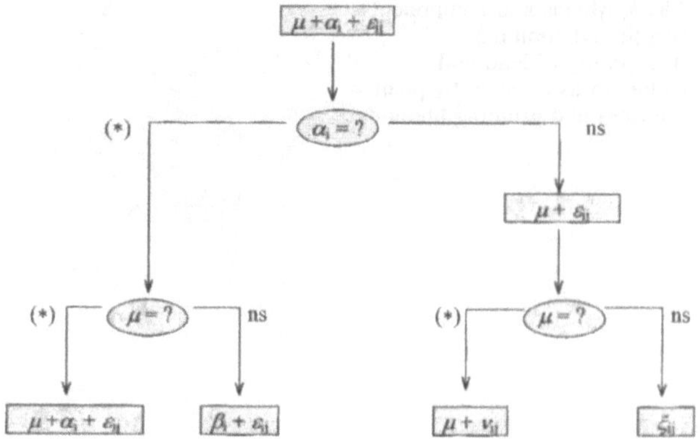

Fig. 2 Statistical analysis based on the difference of the grinding method. Flow diagram for hypothesis testing. (*): significant; ns: not significant

$$H_0: \alpha_i = 0, \ \forall i \quad (3)$$

$$H_0: \mu = 0 \quad (4)$$

The estimates of the above parameters are obtained as in the former case. The hypothesis testing proceeds along the lines of the flow chart represented in Fig. 2.

Thus it is possible to estimate the variance components and hence for the first three models the combined uncertainty and for the last the simple uncertainty $u(y)$.

Final models	Variance components	
$y_{ij} = \mu + \alpha_i + \varepsilon_{ij}$	$s_e^2 = \text{MSE}$	
	$s_a^2 = (\text{MSA} - s_e^2)/n_s$;	$n_a = (n - n_m)/(a-1)$
	$s_y^2 = (\text{MSM} - n_m s_a^2 - s_e^2)/n$;	$n_m = \Sigma n_i^2/n$
	$u(y) = \sqrt{s_e^2 + s_a^2 + s_y^2}$	
$y_{ij} = \beta_i + \varepsilon_{ij}$	$s_e^2 = \text{MSE}$	
	$s_b^2 = s_a^2 + s_y^2$	
	$u(y) = \sqrt{s_e^2 + s_b^2}$	
$y_{ij} = \mu + \nu_{ij}$	$s_\nu^2 = (\text{SSE} + \text{SSA})/(n-1)$	
	$s_y^2 = (\text{SQM} - s_\nu^2)/n$	
	$u(y) = \sqrt{s_\nu^2 - s_y^2}$	
$y_{ij} = \xi_{ij}$	$s_\xi^2 = \Sigma\Sigma y_{ij}^2/n$	
	$u(y) = s_\xi$	

Reproducibility

Following the procedure shown in Stringari et al. [7], q repeatitions of the determinations are considered. Thus the model of Eq. 1 generalises to

$$y_{ijk} = \mu + \alpha_i + \varepsilon_{ij} + \delta_{ijk} \quad (5)$$

and the variance components of the model without δ_{ijk} are multiplied by q.

Further to all mean SS expected values, the component of the reproducibility variance, σ_r^2, is added, thus leaving the estimates of the variance components s_a^2 and s_m^2 unchanged, while s_e^2 is now diminished by s_r^2/q, where s_r^2 is the estimate of the reproducibility variance.

If the experiment is not designed for the model of Eq. 5 but the estimate of s_r^2 is obtained by a separate experiment, this independent estimate is substituted in the above formulas, and also the denominators of the F-test are increased by the same value, thus obtaining new (approximate) F-tests.

Following Stringari et al. [7], these estimates of the reproducibility (s_r) were obtained (in ppm): Pb 0.0822, Cd 0.0067, Ni 0.0456, Cr 0.0628.

Results

The results of the above analyses are reported in the Fig. 4. The horizontal length of the bars indicate the average ppm content for all elements and species for each grinding method.

The total horizontal length of each bar has been divided into portions proportional to the variation coefficients based on the variance components, to which the reproducibility variance has been added.

Lead

Ssg grinding methods gave higher mean values in all the plant species (from 7% to 52%), and these differences were significant for cucumber and apple trees leaves. Moreover, it allowed us to highlight significant effects due to the field treatments in four of the five species, while with pbm methods there were significant effects only for apple trees and grapevines.

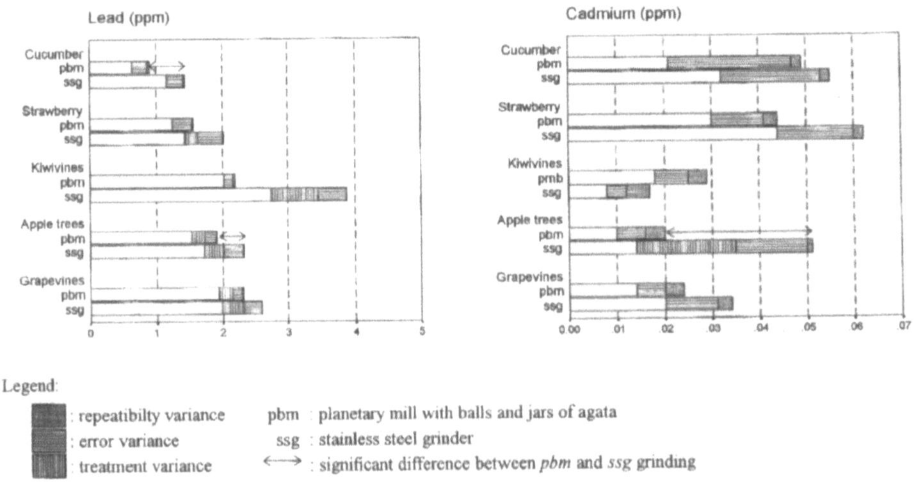

Fig. 3 Means and components of squared combined uncertainty of lead and cadmium as affected by plant matrices and grinding methods

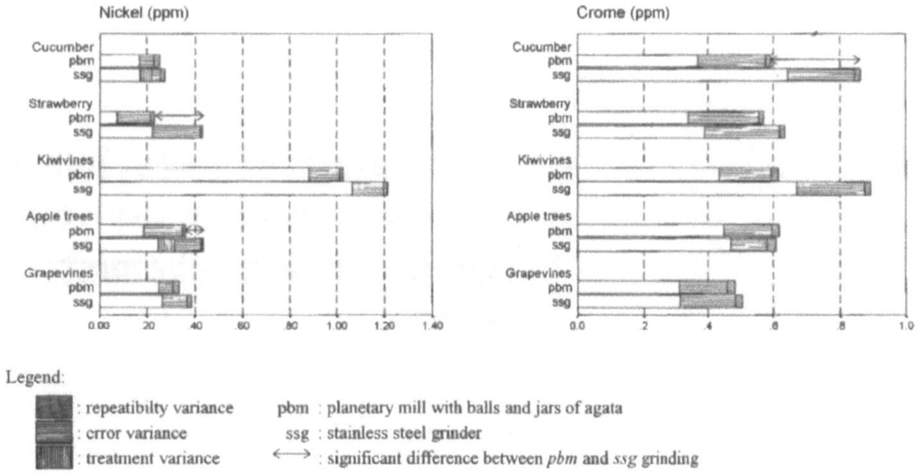

Fig. 4 Means and components of squared combined uncertainty of nickel and chromium as affected by plant matrices and grinding methods

Cadmium

In general, this metal was associated with a high combined uncertainty due to the low analytical values, very close to the analytical limits. With the exception of kiwi vines, *ssg* grinding methods gave higher mean values, with significantly higher mean values in apple trees (from 12% to 67%), and highlighted a significant effect of the field treatments for apple trees.

Nickel

For this metal also, the *ssg* method gave higher values for all the species (from 1% to 63%). The differences were significant for strawberry and apple trees. The *ssg* method allowed to put in evidence the field treatment effects in cucumber and apple trees.

Chromium

In three species out of five, the *ssg* method still determined higher values (from 7.5% to 28%), while for the other two the differences were below 1%. In cucumber the difference was significant.

This metal also showed a high combined uncertainty, which in this case probably was not due to low levels but to the greater difficulties for the instrumental measurement of this element.

Conclusion

With the stainless steel grinder, on average higher levels of the considered heavy metals were obtained (up to 67% of the mean values). On average, the increments were similar for metals contained in steel (Ni and Cr) and those not contained (Pb and Cd). The true causes of these differences need further investigation to determine whether the higher metal detection is due to possible contamination, to a different grinding quality or to other reasons. Further, the stainless steel grinder permitted us to detect more significant effects due to the field treatments. The grinding methods did not seem to affect the combined uncertainty of the analyses.

References

1. C.I.I.-Comité Inter-Instituts d'étude des techniques analytiques (1993–1997) Compte Rendu de 67e, 68e, 69e, 70e, 71e, 72e Réunion
2. Martin-Prével P, Gagnard J.,Gautier P (1984) In: Martin-Prével P, Gagnard J, Gautier P (eds) Plant analysis. Lavoisier, New York
3. BCR Catalogue. BCR Reference Materials. Community Bureau of Reference (BCR) Commission of European Communities, Brussels
4. Slavin W (1984) Graphite furnace AAS a source book. Perkin-Elmer, Ridgefield, Conn
5. Hoehig M, de Kersabiec AM (1990) L'atomisation électrothermique en spectrometrie d'absorption atomique. Masson, Paris
6. Hoening M, Baeten H, Vanhentenrÿk S (in press) Anal Chim Acta
7. Stringari G, Möller F, Ceschini A, Failla O (1996) Comm Soil Sci Plant Anal 27, 5–8:1403–1416

Accred Qual Assur (1998) 3:328–334

Mirella Buzoianu

Measurement uncertainty and its meaning in legal metrology of environmental chemistry and public health

Abstract The need for reliability of measurements supporting legal decisions in environmental policy or medical diagnosis and treatment is well known and widely accepted. This prerequisite can be met only by ensuring that legal measurements are accurate and traceable to national or international standards. Consequently, an outline of the organizational structure of the Romanian National Institute of Metrology (INM) for ensuring uniformity, consistency and accuracy of all measurements including legal measurements performed in chemical laboratories is presented. Since reliable measurements can only be accomplished within an appropriate traceability chain, the experience of the INM in identification and evaluation of measurement uncertainty in legal activities concerning the environment and health is reviewed. Practical examples of measurement uncertainty evaluation in spectrophotometric determination of five analytes, commonly determined in environmental and clinical chemistry are described. The implications of measurement uncertainty for interpretation of regulatory compliance are discussed.

Key words Measurement uncertainty · Analytical chemistry · Environment · Clinical chemistry

M. Buzoianu (✉)
National Institute of Metrology,
Sos. Vitan-Bârzesti No. 11,
75669 Bucharest, Romania
Tel.: +40-1-6344030
Fax: +40-1-3301533

Introduction

Traditional metrological activities in Romania have concentrated on legal chemical measurements performed in trade, environmental chemistry and public health, in conjunction with the implementation of quality assurance system in these fields. In this paper only measurements performed in laboratories from the environment and public health sectors are considered.

The main problem faced by analysts is whether or not they have the methodology to provide a result of the required accuracy and precision. However, after carrying out an analysis, it is very unusual for analysts to give any indication of the measurement uncertainty, or information on the traceability of the reported results. This means that the user of the measurement information is unable to make any judgment on the confidence to be placed in it, nor it is possible to compare, in a rational way, the results of independent analyses of the same sample. Therefore, concepts of measurement uncertainty and traceability are continuously developing in legal activities.

In this framework, the experience of the National Institute of Metrology (INM) on the evaluation of measurement uncertainty of spectrophotometric analyses performed in legal activities, as well as some results on comparability studies using certified reference materials (CRMs) are presented.

Many important decisions are based on the results of spectrophotometric analysis. The extent to which the quality of these results (i.e. measurement uncertainty) is reflected in regulatory compliance against limits is also discussed.

Outline of metrological assurance of legal measurements

In accordance with the Romanian Law of Metrology (issued in 1992), all measurements performed in production and testing of pharmaceuticals, in trade or in the fields of health, safety and environmental chemistry should be traceable to national or international standards, by the proper use of legal instruments, reference materials (RMs), and adequate methods of measurements. Consequently, the necessary metrological activities for legal measurements are:

1. the assurance of the legality of all instruments used by pattern tests and initial or periodical verification;
2. the development of RMs required by legal metrological norms;
3. the assessment of measurement uncertainty and the achievement of traceability.

In this respect, all instruments used in legal activities are subject to pattern approval of each model and any variants of that model. The performances of these instruments are evaluated and verified, using legal metrological norm (NML) methods and appropriate CRMs. Note that various types of CRMs developed, recognized and accepted for use for spectrophotometric systems are presented in Ref. [1]. Metrological assurance of uniformity and traceability of measurements in legal activities is coordinated and supervised by the Romanian Bureau of Legal Metrology (BRML), and carried out by the INM, 14 area-organized metrological inspectorates (IIJM) and a number of accredited metrological laboratories.

Founded in 1951, the INM's mission is to ensure a valid scientific background for uniformity, consistency and accuracy of all measurements in Romania, regardless of their field of application. The main activities of the INM are shown in Fig. 1.

Measurement uncertainty and traceability are very important for regulatory compliance against limits, when a good reliability of the analytic results and/or monitoring of toxic pollutants is needed. Therefore, much is being done by the INM to improve matters in the specific legal metrology of environmental chemistry and public health. Also, for comparability purposes, the INM organized several inter-laboratory studies using appropriate CRMs (single or multielement). The re-

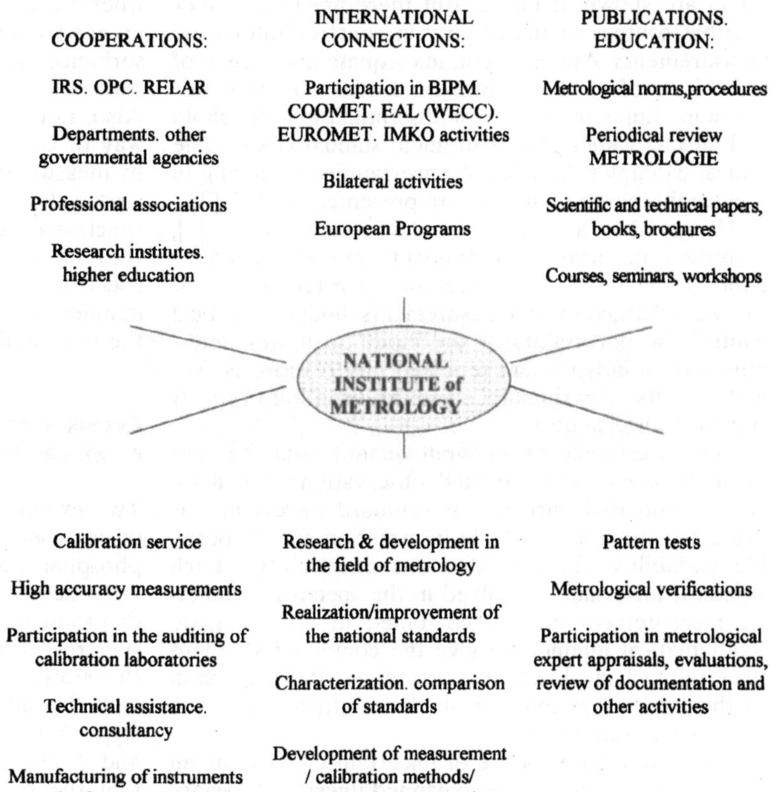

Fig. 1 Activities of the Romanian National Institute of Metrology for ensuring uniformity, consistency and accuracy of all measurements in Romania

sults on comparability of concentration measurements in legal activities are used to support future metrological activities related to NML procedures and metrological training.

Experience of the INM on evaluation of measurement uncertainty in legal metrology

Detailed evaluation of the measurement uncertainty is carried out as common practice by the INM on the realization of base and derived units. The techniques used rely on assessing and determining the correction for each cause, and building up an uncertainty budget.

Among the standardized methods currently performed in chemical laboratories, spectrophotometric ones are routinely used to determine the concentration of analytes using a variety of equipment, starting from discontinuously wide-band instruments to automated devices with a narrow band detection range. For a general photometric system, illustrated in Fig. 2, the evaluation of measurement uncertainty starts with the identification of the sources of errors and uncertainty components. Measurement uncertainties due to sampling (U_s), sample preparation (u_P), the photometric system (u_M), calibration of the system (u_R), the RMs/CRMs used for calibration (u_{RM}) and the data treatment (u_{DA}) are shown in Fig. 2. But there are many other possible sources of uncertainty in spectrophotometric measurements. Among them, inadequate knowledge of the effects of environmental conditions on the measurement, finite resolution or discrimination threshold and inexact values of measurement standards are some typical examples. Sources of uncertainties occurring in spectrophotometric analysis are presented in Ref. [2].

Using RMs and experimental quantification [3], identified uncertainty components are evaluated as either Type A or Type B standard uncertainties. Thus, for well-characterized measurements under statistical control the uncertainty of the input quantities determined from independent repeated observations is estimated as the experimental standard deviation (Type A standard uncertainty).

For an estimate of an input quantity that has not been obtained from repeated observations, the associated estimated variance or standard uncertainty is evaluated using all available information and its possible variability (Type B standard uncertainty). Each standard uncertainty involved in the spectrophotometric measurement is then combined in an 'adequate mathematical manner' to give the combined standard uncertainty [u_c^2(c)], which characterizes the dispersion of the values that can reasonably be attributed to the considered concentration.

The additional measure of uncertainty providing an interval of confidence, the expanded uncertainty, is ob-

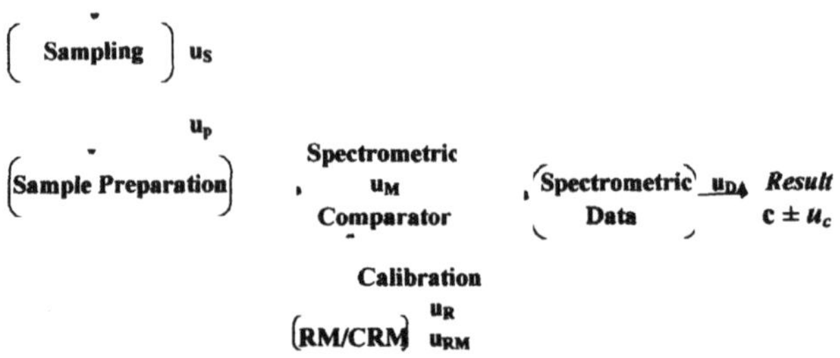

Fig. 2 Schematic diagram of an analytical photometric system

tained by multiplying the combined standard uncertainty by a coverage factor k (for legal metrology applications K is usually taken as 2).

The above-mentioned 'adequate mathematical manner' takes into consideration the function describing the concentration, of a general form:

$$c = c_{rec} \cdot f \cdot V \cdot K \quad \text{(in environmental analyses)} \qquad (1)$$

where: c_{rec} is the recalculated concentration from the calibration curve, f is the dilution factor, V is the sample volume and K is a proportionality factor; or

$$c = c_{standard} \cdot A_{sample} / A_{standard} \quad \text{(in clinical analyses)} \qquad (2)$$

where: $c_{standard}$ is the concentration of the standard reference solution used for comparison, A_{sample} is the absorbance measured for the sample and $A_{standard}$ the absorbance measured for the standard reference solution. Also, note that Eqs. (1) and (2) reflect two different way of calibrating the spectrophotometric system: (a) by measurement of a CRM, and (b) by measurement of a pure standard of the analyte used to calibrate just the spectrometric comparator. For traceability purposes these situations introduce the following points: the spectrophotometer should be calibrated in a traceable manner, and RMs used for its calibration should assure the traceability to SI units.

Examples of evaluation of measurement uncertainty in environmental analyses

Two examples of measurements frequently performed in environmental analyses to determine cadmium and phosphates in waste water will be discussed.

In accordance with SR ISO 5961 'Quality of water: determination of cadmium by flame atomic absorption spectrometry (FAAS)', synthetic water containing (0.500 ± 0.015) mg/l Cd was prepared under well-controlled conditions and measured on five spectrometers (type AAS 1, instrument 5; type AAS 3, instruments 1 and 2, and type AAS 30, instruments 3 and 4). Note that the performance of each instrument was tested against single element CRMs (code 13.01), as indicated in NML 9-02-94 'Atomic absorption spectrometers for water pollution measurements'. A summary of the parameters of each calibration curve is presented in Table 1. An estimated standard uncertainty was evaluated starting from the linear calibration of the instrument (uncertainty of regression, residual standard deviation and uncertainty of calibration curve included) [4]. Then a standard uncertainty was determined, combining standard deviation of repeated measurements, correction of the calibration curve and the uncertainty of RMs. The results of the evaluation of these uncertainties are also presented in Table 1. A good agreement between the two standard uncertainties is observed. Limit ratios of 0.73 and 2.47 were calculated from the determined uncertainty and u_{RM}.

The concentration of phosphates in waste water was determined according to a national standard STAS 10064 'Surface and waste waters: determination of phosphates' by measuring the absorbance of the blue colour of a reduced phosphomolybdate complex. Several types of molecular absorption spectrophotometers (SPECORD M40, instrument 1; DR 2000, instrument 4 and CADAS 100, instrument 5), and photometers of different bandwidth were used. Instrument 2 and 3 were specialized for water measurements (AQUANAL type). The unknown sample of 0.250±0.010 mg/l, was prepared under well-controlled conditions. The measurement conditions and evaluation of measurement uncertainty are presented in Table 2. Starting from the experimental steps involved in each measurement method used, an estimated measurement uncertainty was calculated as the square sum of partial uncertainties for volume and absorbance measurements, preparation of the calibration standards and the calibration curve [5].

By statistical analysis of the results obtained on control RMs or CRMs, an observed measurement uncertainty was evaluated (taking into account repeated measurements, correction of the calibration curve, the calibration curve and the uncertainty of the RMs). A quite good agreement between the two values of measurement uncertainty evaluated starting from two different approaches was accomplished. Furthermore, the experimental standard deviation of the mean value of concentration was determined using the analysis of variance of individual random effects according to [3]. Experimental variances of individual values, mean values and within parallel measurements for cadium are

Table 1 Results on evaluation of the measurement uncertainty on cadmium determination in waste water

	Instrument 1	Instrument 2	Instrument 3	Instrument 4	Instrument 5
Correlation coefficient, r	0.9995	0.9999	0.9995	0.9990	0.9995
Intercept of the regression line, a	−0.0044	−0.0014	0.0060	0.0038	−0.0003
Slope of the regression line, b Standad deviation, s_b	0.1416 (0.0013)	0.0694 (0.0007)	0.2596 (0.0041)	0.1433 (0.0023)	0.1049 (0.0021)
Standard deviation of residuals, s_0	0.005	0.003	0.009	0.005	0.003
Number of calibration points, N	4	4	4	4	5
Number of replicate measurements, n	3	3	3	3	3
Mean of all the absorbance A, values in the calibration	0.296	0.146	0.217	0.145	0.088
Mean value of the absorbance measured on the sample, $A_{(cx)}$	0.067	0.032	0.133	0.075	0.049
Predicted concentration, c_x (mg/l)	0.502	0.481	0.491	0.497	0.470
Standard uncertainty estimated, rel	0.021	0.033	0.026	0.027	0.021
Standard uncertainty determined, rel (mg/l)	0.011 0.014	0.037 0.018	0.032 0.016	0.028 0.014	0.034 0.016
Confidence interval (mg/l)	0.488 ... 0.516	0.463 ... 0.499	0.475 ... 0.507	0.483 ... 0.511	0.446 ... 0.486

Table 2 Results on evaluation of measurement uncertainty of phosphate concentration in waste water

Measurement method	O-phosphate reacts with ammonium molybdate in acidic medium to produce a phosphomolybdate complex. This complex is then reduced to an intense molibden blue colour					
Instrument	1	2	3	4	5	6
Steps considered:						
1. sampling (ml)	50	10	50	25	2	50
2. methods of measurement	STAS[a] 10064	As indicated by the maunfacturer	STAS[a] 10064	As indicated by the manufacturer	As indicated by the manufacturer	STAS 10064
3. volume of reagent (ml)	10	6 drops R_1 2 drops R_2	10		0.2	10
4. final volume (ml)	100	10	100	25	2.2	100
5. calibration curve						
r	0.9989	0.9988	0.9988	$A \cdot 0.5722$	Linear	0.9989
a	0.0168	0.0018	0.0863		−0.129 (k)	0.0933
b	0.8528	0.2589	1.2746		1.423 (F)	1.2086
s_0	0.0450	0.0267	0.0660		–	0.0610
6. measurement conditions; λ (nm),	700	635	650	890	890	660
time (min),	30	5	30	2	10	30
pathlength (mm)	10	15	10	23.5	25	25
$(\Delta\varepsilon/\varepsilon)_{time}$	0.016	0.000	−0.04	+0.03	−0.01	−0.04
$(\Delta\varepsilon/\varepsilon)_{temp\,25°C}$	0	0	0	0	0	0
Accuracy of the method, rel	0.05	0.05	0.05	0.01	0.04	0.05
Mathematicl equation	$(A_x - a) \cdot f_{timp}/b$	$(A_x - a) \cdot f_{timp}/b$	$(A_x - a) \cdot f_{timp}/b$	$0.5722 \cdot A_x$	$F \cdot A_x - k)$	$(A_x - a) \cdot f_{timp}/b$
Validation of the instrument with:						
neutral filters	Yes	Yes	Yes	No	No	No
CRM 1 mg/l PO_4	0.984	1.03	1.05	1.05	0.96	1.01
Absorbance measured on sample	0.246	0.069	0.424	0.472	0.275	0.392
Concentration of the sample (mg/l)	0.246	0.275	0.265	0.270	0.262	0.247
Estimated standard incertainty, rel	0.051	0.065	0.059	0.032	0.022	0.046
Determined standard uncertainty, rel	0.047	0.051	0.060	0.030	0.022	0.044

[a] STAS: National standard STAS 10064 'Surface and waste waters: determination of phosphates'

shown in Table 3. Since $F_{calc}(1.89) < F_{cr(3,10,0.95)}$ (3.71), there is no statistical significance between instrument effects at the 5% level of significance. Also, note the agreement between the standard deviation of the mean values around the known value of RM (0.020 mg/l), and the uncertainty assigned to the RMs used (0.015 mg/l).

The difference between the extreme values of the concentration corresponding to the confidence level reported for the CRMs was of 13.3% in the case of cadmium and of 22% for phosphates.

Practical considerations of measurement uncertainty in photometric clinical analyses

Methods most commonly used in clinical laboratories using photometric systems, as well methods to evaluate measurement uncertainty have been fully described in [1] and [6]. But the concept of measurement uncertainty is still poorly understood in clinical laboratories.

In clinical chemistry, activity has concentrated so far on the evaluation of measurement uncertainty of six analyses (Na, K, Ca, Mg, urea and glucose), following the international ISO Guide [3]. The main components of uncertainty considered for three typical examples of

Table 3 Summary of cadmium concentration values obtained with different instruments

	Instrument 1	Instrument 2	Instrument 3	Instrument 4	Instrument 5
Mean value, $\bar{c}$ (mg/l)	0.502	0.481	0.491	0.497	0.470
Standard deviation (mg/l)	0.014	0.018	0.016	0.014	0.016
Overall mean, $\bar{\bar{c}}$ (mg/l)			0.488		
Experimental variance: of the individual values (around the overall mean)			$\Sigma\,\Sigma\,(c_{ij}-\bar{c})^2/14=0.0005$		
within the parallel measurements			$\Sigma\,\Sigma\,(c_{ij}-\bar{c})/2=0.0071$		
of mean values (around the overall mean)			$\Sigma\,(\bar{c}-\bar{\bar{c}})^2/4=0.0003$		
of the overall mean around the known value of RM			$(\bar{\bar{c}}-c_{RM})^2=0.0002$		
of mean values around the known value of RM			$\Sigma\,(\bar{c}-c_{RM})^2/3=0.0004$		

end-point determination, and their values are indicated in Table 4. A relative measurement uncertainty of 0.058 has been obtained for glucose determination, 0.128 for urea and 0.025 for calcium. Note that the uncertainty of the CRMs used are indicated in parentheses in the table. The ratio between the uncertainty of CRMs and the measurement uncertainty evaluated for the above described analyses varies from 1.03 to 2.78, which is acceptable agreement with the typically recommended value of 3.

Measurement uncertainty meaning in legal metrology

Measurement uncertainty is significant when interpreting an analytical result of a toxic substance concentration lying within particular limits. Unfortunately few legal limits are set with allowance for uncertainty. Several studies of comparability performed in the national area showed a quite large spread of the results obtained in legal activities. For instance spreads of 35% for Cd and Zn, and 25% for Cu and Cr in waste water have been reported [7]. Also, in clinical laboratories under routine conditions, the spread was lower than 4.9% for Na, 19% for K, 26.1% for Ca, 18.6% for Mg and 15.6% for glucose, asymmetrically distributed around the assigned values [1]. Most outliers were obtained in the absence of a reliable uncertainty budget and insufficient quality assurance procedures. Nevertheless, limit results do not necessarily mean a higher measurement uncertainty. For instance, seven photometric systems of different photometric accuracy were used to determine nitrite

Table 4 Uncertainty components for three examples of end-point determination

Uncertainty components	Evaluation of the individual component	Relative measurement uncertainty		
		Glucose (0.056)	Magnesium (0.023)	Urea (0.060)
Due to photometric system	As Type A standard uncertainty (run-to-run vairation) and Type B standard uncertainty (certificate of calibration)	0.018	0.017	0.060
Due to CRM	As Type B standard uncertainty (certificate of CRM)	0.026	0.025	0.022
Due to volume of the pipette	As Type A standard uncertainty (run-to-run variation) and Type B standard uncertainty (manufacturer's specification)	0.002	0.002	0.002
Due to calibration	As described in [2]	0.002	0.011	0.002
Combined uncertainty	Square sum of individual standard uncertainties of above components	0.029	0.032	0.064
Overall uncertainty	$k=2$	0.058	0.064	0.128

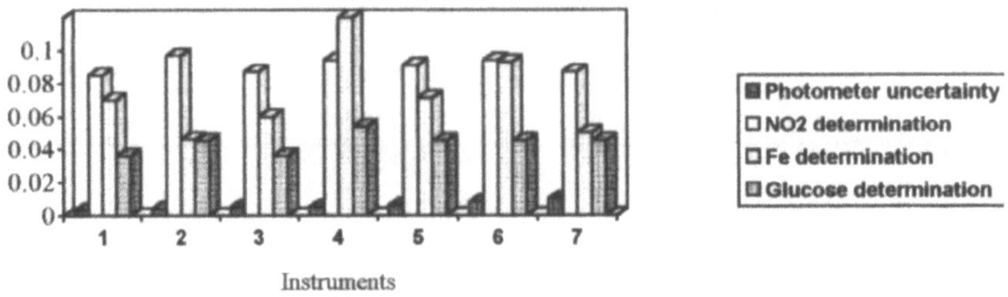

Fig. 3 Comparison of standard measurement uncertainties (rel) evaluated for nitrite, iron and glucose determination, using different photometric systems

and iron in water, and glucose in human serum. In each case a standard measurement uncertainty was evaluated as described above, the results are illustrated in Fig. 3 (light-grey columns). The left-hand column in each group showns the photometric uncertainty, evaluated from the manufacture's specifications. Note that for Fe determination using instruments of the same photometric accuracy, the standard measurement uncertainty (rel) varied from 0.060 to 0.120.

In addition, note how important the confidence interval from Table 2 is when judging the compliance with limits. Measurement results from instruments 1, 3 and 5 need individual consideration if the limit is set with some allowance for measurement uncertainty.

Measurement uncertainty also has a major influence on the traceability chains related to legal spectrophotometric measurements. In such situations both the spectrophotometers and CRMs should be traceable, i.e. they need to be calibrated in a proper manner and with an adequate uncertainty. In this respect the ratio between the uncertainty of upper and lower measurements is very important. For physical standards used to calibrate photometric systems in legal metrology the ratio of 3 is most commonly followed. For concentration calibrations this ratio usually does not exceed 1 or 2.

Conclusions

This paper has examined the importance and legal implications of measurement uncertainty statements in environmental chemistry and in the public health sector.

It is now accepted that the quality of an analytical result relies on the uncertainty of the quoted value, evaluated mainly from the calibration and reproducibility of the measurement system, and from the uncertainty of calibration standards. But, evaluation of the overall uncertainty follows a complex procedure, which is influenced by the skill of the analyst.

References

1. Buzoianu M (1998) Fresenius J Anal Chem 360:479–485
2. Buzoianu M, Aboul-Enein H-Y (1997) Accred Qual Assur 2:11–17
3. Guide to the expression of uncertainty in measurements, ISO (1993), Geneva
4. ISO 8466 (1990) 1 Qualité de l'eau – Étalonage et évaluation des méthodes d'analyse et estimation des caractères de performance. Evaluation statistique de la fonction linaire d'étalonage. ISO, Geneva
5. EURACHEM Guide: Quantifying uncertainty in analytical measurement, 1st edn (1995) Laboratory of the Government Chemist, London
6. Buzoianu M, Aboul-Enein H-Y (1997) Accred Qual Assur 2:186–192
7. Duta S, Buzoianu M (1996) Comparability of spectrophotometric measurement results in the Romanian Institute of Metrology. Proceedings of Central European Conference on Reference Materials, CERM '96', Slovakia

Accred Qual Assur (2001) 6:160–163

Adriaan M.H. van der Veen

Uncertainty evaluation in proficiency testing: state-of-the-art, challenges, and perspectives

Presented at the EURACHEM/EQUALM Workshop "Proficiency Testing in Analytical Chemistry, Microbiology and Laboratory Medicine", 24–26 September 2000, Borås, Sweden

A.M.H. van der Veen
Nederlands Meetinstituut,
Schoemakerstraat 97,
2628 VK Delft,
The Netherlands
e-mail: avdveen@nmi.nl
Tel.: +31-15-269-1733
Fax: +31-15-261-2971

Abstract The evaluation of measurement uncertainty, and that of uncertainty statements of participating laboratories will be a challenge to be met in the coming years. The publication of ISO 17025 has led to the situation that testing laboratories should, to a certain extent, meet the same requirements regarding measurement uncertainty and traceability. As a consequence, proficiency test organizers should deal with the issues measurement uncertainty and traceability as well. Two common statistical models used in proficiency testing are revisited to explore the options to include the evaluation of the measurement uncertainty of the PTRV (proficiency test reference value). Furthermore, the use of this PTRV and its uncertainty estimate for assessing the uncertainty statements of the participants for the two models will be discussed. It is concluded that in analogy to Key Comparisons it is feasible to implement proficiency tests in such a way, that the new requirements can be met.

Keywords Proficiency testing · Measurement uncertainty · Reference value · Consensus value · Assessment of laboratories

Introduction

The current practice in proficiency testing differs considerably from the practice in comparisons in the calibration area. This is not caused by differences between calibration and testing; it finds its origin in the fact that most test results are at best accompanied by an indication of their repeatability, whereas calibration results come with an uncertainty statement. In view of the new ISO 17025 [1], this difference will disappear, leaving a task for the proficiency testing providers in redesigning their services. Uncertainty calculations will play a dominant role at all levels of proficiency testing in the near future. The laboratories are required to express their uncertainty, and the organizer of the proficiency tests will be required to evaluate the uncertainty statements delivered.

This paper aims to set the frame for the newly designed proficiency tests. Furthermore, it will compare the proposed new practices with classical proficiency testing as it is carried out today. An important aspect of this comparison is to see how these developments affect the assessment of the participants' results. Obviously, if a laboratory performs well today, it should also do so tomorrow. This holds only for the reported value; as the laboratory has to deliver an uncertainty statements there is still the option of performing unsatisfactorily on this part.

The "Guide to the expression of uncertainty in measurement" (GUM) [2] provides the framework for doing uncertainty calculations. It does not distinguish between physics, chemistry, or biology; neither does it between calibration and testing. This observation is very important, as it allows the option of using well designed approaches of key comparisons, or other comparisons in the calibration area. The nature of the problems in designing a proficiency test does not differ from that in the calibration area. Problems like obtaining a reference value, expressing its uncertainty, and dealing with covariances and correlations are all the same.

Basic considerations for evaluating measurement uncertainty

The basis for proficiency testing is described in ISO Guide 43–1:1997 [3]. One of the tools necessary to assess the performance of the participating laboratories is an assigned value, which is used as reference point. In this paper, the abbreviation PTRV (proficiency test reference value) will be used for this purpose. Classically, there are two ways to obtain a PTRV:

1. By prior measurement ("reference value")
2. From the participants' results ("consensus value")

Irrespective of the model chosen, the GUM [2] provides a framework for the evaluation of the measurement uncertainty with respect to the PTRV. From a fundamental point of view, there is no difference between the two ways of obtaining a PTRV. A practical example of working out the establishment of a PTRV using prior measurement is given elsewhere [4]. Although the process is not uncomplicated, the estimation of measurement uncertainty is certainly well feasible.

When working with a consensus value, the philosophy is not different: the GUM can be implemented straightforwardly, as soon as the establishment of the consensus value is defined appropriately. There are however some practical difficulties to be overcome, which have mainly to do with the quality of the participants' data. It should be noted first that the quality of the PTRV is directly dependent on the quality of the participants' data. This will be reflected by the uncertainty of the PTRV as well. A further problem is the presence of suspicious results (e.g., outliers). It is not acceptable in a proficiency test to work without some policy to treat outliers.

In this paper, the establishment of a PTRV through consensus among participants will be revisited. There are a few different cases to be considered, in fact:

1. Results with credible uncertainty statement
2. Results with non-credible uncertainty statements
3. Results without uncertainty statements

The other mainstream design, in fact with a PTRV based on prior measurement, is always easier to implement. The understanding of what is going on during the establishment of reference values is usually better than in the case of consensus values: consensus values are often used in cases too complex to be handled by reference values. This is often a result from a lack of understanding, in terms of modeling, of the measurement problem. Properties of the sample, matrix effects, extraction/destruction yields, etc. all contribute greatly to this lack of understanding. All these aspects, that may greatly influence the measurement results and therefore also their uncertainty, may lead to the conclusion that working with a consensus value is inevitable. So, this lack of understanding has more to do with the state-of-the-art in measurement science than with the skills of the team operating the proficiency test.

The topic of correlation between measurement results is a very critical one, and it is gaining more and more interest. The assumption of IID-data (independent, identically distributed data) is easily made, but difficult to verify, and in most cases highly critical. If data are not IID, most of the statistics known do not work. Often, the problem is not so much in the distribution, it is more in the (in)dependence. Dependent data can already be observed in cases where all laboratories use the same pure substances for their calibration, for instance. This happens, for example, in PAH-analysis, where there is only one series of certified pure substances available. Obviously, the purity data of these substances cannot be treated as being independent.

Both in testing and in calibration, correlation of data plays an important role. The consequence of data being correlated and disrespecting this leads to wrong uncertainty estimates. The worst part of the message is that it is even not known whether this leads to over- or underestimation problems. As a result, it will just not work to ignore correlations. A safe practice is to drop the assumption of independence, and to work from there. It does make life somewhat more complicated under certain circumstances, but underestimation problems will be avoided.

PTRV through consensus

Establishment of a PTRV through consensus is more complicated than through prior measurement. The reason for this is that it is more difficult to develop a set of assumptions and assertions that is in compliance with the data obtained, and a sufficient basis on which to develop an algorithm at the same time. The days are gone when all data from all participants could be thrown into a big "hat" and that automatically the consensus value would come out. Building consensus values is probably one of the most complex tasks to be carried out by the organizer.

Case of credible uncertainty statements

The first important case to be considered is the case of credible uncertainty statements. The development of a procedure for the calculation of the consensus value does not differ from an approach suggested for evaluating key comparisons [5] which has also been demonstrated to work for the certification of reference materials [6]. In a recent paper by this author, an implementation of this recipe has been given for the case of reference materials. A disadvantage of the method is that a full description of all measurement models is required. This is – apart from the considerable extra effort – undesirable for another reason: it is far away from the present philosophy of proficiency testing as it violates the principle to work under "normal conditions".

The crux in designing an evaluation method is in the treatment of the data from the laboratories, in relation to the issue of correlations between results. In principle, for each laboratory pair in the proficiency test, the covariance should be computed. To the full extent, this has been established elsewhere [5, 6]. Here, a simpler method will be proposed. The task for the statistician responsible for the evaluation of the proficiency test is to make a fair estimate of the degree of correlation between two laboratory results. In order to make such an estimation, the organizer should have some insight in the methods, chemicals, and standards used, etc. In most proficiency tests, such information is obtained through an inquiry and/or regular participant-organizer communication.

Instead of requesting all measurement models from all laboratories to be reported like in the case of reference materials [6], the statistician should make a conservative estimate of the (possible) degree of correlation of results. This conservative value should flow in into the evaluation method as proposed for the reference materials, and the calculation can be started. Using the methodology of looking at the degrees of equivalence [5, 6], the unsatisfactory results can be removed and the consensus value can be established. Then, with the consensus value after removal of unsatisfactory results, the results of the laboratories can be assessed.

Case of non-credible uncertainty statements

This case cannot be compared with the case of credible uncertainty statements. The problem is that the organizer of the proficiency test gets a lot of information, but the value of this information is to a certain degree questionable. Obviously, the judgment as to whether information is credible or not is something that must be decided from case to case, but always beforehand. If, during a proficiency test, it appears that the wrong decision has been taken, then it is not an easy task to do a repair: the danger of violating other assumptions is great. Furthermore, it leaves the participants in doubt about the outcome of the proficiency test, something to be avoided at all cost.

If the uncertainty statements are not credible, it is better to refrain from using the uncertainty information at all for the establishment of the consensus value. It is better practice to use some kind of approximation, like for instance the following formula:

$$u^2(m) = \frac{s^2}{p} + \sum_{i=1}^{L} u_{i,other}^2 \qquad (1)$$

where the last term reflects those uncertainty sources other than those randomized in the proficiency test. These uncertainties are considered to be more or less the same for all participants. The standard deviation s is the just the standard deviation of the means of the laboratory means, whereas m is the mean of these laboratory means. p denotes the number of laboratories. Further treatment of data can take place as usual, including outlier/straggler testing and/or removal if considered appropriate. It should be noted that the larger the proficiency test (p), the smaller the first term in the expression for the uncertainty, so the more important the second term becomes. This is a serious disadvantage of the approach, and cannot be solved easily, due to apparent problems in the uncertainty estimation.

The method can obviously also be carried out with robust estimation techniques, like for instance the use of the median and the (normalized) median of absolute deviations, MAD_e. The procedure remains the same, and usually the results from robust estimation techniques do not differ significantly from those after an evaluation using classical statistical techniques [7, 8].

The evaluation of the performance of the laboratories can now take place as in the case of the credible uncertainty statements, as the uncertainty of the consensus value is now available, and so are all uncertainty statements from the laboratories.

No uncertainty information available

In several cases it may still be impossible to come up with an uncertainty statement. This is probably the worst situation, as the customer of the laboratory does not have any indication about the reliability of the reported data. In the absence of uncertainty data it is obviously impossible to work with anything else than the reported laboratory averages. It still leaves the organizer of the proficiency test with the task of estimating the uncertainty of the consensus value. Typically, one could proceed as follows. The uncertainty at the level of a laboratory can be computed from

$$u^2(y) = s^2 + \sum_{i=1}^{L} u_{i,other}^2 \qquad (2)$$

where all symbols have the same meaning as in the previous case. The major difference is that the division by p has vanished. This is a necessity, as only the reported value of the laboratory (y) can be assessed (there is no uncertainty information).

In this case, the well known Z-score can still be used:

$$Z = \frac{m - y}{u(y)} \qquad (3)$$

to assess the performance of the laboratories. The estimation of the uncertainty of a "typical" laboratory is a real burden, as the organizer must find ways to come up with an uncertainty statement in a complete lack of information. This situation should be avoided, or circumvented by working with fixed limits in the performance characteristics. This is a completely different philosophy, and outside the scope of this paper.

Role of homogeneity and stability of PTMs

Similarly to the uncertainty of the property values of (certified) reference materials, the uncertainty of the property values of PTMs (proficiency test materials) should also include the between-bottle homogeneity [9] and short- and long-term stability [10]. It should be noted that (1) the stability of the material is only of concern as long as the comparison is ongoing and (2) short-term stability might impose even greater problems than in the case of CRMs. This is due to the fact that PTMs are often more like "real-world" samples, in a sense that the measures taken to improve stability are less severe than for several groups of CRMs. The inclusion of these uncertainty components in the uncertainty of the PTM is analogous to the uncertainty model established for reference materials and is described elsewhere [6, 11].

Conclusions

In conclusion, it is demonstrated that practical approaches are at hand to run proficiency tests in the testing area in the same way as comparisons in the calibration area. The nature of the two comparisons is exactly the same: the problems of credible uncertainty statements as well as that of correlated variables also exist in both cases. The outcome of the restyled proficiency test must not differ from the classical approach, provided that the same assumptions are used and that they are "translated" correctly in the model.

Uncertainty calculations in the testing area are no longer completely different from those in the calibration area. There are differences, and both areas have their specific problems. There is a big task ahead for proficiency testing organizers in adapting to the new situation, but they can borrow a lot from existing techniques made available in comparisons in the calibration area. It will bring probably the science of experimental measurement and the science of uncertainty evaluation more closely and more consistently together, which will improve the learning cycle in proficiency testing considerably. It will give a boost to the understanding of how measurement systems behave, and this will allow for more direct and better heading actions if method improvement is necessary.

References

1. ISO (1999) International Organization for Standardization ISO 17025: General requirements for the competence of testing and calibration laboratories. ISO Geneva
2. ISO (1995) BIPM, IEC, IFCC, ISO, IUPAC, IUPAP, OIML: Guide to the expression of uncertainty in measurement, 1st edn, 2nd corrected print. ISO Geneva
3. ISO (1997) International Organization for Standardization: ISO/IEC Guide 43-1:1997: Proficiency testing by interlaboratory comparisons – Part 1: Development and operation of proficiency testing schemes. ISO Geneva
4. Van der Veen AMH, Horvat M, Milačič R, Buačr T, Repinc U, Ščančar J, Jaćimović R (2001) Operation of a proficiency test of trace elements in sewage sludge with reference values. Accred Qual Assur (submitted for publication)
5. Nielsen L (1999) Evaluation of measurement intercomparisons by the method of least squares. DFM Rep 99-R39, presented at the EUROMET workshop on uncertainty calculations in key comparisons, Teddington, Nov 1999
6. Van der Veen AMH (2000) Determination of the certified value of a reference material appreciating the uncertainty statements obtained in the collaborative study. Presented at AMCTM 2000, Monte de Caparica, May 2000
7. Van der Veen AMH, Broos AJM (1996) Preparation and characterisation of coal samples and maceral concentrates for studies on gasification and combustion reactivity of coals in combined cycle processes. Draft Final Rep ECSC 7220/EC-036, Eygelshoven, NL
8. Cox MG (1999) A discussion of approaches for determining a reference value in the analysis of key-comparison data. NPL Rep CISE 42/99, Teddington, UK
9. Van der Veen AMH, Linsinger TPJ, Pauwels J (2001) Uncertainty calculations in the certification of reference materials. 2. Homogeneity study. Accred Qual Assur 6:26–30
10. Van der Veen AMH, Linsinger TPJ, Lamberty A, Pauwels J (2001) Uncertainty calculations in the certification of reference materials. 3. Stability study. Accred Qual Assur (in press)
11. Van der Veen AMH, Linsinger TPJ, Schimmel H, Lamberty A, Pauwels J (2001) Uncertainty calculations in the certification of reference materials. 4. Characterisation and certification: Accred Qual Assur (in press)

Accred Qual Assur (1998) 3:69–78

Michel Gerboles
Elias Diaz
Alberto Noriega-Guerra

Uncertainty calculation and implementation of the static volumetric method for the preparation of NO and SO_2 standard gas mixtures

Abstract The European Reference Laboratory of Air Pollution implements the static volumetric method for the preparation of nitrogen monoxide and sulphur dioxide reference standard gas mixtures. According to the new ISO guide for the expression of uncertainty, the uncertainty of these standards is up to 0.8% for nitrogen monoxide in the range 100 to 600 ppbv, and up to 0.4% for sulphur dioxide in the range 200 to 400 ppbv. The values presented in the present paper suggest that there is a 95% probability of the true value lying within the interval specified. To attain such low uncertainty values, the standard procedure for the implementation of the static volumetric method must be rigorously followed, and instruments must be carefully maintained.

Key words Uncertainty calculation · Calibration · Static volumetric method · Air pollution monitoring

M. Gerboles (✉) · A. Noriega-Guerra
European Reference Laboratory of Air Pollution (ERLAP), Commission of the European Communities, Joint Research Centre, I-21020 Ispra, Italy
Tel.: +39-332-785652; Fax: +39-332-78
e-mail: michel.gerboles@jrc.it

E. Diaz
Beca de Ampliación de Estudios del F.I.S., Instituto de Salud Carlos III, Madrid, Spain

Glossary of symbols

C_1 Volume concentration of the diluted standard gas mixture
C_0 Volume concentration of the pure component ≈1
p Pressure of the injected pure component (NO or SO_2)
v Volume of the injected pure component (NO or SO_2)
P Pressure in the vessel
V Volume of the vessel
U_{C0} Standard uncertainty of type B error of the purity of the gas
U_{v1} Standard uncertainty of type A error of the volume of the syringe (balance repeatability)
U_{v2} Standard uncertainty of type B error of the volume of the syringe (balance linearity)
U_{v3} Standard uncertainty of type B error of the volume of the syringe (pure gas diffusion)
U_{p1} Standard uncertainty of type A error due to the pressure sensor of the transferred component
U_{p2} Standard uncertainty of type B error due to over-pressure of the transferred component
U_V Standard uncertainty of type B error of determination of the volume of the vessel
U_{P1} Standard uncertainty of type A error of the pressure sensor in the vessel
U_{P2} Standard uncertainty of type B error of the pressure sensor in the vessel
U_{P3} Standard uncertainty of type B error of the pressure in the vessel due to the lack of temperature equilibrium
U_c Combined standard uncertainty on the volume concentration of the diluted standard gas mixture
k Coverage factor ($k = 2$)
U Expanded standard uncertainty on the volume concentration of the standard gas mixture

Introduction

The monitoring of atmospheric sulphur dioxide and nitrogen dioxide is regulated by the European directives 80/779/CEE [1] and 85/203/CEE [2]. In these, the tetrachloromercurate (TCM)/pararosaniline [3] method is proposed as the reference methods for sulphur dioxide determination, and chemiluminescence [4] as the reference method for the determination of nitrogen oxides. However, technological progress has meant that, throughout the European air pollution monitoring network, UV fluorescence [5] has virtually replaced TCM for the analysis of atmospheric sulphur dioxide.

The European Reference Laboratory of Air Pollution (ERLAP) is the reference laboratory for atmospheric pollution serving the European Commission. One of its duties is to maintain European standards for the calibration of NO_2 and SO_2 methods of analysis. ERLAP has chosen the permeation method [6] evaluated by gravimetry to produce reference standards for both NO_2 and SO_2 methods, but uses the static volumetric method [7] to cross check the permeation method.

This paper describes ERLAP's implementation of the static volumetric method and deals with the uncertainty of standards generated by this method. Practical examples of the calculation of uncertainty are given.

Principle of the method

General principles

At atmospheric pressure p and room temperature, a known volume v of the pure component to be analysed ($C_0 \cong 1$) is transferred with a syringe to a large borosilicate vessel of known volume V filled with a selected carrier gas. The vessel is then filled with the selected carrier gas to pressure P, which is usually about 1.5 atm to facilitate use of the mixture. The mixture can be used once temperature has returned to ambient temperature. Under these conditions, the volume concentration of the component C_1 is practically equal to the molar fraction concentration and can be expressed by the formula:

$$C_1 = C_0 \frac{pv}{PV} \tag{1}$$

Implementation at the ERLAP laboratory

The static volumetric method is described in detail in the Guidelines of the VDI 3490 Blatt 14 [8], and it has been successfully tested for over 20 years at the UBA-Pilot Station of the Federal Environmental Agency of Germany. The static volumetric system implemented by ERLAP was devised and developed at the UBA Pilot Station and has been tested at the ERLAP laboratory for more than 3 years. The ERLAP laboratory uses this method for the preparation of SO_2 and NO standard calibration gas mixtures. The static volumetric system used for this purpose is shown in Fig. 1.

Mode of operation

Borosilicate glass mixing vessel

Experiments at the UBA Pilot Station have clearly shown that, for components such as SO_2 and NO, wall effects from borosilicate glass were negligible when preparing concentrations of 100 ppbv and above in dry carrier gases [9]. The volume of the vessel was determined as 0.11184 m3 ± 0.1% by a replicated process of filling with water and deriving volume from weight of

Fig. 1 Static volumetric system

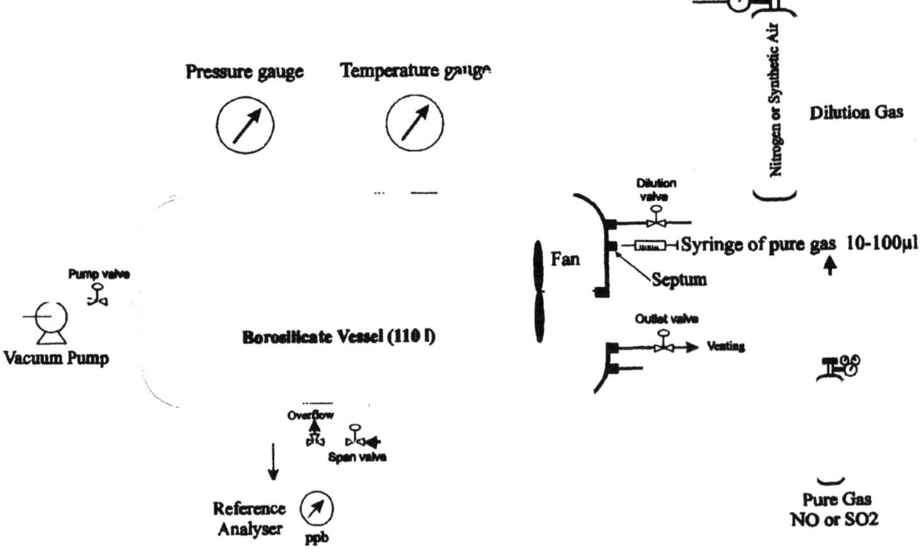

liquid contained within the vessel. To check for leaks in the vessel, pressure stability was periodically determined at 10^{-2} mbar and 1600 mbar.

Fan

A fan mounted inside the vessel to help mix the pure gas and carrier gas comprised a stainless steel propeller mounted on a stainless steel axis. The fan was powered by an electric motor.

Vacuum pump

This was an oil-free Alcatel molecular pump model DRYTEL 100C, capable of creating a residual pressure in the vessel of less than 10^{-2} mbar.

Septum and other external connections

The septum was a silicone rubber disk. Other external valves were made of stainless steel or polytetrafluoroethane (PTFE).

Temperature

Temperature within the vessel was measured with a PT100 stainless steel probe connected to a Thesto Therm 9000 process unit. The overall uncertainty of the system was better than 0.1 K.

Pressure

The pressure sensor was a Druck model DPI 510, with a precision of 0.025% and an accuracy of 0.04%.

Pure gas

The pure gases used for the preparation of NO and SO_2 standard mixtures were manufactured by Messer Griesheim, with a purity better than 99.5% for NO and 99.98% for SO_2. NO with purity better than 99.8% may be available in the future.

Carrier gas

For the preparation of NO standards, cylinders of chromatography-grade N_2 (NO free) were used. For the preparation of SO_2 standards, zero air produced by the ERLAP zero air generator (SO_2 free) was used.

Syringes

Pure gases were injected into the vessel with 100-, 50- and 25-µl Hamilton series 1800 syringes. Special ERLAP mechanisms were used to improve the repeatability of the filling process. The volume of each syringe was determined by filling with water and deriving volume gravimetrically (as described above for the borosilicate glass mixing vessel).

Procedure for the preparation of a mixture of NO with nitrogen

Cleaning the vessel

Check that the contents of the vessel are at ambient pressure, and, if not, open the outlet valve to allow the pressure to equilibrate. Start the molecular pump and open the pump valve sufficiently to establish a pressure inside the vessel of about 10^{-2} mbar.

Filling the vessel with carrier gas

Open the reducing valve on the carrier gas (N_2) followed by the dilution valve until the pressure inside the vessel reaches about 1050 mbar. Open the outlet valve to allow the chamber to return to ambient pressure and then close it. Wait 10 min for the temperature and pressure to stabilise, and then open the outlet valve to allow excess N_2 to escape. Record the values of temperature T_i and pressure p.

Filling the syringe with pure gas

This involves the use of a small stainless steel container (the septum chamber, see Fig. 2), the integrity of which is maintained by two manual stainless steel valves. One valve connects directly to the reducing valve of the pure NO cylinder and the other to a vacuum pump.

Turn on the vacuum pump after first checking that all valves between the pump and the NO cylinder are closed. Introduce an empty syringe into the chamber (through a septum similar to that in the mixing vessel) and carry out the following sequence to ensure that the syringe is filled with pure NO.

(a) Close both septum chamber valves, the reducing valve and the NO cylinder.

(b) Open the septum chamber valves for 20 s to "clean" the septum chamber and tubes.

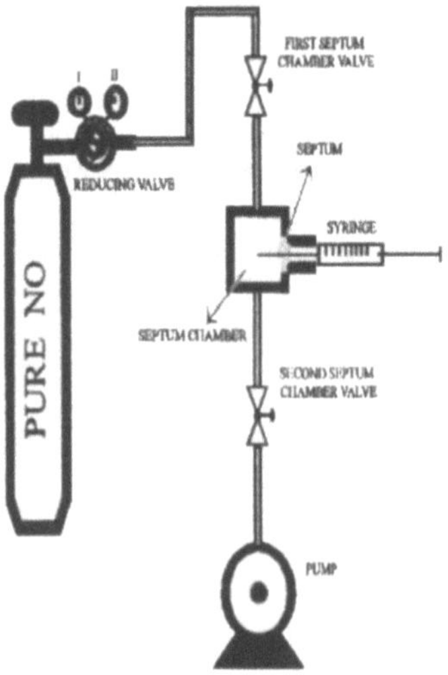

Fig. 2 Syringe filling system

(c) Close both septum chamber valves and open the NO cylinder.
(d) Close the NO cylinder and open the reducing valve until a pressure of 2 bar is established (read on indicator II of the reducing valve).
(e) Close the reducing valve and open the first septum chamber valve to send pure NO into the septum chamber. Close the first septum chamber valve.
(f) Fill the syringe with pure NO and drain the syringe. Repeat three times.
(g) Fill the syringe with pure NO and open the second septum chamber valve to clean the septum chamber and syringe. Drain the syringe and close the second septum chamber valve.
(h) Repeat steps c, f, g.
(i) Repeat steps c, d and c.
(j) Slowly fill the syringe completely with pure NO and wait for one min before removing from the septum chamber.
(k) Switch off the vacuum pump after first opening its venting valve (to avoid oil entering the tubing).

During these operations, it is important not to touch the glass of the syringe or the septum chamber to ensure that these remain at ambient temperature.

Injecting the pure NO into the mixing vessel

Take the syringe filled with pure NO out of the septum chamber, and, 10 s after adjusting it to the required volume (by way of the ERLAP mechanism), slowly inject the pure NO into the mixing vessel. Once the syringe is empty, quickly remove it (to avoid possible loss of pure NO along the surface of the syringe needle) and turn on the fan for 2 min to aid mixing.

Increasing pressure in the mixing vessel

Open the reducing valve on the cylinder of carrier gas (N_2), and the dilution valve to establish a pressure of about 1550 mbar (read on the Druck model 510 instrument). Wait 10 min for the temperature in the mixing vessel to stabilise at the initial value T_i. (read on the Thesto Thern 9000 instrument) and at this point record the value of pressure P.

Calculating the theoretical value – The theoretical value is calculated using Eq. 1.

Procedures for SO_2

Apart from the use of zero air as carrier gas instead of N_2, the procedure for the preparation of SO_2 standard gas mixtures is exactly the same as for the preparation of NO standard gas mixtures.

Correction for the level of purity of the pure gases

The pure gases (NO and SO_2) are supplied by Messer Griesheim Italia. The NO cylinder is of the Nitric Oxide 2.5 F1S type with a certified purity ≥ 99.5%. The SO_2 cylinder is of the Sulphur Dioxide 3.8 F1S type with a certified purity ≥ 99.98%. It was decided to apply a correction factor to the reference value of the NO standards of –0.25%, with lower and upper limits of –0.5 and 0%. For SO_2 standards, the correction factor was –0.01%, with lower and upper limits of –0.02 and 0%. The purity certified by the manufacturer was verified by FT-IR spectrometry.

The new ISO method for calculating uncertainty

In general, an analytical measurement provides only an estimation of the value of a determinant, and must be accompanied by a quantitative statement of the uncertainty attached to the estimate.

In 1993, ISO published a new guide to the expression of uncertainty [10], in which each component contributing to the uncertainty of a measurement is allotted an estimated uncertainty, termed standard uncertainty (u_i) equal to the positive square root of the estimated variance u_i^2.

The uncertainty associated with a measurement generally consists of several components, which may grouped into two categories:

A. Those which are evaluated by statistical methods

B. Those which are evaluated by other means.

Type A evaluation of standard uncertainty

Type A evaluation of uncertainty may be based on any valid statistical method for treating data. For example:
- Calculating the standard deviation of the mean of a series of independent observations
- Using the method of least squares to fit a curve
- Carrying out an analysis of variance ANOVA to quantify random effects.

As an example of Type A evaluation, consider an input quantity X_i whose value is estimated from n independent observations $X_{i,k}$ obtained under identical conditions of measurement. In this case, the estimated standard deviation of the mean is the positive square root of:

$$u^2(x_i) = \left(\frac{1}{n(n-1)}\right) \sum_{k=1}^{n} (X_{i,k} - X_1)^2 \tag{3}$$

Type B evaluation of uncertainty

Type B evaluation of standard uncertainty is usually based on scientific judgement and may make use of all available relevant information, including:
- Previously measured data
- Available information concerning the behaviour and properties of materials and instruments involved
- Manufacturer's specifications
- Calibration data and other available information.

As an example of Type B evaluation, consider an input quantity X_i whose value is estimated from an assumed rectangular probability distribution with lower limit $a-$ and upper limit $a+$. In this case the input estimate is usually expressed by:

$$x_i = (a_+ + a_-)/2 \tag{4}$$

and the standard uncertainty associated with x_i is:

$$u(x_i) = a/\sqrt{3} \text{ where } a = (a_+ - a_-)/2 \tag{5}$$

Combined standard uncertainty

The combined standard uncertainty of a measured value (u_c,) is assumed to correspond to the estimated standard deviation of the result. In the case of non-correlated components, it is derived by combining individual standard uncertainties u_i, which may arise either from type A or type B evaluations. The method is often referred to as the law of propagation of uncertainty, and is expressed as:

$$u_c^2 = \sum_{i=1}^{n} \left(\frac{\partial f}{\partial x_i}\right)^2 u^2(x_i) \tag{6}$$

The partial derivatives are referred to as sensitivity coefficients.

It is assumed that corrections have been applied to compensate for each systematic effect which significantly influences the measured value, and that every effort has been made to identify such effects.

Expanded uncertainty

What is often required is an expression of uncertainty to define the limits associated with a measured value y within which the value Y is confidently believed to lie. The measure of uncertainty intended to meet this requirement is termed the expanded uncertainty, suggested symbol U, and is obtained by multiplying $u_c(y)$ by a coverage factor, suggested symbol k. Thus $U = k\, u_c(y)$ and is confidently believed that $y - U \leq Y \leq y + U$, which is commonly written $Y = y \pm U$. In general, the value of the coverage factor k is chosen on the basis of the desired level of confidence to be associated with the interval defined by $U = k\, u_c$. Typically k is in the range 2 to 3. When the normal distribution applies, $U = 2\, u_c$ defines an interval having a level of confidence of approximately 95% and which is consistent with current international practice.

Uncertainty budget

Purity of the gases C_0

As described in "Correction for the level of purity of the pure gases", for NO, the lower and upper limits of the purity correction are 99.5–100%. The probability that the purity lies in this interval is 100% (rectangular distribution). The best estimate of the standard uncertainty of the quantity is then the positive square root of:

$$u_{C0}^2 = \frac{(a_+ - a_-)^2}{12} = \frac{(1-0.995)^2}{12} = 2.08310^{-6} \tag{7}$$

For SO_2, the lower and upper limits of the purity correction are 99.8–100. The best estimate of the standard uncertainty of the quantity is then the positive square root of:

$$u_{C0}^2 = \frac{(a_+ - a_-)^2}{12} = \frac{(1-0.998)^2}{12} = 3.33310^{-7} \tag{8}$$

Volume of the syringe v

The volume of each syringe (with ERLAP mechanism) was determined by filling with water and deriving the volume from the weight of liquid (measured with a

Table 1 Measured volume of the syringes. Average and standard deviation of 15 replicate measurements, linearity balance deviation u_{v2}^2 and diffusion deviation through the needle u_{v3}^2

Syringes	Average (l)	u_{v1}^2 (l^2)	u_{v2}^2 (l^2)	u_{v3}^2 (l^2)
1 (NO)	$24.71\ 10^{-6}$	$2.19\ 10^{-15}$	$1.33\ 10^{-16}$	$5.09\ 10^{-17}$
2 (NO)	$39.86\ 10^{-6}$	$2.08\ 10^{-15}$	$1.33\ 10^{-16}$	$1.32\ 10^{-16}$
3 (NO)	$78.43\ 10^{-6}$	$8.91\ 10^{-15}$	$1.33\ 10^{-16}$	$5.13\ 10^{-16}$
4 (NO)	$99.23\ 10^{-6}$	$5.11\ 10^{-15}$	$1.33\ 10^{-16}$	$8.21\ 10^{-16}$
5 (SO_2)	$39.65\ 10^{-6}$	$2.16\ 10^{-15}$	$1.33\ 10^{-16}$	$1.31\ 10^{-16}$
6 (SO_2)	$49.16\ 10^{-6}$	$2.44\ 10^{-15}$	$1.33\ 10^{-16}$	$2.01\ 10^{-16}$
7 (SO_2)	$69.29\ 10^{-6}$	$2.36\ 10^{-15}$	$1.33\ 10^{-16}$	$4.00\ 10^{-16}$

Mettler AT201 balance) contained in the syringe (corrected for water density of 0.998 g/cm3at 22 °C). Several replicate volume measurements were carried out in a hysteresis cycle, and the results are shown in Appendix 1. Variations in the measured volume reflected variability in the filling and emptying processes and in the performance of the balance. The standard deviation of the various estimates of v is the square root of $u_{vl}{}^2$ and is given in Table 1.

The balance was also subject to a linearity deviation from the true value, evaluated by the manufacturer as being ± 0.02 mg (0.02 ml) in the range 0–5 g. Assuming a rectangular distribution, the best estimate of the standard uncertainty of v is the square root of $u_{v2}{}^2$, given in Table 1.

The syringe is filled with pure gas (NO or SO_2) at a pressure of about 2 bar. When the pure gas is injected into the glass vessel, it must be returned to the ambient pressure without having reacted and undergone any transformation. In fact, the absorption of NO and SO_2 on the glass walls of the syringe has never been observed and is not likely to produce a relevant reduction of the injected volume. No reaction or transformation of the pure gases has been evidenced so far. However, diffusion of the pure gas out of the syringe through the syringe needle is observed after the transient period needed for the syringe pressure to be adapted to the ambient pressure. The pure gas may leave the syringe chamber by diffusion through the needle before injection into the glass vessel. This diffusion has been investigated in Appendix 2 and Fig. 3. With a time interval of 10 s (+ 15 s of tolerance) before injection, the pure gas represents only 99.9–100% of the syringe volume that is injected into the glass vessel. Assuming a rectangular distribution, the best estimate of the standard uncertainty of v is the square root of $u_{v3}{}^2$, given in Table 1.

Obviously, there are differences between injecting a gas with the syringe and injecting a liquid, but good estimates of volume are possible provided the precautions outlined in "Procedure for the preparation of a mixture" are followed (especially with respect to the duration of injection, see Appendix 2).

Pressure of the pure gas p

It is important that pure gas is injected at room temperature following the procedure described in "Mode of operation" and paying special attention to the precautions relating to pressure.

Room pressure was measured with a barometer manufactured by Lambrecht Klimatologish Messtechnik (Götlingen) model 00.06040.100000, the specified uncertainty u_{p1} of which is ± 0.25 mbar. Pressure is influenced by the time interval between extracting the syringe from the septum chamber and injecting into the vessel. Appendix 2 shows the results of SO_2 measurements made with different time intervals between withdrawal of the syringe from the septum chamber and injection of pure SO_2 into the mixing vessel. The results show that there were few differences between injections after 5- or 15-s intervals, although an transient over-pressure of 0–0.2% cannot be ruled out. Assuming a rectangular distribution, the best estimate of the standard uncertainty of p is the square root of $u_{p2}{}^2$ (in mbar2):

$$u_{p2}^2 = \frac{(a_+ - a_-)^2}{12} = \frac{(4)^2}{12} = 1.333 \tag{9}$$

Volume of the vessel V

Total volume was determined by filling with water as described in "Mode of operation". Assuming rectangular distribution, the best estimate of the standard uncertainty of V is the square root of $u_v{}^2$ (in l^{-2}) with:

$$u_V^2 = \frac{(a_+ - a_-)^2}{12} = \frac{(111.95 - 111.73)^2}{12} = 4.0310^{-3} \tag{10}$$

The 4-cm wide borosilicate glass vessel is not expected to undergo increases in volume at internal pressure up to 1.7 bar.

Pressure in the vessel P

Pressure was determined as described in in "Mode of operation". For pressures up to 1500 mbar, the manufacturer claimed a precision of 0.375 mbar, although the precision of the digital display might be thought to be at least one digit (i.e. 1 mbar). Assuming rectangular distributions, the best estimate of the standard uncertainty of P is the sum of the square root $u_{p1}{}^2$ and $u_{p2}{}^2$ (in $mbar^2$):

$$u_{P1}^2 = \frac{(a_+ - a_-)^2}{12} = \frac{(0.75)^2}{12} = 0.047 \tag{11}$$

$$u_{P2}^2 = \frac{(a_+ - a_-)^2}{12} = \frac{(2)^2}{12} = 0.333 \tag{12}$$

If room temperature were not reached after dilution had established a pressure of 1.5 atm, P would not be correctly estimated, as temperatures exceeding ambient by more than 0.5 °C have been observed to produce deviations of ± 1 mbar in the final pressure P. Assuming rectangular distributions, the best estimate of the standard uncertainty of P is (in $mbar^2$):

$$u_{P3}^2 = \frac{(a_+ - a_-)^2}{12} = \frac{(2)^2}{12} = 0.333 \qquad (13)$$

Uncertainty calculation

Tables 2 and 3 below shows four NO calculations and three SO_2 calculations according to Eq. 1 based on experimental data involving different operators.

The uncertainty calculation uses Eq. 2. For non-correlated variables, the combined standard uncertainty is:

$$u_c^2 = \left(\frac{pv}{PV}\right)^2 u_{C0}^2 + \left(\frac{C_0v}{PV}\right)^2 u_p^2 + \left(\frac{C_0p}{PV}\right)^2 u_v^2$$

$$+ \left(\frac{C_0pv}{P^2V}\right)^2 u_P^2 + \left(\frac{C_0pv}{V^2P}\right)^2 u_V^2$$

For our examples, the calculations are summarised in Table 4.

Table 2 Examples of calculation of NO

Experiment	C_0	v µl	V l	p mbar	P mbar	C_1 ppbv
1	99.75%	24.76	111.84	993	1589	138
2	99.75%	39.93	111.84	1000	1548	230
3	99.75%	78.59	111.84	996	1508	463
4	99.75%	99.43	111.84	995	1459	605

Table 3 Examples of calculation of SO_2

Experiment	C_0	v µl	V l	p mbar	P mbar	C_1 ppbv
1	99.99%	39.73	111.84	995	1558	227
2	99.99%	49.26	111.84	992	1595	274
3	99.99%	69.43	111.84	999	1557	398

Table 4 Uncertainty calculation

$\left(\frac{\partial f}{\partial x_i}\right)^2 U_C^2$ components		NO syringe 1	NO syringe 2	NO syringe 3	NO syringe 4	SO_2 syringe 1	SO_2 syringe 2	SO_2 syringe 3
C_1 in ppbv		138	230	463	605	227	274	398
$\left(\frac{pv}{PV}\right)^2 u_{C_0}^2$	Gas purity	$4.0\ 10^{-20}$	$1.1\ 10^{-19}$	$4.5\ 10^{-19}$	$7.6\ 10^{-19}$	$1.7\ 10^{-20}$	$2.5\ 10^{-20}$	$5.3\ 10^{-20}$
$\left(\frac{C_0p}{PV}\right)^2 u_v^2$	Syringe volume	$7.4\ 10^{-20}$	$7.0\ 10^{-19}$	$3.3\ 10^{-19}$	$2.2\ 10^{-19}$	$7.9\ 10^{-20}$	$8.6\ 10^{-20}$	$9.5\ 10^{-20}$
"... u^2v_1	Repetition of measurements	$6.8\ 10^{-20}$	$6.9\ 10^{-19}$	$3.1\ 10^{-19}$	$1.9\ 10^{-19}$	$7.0\ 10^{-20}$	$7.5\ 10^{-20}$	$7.8\ 10^{-20}$
"... u^2v_2	Linearity of the balance	$4.1\ 10^{-21}$	$4.4\ 10^{-21}$	$4.6\ 10^{-21}$	$4.9\ 10^{-21}$	$4.3\ 10^{-21}$	$4.1\ 10^{-21}$	$4.4\ 10^{-21}$
"... u^2v_3	Diffusion of pure gas	$1.6\ 10^{-21}$	$4.4\ 10^{-21}$	$1.8\ 10^{-20}$	$3.0\ 10^{-20}$	$4.3\ 10^{-21}$	$6.2\ 10^{-21}$	$1.3\ 10^{-20}$
$\left(\frac{C_0v}{PV}\right)^2 u_p^2$	Syringe	$2.7\ 10^{-20}$	$7.4\ 10^{-20}$	$3.0\ 10^{-19}$	$5.2\ 10^{-19}$	$7.3\ 10^{-20}$	$1.1\ 10^{-19}$	$2.2\ 10^{-19}$
"... u^2p_1	Barometer repeatability	$1.2\ 10^{-21}$	$3.3\ 10^{-21}$	$1.4\ 10^{-20}$	$2.3\ 10^{-20}$	$3.2\ 10^{-21}$	$4.8\ 10^{-21}$	$9.9\ 10^{-21}$
"... u^2p_2	Over pressure in the syringe	$2.6\ 10^{-20}$	$7.1\ 10^{-20}$	$2.9\ 10^{-19}$	$4.9\ 10^{-19}$	$6.9\ 10^{-20}$	$1.0\ 10^{-19}$	$2.1\ 10^{-19}$
$\left(\frac{C_0pv}{V^2P}\right)^2 u_V^2$	Vessel volume	$6.1\ 10^{-21}$	$1.7\ 10^{-20}$	$6.9\ 10^{-10}$	$1.2\ 10^{-19}$	$1.7\ 10^{-20}$	$2.4\ 10^{-20}$	$5.1\ 10^{-20}$
$\left(\frac{C_0pv}{P^2V}\right)^2 u_P^2$	Vessel pressure	$5.4\ 10^{-21}$	$1.6\ 10^{-20}$	$6.7\ 10^{-20}$	$1.2\ 10^{-20}$	$1.5\ 10^{-20}$	$2.1\ 10^{-20}$	$4.7\ 10^{-20}$
"... u^2P_1	Random effect of the sensor	$3.5\ 10^{-22}$	$1.0\ 10^{-21}$	$4.4\ 10^{-21}$	$8.1\ 10^{-21}$	$1.0\ 10^{-21}$	$1.4\ 10^{-21}$	$3.1\ 10^{-21}$
"... u^2P_2	Systematic effect of the sensor	$2.5\ 10^{-21}$	$7.3\ 10^{-21}$	$3.1\ 10^{-20}$	$5.7\ 10^{-20}$	$7.0\ 10^{-21}$	$9.8\ 10^{-21}$	$2.2\ 10^{-20}$
"... u^2P_3	Δ pressure due to the room temperature	$2.5\ 10^{-21}$	$7.3\ 10^{-21}$	$3.1\ 10^{-20}$	$5.7\ 10^{-20}$	$7.0\ 10^{-21}$	$9.8\ 10^{-21}$	$2.2\ 10^{-20}$
u_c	Combined standard uncertainty	$3.9\ 10^{-10}$	$9.6\ 10^{-10}$	$1.1\ 10^{-9}$	$1.3\ 10^{-9}$	$4.5\ 10^{-10}$	$5.1\ 10^{-10}$	$6.8\ 10^{-10}$
$U=2u_c$ in ppbv and %	Expanded uncertainty	0.8 0.56%	1.9 0.83%	2.2 0.47%	2.6 0.43%	0.9 0.39%	1.0 0.37%	1.3 0.34%

Conclusions

The uncertainty associated with the reference value of NO and SO_2 standards derived by ERLAP using the static volumetric method have been evaluated. The value was calculated to be about ± 0.5% with 95% probability of the true value lying within this interval. The major component contributing to uncertainty was the determination of syringe volume, the most important aspects of which were the filling and transferring of gas. This was the case except at high concentrations, when NO gas purity became the most important component of uncertainty.

Table 5 Syringe volume

NO Syringe 1 µl	NO Syringe 2 µl	NO Syringe 2 µl	NO Syringe 3 µl	NO Syringe 4 µl	SO_2 Syringe 1 µl	SO_2 Syringe 2 µl	SO_2 Syringe 3 µl
24.71	39.9	40.0	78.49	99.37	39.72	49.22	69.47
24.80	39.6	39.9	78.40	99.30	39.81	49.28	69.52
24.69	39.6	40.0	78.48	99.50	39.67	49.31	69.42
24.80	39.9	40.0	78.60	99.50	39.76	49.24	69.45
24.77	40.0	40.0	78.58	99.45	39.72	49.20	69.44
24.76	39.9	40.1	78.65	99.43	39.66	49.37	69.36
24.74	39.9	39.9	78.69	99.47	39.74	49.23	69.37
24.76	39.8	40.1	78.63	99.44	39.74	49.32	69.39
24.85	39.9	40.0	78.52	99.33	39.68	49.22	69.43
24.73	39.9	39.9	78.56	99.47	39.72	49.27	69.37
24.73	40.2	39.7	78.60	99.55	39.69	49.28	69.47
24.74	39.9		78.58	99.48	39.81	49.24	69.47
24.74	40.1		78.65	99.46	39.69	49.27	69.46
24.71	40.0		78.79	99.36	39.76	49.25	69.44
24.84	40.1		78.63	99.36	39.76	49.18	69.36

Table 6 Static volumetric experiments with various time intervals before injection

Date	Experiment	Time interval before injection	p (initial) mbar	P (final) mbar	Theoretical value (A) ppbv	SO_2 (AF 21M) (**B**) ppbv	$(B/A)^{-1}$ %
3/12/96	2	5	1001	1525	458.6	458.7	0.03
3/12/96	3	5	1000	1523	458.5	458.7	0.03
3/12/96	4	30	999	1530	455.4	453.2	−0.49
3/12/96	5	30	999	1538	453.2	452.2	−0.20
3/12/96	6	30	999	1530	455.7	455.9	0.05
3/12/96	7	5	1000	1532	455.5	456.8	0.28
3/12/96	8	30	1000	1535	454.7	455.9	0.28
4/12/96	9	5	1002	1529	457.8	456.3	−0.34
4/12/96	10	60	1002	1530	457.8	454.4	−0.74
4/12/96	11	5	1002	1570	445.6	446.2	0.14
4/12/96	12	60	1002	1532	456.3	452.6	−0.81
4/12/96	13	5	1002	1532	456.3	457.2	0.19
4/12/96	14	60	1002	1530	456.9	456.3	−0.14
4/12/96	15	60	1002	1581	442.2	441.6	−0.12
5/12/96	16	5	1004	1528	458.7	460.8	0.46
5/12/96	17	120	1004	1537	455.9	449.0	−1.51
5/12/96	18	5	1003	1528	458.0	457.2	−0.17
5/12/96	19	120	1001	1521	459.8	448.1	−2.54
5/12/96	20	120	1001	1528	457.2	447.2	−2.19
5/12/96	21	5	1001	1526	457.6	456.3	−0.30
6/12/96	22	5	1004	1532	457.2	456.3	−0.20
6/12/96	23	15	1004	1573	445.4	446.3	0.18
6/12/96	24	15	1005	1526	459.8	459	−0.17
6/12/96	25	5	1004	1536	456.9	457.2	0.05
6/12/96	26	15	1004	1519	461.0	461.8	0.17
6/12/96	27	5	1004	1525	459.3	460.0	0.13
6/12/96	28	15	1005	1525	459.8	460.0	0.03

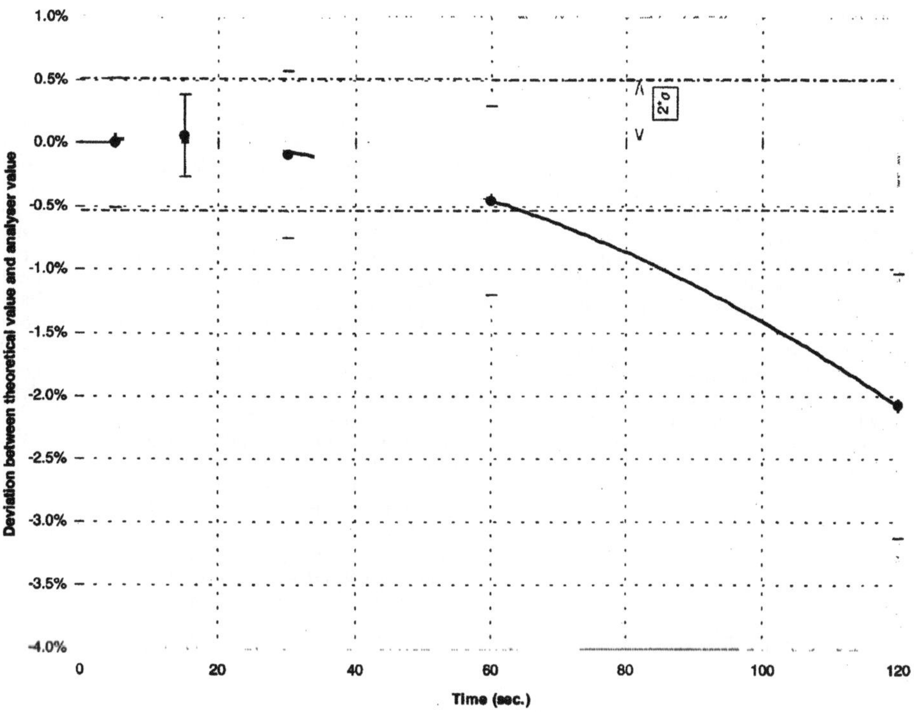

Fig. 3 Losses of pure SO_2 by diffusion through the needle of the syringe

To attain such high levels of confidence on the reference value of NO and SO_2 standards, the following should be taken into account:

- The complex procedure for manipulating the dosing syringe must be carefully followed.
- Values specified in the pure gas manufacturer's certificate must be periodically verified.
- The balance used to weigh the syringe and the pressure sensor serving the mixing vessel must be well maintained to ensure accurate and precise measurements. Traceablility certificates for these instruments must be available.
- Room temperature must remain constant between injection and dilution with carrier gas.

The uncertainty associated with the reference value of NO_2 (obtained by ERLAP using the permeation method) has been evaluated previously [11] as about 1% with 95% confidence limits. This is slightly greater than that of an NO standard prepared using the static volumetric method.

Appendix 1: Volumes dispensed with the syringe

The volumes of the syringes (with ERLAP mechanism) were determined by filling with water and deriving volume from the weight of liquid (measured with a Mettler AT201 balance) contained in the syringe (corrected for water density of 0.998 g/cm^3 at 22 °C). The volumes are reported in table 5.

Appendix 2: SO_2 measurements plotted against time interval before injection

In the present experiments, a UV fluorescence analyser manufactured by Environnement SA model AF21 M was used. A Hamilton syringe series 1800 with a needle series n°80451 was used and a volume of 78 μl (v) was injected for all experiments.

Table 6 shows a series of experiments carried out to determine the maximum tolerable time delay before injection (see "Procedure for the preparation of a mixture"). This time interval depends on the diffusion rate of the SO_2 through the syringe needle and must be established for the methodology to be viable. The results are presented in the Fig. 3, and show that the performance of the methodology is not compromised provided the interval is kept below 30 s.

References

1. Directive du Conseil du 15 juillet 1980 concernant des valeurs limites et des valeurs guides de qualité atmosphérique pour l'anhydride sulfureux et les particules en suspension. Journal officiel des Communautés
2. Directive du Conseil du 7 mars 1995 concernant les normes de qualité de l'air pour le dioxyde d'azote (85/203/CEE). Journal officiel des Communautés
3. Norme internationale ISO 6767 (F) (1990) Air ambiant – Détermination de la concentration en masse du dioxyde de souffre – Méthode au tétrachloromercurate (TCM) et a la pararosaline
4. Norme internationale ISO/DIS 7996 (F)/TC 146 (1984) Qualité de l'air – Détermination des oxydes d'azote dans l'air ambiant – Méthode par chimiluminescence
5. Norme internationale ISO/CD 10498 (F)/TC 146 (1984) Air ambiant – Dosage de souffre – Méthode par fluorescence dans l'ultraviolet
6. Norme internationale ISO 6349 (F) (1979) Analyse des gaz – Préparation des mélanges de gaz pour étalonnage – Méthode par perméation
7. Norme internationale ISO 6144 (F) (1981) Analyse des gaz – Préparation des mélanges de gaz pour étalonnage – Méthode volumétriques statique
8. Verien Deutscher Ingenieure, "Messen von Gasen, Prüfgase – Herstellen von Prüfgasen nach der Volumetrisch-Statischen Methode unter Verwendung von Glasbehältern", VDI 3490 Blatt 14, November 1985
9. Rudolf W, "Implementation of the Static Injection Method", EEC contract n° 4108-90-10-ED ISP D
10. Guide to the expression of uncertainty in measurement, ISBN 92-67-10188-9, copyright International Organisation for Standardisation, Printed in Switzerland
11. Gerboles M, Manalis N, De Saeger E, Payrissat M (1996) Report EUR 16432 EN "Study of the long term stability of NO2 Permeation Sources and the efficiency of Gravimetry in determining their permeation rate"

Accred Qual Assur (2000) 5:280–284

Daniela Kruh

Assessment of uncertainty in calibration of a gas mass flowmeter

D. Kruh (✉)
Rafael Calibration Laboratories,
P.O. Box 2250, 31021 Haifa, Israel
e-mail: danielak@rafael.co.il
Tel.: +972-4-8794494
Fax: +972-4-8794218

Abstract A primary calibration system was set up in Rafael some years ago, based on volumetric flow rate. The primary standard measures volumetric flow by means of the volume change of a dual piston over a specific time interval. This system serves to calibrate secondary standards of the thermal mass flowmeter type. Calibration procedures were prepared and validated. The paper describes the tests and calibration procedure conducted for the uncertainty assessment, the different components contributing to the measurement uncertainties, and the formulas involved with volumetric flow fates and with thermal mass flowmeters.

Key words Volumetric gas flow rate · Mass flowmeter · Calibration · Uncertainty

Introduction

Flowmeters are widely used in analytical chemistry. Examples of some chemical methods which use flowmeters are: gas chromatography, flow injection analysis, gas analysis, monitoring of quality environment, etc.

There are a number of methods available for performing calibration, which may be categorized as primary or secondary (transfer) techniques.

Primary calibration system

In a primary system the measurement is based on fundamental units such as length, mass, temperature, and time. The basic flow input for the instrument to be calibrated is determined through measurement of time and gas volume. Our laboratory is equipped with a volumetric system MKS Califlow A150 (Andover, Mass., USA)based on two pistons in the range of 1 standard cubic centimeters per minute (SCCM) to 50 standard liters per minute (SLPM). The distance the piston travels over a time interval is measured. Due to the fixed cross sectional area of the piston, area times length equals the volume of the cylinder in which the gas is collected. This volume divided by the filling time yields the volume flow rate. Correcting for the absolute pressure and temperature of the gas gives the mass flow rate.

The primary system uses the ideal gas law, Eq. 1, as the basis for determining mass flow rates:

$$PV = n^*R^*T \qquad (1)$$

where:
P = absolute pressure (mmHg),
V = volume of gas (l),
n = moles of gas (mol),
R = universal gas constant (62.364 mmHg·l/K l mol),
T = absolute temperature (K).

Assuming constant temperature and constant pressure, by measuring the change in volume over time the flow rate is attainable. Methods for measuring the volume displaced by a gas are based on Eq. 2.

$$Q = V/(t_2 - t_1) \qquad (2)$$

where:
Q = the volumetric flow rate (l/min),
$(t_2 - t_1)$ = the elapsed time (min).

Generally, one uses Qs, the volumetric flow rate, which refers to standard conditions, standard temperature and pressure (STP), which are given as 0 C and 760 mm of mercury. The subscript "S" denotes standard conditions.

$$Q_S = \frac{P}{P_S} * \frac{T_S}{T} * \frac{V}{(t_2 - t_1)} \tag{3}$$

Since allowances must be made for changes in temperature and pressure, the parameters of the perfect gas law must apply. The precise calculation of the volumetric flow rate through a piston meter of this type is based on Eq. 4:

$$Q_S = \frac{C * T_S * P_m}{K * t * T_m * P_S} \tag{4}$$

where:

Q_S = the gas flow rate corrected to STP (l/min);
T_S = standard temperature: 273.15 K;
P_S = standard pressure: 760 mmHg;
T_m = the temperature of the test gas in the cylinder (K);
P_m = the pressure of the test gas in cylinder (barometric pressure + cylinder pressure) (mmHg);
C = total number of counts accumulated by a shaft encoder synchronized to a crystal clock (counts);
t = total time to accumulate C counts (min);
K = number of counts per liter (calculated for each cylinder at the calibration time).

Secondary standards

As secondary standards our laboratory uses thermal mass flowmeters and controllers in the range of 100 SCCM to 500 SLPM.

Mass flowmeters use the thermal properties of a gas to measure flow rate directly [1]. Mass flow rates are determined by measuring the heat required to maintain an elevated temperature profile along a laminar flow sensor tube. For a specific flowmeter range and gas species, flow is proportional to the voltage necessary to maintain a constant temperature profile. The sensor in a mass flowmeter is a long, thin stainless steel tube, often called a capillary tube because of its shape (see Fig. 1).

Coils wrapped around the midpoint of the capillary tube serve two functions: first as heaters and second as temperature sensors. Since the resistance of the coils varies with temperature, they function as temperature detectors, or resistance temperature detectors (RTDs), which measure the temperature of the gas. The heaters create a known temperature profile along the sensor tube and then maintain the profile during gas flow by means of an autobalancing bridge circuit. As gas flows through the sensor, the gas flow convects heat and the temperature difference is converted into a flow reading.

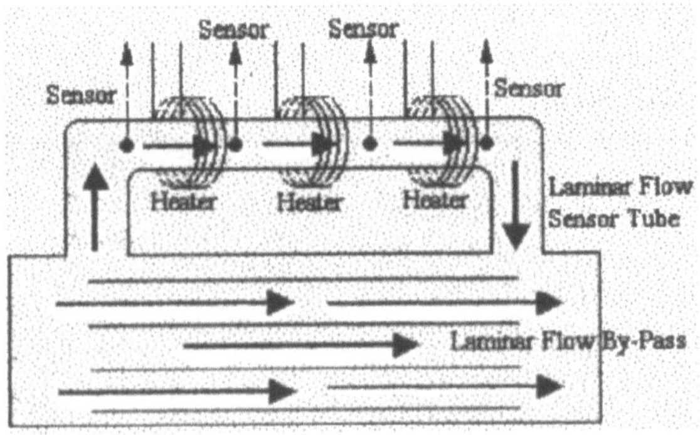

Fig. 1 Schematic structure of a typical thermal mass flowmeter

Thermal mass flowmeters have the capability of giving accurate measurements over a fairly wide range of temperature and pressure without the need to enter pressure or temperature corrections into the calculations. This feature is due to the units that are calibrated with reference to standard conditions and which are stable from about 15–32 °C and from about atmospheric pressure to about 30 PSI for the large volume flowmeters (about 25–50 SLPM). Thermal mass flowmeters are always calibrated for a particular type of gas. All manufacturers list gas conversion factors in their manuals for corrections to be made when measuring different gasses. These figures are mostly theoretically computed based on densities, specific heats, and atomic weights of the gases.

Calibration methods

Standard thermal mass flowmeters up to 50 SLPM are calibrated by comparison to the primary system, Califlow A150 and those used for higher range are sent abroad for calibration.

The secondary thermal mass flowmeters are used to calibrate all kinds of flow devices in accordance to the required range and uncertainty.

A calibration procedure was written by our calibration laboratory, which describes in detail how the two types of calibrations mentioned above are performed.

Uncertainty calculation

The uncertainty evaluation was done according to the recommendations in the ISO Guide for the Expression of Measurement Uncertainty [2]. This procedure was performed on both previously mentioned calibration methods. According to the Guide, we followed the steps described further on.

Calibration vs. Califlow A150

Determination of the measurement model

In our case the physical model is described by Eq. 5:

$$Q_S = \frac{C^*T_S^*P_m}{K^*t^*T_m^*P_S} \tag{5}$$

Sensitivity coefficients calculation

In order to determine the weight of each uncertainty component the partial derivatives are calculated for each of them, resulting in 1 for each of them.

Identification of standard uncertainties components

The various components of the uncertainty and their contribution as required by the ISO Guide for the Expression of Uncertainty [2] appear in Table 1. The drift, temperature measurement and repeatability measurements are the most influential factors.

The sources of uncertainty have to be determined by experiment, or by using figures that are widely accepted and their weighted contribution has to be considered by their sensitivity coefficients.

The product of these two values yields the weighted uncertainty, which after summation constitute the combined uncertainty. The expanded uncertainty at 95% confidence level is obtained by multiplying it by a coverage factor, k, which depends on the number of degrees of freedom, $\nu_{eff,}$ determined by Welch-Satterthwaite [2].

In our case most of the components are of Type B and each repeatability test consisted of ten measurements. The computed effective degrees of freedom is large enough, hence $k=2$.

The components of the uncertainty computations have been divided into three groups:

1) Reference standard uncertainty:
   - Volumetric factors, such as the cylinder area, the length of the stroke, etc.
   - Thermal and pressure influence on the measurements and system
   - Measuring instruments resolutions and variations during the calibration
   - Electronic uncertainty, mainly caused by the encoder and time measurement
2) Unit under test (UUT) uncertainty:
   Measuring instrument calibration uncertainty, which in our case was a Wavetek 1271 DMM.
3) Repeatability measurements which evaluate the closeness of sequential measurement results of the same parameter at the same experimental conditions.
4) Drift of the readings.

The drift was evaluated from the deviations between the readings obtained in different runs, of ten measurements each, at the same flow.

Expanded uncertainties for calibration vs. Califlow A150 of additional flowmeters are presented in Table 2, using the same method as described above.

Table 1 Uncertainty budget for the calibration of a 10 standard liters per minute (SLPM) mass flowmeter vs. the Califlow

Source of uncertainty	Standard uncertainty $u(x_i)$ (%)	Sensitivity coeff. symbol and value C_i	Weighted uncertainty $u(x_i)^*\ C_i$ (%)
Reference standard uncertainty			
Volume measurement collective uncertainty (Type B)	0.0076000	1	0.0076000
Barometric pressure repeatability (Type A)	0.0009671	1	0.0009671
Barometric pressure resolution (Type B)	0.0037984	1	0.0037984
Density temperature effect (Type B)	0.0115470	1	0.011547
Time reading (Type B)	0.0002887	1	0.0002887
Temperature measuring RTD calibration (Type B)	0.0307000	1	0.0307000
Temperature reading resolution (Type B)	0.0577350	1	0.0577350
Encoder calibration uncertainty (Type B)	0.0138000	1	0.0138000
Encoder resolution (Type B)	0.0048113	1	0.0048113
Nitrogen purity uncertainty (Type B)	0.0028752	1	0.0028752
System effects uncertainty			
Mass flowmeter calibration repeatability (Type A)	0.0770000	1	0.0770000
Voltmeter calibration uncertainty (Type B)	0.0003600		0.0003600
Voltmeter manufacturer uncertainty per year (Type B)	0.0000500		0.0000500
Voltmeter reading repeatability (Type A)	0.0230940		0.0230940
Calibrated mass flowmeter zeroing repeatability (Type A)	0.0115470		0.0115470
Drift of readings during different runs (Type B)	0.1443376	1	0.1443376
Combined Uncertainty of the Calibration Process (u_c)			0.2% F.S.
Expanded uncertainty of the calibration process (U)			0.4% F.S.

Table 2 Calculated expanded uncertainties for calibrated flowmeters vs. Califlow A150

Description of flowmeter	Calculated expanded uncertainty (U) (%)
0.5 SLPM	0.35
5 SLPM	0.33
10 SLPM	0.36
30 SLPM	0.34

Calibration vs. a secondary standard mass flowmeter

In this kind of calibration a precise mass flowmeter previously calibrated by the A150 Califlow serves as a secondary standard. Using the same method for determining the measurement uncertainty as described in the section on Calibration vs. Califlow A150 we proceeded with the following steps:

Determination of the measurement model

In this case the physical model is described by the Eq. (12):

$$D = Q_m - Q_R \tag{6}$$

where:

Q_m = the gas flow rate measured by the tested flowmeter (SLPM),
Q_R = the gas flow rate measured by the reference flowmeter (SLPM).

Sensitivity coefficients calculation

Sensitivity coefficients for Q_m and Q_R were calculated and resulted in 1 for each of them.

Table 4 Calculated expanded uncertainties for calibrated flowmeters vs. flowmeters

Description of flowmeter	Calculated expanded uncertainty (U) (%)
0.1 SLPM vs. 0.5 SLPM	0.47
5 SLPM vs. 5 SLPM	0.46
10 SLPM vs. 10 SLPM	0.48
10 SLPM vs. 30 SLPM	0.46

Identification of standard uncertainty components

Standard uncertainty components are described in Table 3. Both flowmeters were assumed to be calibrated by the same type of gas and were exposed to the same temperature and pressure conditions.

Expanded uncertainties for calibration of flowmeters are presented in Table 4 using the same method as described above. In this case, the reference mass flowmeter, the drift and the repeatability measurements contribute mostly to the uncertainty.

Summary and conclusions

Descriptions are given of the Rafael Calibration Laboratory's facilities for calibrating flowmeters.

Methods of operation are given together with the measurement uncertainties obtained using the "ISO Guide for the Expression of Uncertainty" recommendations.

The first method uses a primary standard of the piston prover type, Califlow A150. The uncertainty estimation in this case is based on the manufacturer's uncertainties, experienced judgment, and propagation uncertainty techniques. Typical values of uncertainty us-

Table 3 Calibration uncertainty budget for 10 SLPM flowmeter vs. secondary standard flowmeter

Source of uncertainty	Standard uncertainty $u(x_i)$ (%)	Sensitivity coeff. symbol and value C_i	Weighted uncertainty $u(x_i)^* C_I$ (%)
Reference standards uncertainty			
Reference mass flowmeter uncertainty (Type B)	0.179257	1	0.179257
Voltmeter calibration uncertainty (Type B)	0.000360	1	0.000360
Voltmeter manufacturer uncertainty per year (Type B)	0.000050		0.000050
Voltmeter reading repeatability (Type A)	0.023094	1	0.023094
Reference mass flowmeter zeroing repeatability (Type A)	0.011547	1	0.011547
System effects uncertainty			
Voltmeter calibration uncertainty (Type B)	0.000360	1	0.000360
Voltmeter manufacturer uncertainty per year (Type B)	0.000050	1	0.000050
Voltmeter reading repeatability (Type A)	0.023094	1	0.023094
Calibrated mass flowmeter zeroing repeatability (Type A)	0.011547	1	0.011547
Drift of readings during different runs (Type B)	0.144338	1	0.144338
Mass flowmeter calibration repeatability (Type A)	0.057000	1	0.057000
Combined uncertainty of the calibration process (u_c)			0.24% F.S.
Expanded uncertainty of the calibration process (U)			0.5% F.S.

ing this method are around 0.3% F.S., with the drift and repeatability measurements as the most significant contributors to the total budget. The most significant contributor to Califlow A150 uncertainty is the temperature measurement uncertainty, followed by that of the encoder and the volume.

The second method is based on a comparison of the UUT to a secondary standard calibrated by the first method. The uncertainty budget in this case is determined equally by the drift and the reference standard used for the calibration. Typical values received in that case, are around 0.5%.

References

1. Hinkle LD, Marino CF (1990) Towards understanding the fundamental mechanism and properties of the thermal mass flow controller. MKS Instruments, Andover, Mass., USA
2. ISO (1995) Guide to the expression of uncertainty in measurement. ISO, Geneva, Switzerland

Accred Qual Assur (1998) 3:231–236

Paul Willetts
Roger Wood

Measurement uncertainty – a reliable concept in food analysis and for the use of recovery data?

Presented at: 2nd EURACHEM Workshop on Measurement Uncertainty in Chemical Analysis, Berlin, 29–30 September 1997

P. Willetts · R. Wood (✉)
Food Labelling and Standards Division, Ministry of Agriculture, Fisheries and Food, CSL Food Science Laboratory, Norwich Research Park, Colney, Norwich NR4 7UQ, UK
Tel.: +44-1603-259350
Fax: +44-1603-501123
e-mail: r.wood@fscii.maff.gov.uk

Abstract Steps which are taken to implement the concept of measurement uncertainty in analytical chemical laboratories should take full account of existing internationally agreed protocols for analytical quality assurance and reflect the needs of particular analytical sectors. For the food sector this may mean that for official purposes the use of the term measurement uncertainty is replaced by the term measurement reliability and that a quantitative estimation of this is made based on existing collaborative trial data. In many analytical sectors, the differing strategies currently followed for the determination and use of recovery information are an important cause of the non-comparability of analytical results. Guidelines which are being prepared for the estimation and use of recovery information in analytical measurement may provide a more unified approach which includes measurement uncertainty as a key concept in the use of recovery data.

Introduction

Recent years have seen the issue of the quality and reliability of data become of paramount importance in all analytical sectors. In order to address this matter, analysts from the different analytical sectors have worked together under the sponsorship of ISO, IUPAC and AOAC INTERNATIONAL, to produce International Harmonised Protocols on the subjects of the collaborative testing of analytical methods [1], proficiency testing [2] and the use of internal quality control in analytical chemistry laboratories [3]. In addition, the use of certified reference materials is increasingly being advocated with respect to the traceability of analytical data [4], and laboratory accreditation schemes are being widely implemented.

Each of these components of analytical quality assurance concerns a different aspect of data reliability, namely the external testing of methods and laboratory performance, internal data quality, trueness and the auditing of procedures and records. With the exception of the latter which is administratively based, in each case reliability is limited by either, or in many cases both, systematic or random experimental 'inaccuracies', quantities now being embraced by the concept of measurement uncertainty.

Requirements and initiatives in the food sector

In introducing this concept to the analytical chemical community there is a need to ensure that steps taken to implement measurement uncertainty are made in the context of the existing protocols and strategies in analytical quality assurance. Moreover, the needs and actions of particular analytical sectors must also be recognised. In the food sector a number of initiatives have been advanced recently which affect specifically the issue of data quality, and therefore reliability, in this area of analysis. Firstly, in the EU, there is a tendency in the food analysis sector to not prescribe specific methods of analysis but to adopt a "criteria of methods approach" whereby analysts may use the method of their choice provided it meets certain prescribed quality criteria. This flexibility of approach, to take advantage of the developments of new techniques and procedures as they occur in analytical chemistry, clearly has consequences for the comparability and measurement uncertainty of reported data. Secondly, there is a requirement in the food sector, as set out in EC Directive 93/99, that methods of analysis for food control purposes should wherever possible be formally validated by collaborative trial [5]. Thirdly, there have been discussions on measurement uncertainty within the Codex Committee on Methods of Analysis and Sampling. The Report of the March 1997 Session of that Committee states that with regard to measurement uncertainty [6]:

1. The Committee will develop for Codex purposes an appropriate alternative term for measurement uncertainty, e.g. measurement reliability.
2. The precision of a method may be estimated through a method-performance study, or where this information is not available, through the use of internal quality control and method validation.
3. Consideration should be given as to whether it is necessary to undertake an additional formal evaluation of a method of analysis using the ISO approach [7] in addition to using information obtained through a collaborative trial.
4. Governments should advise accreditation agencies that for national and Codex purposes the measurement uncertainty result need not be calculated using the ISO approach [7] providing the laboratory is complying with the appropriate Codex principles.

Discussions are on-going in Codex. However, if these proposals are accepted, it is likely that the term 'measurement reliability' rather than measurement uncertainty will be adopted and that estimates of this will be made from collaborative trial data if such data are available. In a recent study, carried out in the UK, which compared 'top-down' (collaborative trial) and 'bottom-up' (ISO) approaches to the estimation of measurement uncertainty, it was concluded that for comparable matrix/analyte combinations these approaches gave not dissimilar results in the limited number of cases studied [8]. It should be noted that, in recognising the importance of the concept of measurement uncertainty in underpinning the reliability of analytical data, the Codex recommendations and discussions are in accordance with statements on uncertainty in ISO Guide 25 and EN 45001, which require accreditation agencies to ensure that measurement uncertainty estimations are carried out as part of the accreditation process [9].

Recovery and analyte losses

One aspect of analytical chemistry where, for all analytical sectors including the food sector, current practice continues to have important consequences in terms of the non-comparability and uncertainty or reliability of reported data, is that of the use of recovery information. This arises because of the different strategies for dealing with recovery assessment and the effect these may have on the variability of the analytical results reported.

Recovery studies are an essential component of quality assurance systems in analytical measurement. Their use, particularly in the trace analyte area, to assess the efficiency of the removal of the measurand from the sample matrix and its transfer prior to detection is widely quoted in the scientific literature. Although they thus provide an important indication of the reliability of these steps in the measurement process, there generally has been no consistent approach to the way in which recovery information is derived and used in analytical data. In particular, in the case of recovery factors calculated and applied to analytical data to correct for displacement or bias, the absence of accepted strategies for the determination and use of these factors has meant that it frequently has been difficult to make comparisons between analytical results produced in different laboratories or verify the suitability of that data for the intended purpose. This is particularly marked in the case of complex matrices, such as foodstuffs, where the difficulties of completely extracting the analyte are most pronounced. Quite commonly in such procedures a substantial proportion of the analyte remains in the matrix after extraction, so that the transfer is incomplete, and the subsequent measurement is lower than the true concentration in the original test material. If no compensation for these losses is made, then markedly discrepant results may be obtained by different laboratories. Even greater discrepancies are likely to arise if some laboratories compensate for losses and others do not. These considerations are especially important in legislative/enforcement situations where for instance the difference between applying or not applying a recovery factor to correct for the incomplete removal of the analyte may mean respectively that a legislative limit is exceeded or that a result is in compliance with the limit.

Recovery correction factors

Thus, where an estimate of the *true concentration* is required, there is a compelling case for including a compensation for losses in the calculation of the reported analytical result, provided that the correction factor can be estimated reliably. In the case of an empirical method, where the measurand is defined in terms of the method used and no attempt is being made to estimate the amount of analyte actually present in the sample matrix, the question whether or not a correction is applied is a matter for the definition of the empirical method.

The four most common approaches which typically have been taken by analysts in respect of the application of recovery factors are shown in Table 1.

Table 1 Typical approaches relating to the application of recovery factors

a	The reporting of an analytical result without correcting for bias by the application of a recovery factor, no accompanying statement being given of the level of recovery achieved
b	The reporting of an analytical result without correcting for bias by the application of a recovery factor, together with a statement of the level of recovery achieved
c	The reporting of an analytical result corrected for bias by the application of a recovery factor, without an accompanying statement of the level of recovery
d	The reporting of an analytical result corrected for bias by the application of a recovery factor, together with a statement of the level of recovery achieved

Reference materials and spiking experiments

Quite apart from the variation which can arise from laboratories adopting different practices in respect of whether a correction factor is applied or is not applied to an analytical result, a further aspect which can hinder data comparison is the fact that 'recovery' information may be derived either from the inclusion of reference materials or the use of spiked samples.

In the case of reference materials, the analyte is usually integrated or incorporated into the matrix, whereas in the case of spiked samples the analyte is merely added to the matrix. Potentially different information relating to the behaviour of the native analyte to be measured may be derived from each type of recovery measurement. Moreover, the regularity and pattern of use of these recovery materials may affect the recovery information produced. In the case of spiking, for example, the different ways in which the recovery factor may be determined include those shown in Table 2.

Table 2 Examples of ways in which the recovery factor may be determined with spiking

a	Basing a recovery correction factor on the recovery of the analyte from a spiked sample in the batch
b	Basing a recovery correction factor on the mean value obtained for the recovery of the analyte spiked into a sample in each of a number of batches
c	Basing a recovery correction factor on the recovery of a chemically similar internal standard added to the test material
d	Basing a recovery correction factor on the recovery of an isotopic form of the analyte added as internal standard to the test material

Each of these approaches differs in the representativeness it provides of the actual extraction of the analyte itself, the basis of the representation being different in each case. While it is generally agreed that, of these four alternatives, the use of an isotopic internal standard is the preferred approach since the recovery of the auxiliary analyte equates most closely to being 'fully equivalent' to that of the target analyte, this option is often not possible. As a consequence one of the other alternatives is often followed in spiking experiments.

When a reference material is used rather than spiking, then it will be included at a different position in the batch to the test material itself. In this respect the use of a reference material is akin to options a or b for spiking (see Table 2).

Consideration of these different strategies has led analytical chemists to recognise the desirability of using a more uniform approach when dealing with the topic of recovery measurements in order to facilitate the comparability of data.

Guidelines for using recovery information

Following the circulation to a broad cross section of the analytical community world-wide of a questionnaire on the determination and use of recovery measurements in 1995, background information was obtained which enabled further consideration to be given to the role of recovery studies in chemical analysis [10]. The main questions addressed the issues shown in Table 3.

As expected, the differing answers given to the questions posed revealed considerable variation in the ways in which analysts deal with recovery measurements. In particular, the question on measurement uncertainty itself produced more differences than any of the other questions, perhaps suggesting a lack of appreciation of either the need for or the means of calculating this value. The findings of this survey were presented at the

Table 3 Outline of questions included in the recovery factors questionnaire

Question number	Question	Question number	Question
1	Meaning of recovery	12	Recovery of analyte and internal standard
2.1	Purpose/use of recovery measurements	13	Spiking procedure
2.2	Reporting results	14	Blank material
3	Recovery frequency in time	15	Spiking level
4	Recovery frequency in batch	16	Spiking concentration
5	Recovery level	17	Carrier solvent for analyte
6	Acceptable recovery levels	20	Recovery solution
7	Assessment of acceptable recovery	21	Time of sample preparation
9	Multi-analyte determinations	23	Precision
10	Procedure for the determination of recovery	24	Measurement uncertainty
11	Matrices used		

Symposium on Harmonisation of Quality Assurance Systems in Chemical Analysis held in Orlando, USA, in 1996. From the deliberations of that meeting, harmonised guidelines for the use of recovery information in analytical measurement are being prepared under the sponsorship of IUPAC, ISO and AOAC INTERNATIONAL [11]. The guidelines, the main points of which are summarised in Table 4, refer to uncertainty as being a key concept in formulating an approach to the estimation and use of recovery information.

Uncertainty and recovery correction

Although the estimation of uncertainty in recovery has yet to be studied in detail, the guidelines list some sources of the uncertainty in measured recovery (Table 5) and include a treatment which considers the uncertainty estimation in cases of incomplete recovery where either a correction is or is not applied to an analytical result [12]. In this treatment, the difference in the measured recovery (R) from the value of unity, representing total recovery, is compared to the uncertainty in the determination of R.

The comparison is made using a significance test to assess whether $|R-1|$ is greater than the uncertainty

Table 4 A summary of guidelines for the use of recovery information

1. A distinction is recognised between:
   surrogate recovery (recovery of a pure compound or element specifically added to the test portion or test material as a spike – sometimes called "marginal recovery")
   the recovery of native analyte incorporated into the test material by natural processes and manufacturing procedures – sometimes called "incurred analyte".
2. It is recognised that there is a dual role for recovery determinations in analytical measurement, that is, (a) for quality control purposes and (b) for deriving values for recovery factors. In the latter application, more extensive and detailed data are required.
3. Variable practice in handling recovery information is an important cause of the non-equivalence of data. To mitigate its effects, in general, results should be corrected for recovery, unless there are overriding reasons for not doing so. Such reasons would include the situation where a limit (statutory or contractual) has been established using uncorrected data, or where recoveries are close to unity.
4. It is of over-riding importance that (a) all data, when reported, should be clearly identified as to whether or not a recovery correction has been applied and (b) if a recovery correction has been applied, the amount of the correction and the method by which it was derived should be included with the report. This will promote direct comparability of data sets. Thus, in all situations, correction functions should be established based on appropriate statistical considerations, documented, archived and available to the client.
5. Recovery values should always be established as part of method validation, whether or not recoveries are reported or results are corrected, so that measured values can be converted to corrected values and vice versa.
6. When the use of a recovery factor is justified, the method of calculation should be given in the method.
7. IQC control charts for recovery should be established during method validation and used in all routine analysis. Runs giving recovery values outside the control rånge should be considered for re-analysis in the context of acceptable variation, or the results should be reported as semi-quantitative.
8. Uncertainty is a key concept in formulating an approach to the estimation and use of recovery information. Although there are substantive practical points in the estimation of uncertainty that remain to be settled, the principle of uncertainty is an invaluable tool in conceptualising recovery issues.

Table 5 Sources of uncertainty in recovery estimation

1	Repeatability of the recovery experiment
2	Uncertainties in reference material values
3	Uncertainties in added spike quantity
4	Poor representation of native analyte by the added spike
5	Poor or restricted match between experimental matrix and the full range of sample matrices encountered
6	Effect of analyte/spike level on recovery and imperfect match of spike or reference material analyte level and analyte level in samples

(u_R) in the determination of R, at some level of confidence. The significance test takes the form

$|R-1|/u_R>t$: R differs significantly from 1
$|R-1|/u_R\leq t$: R does not differ significantly from 1

where t is a critical value based either on a 'coverage factor' allowing for practical significance or, where the test is entirely statistical, $t_{(\alpha/2,\ n-1)}$, being the relevant value of Student's t for a level of confidence $1-\alpha$.

Following this assessment, for a situation where incomplete recovery is achieved, four cases can be distinguished, chiefly differentiated by the use made of the recovery R.

(a) R is not significantly different from 1. No correction is applied.
(b) R is significantly different from 1 and a correction for R is applied.
(c) R is significantly different from 1 but, for operational reasons, no correction for R is applied
(d) An empirical method is in use. R is arbitrarily regarded as unity and u_R as zero. (Although there is obviously some variation in recovery in repeated or reproduced results, that variation is subsumed in the directly estimated precision of the method.)

In the first case, where R is not significantly different from 1, the recovery can be viewed as being equal to unity, no correction being applied. There is still an uncertainty, u_R, about the recovery that contributes to the overall uncertainty of the analytical result.

In the cases where R is significantly different from 1, the loss of analyte occurring in the analytical procedure is taken into account, and two uncertainties need to be considered separately. First, there are the uncertainties associated only with the determination, namely those due to gravimetric, volumetric, instrumental, and calibration errors. That relative uncertainty u_x/x will be low unless the concentration of the analyte is close to the detection limit. Second, there is the uncertainty u_R on the estimated recovery R. Here the relative uncertainty u_R/R is likely to be somewhat greater. If the raw result is corrected for recovery, we have $x_{corr}=x/R$ (*i.e.*, the correction factor is $1/R$). The relative uncertainty on x_{corr} is given by

$$\frac{u_{corr}}{x_{corr}}=\sqrt{\left(\frac{u_x}{x}\right)^2+\left(\frac{u_R}{R}\right)^2}$$

which is necessarily greater than u_x/x and may be considerably greater. Hence correction for recovery seems at first sight to degrade, perhaps substantially, the reliability of the measurement.

It is stated that such a perception is incorrect. Only if the method is regarded as empirical, and this has drawbacks in relation to comparability as already discussed, is u_x the appropriate uncertainty. If the method were taken as rational, and the bias due to loss of analyte were not corrected, a realistic estimate u_x would have to include a term describing the bias. Hence u_x/x would be at least comparable with, and may be even greater than, u_{corr}/x_{corr}.

These approaches to the estimation of the uncertainty of a recovery are necessarily tentative. Nevertheless, the following important principles of relevance to the conduct of recovery experiments are demonstrated.

(a) The recovery and its standard uncertainty may both depend on the concentration of the analyte. This may entail studies at several concentration levels.
(b) The main recovery study should involve the whole range of matrices that are included in the category for which the method is being validated. If the category is strict (e.g., bovine liver) a number of different specimens of that type should be studied so as to represent variations likely to be encountered in practice (e.g., sex, age, breed, time of storage etc.). Probably a minimum of ten diverse matrices are required for recovery estimation. The standard deviation of the recovery over these matrices is taken as the main part of the standard uncertainty of the recovery.
(c) If there are grounds to suspect that a proportion of the native analyte is not extracted, then a recovery estimated by a surrogate will be biased. That bias should be estimated and included in the uncertainty budget.
(d) If a method is used outside the matrix scope of its validation, there is a matrix mismatch between the recovery experiments at validation time and the test material at analysis time. This could result in extra uncertainty in the recovery value. There may be problems in estimating this extra uncertainty. It would probably be preferable to estimate the recovery in the new matrix, and its uncertainty, in a separate experiment.

References

1. Horwitz W (1988) Pure Appl Chem 60:855–864
2. Thompson M, Wood R (1993) Pure Appl Chem 65:2123–2144 [Also published in J. AOAC Int (1993) 76:926–940]
3. Thompson M, Wood R (1995) Pure Appl Chem 67:649–666
4. Thompson M (1996) Analyst 121:285–288
5. Official Journal of the European Communities (1993) L290/14 Council Directive 93/99/EEC
6. Codex Alimentarius Commission (1997) Codex Committee on Methods of Analysis and Sampling 21st Session

7. ISO (1993) Guide to the Expression of Uncertainty in Measurement, Geneva
8. Brereton P, Anderson S, Willetts P, Ellison S, Barwick V, Thompson M, Wood R (1997) CSL Report FD 96/103
9. Draft - ISO/IEC Guide 25 (1996) General requirements for the competence of testing and calibration laboratories
10. Willetts P, Anderson S, Wood R (1998) CSL Report FD 97/65
11. International Union of Pure and Applied Chemistry (1997) Draft Harmonised Guidelines for the use of Recovery Information in Analytical Measurement
12. Ellison SLR, Williams A (1996) In: Parkany M (ed) Proceedings of the Seventh International Symposium on the Harmonisation of Quality Assurance Systems in Chemical Analysis. Royal Society of Chemistry, London

Accred Qual Assur (1998) 3:127–130

André Henrion

In- and off-laboratory sources of uncertainty in the use of a serum standard reference material as a means of accuracy control in cholesterol determination

Presented at: 2nd EURACHEM Workshop on Measurement Uncertainty in Chemical Analysis, Berlin, 29–30 September 1997

A. Henrion (✉)
Physikalisch-Technische Bundesanstalt, Bundesallee 100, D-38116 Braunschweig, Germany
Tel.: +49 531 592 3321;
Fax: +49 531 592 3015;
e-mail: andre.henrion@ptb.de

Abstract Repeated subsampling or a hierarchical design of experiments combined with an analysis of variance (ANOVA) is demonstrated to be a useful tool in the determination of uncertainty components in amount-of-substance measurements. With the reference material of human serum as investigated here for total cholesterol, besides several in-laboratory sources of uncertainty, a vial-to-vial effect which can be regarded as an off-laboratory source was found to be significant. This knowledge might be essential when the material is used for calibration and for the self-assessment of a laboratory.

Key words Experimental design · Analysis of variance (ANOVA) · Determination of uncertainties · Human serum · Cholesterol · Standard reference material

Introduction

For an amount-of-substance measurement to be regarded as under control, it is necessary to be able to clearly state its uncertainty. Under many circumstances, knowledge of particular uncertainty components to be assigned to particular steps of the whole procedure would also be of valuable assistance. This would allow future work to be focussed on the points requiring improvement. A carefully designed series of repeated measurements along with the evaluation of the results by analysis of variance (ANOVA) offers a powerful tool for obtaining this information. This will be demonstrated by the example of the determination of total cholesterol in a human serum reference material.

Material and method

The material investigated was the NIST Standard Reference Material 1952a [1]. This is a freeze-dried serum certified for its concentration of cholesterol.

The amount-of-substance measurement would comprise the following steps:

- Reconstitution of the serum by addition of water
- Separation of the analyte from the matrix (saponification of cholesterol fatty acid esters, extraction into hexane)
- Derivatization of the cholesterol (for preparation of gas chromatography)
- Quantification by gas chromatography/mass spectrometry (GC/MS)

Details of these manipulations are given elsewhere [2].

An experiment was set up according the strategy of *repeated subsampling,* also called *a nested* or *hierarchical* design. This is sketched in Fig. 1. Two vials had been drawn by the manufacturer as samples out of each of three different serum pools. After reconstitution, the contents of each of the vials were subdivided into three aliquots. Then, after addition of an internal standard (spike), the cholesterol was separated from the matrix and extracted for each aliquot independently of the others. Subsamples of the extracts were derivatized on three different occasions to obtain three solutions ready for analysis, each of which was finally subjected to three different GC/MS runs.

It should be noted that it would also have been possible to carry out this survey with two subsamples on every tier, instead of three. This often might be more advantageous because of the cost reduction involved, though, of course, degrees of freedom in the ANOVA would be sacrificed.

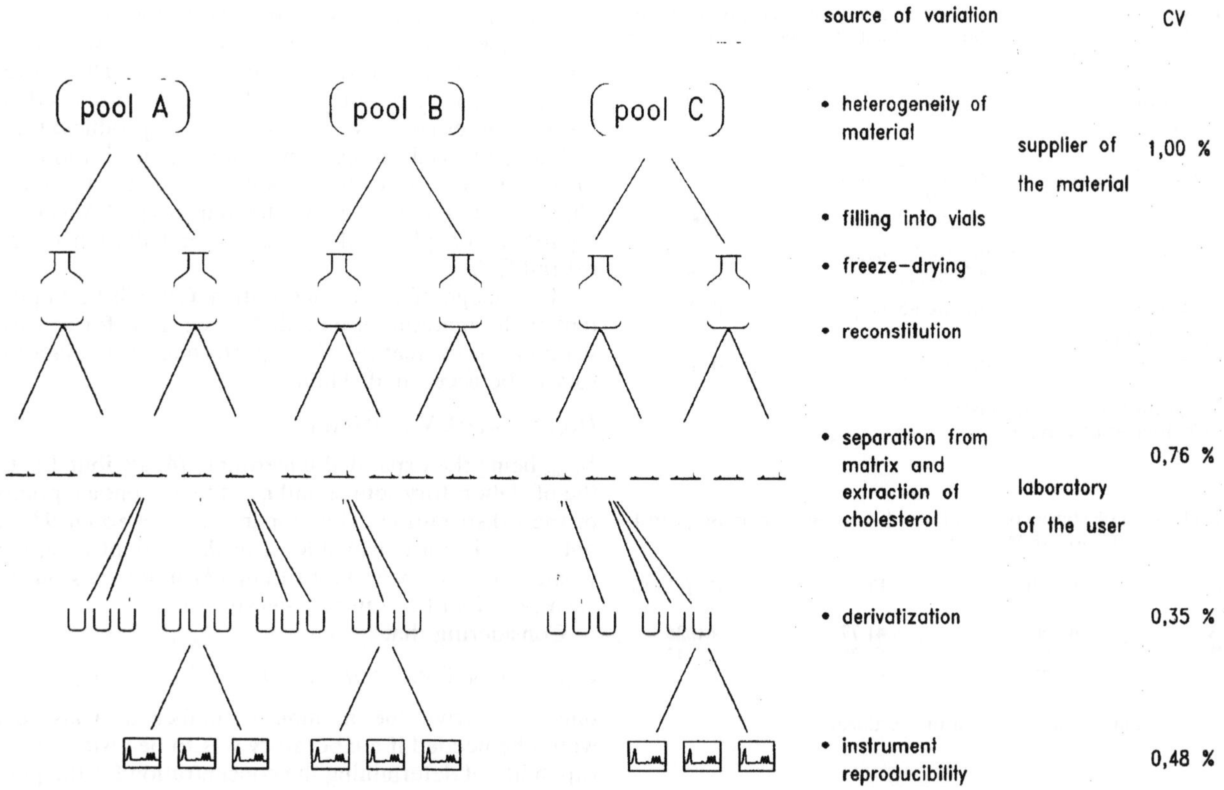

Fig. 1 Sampling scheme and sources of variation attributable to the tiers

For particulars of experimental designs and ANOVA, the reader not familiar with it is referred to the multitude of textbooks on this subject, e.g. Anderson and Bancroft [3].

Results

The direct results of the ANOVA are compiled in Table 1. Since all of the F ratios exceed the tabulated critical value corresponding to 5% risk of error, all of the mean squares can reasonably be assumed to represent significant contributions to the total sum of squares apart from the sole GC/MS sum of squares. Otherwise, the model would have to be recalculated, omitting the insignificant source(s).

The mean squares given in Table 1 do not yet represent the variances attributable to the individual sources, as each of these is actually a weighted sum of contributions of all sources below it in the hierarchy. This becomes evident from the calculated expected mean squares derived from theory for this nested model [3]. These are given in Table 2. The coefficients reflect the

Table 1 ANOVA of the amount-of-cholesterol data

Source of variation	DF[a]	SS[b]	MS[c]	F ratio
Inter vials	3	99.067	33.022	147.595
Inter extr.	12	70.068	5.839	26.098
Inter deriv.	36	21.296	0.592	2.644
Inter GC/MS runs	108	24.164	0.224	

[a] Degrees of freedom
[b] Sum of squares
[c] Mean square

numbers of subsamples drawn on each of the tiers and would change when other sampling schemes were used.

Equating the mean squares in Table 1 with the corresponding expected values furnishes the sought for estimates of variances of the individual sources. Their square roots, which can be interpreted in terms of uncertainties caused by them, are compiled in the third column of Table 2. These figures at the same time represent the coefficients of variation (CV), since, in this example, prior to the ANOVA, all data had been standardized to yield mean values of 100 for each of the pools.

Table 2 Expected mean squares and derived coefficients of variation for the amount-of-cholesterol data (σ: standard deviations of the individual sources)

Source of variation	E(MS)[a]	CV[b] in %
Inter vials	$\sigma^2_{GC/MS}+3\sigma^2_{derivs.}+3\cdot3\sigma^2_{extr.}+3\cdot3\cdot3\sigma^2_{vials}$	1.00
Inter extr.	$\sigma^2_{GC/MS}+3\sigma^2_{derivs.}+3\cdot3\sigma^2_{extr.}$	0.76
Inter deriv.	$\sigma^2_{GC/MS}+3\sigma^2_{derivs.}$	0.35
Inter GC/MS Runs	$\sigma^2_{GC/MS}$	0.48

[a] Expectation of mean square
[b] Coefficient of variation

Table 3 Cholesterol concentrations [in µmol/g (dry mass)] certified and found for SRM 1952a

Pool	Certified[a]	Found	Diff. in %
A	41.66	41.77	+0.25
B	63.85	64.76	+1.42
C	86.72	85.50	−1.40

[a] Calc. from the data given in the certificate

A comparison of the mean concentrations found for the pools with the data stated in the manufacturer's certificate is given in Table 3.

Discussion

The CVs (Table 3) can be combined to obtain an estimate of the CV of a single measurement: $CV=(CV^2_{vials}+CV^2_{extr.}+CV^2_{deriv.}+CV^2_{GC/MS})^{1/2}=1.39\%$. This is characteristic of the one-time drawing of a random vial out of a pool, separating from the matrix and extracting the cholesterol, derivatizing it and finally detecting the concentration by GC/MS. It is, of course, equivalent to the CV calculated with the whole measurement repeated several times in a straightforward way (i.e. formally with only one element on each of the tiers).

Knowledge of the individual variance contributions allows a further interpretation. For illustration, see the right-hand part of Fig. 1. Instrument reproducibility (GC/MS) is not a critical source of uncertainty, as possibly could have been assumed, and neither is the derivatization step. Among the manipulations in the laboratory of the user of the material, the separation from the matrix and the extraction into solvent might be reviewed for improvement. The in-laboratory CVs can be combined to form a joint CV of about 1.0%.

In addition to this, a vial-to-vial effect is observed. It is significant though it was characterized with only three degrees of freedom (see Table 1). The corresponding CV is about 1.0%. It can be discussed in terms of heterogeneity of the material, reproducibilities of filling into vials, freeze-drying and reconstitution prior to use. The author believes the last source to be negligible in magnitude. If so, the vial-to-vial CV can be regarded as a plain off-laboratory contribution to the overall CV.

Knowledge of the off-laboratory CV will be important if the material is intended to be used for self assessment or evaluation of the performance of a laboratory to be accredited. Then

$$U^2_{vials}=(t_{p,df}\cdot CV_{vials})^2/n_{vials},$$

U_{vials} being the expanded uncertainty [4] attributable to the off-laboratory source and $t_{p,df}$ the percentage points of the t distribution. This, for instance, is about 9% if only two vials are available as in the present case, but would not have been more than 1% if six vials out of each pool had been under investigation.

Considering that

$$U^2_{total}=(t_{p,df}\cdot CV_{vials})^2/n_{vials}+(t_{p,df}\cdot CV_{in\text{-}lab.})^2/n_{in\text{-}lab.}$$

one can derive the minimum number of vials that would be needed if a laboratory was to be tested for its capability of determining the concentration for the pool with a given uncertainty $U_{in\text{-}lab.}$.

In the example discussed here, U_{vials}, as already mentioned, is estimated to be at as high as 9%. However, the mean concentrations of cholesterol found for the pools happen to be pretty close to those certified (see Table 3). Therefore, at this level of uncertainty, there is no reason to suspect that the results are biased.

Conclusions

A carefully designed survey combined with ANOVA is a powerful tool for providing the experimenter with knowledge of particular components of the total variance. In many instances, as in the example presented here, no other way of detecting them is conceivable. The variance components furnish valuable information as to what steps of the whole procedure need to be checked for possible improvement. These at the same time would be the steps which perhaps would require further subsampling and averaging of the results in order to keep the uncertainty of the final result low.

References

1. Certificate of Analysis for SRM 1952a, National Institute of Standards & Technology, Gaithersburg, MD 20899, January 8, 1990
2. Henrion A, Dube G, Richter W (1997) Fresenius J Anal Chem 358:506–508
3. Anderson RL, Bancroft TA (1952) Statistical Theory in Research. McGraw-Hill Book Company, New York
4. Guide to the Expression of Uncertainty in Measurement (1st edn) (1993) International Organization for Standardization, ISBN 92-67-10188-9

Accred Qual Assur (1999) 4:124–128

Ilya Kuselman
Felix Sherman

Assessment of limits of detection and quantitation using calculation of uncertainty in a new method for water determination

I. Kuselman (✉) · F. Sherman
The National Physical Laboratory of Israel, Givat Ram, Jerusalem 91904, Israel
Tel.: +972-2-6536-534
Fax: +972-2-6520-797
e-mail: kuselman@netvision.net.il

Abstract An approach to the assessment of the limit of detection and the limit of quantitation using uncertainty calculation is discussed. The approach is based on the known evaluation of the limits of detection and quantitation as concentrations of the analyte equal to three and ten standard deviations of the blank response, respectively. It is shown that these values can be calculated as the analyte concentrations, for which relative expanded uncertainty achieves 66% and 20% of possible results of the analyte determination, correspondingly. For example, the calculation is performed for the validation of a new method for water determination in the presence of ene-diols or thiols, developed for analysis of chemical products, drugs or other materials which are unsuitable for direct Karl Fischer titration. A good conformity between calculated values and experimental validation data is observed.

Key words Limit of detection · Limit of quantitation · Uncertainty of measurements · Analytical method validation · Water determination · Karl Fischer titration

Introduction

Validation of analytical methods used for the enforcement of regulations, in particular in the pharmaceutical industry, has become obligatory in last few years [1]. Obviously, different methods should be validated using different validation parameters. For example, Category I methods for the quantitation of major components of bulk drug substances or active ingredients in finished pharmaceutical products do not require evaluation of the limit of detection (LOD) and the limit of quantitation (LOQ). In contrast, validation of methods from Category II for the determination of impurities in bulk drug substances or degradation compounds in finished products is incomplete without these parameters [2, pp. 1982–1984]. Experimental design for LOD and LOQ assessment is complicated, and expensive [3].

On the other hand, quantifying uncertainty in analytical measurement is not included in the list of validation parameters [1, 2] but is required during laboratory accreditation according to the guides of the International Organization for Standardization (ISO) [4], CITAC [5], EURACHEM [6] and other regulations. Uncertainty values and values of LOD and LOQ describing the same analytical method are interdependent [7–9]. When this interdependence is described in an analytical (mathematical) form, the design of the experiment for validation can be based on the uncertainty calculation and prediction of the parameters.

For example, Karl Fischer titration of water in solid, liquid or gaseous samples is a routine analytical method used today for quality control of many products. The method is relevant to ISO 9001–9003, Good Manufacturing Practice, Good Laboratory Practice, and Food and Drug Administration (FDA) guidelines [10]. Therefore, different aspects of the measurement uncertainty in this method were studied in a number of publications [10–13]. In particular, the work concerning the analysis of the uncertainty budget for $Ca(OH)_2$ and $Mg(OH)_2$ determination in CaO and MgO is interesting [14, 15]. In this method the Karl Fischer titration of water produced by the analytes at high temperatures is the second final step of the determination.

One of the most important sources of uncertainty is the presence of materials in the sample (components of the matrix) other than water that react with the Karl Fischer reagent (KFR). If the sample includes such components in significant concentrations, direct Karl Fischer titration will be impossible. This is a problem, for example, for ene-diols (such as ascorbic acid and its preparations) and thiols which are used in the pharmaceutical, perfumery, food and other industries.

We have developed a new method for simultaneous determination of water and ene-diols or thiols in samples unsuitable for direct Karl Fischer titration [16]. The method is based on the consecutive titration first of ene-diol or thiol against a novel reagent [17] and then of water against a conventional KFR in the same test portion and the same cell for electrometric location of the end-point in both titrations. For ene-diol or thiol the method is classified as a Category I method and for water a Category II method. Therefore LOD and LOQ evaluation is only necessary for water.

In the present paper, the dependence of LOD and LOQ on uncertainty in analytical measurement is discussed and used in the design of an experiment for the assessment of LOD and LOQ in the new method for water determination.

Dependence of LOD and LOQ on uncertainty in analytical measurement

LOD is the lowest concentration of an analyte in a sample that can be detected, but not necessarily quantified, by the analytical method [2, p.1983]. Wegscheider has shown [9], that in order to define this concentration C as corresponding to three standard deviations of the blank response, it would mean that the relative standard measurement uncertainty (relative standard deviation in the concentration domain) is $u(C)/C=33\%$ at $C=LOD$. Hence, using the coverage factor 2, the relative expanded uncertainty at $C=LOD$ is $U(C)/C=2u(C)/C=66\%$ and vice versa:

$$LOD=(100/66)\,U(C)=1.5\,U(C), \quad (1)$$

where U(C) is the expanded uncertainty (absolute value).

The analogous dependence is also valid for LOQ. By definition, LOQ is the lowest concentration of an analyte in a sample which can be determined by the analytical method with acceptable precision and accuracy; LOQ is usually equal to ten standard deviations of the blank response [2, p.1983]. In other words, the relative standard measurement uncertainty is $u(C)/C=10\%$ at $C=LOQ$. Therefore, the relative expanded uncertainty with the coverage factor 2 at the LOQ is $U(C)/C=20\%$ and

$$LOQ=(100/20)\,U(C)=5\,U(C). \quad (2)$$

Since uncertainty in analytical measurement can be calculated with "pen and paper only" [18], the Eqs. (1) and (2) also allow one to calculate or predict LOD and LOQ before an experiment, as it is shown below for water determination in presence of ene-diols or thiols.

Uncertainty in water determination

The analytical procedure begins with the titration of the sample test portion against the novel reagent [17]. This reagent includes iodine in non-aqueous solvents, which oxidizes ene-diol or thiol to diketones or sulphide derivatives, respectively, which do not interfere with the next titration of water by KFR.

After the first titration (assay) the total water content in the flask consists of the original amount of water in the test portion and that introduced with the novel reagent during titration. This total water content is titrated against KFR (second titration).

The original water content in the sample (C_w, % mass) is calculated from the equation:

$$C_w=[V_{KFR}-(F\times V_{r\text{-}a})]\times T_{KFR}\times 100/m, \quad (3)$$

where m is the mass of the sample test portion in mg; V_{KFR} is the volume of the KFR spent for titration of the solution formed after the first titration in ml; T_{KFR} is the titre of the KFR in mg H_2O/ml. Then

$$T_{KFR}=m^T_{H2O}/V^T_{KFR}, \quad (4)$$

where m^T_{H2O} is the mass of water test portion used for determination of the KFR titre in mg; V^T_{KFR} is the volume of the KFR spent for the titration of m^T_{H2O} in ml; $V_{r\text{-}a}$ is the volume of the reagent spent for titration of the sample test portion in ml, and F is the factor that corresponds to the volume of the KFR in ml spent for titration of water traces in 1 ml of the novel reagent. The F value is calculated from two consecutive titrations of one and the same dry $SnCl_2$ sample by the novel reagent and then by the KFR [16]:

$$F=V^0_{KFR}/V^0_r, \quad (5)$$

where V^0_r is the volume of the novel reagent spent for the first $SnCl_2$ titration in ml, and V^0_{KFR} is the volume in ml of the KFR spent for the second titration of the solution formed after the first titration.

Masses of the sample test portion ($m \approx 50$ mg) and even of the water test portion for determination of the KFR titre ($m^T_{H2O} \approx 5$ mg) are measured with negligible uncertainty (for example, with Mettler AT 201 balance it is 0.015 mg) in comparison to the uncertainty of Volumes [6, 8]. Therefore, it is desirable to transform Eqs. (3–5) in the following way:

$$C_w = [V_{KFR} - (V_{r\text{-}a} \times V^0_{KFR}/V^0_r)] \times [(m^T_{H2O} \times 100/m)/V^T_{KFR}], \quad (6)$$

where $(m^T_{H2O} \times 100/m) = K$ is a negligible source of uncertainty (designated as K for convenience).

The standard combined uncertainty of the water determination can be evaluated by the partial differentiation of Eq. (6) taking into account the value K:

$$u(C_w) = \{(K/V^T_{KFR})^2 \times [(u(V_{KFR})/V_{KFR})^2 + (V_{r\text{-}a}/V^0_r)^2 \times (u(V^0_{KFR})/V^0_{KFR})^2 + (V^0_{KFR}/V^0_r)^2 \times (u(V_{r\text{-}a})/V_{r\text{-}a})^2 + (V_{r\text{-}a} \times V^0_{KFR}/(V^0_r)^2)^2 \times u(V^0_r)/V^0_r)^2] + [(V_{KFR} - (V_{r\text{-}a} \times V^0_{KFR}/V^0_r)) \times K/(V^T_{KFR})^2]^2 \times (u(V^T_{KFR})/V^T_{KFR})^2\}^{1/2}. \quad (7)$$

All the volumes were measured with a 10-ml burette graduated in 0.02 ml divisions (Bein Z.M., Israel). The manufacturer specifies a calibration accuracy of ± 0.02 ml, which can be converted using rectangular distribution to a standard deviation $u_c(V) = 0.02/\sqrt{3} = 0.012$ ml. The standard deviation of the burette filling obtained was equal to $u_f(V) = 0.013$ ml. Since the volumes are spent for titrations, they also depend on the standard deviation of end-point detection. Bipotentiometric location of the end-point by the direct dead-stop technique is very precise [19]. Therefore, in both titrations the main deviation in end-point detection arises due to the drop size of the burette. In our case the drop is 0.013 ml and the corresponding standard deviation is $u_e(V) = 0.013/\sqrt{3} = 0.0075$ ml. The temperature uncertainty source is negligible here, therefore the standard uncertainty of each volume spent for titration using the burettes described above is $u(V) = [(u_c(V))^2 + (u_f(V))^2 + (u_e(V))^2]^{1/2} = 0.019$ ml. Volumes V_{KFR}, $V_{r\text{-}a}$, V^0_r and V^T_{KFR} are equal of ~5 ml, while the volume V^0_{KFR} is 1 ml. Therefore the relative standard deviations are different: $u(V_{KFR})/V_{KFR} = u(V_{r\text{-}a})/V_{r\text{-}a}) = u(V^0_r)/V^0_r) = u(V^T_{KFR})/V^T_{KFR} = 0.0038$, whereas $u(V^0_{KFR})/V^0_{KFR} = 0.019$. By substituting these values into Eq. (7) one can calculate the combined uncertainty $u(C_w) = 0.037$. The expanded combined uncertainty in the water determination with the coverage factor 2 is $U(C_w) = 0.074$% mass, and its corresponding relative value is $U(C_w)/C_w = 0.074/C_w$ or $7.4/C_w$% rel. Calculated values of the relative combined uncertainty for water concentrations in the range 0.1–1.0% mass are shown in Fig. 1.

LOD and LOQ prediction and design of the experiment

According to Eqs. (1) and (2) and the calculated value $U(C_w)$, the predicted values of LOD and LOQ for the water determination are as follows:

$$LOD_{wp} = 1.5\ U(C_w) = 1.5 \times 0.074 = 0.11\%\ \text{mass} \quad (8)$$

and

$$LOQ_{wp} = 5\ U(C_w) = 5 \times 0.074 = 0.37\%\ \text{mass}. \quad (9)$$

Corresponding values of the relative combined uncertainty (66% rel. and 20% rel.) are indicated in Fig. 1 by empty circles.

Based on this prediction, the experiment for evaluation of LOD and LOQ was designed to analyse a sample with a water concentrations close to LOD_{wp} and two samples with water concentrations, one a little less and one a little more than LOQ_{wp}. A purchased ascorbic acid powder containing 0.15% mass of water, purchased α-monothioglycerol with 0.24% mass of water and a fortified sample of α-monothioglycerol with 0.53% mass of water were used for this purpose.

Experimental data

The "true" values of water concentration, C_{tr}, in these samples, shown in Table 1, were obtained as a difference between two titrations of two independent test

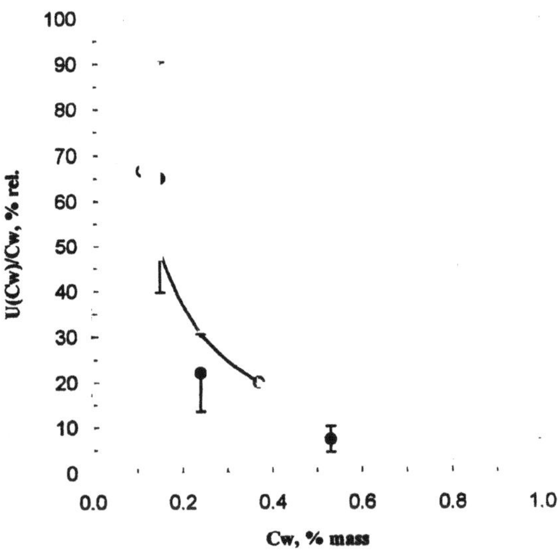

Fig. 1 Dependence of relative expanded uncertainty in water determination (% rel.) on the water concentration in a sample (% mass). *Empty circles* correspond to predicted LOD_{wp} and LOQ_{wp}, *full circles* and *bars* to experimental data

Table 1 Results of the water determinations

Sample	C_{tr}, % mass	C_{av}, % mass	S_r, parts of 1	S_b, parts of 1
Ascorbic acid (purchased)	0.15	0.18	0.630	0.167
α-monothioglycerol (purchased)	0.24	0.25	0.215	0.042
α-monothioglycerol (fortified)	0.53	0.55	0.064	0.036

portions (each with ten replicates). First, by the Karl Fischer method and then by the pharmacopoeial method for the determination of ascorbic acid [2, p. 131] or of α-monothioglycerol [2, p. 2271]. The average of 20 replicate water determinations by the new method, C_{av}; the corresponding relative standard deviation of replicates, S_r, and relative bias $S_b=(C_{av}-C_{tr})/C_{av}$ are shown in Table 1.

The relative standard uncertainty of the replicates $u_{exp}(C_w)/C_w=100(S_r^2+S_b^2)^{1/2}$ is shown (% rel.) for values $C_w=C_{tr}$ in Fig. 1 by full circles. Relative expanded uncertainty in the uncertainty $u_{exp}(C_w)/C_w$ by [18] for the level of confidence 0.95 and 20–1=19 degrees of freedom is $U_{exp}=2.09\times100/(2\times19)^{1/2}=39\%$ rel., where 2.09 is the corresponding two-tailed percentile of the Student's *t-test* distribution. The U_{exp} values (39% rel. of $u_{exp}(C_w)/C_w$) are shown in Fig. 1 by the bars to the full circles.

Discussion

From Fig. 1 one can see, that the calculated relative uncertainty $U(C_w)/C_w$ based on Eq. (7), being a hyperbola, depends on each 0.01% mass of water concentration at C_w close to LOD_{wp}. At $C_w<LOD_{wp}$ the values of $U(C_w)/C_w$ quickly tend to infinity. On the other hand, for $C_w>LOQ_{wp}$ the relative uncertainty asymptotically draws nearer to zero.

Note that another definition of LOD and LOQ (for example, with requirements to both Type I and Type II errors in decisions [20–22]) leads to another correlation between these parameters and the uncertainty. It is clear also that the model of uncertainty given in Eq. (7) is not absolutely complete, but predicted parameters of the method correspond rather well to the experimental data. Maybe $U(C_w)/C_w$ values calculated for concentrations close to LOQ_{wp} and above, are excessively pessimistic (high). However, it is obvious that the experimental data (even obtained from 20 replicates) have a wide uncertainty range: sample statistical values can be significantly different from the population ones.

Because the whole study was performed within the framework of the new method validation according to the AOAC Peer-Verified Methods Program [1] which defines LOD and LOQ as experimental values, finally $LOD_w=0.2\%$ mass and $LOQ_w=0.5\%$ mass were accepted. These values are sufficient for the purposes of the method. It has been adopted as an AOAC Peer-Verified Method with the assigned number PVM 1:1998.

Conclusions

The approach to the assessment of the limits of detection and quantitation using uncertainty calculation can be helpful for the prediction of the former and experimental design.

The calculation performed for the new method of water determination in samples which are unsuitable for direct Karl Fischer titration is in good conformity with the experimental validation data.

Acknowledgements The authors thank Prof. E. Schoenberger for helpful discussions.

References

1. Lauwaars M (1998) Accred Qual Assur 3:32–35
2. USP 23 (1995) The United States Pharmacopeia (USP). The National Formulary. United States Pharmacopeial Convention Inc, Md., USA
3. Kuselman I, Shenhar A (1995) Anal Chim Acta 306:301–305
4. ISO/IEC Guide 25 (1990) General requirements for the competence of calibration and testing laboratories, 3d edn. International Organization for Standardization (ISO), Geneva,
5. CITAC Guide 1 (1995) International guide to quality in analytical chemistry: An aid to accreditation, 1st edn. Teddington, UK
6. EURACHEM (1995) Quantifying uncertainty in analytical measurement, 1st edn. EURACHEM, Teddington, UK
7. Ellison SLS, Williams A (1998) Accred Qual Assur 3:6–10
8. Kuselman I, Shenhar A (1997) Accred Qual Assur 2:180–185
9. Wegscheider W (1997) The Proceedings of the 2nd EURACHEM Workshop "Measurement Uncertainty in Chemical Analysis. Current Practice and Future Directions", 29–30 September, Berlin. EURACHEM

10. Dietrich A (1994) American Lab 26 (5): 33–39
11. Mitchell J Jr, Smith DM (1980) Aquametry, Part III (the Karl Fischer reagent). A treatise on methods for the determination of water. Willey-Interscience, N.Y., USA
12. Margolis SA (1995) Anal Chem 67:4239–4246
13. Margolis SA (1997) Anal Chem 69:4864–4871
14. Zeiler HJ, Heindl R, Wegscheider W (1996) Veitsch-Radex Rundschau 2:48–55
15. Wegscheider W, Zeiler HJ, Heindl R, Mosser J (1997) Ann Chim 87:273–283
16. Sherman F, Kuselman I, Shenhar A (1996) Talanta 43:1035–1042
17. Sherman F, Kuselman I, Shenhar A (1998) Reagent for determining water and ene-diols or thiols. USA Patent No. 5,750,404, 12.05.98
18. Kuselman I (1998) Accred Qual Assur 3:131–133
19. Cedergren A (1996) Anal Chem 68:3679–3681
20. American Public Health Association, American Water Works Association, Water Environment Federation (1995) Standard methods for the examination of water and wastewater, 19th edn. American Public Health Association, Washington, USA, pp 1-10, 1-11
21. Kaus R (1998) Accred Qual Assur 3:150–154
22. Vogelgesang J, Hadrich J (1998) Accred Qual Assur 3:242–255

Accred Qual Assur (2002) 7:115–120
DOI 10.1007/s00769-002-0442-6

Yunqiao Li
Guanghui Tian
Naijie Shi
Xiaohua Lu

Study of the uncertainty in gravimetric analysis of the Ba ion

Abstract The determination of barium by the gravimetric method, in which the precipitation of $BaSO_4$ was formed and weighed, coupled with instrumental measurement of trace constituents was studied. The analyte's remaining in the filtrate and washes, mechanical loss, contaminants in the precipitate are the main influencing factors of uncertainty. A series of condition tests have been done, to reduce the effect of the factors mentioned above and the optimum test condition was found. The determination was carried out with a strictly defined operational procedure. The trace amounts of barium in the filtrate, washes and mechanical loss were determined by ICP-AES, the chloride occluded in the precipitate was determined by ion chromatography (IC), calculated as $BaCl_2$ and barium, and sodium by FAAS, calculated as Na_2SO_4. The average mass of barium in the filtrate contributes about 0.06% relative to that of the total barium, in washes about 0.09%, mechanical loss about 0.06%, contaminants of $BaCl_2$ about 0.08% and Na_2SO_4 about 0.05%. All the trace constituents were determined and corrected on a sample-by-sample basis. Sources of uncertainty were assessed thoroughly. The uncertainty of this combined gravimetric-instrumental method was improved remarkably compared with that of gravimetric method alone. The expanded uncertainty (k =2) is 0.08%.

Keywords Gravimetry · Uncertainty · Barium · Traceability

Y. Li (✉) · G. Tian · N. Shi · X. Lu
National Research Center
for Certified Reference Material,
No.18 Bei San Huan Dong Lu,
Chaoyang Qu, Beijing, 100013,
P. R. China
e-mail: nrccrm@public3.bta.net.cn
Tel.: +86-10-64228404
Fax: +86-10-64228404

Introduction

Gravimetric analysis is a classical chemical determination method, which has been developed for more than one hundred years [1]. The precipitation method is of the greatest importance because often it is more or less specific for the constituent being determined and is of general applicability, especially for the determination of alkali metals, alkaline earth metals, sulfate, phosphate and so on. The analyte is precipitated as a very slightly soluble compound, and separated from the solution, ignited and weighed, and then the content of the constituent is found. The results can be traced to SI units without any reference materials. This method is traditionally used for the measurement of "percentage-level" concentrations and the results are often unsurpassed. The previous research work done on the method was on the repeatability. The uncertainty of the method is not obvious, the main influencing factors on uncertainty and their magnitudes are not clear. Thomas W. Vetter et al [2] used "instrumental-enhanced" gravimetric analysis for determination of sulfate. Sulfate in the filtrate, contaminants in the precipitate, and volatilized sulfate was quantified with instrumental methods. The mechanical loss of sulfate was estimated. The accuracy of the method was increased and the expanded uncertainty (k =2) of the method was 0.16%.

The correction factors

Many variables that influence the contamination of precipitate and loss of barium have been obtained in our experiment by systematic researching work and summarized in Table 1. A variable that decreases the magnitude of a factor is considered favorable and its effect is indi-

Table 1 Effect of variables on factors of uncertainty

Variable	Factor				
	Ba^{2+}in filtrate	Ba^{2+}in washes	mechanical loss of Ba^{2+}	$BaCl_2$ occlusion	Na_2SO_4 occlusion
Regular precipitation	–	–	nd	+	–
Reverse precipitation	++	nd	– –	–	++
Increase excess of precipitant	–	nd	nd	nd	+
Increase adding rate of precipitant	–s	–s	nd	–s	+
Increase concentration of precipitant	–s	+	nd	+	++
Increase acidity of solution	+	–	nd	–s	– –
Increase concentration of analyte	–s	–	nd	+	+
Increase standing time	–	–	nd	–	–
Increase standing temperature	–s	nd	nd	nd	nd
Increase volume of washes	– –	+	–	–	–
Ignition condition	– –	– –	– –	nd	nd
Increase coexisting ion	nd	nd	nd	nd	+

–: factor decreases, +: factor increases, nd: no data available, –s: factor decreases not distinctly, – – : factor not applicable

cated with the minus (–). A variable that increases the magnitude of a factor is indicated with the plus (+). The relative magnitude of the effect is not considered in this table. The influence of these variables is explained below.

1. Barium in the filtrate: Because of the solubility of precipitate, there is always trace amount of barium in the filtrate. A relatively large amount of barium was in the filtrate when reverse precipitation was used. The amount of barium in the filtrate will increase along with the increase of acidity of the precipitating solution. To reduce barium in the filtrate, regular precipitation was used with the proper excess amount and the adding rate of precipitant, so the common-ion effect is dominant. Diluting the precipitating solution, with the proper acidity and temperature of the precipitating solution, standing the precipitate in the filtrate for a time, so that the small and imperfect crystals will form larger and more perfect crystals and the solubility of the precipitate will be decreased. Another important variable is the filtrate volume, for a given condition, the solubility of the precipitate is almost constant, as the filtrate volume decreased, the loss of barium in the filtrate will be reduced.
2. Barium in the washes: There is always a trace amount of barium dissolved in washes by washing the precipitate. Increasing the concentration of the precipitant and the volume of washes, the amount of barium in the washes will be increased. In order to decrease precipitate contamination, thorough rinsing of the precipitate is necessary. Increasing the adding rate of precipitant, the acidity of precipitating solution, the concentration of analyte and the standing time of precipitate, the amount of barium in washes will be decreased.
3. Mechanical loss of barium: Some of the micro crystals of the precipitate still adhere to the beaker side, stir rod and policeman when it is transferred onto the filter. The amount of the mechanical loss depends upon the transferring and washing processes. It can be dissolved in a hot HCl solution and detected.
4. $BaCl_2$ and Na_2SO_4 occlusion: It has been found that all chloride in the precipitate was presented as $BaCl_2$ and the sodium as Na_2SO_4.[2] The variables are sometimes contradictory for the contamination of $BaCl_2$ and Na_2SO_4. Using reverse precipitation, the occlusion of $BaCl_2$ decreased but that of Na_2SO_4 increased markedly. Increasing the excess amount and the adding rate of precipitant, the content of coexisting cation and anion, the occlusion of Na_2SO_4 was aggravated, but almost no data changed for $BaCl_2$. By diluting the solution, prolonging the standing time of precipitate, and increasing the volume of washes, both occlusions of Na_2SO_4 and $BaCl_2$ were minimized.

Experimental

1. Sample preparation

The analytical-reagent grade $BaCl_2$ was weighed and dissolved in 5% HCl, 20 kg of barium solution was produced and the nominal concentration was 20 $mg.g^{-1}$. The solution was mixed and packed in 20 mL ampoules. Some of them were taken for analysis.

2. General procedure

The optimal test conditions shown in Table 2 were used in the following experiment. A hot solution of Na_2SO_4 was added to a pre-weighed hot acidified solution of $BaCl_2$. The precipitated $BaSO_4$ was left overnight on a water-bath at about 94 °C and then filtered and washed. The precipitate and filter were charred, then ignited to constant mass (the difference between two ignitions was less than 0.03 mg) at 800 °C in a platinum crucible. The masses of the

Table 2 Optimal test conditions for the gravimetric analysis of barium sulfate

Precipitation mode	Regular
Excess amount of precipitant	2.5 times
Adding rate of precipitant	10 mL of precipitant is poured at 4 minute intervals
Concentration of precipitant	0.5% (10 mL 5% Na_2SO_4 diluted by water to 100 mL)
Acidity of precipitation solution	pH: 1.8
Concentration of analyte	0.67mg·g^{-1} (about 11 g sample solution diluted by water to 300 mL)
Standing time of precipitate	20 h
precipitating temperature	90–95 °C
Standing temperature of precipitate	94 °C
Volume of washes	150mL
Igniting condition of precipitate	800 °C
Fusing condition of precipitate	1.000 g, K_2CO_3, 910 °C, 35 min
Extracting condition of melts	Hot water, several times, acidified with HNO_3, total volume is 100 mL

Table 3 Gravimetric factors used for calculation

Correction factors	Symbol	Conversion	Coefficient	Calculation
Mass of Sample (g)	m_1			
Mass of Precipitate weighed (mg)	m_2			
Mass of Ba^{2+}in filtrate(mg)	m_3	$Ba^{2+} \rightarrow Ba^{2+}$	1	
Mass of Ba^{2+}in washes (mg)	m_4	$Ba^{2+} \rightarrow Ba^{2+}$	1	
Mechanical Loss of Ba^{2+} (mg)	m_5	$Ba^{2+} \rightarrow Ba^{2+}$	1	
Mass of $BaCl_2$ in Precipitate (mg)	m_6	$Cl^- \rightarrow BaCl_2$	2.9368	IC Cl^-×2.9368
Equivalent mass of Ba^{2+}in $BaCl_2$ (mg)	m_7	$Cl^- \rightarrow Ba^{2+}$	1.9368	IC Cl^-×1.9368
Mass of Na_2SO_4 in Precipitate (mg)	m_8	$Na^+ \rightarrow Na_2SO_4$	3.0893	AAS Na^+×3.0893

precipitate were blank-corrected. The crucibles had been pre-ignited to constant mass. All the weights were corrected for buoyancy. The barium in filtrate, washes, and the mechanical loss were determined by ICP-AES. The precipitate was fused with K_2CO_3 at 910 °C, extracted with hot water and acidified with HNO_3. Some of the aliquots for determining chloride by IC and were calculated as $BaCl_2$ and barium. Other for determining sodium by AAS and was calculated as Na_2SO_4. A schematic diagram of the procedure is shown in Fig. 1 and the gravimetric factors used for calculation are listed in Table 3.

3. Apparatus

All the instruments and the significant experimental parameters are listed in Table 4.

4. Analytical reagents

Na_2SO_4, HNO_3, HCl and K_2CO_3 are all of guaranteed reagent grade, $BaCl_2$ is analytical-reagent grade. All the important impurities in the reagents were measured. The sodium in K_2CO_3 is 28 µg g^{-1} and chloride is 6 µg g^{-1}.

The Na_2SO_4 solution was prepared and at a concentration of 5%, filtered, and left for several weeks (to promote the formation of larger crystals when used to precipitate $BaSO_4$) in a 1000 mL high-pressure polyethylene bottle. The barium, sodium and chloride calibration solutions were made up from GBW(E)080243, GBW(E)080260 and GBW(E)080269, NRCCRM standard solutions.

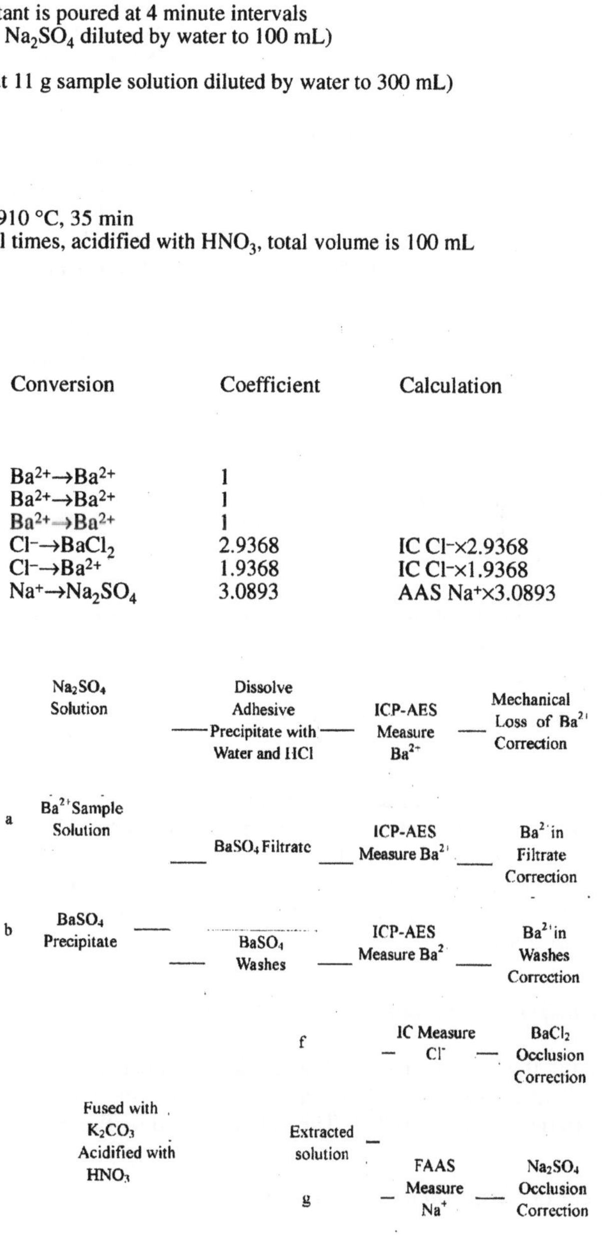

Fig. 1 Gravimetric coupled with instrumental measurement for determined of barium

Table 4 Instruments and parameters used in the determination

Method	Constituents	Condition	Parameter
ICP-AES	barium	Wave length	455.403 nm
		ICP power	1.0 kW
		Plasma flow	12 $L \cdot min^{-1}$
		Nebulizer flow	0.7 $L \cdot min^{-1}$
		Auxiliary flow	1.0 $L \cdot min^{-1}$
		Observation height	7 mm
IC	Chloride	Eluent	25 $mmol \cdot L^{-1} NaHCO_3$/25 $mmol \cdot L^{-1} Na_2CO_3$
		Flow rate	1.5 $mL \cdot min^{-1}$
		Injection Volume	50 µL
		Detector	Conductivity
		Column	SA4 anion-exchange column
FAAS	Sodium	Wavelength	589.0 nm
		Slit	0.4 nm
		Oxidant gas flow rate	air:1.6 $kg \cdot min^{-1}$
		Reducing gas flow rate	C_2H_2: 0.25 $kg \cdot min^{-1}$
		Burner height	7.5 mm
		Background correction	Zeeman mode

Results and discussion

Results

The total concentration of barium in solution is calculated as follows

$$C_{Ba^{2+}}(mg \cdot g^{-1}) = \frac{(m_2 - m_6 - m_8) \times 0.5883993 + m_3 + m_4 + m_5 + m_7}{m_1} \times 10^{-3} \quad (1)$$

The factors used for calculation are given in Table 3. The results for gravimetric coupled with instrumental determination are listed in Table 5. All corrections were applied on a sample-by-sample basis. The expanded uncertainty of the final result was calculated according to GUM guide [3].

Gravimetric determination

Using the Gravimetric method alone, the concentration of Ba^{2+} in solution is 19.2680 mg g^{-1} and the relative standard deviation is 0.089%. They are all listed in Table 5.

Loss of barium and contamination

Loss of barium in filtrate. The mass of barium in the filtrate is about 0.02%–0.09% relative to that of the total barium. This quantity of barium should be added to the total barium. If it is not corrected, it may lead to a negative error of about 0.1% under this condition.

Loss of barium in washes. The mass of barium in washes was about 0.09% except in one sample which was up to 0.3% due to leaking, relative to that of the total barium. This quantity of barium should be added to that of the total barium. If it is not corrected, it may lead to a negative error of about 0.09% under this condition. It can also be seen that the effect of leaking or other factors can be minimized or eliminated by using correction.

Mechanical loss of barium. The mass of the mechanical loss of barium was about 0.03%–0.08% relative to that of the total barium. This quantity of barium should be added to the total barium. If it is not corrected, it may lead to a negative error about 0.03%–0.08% under this condition.

Occlusion of $BaCl_2$. The mass of $BaCl_2$ occluded was about 0.04%–0.1% relative to that of $BaSO_4$, or 0.02%–0.05% relative to that of the total barium. This quantity of $BaCl_2$ should be subtracted from the mass of $BaSO_4$and the relevant mass of barium should be added to that of the total barium. If it is not subtracted, it may lead to a positive error about 0.02%–0.05% under this condition.

Occlusion of Na_2SO_4. The mass of Na_2SO_4 occluded was about 0.03%–0.06% relative to that of $BaSO_4$, or 0.02%–0.04% relative to that of the total barium. This quantity of Na_2SO_4 should be subtracted from the mass of $BaSO_4$. If it is not subtracted, it may lead to a positive error about 0.02%–0.04% under this condition.

Total loss of barium Total loss of barium includes the loss in filtrate and washes, mechanical loss and contained in the occlusion of $BaCl_2$. The mass of total loss of barium was about 0.2%–0.3% relative to that of the total barium. This quantity of barium should be added to that of the total barium. Otherwise, it may lead to a negative error about 0.2%–0.3% under this condition.

Table 5 The results of gravimetric analysis combined with instrumental analysis of Ba^{2+} solution

Sample	No.	Sample Mass (g)	Precipitate Mass (mg)	Ba^{2+} by Gravimetric alone (mg·g^{-1})	Ba^{2+} in filtrate (mg)	Ba^{2+} in washes (mg)	Mechanical loss of Ba^{2+} (mg)	$BaCl_2$ Occlusion (mg)	Na_2SO_4 Occlusion (mg)	Total Loss of Ba^{2+} (mg)	Total Occlusion (mg)	Ba^{2+} after Correction (mg·g^{-1})
4	4–1	11.07213	362.796	19.2798	0.0640	0.1538	0.0848	0.2376	0.1304	0.4593	0.3680	19.3028
	4–2	11.20975	367.192	19.2739	0.0424	0.1886	0.0978	0.3534	0.1597	0.5619	0.5131	19.2971
5	5–1	10.71454	350.963	19.2735	0.0670	0.1784	0.1078	0.3309	0.2153	0.5714	0.5462	19.2968
	5–2	10.79263	353.724	19.2845	0.0956	0.1668	0.0876	0.2264	0.2116	0.4993	0.4380	19.3069
6	6–1	10.64595	348.467	19.2597	0.1289	0.1777	0.1422	0.1557	0.0837	0.5515	0.2394	19.2945
	6–2	10.75338	351.439	19.2299	0.1070	0.6055*	0.1620	0.2135	0.1782	1.0153	0.3917	19.3062
7	7–1	10.93160	357.985	19.2687	0.1775	0.1628	0.0652	0.2547	0.1637	0.5735	0.4184	19.2987
8	8–1	11.00880	360.619	19.2744	0.1740	0.1785	0.0674	0.1768	0.1492	0.5365	0.3260	19.3057
	8–2	10.95523	358.668	19.2639	0.1378	0.1756	0.1840	0.2392	0.1918	0.6552	0.4310	19.3005
Average (mg·g^{-1})				19.2680								19.3010
Standard Deviation				0.018								0.0046
R.S.D%				0.089								0.024

Note: *portion of the precipitate was leaked because the apertures of the filter were not equal

Total occlusion The total occlusion includes Na_2SO_4 and $BaCl_2$. The mass of the total occlusion was about 0.06%–0.15% relative to that of $BaSO_4$, or 0.04%–0.1% relative to that of total barium. This quantity of occlusion should be subtracted from the mass of $BaSO_4$. Otherwise, it may lead to a positive error about 0.04%–0.1% under this condition.

The data listed above show that the effect of individual factors is approximately parts per ten thousand compared to the final results, its magnitude is to some extent in tune with the relative standard deviation of the " gravimetric-instrumental " method. The quantity of total loss of barium is more than that of total occlusion, so if the correction is not made, the results will be on the low side.

Gravimetric-instrumental determination

Using gravimetric-instrumental determination, the concentration of Ba^{2+} in solution is 19.3010 mg·g^{-1} and a relative standard deviation of 0.024% was obtained. The results are list in Table 5 and were calculated with Eq. (1). All the corrections were made on a sample-by-sample basis.

The over-all repeatability of the low-precision instrumental measurement coupled with high-precision gravimetric analysis (relative standard deviation, 0.024%) is three times better than the repeatability of the gravimetric analysis alone and the value (19.3010 mg·g^{-1}) is greater than the gravimetric value (19.2680 mg·g^{-1}). This indicates that an improvement has been made by the coupled instrumental analysis. In addition, a determination made without the instrumental correction would be negatively biased by 0.2%.

Assessment of uncertainty

Instrumental measurement

Barium in the filtrate measured by ICP-AES. There was a lot of NaCl in the filtrate, so the effect of matrix is very high. The barium was measured by means of standard addition. The concentration of barium in the filtrate was about 0.2 mg·L^{-1}, and the repeatability of ICP-AES was less than 2.5%. The combined standard uncertainty was 3.5% and the mass of barium in filtrate was about 100 μg, so the estimated relative standard uncertainty of the correction for the barium in filtrate was 1.8×10^{-5}.

Barium in washes measured by ICP-AES. The barium in washes was determined by ICP-AES with the matrix-matched calibration curve method. The concentration of barium in washes was about 1.5 mg·L^{-1}, and the repeatability was 1.5%. The combined standard uncertainty was 2% and the mass of barium in filtrate was about

180 μg, so the estimated relative standard uncertainty of the correction for the barium in washes was 1.7×10^{-5}.

Mechanical loss of Barium measured by ICP-AES. The barium was determined by ICP-AES with the matrix-matched calibration curve method. The concentration of barium in solution was about 2 mg·L^{-1}, and the repeatability was 1.5%. The combined standard uncertainty was 2% and the mass of mechanical loss of barium was about 100 μg, so the estimated relative standard uncertainty of the correction for mechanical loss of barium was 1.1×10^{-5}.

Chloride measured by IC. There is a lot of KNO_3 in the extract, so the effect of the matrix is high. The chloride is measured by means of standard addition. The concentration of chloride in solution is about 0.2 mg·L^{-1}, and the repeatability of IC is about 5%. The combined standard uncertainty was 10%. The mass of $BaCl_2$ (transferred from chloride) in extract was about 250 μg, so the estimated relative standard uncertainty of the correction for $BaCl_2$ and barium from occlusion was 1.1×10^{-4}.

Sodium measured by FAAS. There was a lot of KNO_3 in the extract, so the effect of the matrix is high. The sodium is measured with a matrix-matched calibration curve method. The concentration of sodium in solution is about 0.3 mg·L^{-1}, and the repeatability of FAAS was about 3%. The combined standard uncertainty was 5% and the mass of Na_2SO_4 (transferred from sodium) in the extract was about 160 μg, so the estimated relative standard uncertainty of the correction for occluded Na_2SO_4 was 2.3×10^{-5}.

Assessment of the combined uncertainty

Uncertainty Assessment of type A. The source of Type A uncertainty is the relative standard deviation of result by gravimetric -instrumental method $u_1=2.4\times10^{-4}$.

Uncertainty Assessment of type B. The sources of Type B uncertainty are as follows:

(1) weighing uncertainty:
  a standard uncertainty for sample weighing $u_2=2\times10^{-6}$
  b standard uncertainty for precipitate weighing $u_3=8\times10^{-6}$
(2) standard uncertainty for correction of barium loss in filtrate $u_4=1.8\times10^{-5}$
(3) standard uncertainty for correction of barium loss in washes $u_5=1.7\times10^{-5}$
(4) standard uncertainty for correction of mechanical loss of barium
  a standard uncertainty for the measurement of mechanical loss of barium $u_6=1.1\times10^{-5}$
  b standard uncertainty caused by the incompletely extraction for the mechanical loss of barium $u_7=1\times10^{-5}$(the extraction rate of barium from the beaker, stir rod and policeman was about 98%)
(5) standard uncertainty for correction of occlusion $BaCl_2$
  a standard uncertainty for the measurement of occlusion $BaCl_2$ $u_8=1.1\times10^{-4}$
  b standard uncertainty caused by the incompletely extraction for occluded $BaCl_2$ in precipitate is $u_9=1.5\times10^{-5}$
(6) standard uncertainty for correction of occlusion Na_2SO_4
  a standard uncertainty for the measurement of occlusion Na_2SO_4 $u_{10}=2.3\times10^{-5}$
  b standard uncertainty caused by the incompletely extraction for occluded Na_2SO_4 in precipitate is $u_{11}=1.0\times10^{-5}$
(7) standard uncertainty of the measurement for atomic weight $u_{12}=1.9\times10^{-4}$
(8) standard uncertainty of the buoyancy modification $u_{13}=5.8\times10^{-6}$

The combined standard uncertainty is 4×10^{-4}
The expanded standard uncertainty is 8×10^{-4}

Conclusion

The combination of a classical gravimetric determination together with instrumental techniques was used to analyze the concentration of barium in solution. Corrections were made to the classical gravimetric method to correct the loss of barium in the filtrate, washes and the mechanical loss, and the contaminants of $BaCl_2$ and Na_2SO_4. Instrumental methods (ICP-AES, IC, FAAS) were used to quantify the loss of barium and the contaminants in the precipitate. The sources of the uncertainties have been assessed thoroughly, and the values were obtained. The uncertainty of the combined method has been improved remarkably and the expanded standard uncertainty (k =2) is 0.08%.

References

1. Kolthoff IM, Sandell EB: Textbook of Quantitative Inorganic Analysis 3rd edn.
2. Vetter TW (1995) Analyst 120:2025–2030
3. GUM (1995) Guide to the expression of uncertainty in measurement"(issued by ISO, IEC, BIPM, IFCC, IUPAC and OIML)

Accred Qual Assur (2000) 5:100–103

Ilya Kuselman

Assessment of permissible ranges for results of pH-metric acid number determinations using uncertainty calculation

Presented at: EURACHEM Workshop on Efficient Methodology for the Evaluation of Uncertainty in Analytical Chemistry, Helsinki, Finland 14–15 June 1999

I. Kuselman
The National Physical Laboratory of Israel (INPL), Givat Ram, Jerusalem 91904, Israel
e-mail: kuselman@netvision.net.il
Tel.: +972-2-6536534
Fax: +972-2-6520797

Abstract An approach to assess the permissible ranges for results of replicate determinations using uncertainty calculation is discussed. The approach is based on the known range distribution for normalized "range/standard deviation" values, which is equivalent to the distribution of the range for normalized results of replicate determinations having an average of 0 and a standard deviation of 1. It is shown that the permissible ranges can be assessed using tabulated percentiles of this distribution and calculated values of the determination (analysis) standard uncertainty. When the standard uncertainty calculation is performed before the analytical method validation, the permissible ranges can be predicted. As an example, the range is predicted for a new pH-metric method for acid number determination without titration in petroleum oils (basic, white and transformer). The results of the prediction are in good conformity with the experimental data.

Key words Range · Prediction · Uncertainty of measurements · Analytical method validation · Acid number determination · pH-metry

Introduction

According to the definition [1] a range w_n is the difference between the highest and lowest values in data consisting of n values. During the use of an analytical method a permissible range $R_{n,s}$ of results of n replicate determinations $x_1, x_2, \ldots, x_n$ obtained under the same conditions should be known (repeatability level [2]). So, each $w_n \leq R_{n,s}$. For example, if $n=2$, the value $R_{2,s}$ is the maximal acceptable difference of the duplicate results x_1 and x_2. A norm $R_{n,d}$ of the difference between two or more results of the analysis of the same sample in the same laboratory but under different conditions (days, analysts, instruments, etc.) and a norm $R_{n,l}$ for n results of different laboratories are also necessary for quality control of the analysis. These are known as the intermediate precision level and reproducibility level, respectively [2].

Usually these parameters are estimated from the data of the method validation and its collaborative study. However, even before validation, an assessment (prediction) of $R_{n,s}$, $R_{n,d}$ and $R_{n,l}$ values can be helpful in deciding whether the method is "fit-for-purpose". Predicted $R_{n,s}$, $R_{n,d}$ and $R_{n,l}$ values are also expedient for the design of the experiment for method validation and diagnosis of outliers during the experiment. Such predictions can be performed using calculation of the analysis uncertainty "with pen and paper", as was done for the assessment of limits of detection and quantitation [3].

In the present paper, the range prediction is discussed and used for statistical analysis of the data obtained during validation of a new method for acid number (AN) determination without titration in petroleum oils which has been developed in our laboratory (INPL) [4].

Assessment of a range using analysis uncertainty

If combined standard uncertainty u_c is the standard deviation quantified using intra-laboratory components for analysis under the same conditions [5], predicted $R_{n,s}=Q_p u_c$, where Q_p is the critical value (limit) of the ratio "range/standard deviation" at the level of confidence P. In essence, Q_p is the limit of the range for normalized values x which have an average of 0 and a standard deviation of 1. The limits are calculated from the function of the range distribution [6]:

$$P\{w_n \le R_n\} = n \int_{-\infty}^{+\infty} [P(x+R_n) - P(x)]^{n-1} dP(x). \quad (1)$$

Q_p values are tabulated for normally distributed x, for example in [6, 7]. Figure 1 gives an overview of the values which are dependent on n for different P. One can see that at $P=0.95$ for $n=2$ the value $Q_{0.95}=2.77$ and for $n=3$ it is $Q_{0.95}=3.31$. Therefore, a range of duplicates, should be no more than $R_{2,s}=2.77u_c$, and a range of three replicates – no more than $R_{3,s}=3.31u_c$.

Since the ratio between intra-laboratory and inter-laboratory standard deviations of analytical results is approximately 0.67 [8], the analogous limit for a range of results obtained in different laboratories is $R_{n,l}=Q_p u_c/0.67$. Therefore, the difference between two such results should be no more than $R_{2,l}=4.13u_c$, and the range for three results – no more than $R_{3,d}=4.94u_c$ at $P=0.95$.

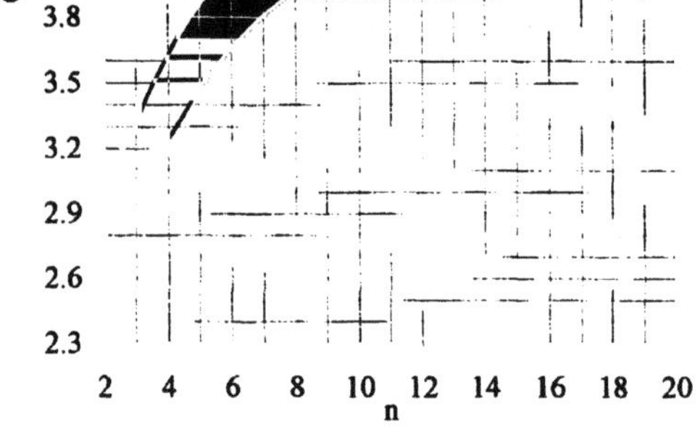

Fig. 1 Dependence of limits Q_p on the number of replicates n at different levels of confidence P. *Curve 1* corresponds to $P=0.90$, *2*–$P=0.95$, and *3*–$P=0.99$

The ratio of intra-laboratory standard deviation for the repeatability level and the corresponding standard deviation for the intermediate precision level is an intermediate between 1 (repeatability level) and 0.67 (reproducibility level). It can be accepted as approximately $(1+0.67)/2=0.83$. In this case $R_{n,d}=Q_p u_c/0.83$ and the difference between the two results should be no more than $R_{2,d}=3.34u_c$, while the range for three results – no more than $R_{3,d}=3.99u_c$ at $P=0.95$.

If a range is assessed for values $X_1, X_2, ..., X_n$, which are the averages from n replicates, the permissible value of the range should be divided by $\sqrt{n}$. The reason is that the standard deviation of the average is $\sqrt{n}$ times less than the standard deviation of the replicate.

Uncertainty in AN determination

The method is proposed for determination of AN<0.1 mg KOH/g oil in such oils as white, transformer and basic oils. The AN is an important characteristic of a petroleum oil's quality because the conductive and corrosive properties, and several other properties of the oil are dependent on AN. The method is based on rapid and complete extraction of acids from an oil test portion into a special reagent and measurement of the conditional pH in the "oil-reagent" mixture before (pH_1') and after (pH_2') standard acid addition. As a standard addition, a solution of hydrochloric acid is used.

The calculation of AN is carried out according to the following formula:

$$AN = 56.11\, N_{st} V_{st}/[m(10^{\Delta pH}-1)], \quad (2)$$

where 56.11 is the molecular mass of KOH, N_{st} and V_{st} are the concentration (M) and volume (ml) of the added HCl standard solution, m is the mass of the oil test portion (g) and $\Delta pH = pH_1' - pH_2'$.

The standard uncertainty of the AN determination can be evaluated according to [9] as

$$u(AN) = 0.032\, AN/(1-1/10^{\Delta pH}). \quad (3)$$

So, for values $\Delta pH = 0.25$–0.40 recommended in the method [4] the standard uncertainty $u(AN)$ is about 0.06 AN.

Design of the experiment for method validation and prediction of ranges

The experiment for the method validation was designed in order to obtain four replicate results of AN determination for each sample daily, over 5 days, in two laboratories (Lab 1 – INPL, as the method originator, and Lab 2 – Bio-Lab Ltd., Israel, as an independent laboratory [8]). Samples of purchased and fortified white, transformer and basic oils were used. The design allowed the assessment of the method precision for the levels of repeatability (intra-laboratory, within a day), intermediate precision (intra-laboratory, between days) and reproducibility (inter-laboratory).

The predicted permissible range at the level of repeatability for four replicates is $R_{4,s} = 3.63\,u(\mathrm{AN}) = 0.22$ AN.

At the level of intermediate precision for five replicates, each an average from four daily results, the permissible range is $R_{5,d} = 3.86\,u(\mathrm{AN})/(0.83\sqrt{4}) = 0.14$ AN.

For the reproducibility level, the predicted permissible range for laboratory results, each an average from $4 \times 5 = 20$ replicates, is $R_{2,I} = 2.77\,u(\mathrm{AN})/(0.67\sqrt{20}) = 0.06$ AN.

Results and discussion

The results of the experiment are shown in Table 1, where $X_i = \sum_{j=1}^{4} x_{ij}/4$ is the daily average result of AN determination (from 4 replicates, $j = 1, 2, \ldots, 4$), and w_i is the range of replicates x_{ij} in i-th day ($i = 1, 2, \ldots, 5$).

Table 2 shows the ranges w_5 between the highest and lowest daily average values X_i, the total laboratory average results $X_{avg} = \sum_{i=1}^{5} \sum_{j=1}^{4} x_{ij}/20$ and the difference w_2 between them, with the predicted permissible ranges: 1) for four replicates during a day – $R_{4,s}$; 2) for five daily average values – $R_{5,d}$ and 3) for two total average results – $R_{2,I}$.

Comparing w_i values with their norms $R_{4,s}$ one can see that all $w_i \leq R_{4,s}$, so the method repeatability is satis-

Table 1 Results of the experiment for method validation

Oil, sample	Parameter	Lab 1 (INPL) Day i/Results (mg KOH/g oil)					Lab 2 (Bio-lab Ltd.) Day i/Results (mg KOH/g oil)				
		1	2	3	4	5	1	2	3	4	5
Basic oil, purchased	X_i	0.0060	0.0062	0.0060	0.0061	0.0058	0.0062	0.0061	0.0062	0.0060	0.0066
	w_i	0.0004	0.0010	0.0009	0.0003	0.0008	0.0006	0.0008	0.0007	0.0004	0.0003
Basic oil, fortified	X_i	0.053	0.055	0.051	0.052	0.052	0.055	0.050	0.051	0.050	0.052
	w_i	0.003	0.003	0.002	0.005	0.001	0.003	0.008	0.005	0.006	0.005
Transformer oil, purchased	X_i	0.0023	0.0022	0.0023	0.0023	0.0023	0.0021	0.0021	0.0021	0.0022	0.0022
	w_i	0.0001	0.0002	0.0001	0.0001	0.0001	0.0002	0.0000	0.0001	0.0001	0.0001
Transformer oil, fortified	X_i	0.052	0.051	0.052	0.051	0.051	0.049	0.052	0.052	0.051	0.052
	w_i	0.001	0.003	0.001	0.003	0.004	0.000	0.000	0.001	0.001	0.001
White oil, purchased	X_i	0.0021	0.0022	0.0021	0.0021	0.0022	0.0020	0.0021	0.0019	0.0019	0.0020
	w_i	0.0003	0.0002	0.0002	0.0002	0.0001	0.0002	0.0002	0.0000	0.0000	0.0000
White oil, fortified	X_i	0.050	0.052	0.052	0.051	0.050	0.051	0.050	0.050	0.049	0.049
	w_i	0.003	0.001	0.002	0.001	0.003	0.002	0.004	0.004	0.002	0.002

Table 2 Results of the range prediction and statistical analysis of the experimental data using Horwitz's norms

Oil, sample	Laboratory	Range calculations (mg KOH/g oil)						Statistical analysis with the Horwitz's norms						
		$R_{4,s}$	w_5	$R_{5,d}$	X_{avg}	w_2	$R_{2,I}$	C, parts of 1	RSD_1, %	RSD_{1N}, %	RSD_2, %	RSD_{2N}, %	RSD_3, %	RSD_{3N}, %
Basic oil, purchased	1	0.0014	0.0004	0.0009	0.0060	0.0002	0.0004	$1.1 \cdot 10^{-5}$	4.29	7.47	1.96	4.50	2.32	2.49
	2		0.0006		0.0062				3.40		2.91			
Basic oil, fortified	1	0.011	0.004	0.007	0.053	0.001	0.003	$9.4 \cdot 10^{-5}$	2.24	5.40	2.77	3.25	1.35	1.80
	2		0.005		0.052				4.28		3.66			
Transformer oil, purchased	1	0.0005	0.0001	0.0003	0.0023	0.0001	0.0001	$4.0 \cdot 10^{-6}$	2.33	8.71	2.22	5.25	3.14	2.90
	2		0.0001		0.0022				2.01		1.05			
Transformer oil, fortified	1	0.011	0.001	0.007	0.051	0.000	0.003	$9.1 \cdot 10^{-5}$	1.77	5.43	0.90	3.27	0.00	1.81
	2		0.003		0.051				0.83		2.17			
White oil, purchased	1	0.0004	0.0001	0.0003	0.0021	0.0001	0.0001	$3.7 \cdot 10^{-6}$	3.47	8.83	2.70	5.32	3.45	2.94
	2		0.0002		0.0020				1.67		4.24			
White oil, fortified	1	0.011	0.002	0.007	0.051	0.001	0.003	$9.0 \cdot 10^{-5}$	1.35	5.43	1.65	3.27	1.40	1.81
	2		0.002		0.050				2.16		1.66			

factory. The intermediate precision is also satisfactory, since $w_5 \leq R_{5,d}$ for all oils in both laboratories. The similar assessment of the method reproducibility by the condition $w_2 \leq R_{2,l}$ is positive too.

The same results can be obtained by statistical analysis of the experimental data using comparison of the relative standard deviations (RSDs) of x_{ij}, X_i and X_{avg} with the corresponding empirical Horwitz's norms [8, 10–12]. For this purpose the following parameters are calculated and shown in Table 2:

1. Values X_{avg} (mg KOH/g oil) expressed as concentrations of naphthenic acid in decimal fractions: $C = X_{avg}$ (100/56.11)/1000, where 100 and 56.11 are the molecular masses of naphthenic acid and KOH, respectively and 1000 is the factor for transformation of mg to g.
2. RSDs of x_{ij} averaged for 5 days (the i-th values were homogeneous):
   $$RSD_1 = 100 \cdot \left\{ \sum_{i=1}^{5} \sum_{j=1}^{4} [(x_{ij} - X_i)^2 / X_i^2] / [5(4-1)] \right\}^{1/2}, \%.$$
3. Norms for RSD_1 by Horwitz:
   $RSD_{1N} = 2^{(1-0.5 \mathrm{Log}\, C)} \times 0.67$, %.
4. RSD of X_i:
   $$RSD_2 = 100 \left\{ \sum_{i=1}^{5} [(X_i - X_{avg})^2] / (5-1) \right\}^{1/2} / X_{avg}, \%.$$
5. Norms for RSD_2 derived from RSD_1:
   $RSD_{2N} = RSD_{1N} / (0.83\sqrt{4})$, %.
6. RSD of X_{avg}:
   $RSD_3 = 100\sqrt{2}\,[(X_{avg1} - X_{avg2})/(X_{avg1} + X_{avg2})]$, %,
   where numbers 1 and 2, as additional indices for X_{avg}, denote Lab 1 and Lab 2, respectively.
7. Norms for RSD_3 by Horwitz:
   $RSD_{3N} = RSD_{1N} / (0.67\sqrt{20})$, %.

All the RSD values are less than their norms, except inter-laboratory RSD_3 for purchased transformer and white oils. These samples have the lowest AN and are, therefore, more difficult to analyse. However, it should be noted, the sample standard deviations RSD_3 have 2(20–1)–1 = 37 degrees of freedom while the number of degrees of freedom of the norms RSD_{3N} can be accepted as infinity. So, correct comparison of the standard deviations with their norms should be based on χ^2 or Fisher's criteria [13]. For example, by Fisher's criterion for transformer oil $F = RSD_3^2 / RSD_{3N}^2 = 3.14^2/2.90^2 = 1.17$. It is less than the critical value $F_{0.95}\{37, \infty\} = 1.54$ at the level of confidence 0.95. Therefore, the population value of RSD_3 for this oil is no more than the Horwitz's norm. The same is true also for white oil, since the corresponding $F = 3.45^2/2.94^2 = 1.38 < F_{0.95}\{37, \infty\} = 1.54$.

From the values above it can be seen that the range prediction and uncertainty calculation (on which the prediction is based) are adequate and in good conformity not only with the experimental data for the method validation, but also with the database used by Horwitz for calculation of his norms.

Conclusions

The approach to evaluate the permissible ranges for results of replicate determinations using uncertainty calculation can be helpful for prediction of ranges and statistical analysis of the data obtained during validation of the analytical (chemical) method.

The range prediction performed for the new method of pH-metric AN determination without titration in petroleum oils is in good conformity with the experimental validation data.

Acknowledgements The author thanks Professor E. Schoenberger, Professor Ya. Tur'yan and Dr. E. Strochkova for helpful discussions.

References

1. Havilcek LL, Crain RD (1988) Practical statistics for the physical sciences, American Chemistry Society, Washington, D.C.
2. United States Pharmacopeia. USP 23 (1995) US Pharmacopeial Convention, Inc., Rockville, Md., pp 1982–1984
3. Kuselman I, Sherman F (1999) Accred Qual Assur 4:124–128
4. Tur'yan YI, Strochkova E, Berezin OY, Kuselman I, Shenhar A (1998) Talanta 47:53–58
5. EURACHEM (1995) Quantifying uncertainty in analytical measurement 1st edn. EURACHEM, p 16
6. Owen DB (1962) Handbook of statistical tables, Addison-Wesley, Reading, Mass., pp 138–139
7. Dixon WJ, Massey FJ Jr (1969) Introduction to statistical analysis, 3rd edn. International Student Edition, New York, Table A-8b
8. AOAC Peer-Verified Methods Program (1993) Manual on policies and procedures. Association of Official Analytical Chemists International, Arlington, p 9
9. Kuselman I, Shenhar A (1997) Accred Qual Assur 2:180–185
10. Horwitz W, Albert R (1987) Anal Proc 24:49–55
11. Thompson M, Fearn T (1996) Analyst 121:275–278
12. King B (1999) Accred Qual Assur 4:27–30
13. Miller JC, Miller JN (1993) Statistics for analytical chemistry, 3rd edn. Ellis Horwood, Bodmin, England

Accred Qual Assur (2002) 7:13–18

Ilya Kuselman
Elena Kardash-Strochkova
Yakov I. Tur'yan

Uncertainty and other metrological parameters of peroxide value determination in vegetable oils

Abstract Measurement uncertainty in the proposed redox-potentiometric methods for peroxide value (PV) determination in vegetable oils is evaluated in comparison with uncertainty in the standard methods. The methods determine all peroxides in oils, in terms of milliequivalents per kg of sample (meq/kg), that oxidize potassium iodide (KI) under the conditions of the test. The standard methods are based on KI oxidation by the oil test portion and volumetric titration of the liberated iodine, while the proposed methods are using redox-potentiometric iodine determination without titration. As far as fresh refined oils have PV≤0.5 meq/kg, the limit of detection (LOD) and limit of quantitation (LOQ) of the methods are important. An approach to assess the LOD and LOQ using uncertainty calculation was applied. It is shown how important is the influence of the solvents purity on the values of LOD and LOQ.

Key words Peroxide value · Vegetable oils · Uncertainty of measurements · Limit of detection · Limit of quantitation

I. Kuselman (✉) · E. Kardash-Strochkova
Y.I. Tur'yan
The National Physical Laboratory
of Israel (INPL), Givat Ram,
Jerusalem 91904, Israel
e-mail: kuselman@netvision.net.il

Introduction

If standard analytical methods are time and labor consuming, new methods are developing instead of them. The aims of the development should be achieved with the condition that the metrological parameters of a new method are "fit for purpose" [1]. The choice of these parameters and their assessment affect the final result of the development. As a universal parameter, the measurement uncertainty can be applied. The others, such as limit of detection (LOD) and limit of quantitation (LOQ), can be expressed through the uncertainty values [2].

In the present paper the measurement uncertainty in the proposed redox-potentiometric methods for peroxide value (PV) determination in vegetable oils, developed by us [3], is evaluated in comparison with the uncertainty in the standard methods [4, 5].

PV is an important characteristic of the oil quality and appears as an indicator of the lipid oxidation and oil properties deterioration [6]. The methods [3, 4, 5] determine all peroxides in oils, in terms of milliequivalents per kg of sample (meq/kg), that oxidize potassium iodide (KI) under the conditions of the test. The standard methods are based on KI oxidation by the oil test portion and volumetric titration of the liberated iodine, while the proposed methods are using redox-potentiometric iodine determination without titration. The major drawback of the standard methods is that the titrimetric determination of low levels of PV is complicated and requires a certain experience of the analyst. Such PV levels (less than 0.5 meq/kg) should be guaranteed, for example, in fresh refined oils [7]. The advantage of the proposed methods is their simplicity for automation [3].

The standard methods are highly empirical, and any variation in the test procedure may lead to erratic results. The difference between standard methods consists only in the kind of solution (acetic acid-chloroform or acetic acid-isooctane) used for the oil dissolution: commercial chloroform is more pure and ensures low blank values, while isooctane is less toxic. Therefore, stages of the KI oxidation in all the discussed methods are the same, but in the new methods they are combined with the iodine redox-potentiometric measurements in the same electrochemical cell. So, the main point is that the uncertainty of iodine determinations by volumetric titration and the uncertainty of iodine redox-potentiometric measurements without titration should be compared.

To compare the iodine measurement uncertainties, they are assessed by identification of the uncertainty sources, quantification of uncertainty components and calculation of combined uncertainties according to the EURACHEM/CITAC Guide [8] as values $u_c(y(x_1, x_2,..., x_n))=[\Sigma(u(y, x_i))^2]^{1/2}$, where $y(x_1, x_2,...,x_n)$ is a function of parameters $x_1, x_2,...,x_n$ and $u(y, x_i)$ is the uncertainty in y arising from the uncertainty in x_i, i=1, 2,...,n.

Uncertainty of measurement results by the proposed methods

After completing the reaction of the KI oxidation by hydroperoxides contained in the oil test portion the equilibrium $I_2+I^-\Leftrightarrow I_3^-$ is established at the KI excess. Thus, redox-potential E_1 caused by the electrochemical reversible couple $I_3^-+2e^-\Leftrightarrow 3I^-$ is measured in the aqueous phase with the Pt indicator and Ag/AgCl, 3 mol/l KCl, 3 mol/l KNO_3 reference electrodes. After E_1 measurement the standard addition of the iodine aqueous solution is introduced into the cell and potential E_2 is measured (for more details see [3]).

PV of the test portion is calculated using the following equation:

$$PV_t = (1000 / m)[(C_{st} \times V_{st}) / (10^{\Delta E/S} - 1)],\ \text{meg / kg}, \quad (1)$$

where m is the mass of the oil test portion [g]; C_{st} is the iodine concentration in the standard solutions for addition, expressed in gramequivalents per liter [N]; V_{st} is the volume of the iodine standard addition [ml]; $\Delta E=E_1-E_2$ is the difference of the potentials [mV]; S is an electrochemical parameter equal to 2.303 RT/2F [mV], where R is the universal gas constant (8.314 J/(K × mol)), T is the temperature [K], and F is the Faraday constant (96485 coulombs).

To obtain PV of the tested oil, PV_t should be corrected for the blank (organic solvent -water system without oil). Blank value PV_0 is calculated by the same formula, as PV_t, for the same mass value m. Finally, PV of the oil is

$$PV = PV_t - PV_0. \quad (2)$$

The main PV uncertainty components following from Eqs. (1) and (2) are shown in Table 1. Note only some details for their evaluation.

The final mass m of an oil test portion is the difference in masses between a beaker with the test portion and the empty beaker (after oil transfer to the solvent). These masses are weighed using the balance (Mettler AE 163, Switzerland) with reading 0.0001 g and calibration expanded uncertainty of ±0.0002 g at the level of confidence 0.95 and coverage factor 2 (normal distribution) in the range up to 100 g.

Uncertainty in iodine concentration C_{st} in the standard solutions is calculated taking into account the manufacturer information on possible deviation of the iodine titer (0.02%/°C) in a Titrisol ampoule (Merck, Germany), as well as the information on the volume uncertainty for volumetric flasks according to DIN, Class A, used for the solutions preparation, and possible temperature variation in the laboratory (in limits of 20±2 °C).

A recommended standard addition for oil samples with different expected PV is 0.1–1 ml of 0.01 N, 0.1 N, or 0.5 N iodine solutions: this volume should be negligible in comparison with the volume of the aqueous phase (70–110 ml). For transfer of the addition to the "oil-organic solvent-water" system, a mechanical hand pipette is used (Gilson, France, calibrated at INPL based on the gravimetric method [9]).

E_1 and E_2 are measured under the same conditions, both within 2–3 min, by the same instrument (a pH/ion-meter PHM 95, Radiometer, France). The expanded measurement uncertainty is ±(0.2+0.0005E) mV according to the Radiometer's information. At the normal distribution it corresponds to the standard uncertainty u(E)=0.1±0.00025E mV. Since the E measurement range

Table 1 Values and uncertainty components in the proposed methods

Symbol of the source	Description	Value x	Standard uncertainty u(x)	Relative standard uncertainty u(x)/x
m	Mass of the oil test portion	5 g	0.00024 g	0.00005
C_{st}	I_2 concentration in a standard addition	0.01–0.5 N	0.000007–0.00017 N	0.0007–0.00034
V_{st}	Volume of the standard addition	0.1–1 ml	0.0002-0.0016 ml	0.0025–0.0016
ΔE	E_1-E_2	7–13 mV	0.3 mV	0.043–0.023
S	2.303×RT/2F	28.9–29.3 mV	0.2 mV	0.007
PV	Peroxide value	0.03–100 meq/kg	0.003–4.7 meq/kg	0.10–0.047

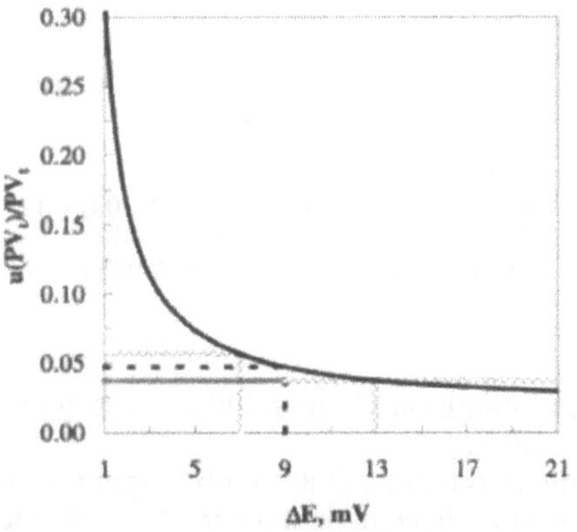

Fig. 1 Dependence of the relative standard uncertainty $u(PV_t)/PV_t$ on the measured difference of potentials ΔE [mV]. The optimal ΔE value is shown by the *dotted line*. *Wavy lines* show the range of recommended ΔE values

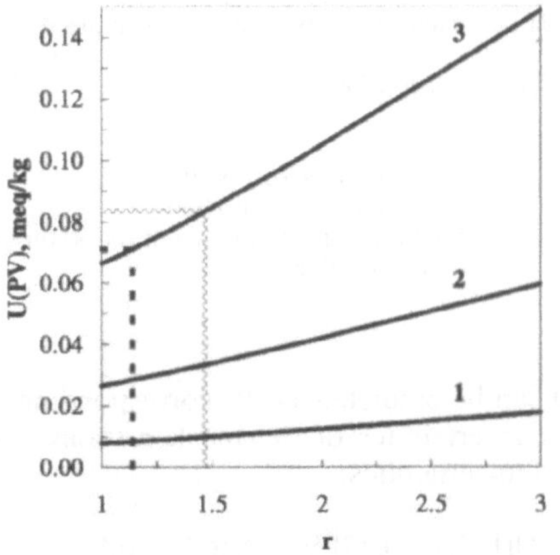

Fig. 2 The expanded uncertainty U(PV) [meq/kg] as function of the ratio $r=PV_t/PV_0$. *Line 1* corresponds to PV_0=0.06 meq/kg, *line 2* to PV_0=0.2 meq/kg, *line 3* to PV_0=0.5 meq/kg. LOD and LOQ for PV_0=0.5 meq/kg are shown by the *dotted and wavy lines*, correspondingly

recommended in the method [3] is 282–333 mV, u(E)=0.2 mV. So, the standard uncertainty of the difference between such two E measurements is

$$u(\Delta E)=\sqrt{2}\times u(E)=0.3\ \text{mV}.$$

The main parameter influencing S value is the temperature. Its variations in the laboratory in the range 291–295 K (18–22 °C) lead to the S changes from 28.87 to 29.27 mV.

As one can see from Table 1, ΔE measurement is the dominant source of the uncertainty in results of PV determination calculated by Eq. (1). So, the relative standard uncertainty of such a result $u(PV_t)/PV_t$ is calculated by the logarithmic partial differentiation of the function (1) concerning ΔE:

$$u(PV_t)/PV_t=[2.303\times 10^{\Delta E/S}\times u(\Delta E/S)10^{\Delta E/S}-1]. \quad (3)$$

Taking into account u(ΔE)=0.3 mV and S=28.87–29.27 mV, Eq. (3) can be simplified:

$$u(PV_t)/PV_t=0.024/(1-1/10^{\Delta E/S}). \quad (4)$$

The dependence of $u(PV_t)/PV_t$ on ΔE in the range 1–21 mV is shown in Fig. 1. From this dependence it follows that ΔE<7 mV leads to an essential increase in the PV uncertainty (ΔE=7 mV and corresponding $u(PV_t)/PV_t$ are shown in Fig. 1 by a wavy line). To achieve value ΔE>13 mV the amount of iodine added with the standard solution may exceed three times the iodine amount formed by the reaction of KI oxidation by hydroperoxides contained in the oil test portion (ΔE=13 mV and corresponding $u(PV_t)/PV_t$ are shown in Fig. 1 by the wavy line also). So, the optimal ΔE is 9 mV leading to $u(PV_t)/PV_t$=0.047 which are shown in Fig. 1 by a dotted line. Moreover, all the range ΔE=7–13 mV can be recommended for practical use as far as it covers uncertainty values close to the optimal one: $u(PV_t)/PV_t$= 0.047±0.010. The same is correct for the blanks.

Uncertainty of the final result of PV determination in the tested oil calculated by Eq. (2) is $u(PV)=[(u(PV_t))^2+(u(PV_0))^2]^{1/2}$. At the optimal ΔE, assuring $u(PV_t)/PV_t=u(PV_0)/PV_0=0.047$, normal distribution and coverage factor 2, the expanded uncertainty U(PV) is

$$U(PV)=2\times(0.047[PV_t^2+PV_0^2]^{1/2})$$
$$=0.094[PV_t^2+PV_0^2]^{1/2},\ \text{meg/kg}. \quad (5)$$

U(PV) as function of the ratio PV_t/PV_0=r is shown in Fig. 2 in the range r=1–3 at PV_0 equal to 0.06, 0.20, and 0.50 meq/kg (lines 1–3, correspondingly). From Eq. (5) the relative expanded uncertainty is the following:

$$U(PV)/PV=0.094(r^2+1)^{1/2}/(r+1). \quad (6)$$

When PV_0 is negligible, $PV=PV_t$, r+1⇒r, $r^2+1 \Rightarrow r^2$ and the relative expanded uncertainty is U(PV)/PV=0.094.

LOD and LOQ prediction

Defining LOD as PV corresponding to the three standard deviations (standard uncertainties) of the blank response or its 1.5 expanded uncertainties [2], the following prediction can be obtained at the optimal ΔE:

$$LOD=1.5U(PV_0)=0.14PV_0,\ \text{meg/kg oil}. \quad (7)$$

In this case PV_t according to Eq. (2) should be equal to 1.14 PV_0, i.e., r=1.14 and U(PV)/PV=0.067 by Eq. (6).

Table 2 Values and uncertainty components in the standard methods

Symbol of the source	Description	Value x	Standard uncertainty u(x)	Relative standard uncertainty u(x)/x
m	Mass of the oil test portion	5 g	0.00024 g	0.00005
C_{th}	Thiosulfate concentration	0.01–0.1 N	0.0000070–0.000034 N	0.00070–0.00034
$V_{th}-V_{th-0}$	Volume of thiosulfate spent for oil titration	0.1–4 ml	0.014 ml	0.14–0.003
PV	Peroxide value	0.2–100 meq/kg	0.02–0.56 meq/kg	0.10–0.0056

LOQ can be predicted as PV corresponding to the ten standard uncertainties of the blank response or its five expanded uncertainties:

$$LOQ = 5U(PV_0) = 0.47PV_0,\ \text{meg / kg oil.} \quad (8)$$

It means PV_t=1.47 PV_0 or r=1.47 and again, practically as for LOD, U(PV)/PV=0.068.

Values r and U(PV) appropriate to LOD and LOQ are shown in Fig. 2 for the case of PV_0=0.50 meq/kg, for example, by the dotted and wavy lines, correspondingly. The limit values here are LOD=0.14×0.50=0.07 meq/kg and LOQ=0.47×0.50=0.23 meq/kg. For more pure solvents, i.e., for blanks with lower PV_0, LOD and LOQ are lower also.

It is clear from Eqs. (5), (6), (7), and (8) and Fig. 2 how the uncertainty of the result of PV determination depends on the solvent purity (blank PV_0), especially for fresh refined oils having PV≤0.5 meq/kg.

Uncertainty of measurement results by the standard methods

According to the standard methods [4, 5], iodine, liberated after completing the reaction of the KI oxidation by hydroperoxides contained in the oil test portion, is titrated against sodium thiosulfate. Final result is calculated as usually in titrimetry:

$$PV = (1000 / m) \times (V_{th} - V_{th-0}) \times C_{th},\ \text{meq / kg.} \quad (9)$$

where V_{th} and V_{th-0} are the volumes of thiosulfate solution spent for titration of the test solution and of the blank, correspondingly [ml]; C_{th} is the thiosulfate concentration [N].

The main PV uncertainty components in this way are shown in Table 2. Some details for their evaluation are discussed below in same manner as previously for the new methods.

The final mass of an oil test portion (m=5 g) is determined as described already: it is not depending on a method for PV determination.

Uncertainty in thiosulfate concentration C_{th} in solutions prepared from a Titrisol ampoule (Merck, Germany) is calculated by analogy with the uncertainty in iodine concentration C_{st} in solutions also prepared from such ampoules.

Volume of the thiosulfate solution spent for titration is the more complicated parameter. A 2-ml microburette (Bein Z.M., Israel) with 0.01-ml divisions and a drop size reduced to 0.008 ml was used for this titration [3]. The manufacturer specifies a calibration accuracy of ±0.01 ml. The standard deviation of the burette filling obtained was equal to the standard deviation of calibration – 0.006 ml. Since the volumes V_{th} and V_{th-0} are spent for titrations, they depend also on the standard deviation of end-point detection. Using "potato starch for iodometry" at the end of titration, as recommended in the standard [5], which produces a deep blue color in the presence of the iodonium ion, location of the end-point (when the blue color just disappears) is precise. It requires an analytical experience, especially at low iodine concentrations (low PV), but for an experienced analyst the main deviation in the end-point detection arises due to drop size of the burette. For the described burette corresponding standard deviation is $0.008/\sqrt{3}=0.005$ ml. The temperature uncertainty source is negligible here, and therefore the standard uncertainty of each volume spent for titration using such a burette is

$$u(V_{th})=u(V_{th-0})=(0.006^2+0.006^2+0.005^2)^{1/2}=0.01 \text{ ml.}$$

Note, the use of 50-ml burette, DIN, Class A (Duran, Germany) with 0.1-ml divisions, calibration accuracy ±0.05 ml and a drop size 0.05 ml, recommended in the standard [10], leads to the standard uncertainty $u(V_{th})$= $u(V_{th-0})$=0.05 ml. So, such a burette is not suitable for fresh refined oils with PV≤0.5 meq/kg, as far as the volume of the 0.01 N thiosulfate solution spent for titration in this case is V_{th}≤0.25 ml.

From comparison of the main components of PV uncertainty in Table 2, one can see that volume components are dominant. So, assuring normal distribution and coverage factor 2, the relative expanded uncertainty of the titration result is

$$U(PV) / PV = 2 \times [(u(V_{th}))^2 + (u(V_{th-0}))^2]^{1/2} / (V_{th} - V_{th-0}) = 0.028 / (V_{th} - V_{th-0}). \quad (10)$$

Dependence of U(PV)/PV on the volume spent for titration is shown in Fig. 3 in the range of $(V_{th}-V_{th-0})$ up to

Table 3 Results of PV_0 determination, meq/kg

Solvents	Proposed technique		Standard titration	
	Mean	Standard deviation	Mean	Standard deviation
Chloroform GR	0.063	0.002	0.037	0.006
Isooctane SP	0.084	0.007	0.080	0.010
Isooctane GR–p	0.114	0.002	0.130	0.010
Isooctane GR	0.30	0.01	0.26	0.02
Isooctane TR	0.45	0.02	0.44	0.03

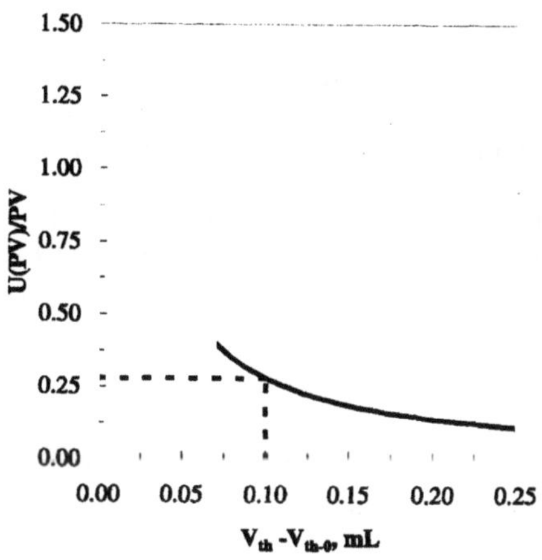

Fig. 3 Dependence of the expanded relative uncertainty U(PV)/PV on the volume spent for titration $V_{th}-V_{th-0}$ [ml]. The case of PV=0.2 meq/kg is shown by the *dotted line*

0.25 ml. It is clear that greater blank response (greater V_{th-0}) leads here to greater relative uncertainty U(PV)/PV as in the proposed methods.

The volume ($V_{th}-V_{th-0}$) for oils with PV>4 meq/kg at recommended m=5 g and C_{th}=0.1 N is more than 0.2 ml. In this PV range the expanded uncertainty is U(PV)= 0.56 meq/kg according to Eqs. (9) and (10), and U(PV)/PV<0.14. For oils with PV≤4 meq/kg and C_{th}=0.01 N the volume is ($V_{th}-V_{th-0}$)≤2 ml, U(PV)= 0.056 meq/kg, and U(PV)/PV≥0.014. The uncertainties for fresh refined oils with PV≤0.5 meq/kg are U(PV)/PV≥0.11. For example, for PV=0.2 meq/kg and ($V_{th}-V_{th-0}$)=0.1 ml it is U(PV)/PV=0.28: see shown in Fig. 3 by the dotted line. So, uncertainties of PV determination by the standard methods are close to the uncertainties by the proposed methods or less of them for oils with PV>0.5 meq/kg. However, for fresh refined oils with PV≤0.5 meq/kg the proposed methods are better, if the criterion is the uncertainty.

LOD and LOQ prediction

For a blank the volume V_{th-0} is relevant only in Eq. (10), so $U(PV_0)/PV_0=0.02/V_{th-0}$. The same in Eq. (9): at C_{th}=0.01 N the blank peroxide value is $PV_0=2V_{th-0}$ meq/kg. Therefore, for the standard methods the limits are

$$LOD = 1.5U(PV_0) = 0.03PV_0 / V_{th-0} = 0.06 \text{ meg / kg.} \quad (11)$$

and

$$LOQ = 5U(PV_0) = 0.10PV_0 / V_{th-0} = 0.20 \text{ meg / kg.} \quad (12)$$

Note, the titrimetric LOD and LOQ are not depending finally on the solvent purity, i.e., on the blank peroxide value PV_0, as the redox-potentiometric ones. However, PV_0 for both standard and proposed methods are the minimal detectable peroxide values and their comparison can be helpful for understanding the methods possibilities.

Solvent analysis

Five kinds of solvents were analyzed. Results of the analysis – means from three replicates and corresponding standard deviations – are shown in Table 3. The solvents listed in the table are chloroform GR purchased from Baker (Holland); isooctane GR, isooctane for organic trace analysis (TR) and isooctane for spectroscopy (SP) purchased from Merck (Germany). Isooctane signed as GR-p is isooctane GR purified by us using ion-exchange resin with active $-SO_3$ groups ("Amberlyst 15" from BDH, England).

The results of the analysis by proposed and standard methods are close enough (not differ more than for one-two standard deviations). The exception is chloroform only: the $U(PV_0)/PV_0$=0.5 for the standard method in this case (volume of 0.01 N thiosulfate solution spent for titration is 0.02 ml, i.e., 3 drops in all). Anyway chloroform is most pure from all available solvents.

Taking into account the chloroform PV_0 obtained by the proposed method, minimal LOD and LOQ can be calculated for this method: LOD=0.14×0.063=0.01 meq/kg and LOQ=0.47×0.063=0.03 meq/kg.

Analysis of oils

To compare the methods at PV levels higher than PV_0, five kinds of oils were analyzed using the purified isooc-

Table 4 Results of PV determination in oils

Oil	Proposed technique		Standard titration		Fisher's ratio F	Student's ratio t	PV_p/PV_s,
	PV_p, meq/kg	S_p, meq/kg	PV_s, meq/kg	S_s, meq/kg			
Canola 1	0.45	0.02	0.44	0.01	1.38	1.12	102.3
Soya	0.42	0.01	0.43	0.01	2.76	1.54	97.7
Sunflower	2.38	0.02	2.41	0.02	0.76	2.07	98.8
Canola 2	7.45	0.14	7.49	0.15	0.84	0.42	99.5
Olive	35.3	1.6	36.0	1.0	2.58	0.81	98.1
Maize	69.8	2.1	68.9	1.9	1.25	0.74	101.3

tane: canola, soya, sunflower, olive, and maize. Table 4 shows the average results PV_p and PV_s obtained by the proposed and standard methods from n=5 replicates for each sample; the standard deviations for these replicates – S_p and S_s, respectively; Fisher's ratio $F=S_p^2/S_s^2$; Student's ratio $t=|PV_s-PV_p|/[(S_s^2+S_p^2)/5]^{0.5}$; and PV_p/PV_s, %.

The critical value for the F-ratio is 6.39 at the 95% level of confidence and the number of the degrees of freedom n–1=4. For the t-ratio the critical value is 2.31 at the 95% level of confidence and the number of the degrees of freedom 2(n–1)=8.

From the comparison of the F-data with the critical value it follows that the differences between repeatability of the results obtained by the standard titration and by the proposed method are insignificant (all F values are less than 6.39), i.e., repeatability of the proposed method is sufficient.

The accuracy of these techniques is approximately the same, since the deviations of the average PV_p results obtained by the proposed method from the average results obtained by the standard titration PV_s are insignificant in comparison with the repeatability deviations (all t values are less than 2.31), i.e., accuracy of the proposed technique is sufficient. Average ratio PV_p/PV_s is 99.6%.

Similar results were obtained in our previous work with the same oils dissolved in chloroform [3]. However, in experiment described in [3] the oils were more fresh, and though were stored in refrigerator, their PV have increased after some months. Therefore, results of PV determinations shown in Table 4 are higher than those in [3].

Since LOD and LOQ should be determined experimentally [11], only LOQ=0.2 meq/kg predicted for the standard methods is approved based on the work [3] data for fresh refined canola. Test of other predictions requires an additional experiment with very good refined fresh oils.

Conclusions

1. Metrological parameters of the new methods for redox-potentiometric PV determination in vegetable oils are fit for purposes (similar to demonstrated by the standard methods).
2. Uncertainties of results of PV determination by the proposed methods are close to those by the standard methods or worse of them for oils with PV significantly more than 0.5 meq/kg. However, for fresh refined oils with PV≤0.5 meq/kg the proposed methods are better than standards ones, if the criterion is the measurement uncertainty.
3. For proposed methods the LOD=0.01 meq/kg and LOQ=0.03 meq/kg are predicted, while for the standard methods there are LOD=0.06 meq/kg and LOQ=0.20 meq/kg.

References

1. EURACHEM (1998) EURACHEM Guide. The fitness for purpose of analytical methods. A laboratory guide to method validation and related topics, 1st edn. Teddington, UK
2. Kuselman I, Sherman F (1999) Accred Qual Assur 4:124–128
3. Kardash-Strochkova E, Tur'yan Ya, Kuselman I (2001) Talanta 54:411–416
4. AOCS (1996) AOCS official methods, vol II, method Cd 8–53: Peroxide value. Acetic acid-chloroform method. AOCS, Champaign, IL, USA
5. AOCS (1996) AOCS Official methods, vol II, method Cd 8b–90: Peroxide value. Acetic acid-isooctane method. AOCS, Champaign, IL, USA
6. Finne G, Ikins WG, Williams J Jr, Welborn JL (1998) Inside Lab Manage 2:24–26
7. Israel Standard No 216 (1994) Edible vegetable oils. Tel Aviv, Israel
8. EURACHEM/CITAC (2000) EURACHEM/CITAC Guide. Quantifying uncertainty in analytical measurement, 2nd edn. Teddington, UK
9. ISO (1999) ISO/DIS 8655–6 Draft. Piston-operated volumetric apparatus. Part 6: Gravimetric test methods. Geneva, Switzerland
10. AOCS (1996) AOCS official methods, vol II, method Ja 8–87: Peroxide value. AOCS, Champaign, IL, USA
11. AOAC (1997) AOAC peer-verified methods program. Manual on policies and procedures. AOAC, Gaithersburg, MD

Accred Qual Assur (1999) 4:504–510

Thomas Anglov
Inge M. Petersen
Jesper Kristiansen

Uncertainty of nitrogen determination by the Kjeldahl method

J. Kristiansen (✉)
The National Institute of Occupational Health, Lersø Parkallé 105, DK-2100 Copenhagen, Denmark
e-mail: jkr@ami.dk
Tel.: +45-39-16 52 00
Fax: +45-39-16 52 01

T. Anglov, I.M. Petersen
Department of Metrology, Novo Nordisk A/S, Krogshøjvej 51, DK-2880 Bagsværd, Denmark

Abstract The uncertainty of the Kjeldahl method for determination of nitrogen in insulin was evaluated according the procedure described in the Guide to the Expression of Uncertainty in Measurement. The relative standard uncertainty of the method was found to be 0.19%, compared to the relative intermediate precision experimentally found to be 0.085%. The uncertainty components were organized in Tables, which allowed an easy overview and evaluation. The largest contribution to the uncertainty came from volumetric equipment. Systematic uncertainty budgets such as the design presented here facilitate the uncertainty evaluation process and makes it easier to compare uncertainty evaluations performed by different analysts.

Key words Uncertainty budget · Uncertainty evaluation · Uncertainty component · Traceability · Reference standards

Introduction

Determination of nitrogen content plays a key role in assigning values to insulin reference materials. The reference materials in question serve as reference standards when measuring insulin in drug products. Thus, the Kjeldahl nitrogen determination is a crucial link in the traceability chain. The uncertainty budget for the Kjeldahl method published in this paper has three objectives: Firstly, to estimate the uncertainty of results obtained by the Kjeldahl method; secondly, to identify steps in the analytical procedure that may be targets for improvement; and thirdly, to contribute a generally applicable procedure for evaluating an uncertainty budget for a chemical analytical method to the current literature. The need for such schemes or procedures is urgent as accredited laboratories in the near future will be required to state their uncertainty of measurement [1]. The uncertainty budget published in this paper is based on existing guidelines [2, 3].

Methods

Kjeldahl method

The nitrogen content of dry insulin was determined by the Kjeldahl method, which consists of three steps, *digestion, distillation* and *titration*. In brief, approximately 50 mg of the sample was weighed accurately and transferred to a digestion test tube. Concentrated sulphuric acid was added and the mixture was heated to 390 °C for 4 h (*digestion*). The tube with the digested sample was placed in the Kjeldahl apparatus, sodium hydroxide and steam was added, and ammonia was distilled off. The ammonia-rich steam was condensed in the receiving flask (*distillation*). The content of ammonia in the receiving flask was determined by end-point titration to pH 4.5 with 0.1 mol/l hydrochloric acid (*titration*). The normality of the acid (N_{HCl}, in mol/l) was determined by titration of tris-(hydroxymethyl)-aminomethane (Tris). Blanks consisted of empty digestion tubes treated as samples. Samples were measured in duplicate and blanks in triplicate.

Let a (mg) denote the amount of sample. If b (ml) and c (ml) denote the volume of hydrochloric acid used for titration of the blanks and the sample, respectively, then the relative nitrogen content of the sample (N_{total}) can be calculated as:

$$N_{total}=\frac{14.01\ \text{g/mol}\cdot(c-b)\cdot N_{HCl}}{a} \tag{1}$$

The method was validated using two insulin drugs. The relative standard deviation under intermediate precision conditions was found to be 0.085%.

Uncertainty budgets

The uncertainty of N_{total} was estimated by combining standard uncertainties of N_{HCl}, a, b and c. Uncertainties were combined by using the rule of "error propagation" [3, 4]. In general, when a result of measurement (y) is determined from other quantities, the relationship between y and the values of these quantities (input estimates) can be expressed by a function, f [2]:

$$y=f(x_1, x_2 \ldots, x_i \ldots, x_N) \tag{2}$$

where $x_1 \ldots x_i \ldots x_N$ represents N input estimates. The uncertainty of y ($u(y)$) is related to the uncertainty of the input estimates by the equation:

$$u(y)^2=\sum_{i=1}^{N}\left(\frac{\partial f}{\partial x_i}\right)^2 u(x_i)^2 \tag{3}$$

where $u(x_i)^2$ is the standard uncertainty variance and $u(x_i)$ the standard uncertainty of the uncertainty component number i, and where $\frac{\partial f}{\partial x_i}$ is the partial derivative of the function, f, with respect to the uncertainty component number x_i. The partial derivative is often called the sensitivity coefficient because it describes how the measurement result varies with changes in the value of the input estimates [2]. It should be noted that Eq. (3) is an approximation that is valid if there are no correlated input estimates [2]. The result of measurement of the Kjeldahl method is N_{total} as given by Eq. (1), and the expression for N_{total} is the function f in Eq. (2). Thus, the following expression for the uncertainty of N_{total} is obtained:

$$\left(\frac{u(N_{total})}{N_{total}}\right)^2=\frac{\left(\frac{u(c)}{\sqrt{2}}\right)^2+\left(\frac{u(b)}{\sqrt{3}}\right)^2}{(c-b)^2}+\left(\frac{u(N_{HCl})}{N_{HCl}}\right)^2+\left(\frac{\frac{u(a)}{\sqrt{2}}}{a}\right)^2 \tag{4a}$$

The factors $1/\sqrt{2}$ and $1/\sqrt{3}$ account for the number of replicate determinations (2 and 3, respectively). The expression may be rearranged in order to emphasize the sensitivity coefficient of each uncertainty component:

$$u(N_{total})^2=\left(\frac{N_{total}}{\sqrt{2}\cdot(c-b)}\right)^2 u(c)^2+\left(\frac{N_{total}}{\sqrt{3}\cdot(c-b)}\right)^2 u(b)^2+\left(\frac{N_{total}}{\sqrt{2}\cdot a}\right)^2 u(a)^2+\left(\frac{N_{total}}{N_{HCl}}\right)^2 u(N_{HCl})^2 \tag{4b}$$

The standard uncertainties $u(N_{HCl})$, $u(a)$, $u(b)$ and $u(c)$ are themselves composed from various uncertainty components. These uncertainties were likewise obtained from uncertainty budgets (Appendix B1–4).

Relative uncertainty variance contributions

In this paper the relative uncertainty variance contributions are used to illustrate the relative significance of different uncertainty components [5]. The relative contribution (r_i) of an uncertainty component x_i to the combined uncertainty, is defined here as the standard uncertainty variance of the component multiplied with its squared sensitivity coefficient divided by the combined standard uncertainty variance [5]:

$$r_i=\frac{\left(\frac{\partial f}{\partial x_i}\right)^2 u(x_i)^2}{u(N_{total})^2} \tag{5}$$

where $x_i = N_{HCl}$, a, b and c. It follows from Eqs. (3) and (5) that $\sum r_i = 1$ (assuming the absence of correlated input estimates).

Results

Uncertainty of N_{total}

The values of N_{HCl}, a, b, c and N_{total} as well as the respective standard uncertainties $u(N_{HCl})$, $u(a)$, $u(b)$, $u(c)$ and $u(N_{total})$ and the corresponding sensitivity coefficients are given in Table 1. The combined standard uncertainty $u(N_{total})$ was estimated to 0.00023, corresponding to a relative standard uncertainty of 0.19%. Using an coverage factor $k=2$ the result of the measurement should be reported as:

$$N_{total} \pm U(N_{total}) = 0.1233\ (\pm 0.00046)$$

where $U(N_{total})$ is the expanded standard uncertainty. Details on the uncertainty budgets for $u(N_{HCl})$, $u(a)$, $u(b)$ and $u(c)$ are given in Appendix B.

Table 1 Values of uncertainty components, their standard uncertainty and sensitivity coefficient

Component	Symbol	Value	Standard uncertainty	Sensitivity coefficient	Reference
Normality of hydrochloric acid	N_{HCl}	0.1 mol/l	1.1×10^{-4} mol/l	$\left(\frac{N_{total}}{N_{HCl}}\right)=1.23$ l/mol	Appendix B1
Amount of sample	a	50 mg	0.059 mg	$\left(\frac{N_{Total}}{\sqrt{2}\cdot a}\right)=1.74\times10^{-3}\ mg^{-1}$	Appendix B2
Volume of titrant used on blank sample	b	0.1 ml	4.0×10^{-3} ml	$\left(\frac{N_{total}}{\sqrt{3}\cdot(c-b)}\right)=1.62\times10^{-2}\ ml^{-1}$	Appendix B3
Volume of titrant used on the sample	c	4.5 ml	7.2×10^{-3} ml	$\left(\frac{N_{total}}{\sqrt{2}\cdot(c-b)}\right)=1.98\times10^{-2}\ ml^{-1}$	Appendix B4
Result of measurement	N_{total}	0.12329	0.00023		Eq. (4b)

Relative contributions to the uncertainty

The relative contributions (r_i) from $u(N_{HCl})$, $u(a)$, $u(b)$ and $u(c)$ to the combined standard uncertainty variance $u(N_{total})^2$, are shown in Fig. 1. The two largest contributions come from $u(N_{HCl})$ and $u(c)$. Both contribute 35–38% to the combined standard uncertainty variance. The uncertainty budget for $u(N_{HCl})$ (Appendix B1) shows that the uncertainty of this component is composed mainly of uncertainty contributions from the temperature of the titrant, the weighing of the Tris base and the volume of the titrant. The primary contributions to $u(c)$ (see Appendix B4) come from the uncertainty of digesting the sample (which consists of the uncertainty of the amount of sample transferred to the digestion tube) and from the uncertainty of the volume of the titrant.

In Fig. 2 the relative contributions to the uncertainty variance are summarized for various types of analytical steps and equipment used in the Kjeldahl method. The contribution from *volumetric* equipment (Fig. 2) includes the contributions from uncertainty of the volume of the titrant and from uncertainty of the temperature of the titrant (Appendix B1, 3 and 4). *Weighing* (Fig. 2) includes first and second weighing of both Tris base and the sample (Appendix B1 and 4). *Digestion* (Fig. 2) includes all uncertainty components denoted in digestion in Appendix B3 and 4. Lastly, *Tris-purity* (Fig. 2) denotes the uncertainty component associated with the purity of the Tris base evaluated in Appendix B1. From Fig. 2 it can be seen, that the largest uncertainty contribution comes from the use of volumetric equipment, i.e. burettes used for titration of hydrochloric acid and for titration of samples and blanks. This uncertainty component is composed of the precision of titration and of the uncertainty of the temperature. The latter factor influences the volume of titrant used for titration. The second largest contribution comes from weighing (of Tris base and of the sample).

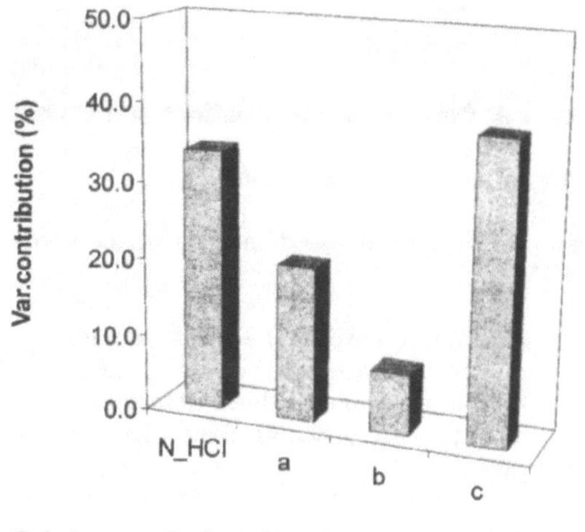

Fig. 1 Relative contributions from N_{HCl}, a, b and c to the combined standard uncertainty variance $u(N_{total})^2$

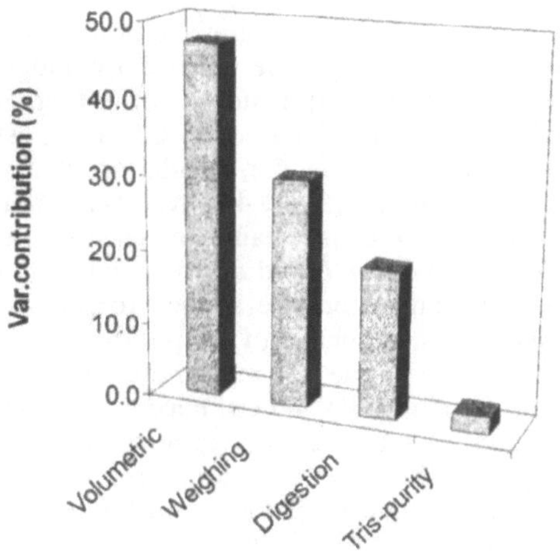

Fig. 2 Relative contributions from volumetric equipment, weighing, digestion and the purity of the Tris base to the combined standard uncertainty variance

Discussion

The uncertainty budget presented in Table 1 indicates that a typical result obtained by the Kjeldahl method (as described in this paper) will have a relative standard uncertainty of around 0.19%, which is approximately twice as large as the experimentally derived relative standard deviation estimated under intermediate precision conditions. This is not surprising as some of the uncertainty components included in the uncertainty budget are difficult to evaluate experimentally, e.g. purity of the Tris base, accuracy of the pH meter, etc.

The uncertainty evaluated for the Kjeldahl method should be propagated to the uncertainty of the insulin reference standards as well as to the analytical methods that are calibrated with these standards. Failure to consider the uncertainty of the reference standards or of the analytical method, or to base the uncertainty on less encompassing estimates (e.g. intermediate precision), may lead to an excess number of nonconforming results when evaluating the fulfilment of acceptance criteria for insulin drug products.

The present uncertainty budget demonstrates that the Kjeldahl method may be improved. The most promising target for improvement is the titration. According to the analysis of the uncertainty contributions, this can be accomplished by improving the temperature control during titration or by improving the accuracy of the volume of titrant. The latter goal may be achieved by using more accurate titration equipment, or, at least partly, by increasing the number of replicate determinations. Increasing the number of replicates will improve the precision, but will not influence the trueness of the volumetric equipment, nor does it compensate for long-term fluctuations that may affect the equipment.

The uncertainty budget in this study was organized to make it easy to review the evaluation of individual uncertainty components. The uncertainty budget is organized in three evaluation steps: In the first step, the uncertainty is evaluated based on the input estimates that are used directly in the calculation of N_{total} (i.e. N_{HCl}, a, b and c in Eqs. 1 and 4b) (results in Table 1). In the second step, the uncertainty of each of the input estimates (i.e. N_{HCl}, a, b and c) is evaluated and calculated. Correction factors are applied to correct for effects such as the influence of temperature on the volume, inaccuracy of the pH meter, etc. (equations and results in Appendix B). Where necessary, a third step of uncertainty evaluation was applied. An example is air bubbles in the burette, water in the Tris base and other effects that are evaluated and included in the uncertainty of q_{contr} (Appendix B1). Generally evaluations in the third step should only be carried out if the uncertainty components contribute significantly to the combined uncertainty.

It was not found necessary to consider the correlation between input estimates because the contribution from covariances was judged to be minor in the uncertainty budget presented above. Examples are c and b in Eq. (1) which are determined using the same burette. This means that uncertainties of c and b originating from, e.g. long-term variations of volume (delivered by the burette), wear of the burette, etc. are positively correlated. In other words, the uncertainty of the difference $(c-b)$ tends to be unaffected by these effects. The same reasoning can be used for m_1 and m_2, which are determined for calculation of a (Appendix B2).

The structure of this uncertainty budget for the Kjeldahl method may be used as a paradigm for other analytical methods. However, it must be understood that an uncertainty budget is designed for a specific method, and is therefore only valid for a set of specified analytical conditions (including the analytical method, type of sample and its concentration level). For highly automated (i.e. computerized) analytical methods, the result of measurement is often calculated by specialized software, and this makes the interdependency of various analytical steps less obvious. For such systems other approaches for making uncertainty budgets must be applied [5, 6].

Conclusions

A structured evaluation of uncertainty components was applied for the Kjeldahl method for nitrogen determination. The design offers a systematic approach for making uncertainty budgets for analytical chemical methods, and it may facilitate comparison of uncertainty budgets made by different analysts or laboratories. The evaluation of the contribution from individual or grouped uncertainty components made it possible to suggest specific improvements of the analytical method.

Appendix A: Definitions, abbreviations and equations

1. Definitions and abbreviations

Terms and definitions used in accordance with GUM and VIM [2, 7].

a:	mg sample used in the measurement
b:	ml titrant (hydrochloric acid) used on the blank sample (average of three determinations)
c:	ml titrant (hydrochloric acid) used on the sample
k:	Correction factor 14.01 g/mol containing the atomic weight of nitrogen
MPE:	Maximum permissible error

Table B1 Estimation of the uncertainty of N_{HCl}

No.	Input quantitiy	Symbol (x_i)	Value of input estimate	Stated or evaluated uncertainty	Type (A/B) and type of distribution	Standard uncertainty $u(x_i)$	Sensitivity coefficient $\frac{\partial f}{\partial x_i}$	Contribution to the standard uncertainty of the output estimate $u(x_i)\frac{\partial f}{\partial x_i}$	Reference
1	various contributions[a]	q_{contr}	1	1×10^{-4} [a]	B rectangular	$\left(\frac{1\times10^{-4}}{\sqrt{3}}\right)=5.8\times10^{-5}$	$\frac{N_{HCl}}{q_{contr}}=0.1$ mol/l	5.8×10^{-6} mol/l	Own estimate
2	MPE of the pH-meter	q_{pH}	1	≈ 0[b]	–	–	–	–	Instrument specification, own estimate
3	Purity of Tris (99.89%)	q_{pur}	0.9989	0.0005[c]	B	$\frac{0.005}{\sqrt{3}}=2.9\times10^{-4}$	$\frac{N_{HCl}}{q_{pur}}=0.1$ mol/l	2.9×10^{-5} mol/l	Supplier certificate
4	First weighing of Tris base	m_1	100 mg	0.1 mg	B rectangular	$\frac{0.1\text{ mg}}{\sqrt{3}}=5.8\times10^{-2}$ mg	$\frac{N_{HCl}}{m_2-m_1}=1.02\times10^{-3}$ mol/(mg·l)	5.9×10^{-5} mol/l	Instrument specification
5	Second weighing of Tris base	m_2	2 mg	0.02 mg	B rectangular	$\frac{0.02\text{ mg}}{\sqrt{3}}=1.2\times10^{-2}$ mg	$\frac{N_{HCl}}{m_2-m_1}=1.02\times10^{-3}$ mol/(mg·l)	1.2×10^{-5} mol/l	Instrument specification
6	Volume of titrant	v	8.06 ml	0.004 ml[d]	A normal	0.004 ml	$\frac{N_{HCl}}{v}=0.012$ mol/(L·ml)	5.0×10^{-5} mol/l	Instrument qualification report
7	Temperature of titrant	q_{temp}	1	0.0012[e]	B rectangular	$\left(\frac{1.2\times10^{-3}}{\sqrt{3}}\right)=6.9\times10^{-4}$	$\frac{N_{HCl}}{q_{temp}}=0.1$ mol/l	6.9×10^{-5} mol/l	Own estimate Laboratory temp. 15–25 °C
8	Molar mass of Tris base	M_{Tris}	121.14 g/mol	≈ 0 g/mol[f]	–	–	–	–	Own estimate

$$u(x_{N_{HCl}})=\sqrt{(5.8\times10^{-6})^2+(2.9\times10^{-5})^2+(5.9\times10^{-2})^2+(1.2\times10^{-5})^2+(5.0\times10^{-5})^2+(6.9\times10^{-5})^2}=1.1\times10^{-4}\text{ mol/l}$$

[a] Contributions from: The water content of the silica gel; lack of complete drying of the Tris base; variation in the time of drying; variation in the time of dissolution of the Tris base; the hygroscopy of the Tris base; air bubbles in the burette: Combined uncertainty estimated to 0.01 mg Tris base or 0.01% of approximately 100 mg Tris base

[b] The pH-meter specifications stipulate a MPE of 0.02 pH units. The form of the titration curve indicate that an uncertainty of 0.02 pH corresponds to a negligible uncertainty in the volume of titrant. Thus, the value of this uncertainty component is not included in the uncertainty budget

[c] The stipulated purity of (hydroxymethyl)-aminomethane (Tris) is 99.89% with an uncertainty of 0.05%

[d] A standard deviation of 0.004 mL was found experimentally

[e] Water density at 15 °C is 0.99913 g/mL, at 20 °C 0.99823 g/ml and at 25 °C 0.99707 g/mL. It is assumed that the temperature in the laboratory (and the temperature of the acid) in average is 20 °C, but in *worst case* may vary ±5 °C. The largest change in water density from 20 to 25 °C is 0.00116 g/ml, or 0.12%

[f] The uncertainty of the molar mass is assumed to give a negligible contribution to the combined uncertainty

Table B2 Estimation of the uncertainty of *a*

No.	Input quantity	Symbol (x_i)	Value of input estimate	Stated or evaluated uncertainty	Type (A/B) and type of distribution	Standard uncertainty $u(x_i)$	Sensitivity coefficient $\frac{cf}{\partial x_i}$	Contribution to the standard uncertainty of the output estimate $u(x_i)\frac{\partial f}{\partial x_i}$	Reference
1	First weighing of sample	m_1	52 mg	0.1 mg	B rectangular	$\frac{0.1\text{ mg}}{\sqrt{3}}$	1	5.8×10^{-2} mg	Instrument specifications
2	Second weighing of sample	m_2	2 mg	0.02 mg	B rectangular	$\frac{0.02\text{ mg}}{\sqrt{3}}$	1	1.2×10^{-2} mg	Instrument specifications

$$u(a)=\sqrt{(5.8\times10^{-2})^2+(1.2\times10^{-2})^2}=0.059\text{ mg}$$

Table B3 Estimation of the uncertainty of *b*

No.	Input quantity	Symbol (x_i)	Value of input estimate	Stated or evaluated uncertainty	Type (A/B) and type of distribution	Standard uncertainty $u(x_i)$	Sensitivity coefficient $\frac{\partial f}{\partial x_i}$	Contribution to the standard uncertainty of the output estimate $u(x_i)\frac{\partial f}{\partial x_i}$	Reference
1	Digestion[a]	q_{digest}	1	N/A	N/A	N/A	N/A	N/A	Digestion not performed
2	Distillation[b]	q_{distil}	1	–	–	–	–	–	Own estimation
3	Volumen of titrant	v	0.1 ml	0.004 ml	A normal	0.004 ml[c]	$\frac{b}{v}=1$	4×10^{-3} ml	Instrument qualification report
4	Temperature of titrant	q_{temp}	1	1.2×10^{-3} [d]	B rectangular	$\left(\frac{1.2\times10^{-3}}{\sqrt{3}}\right)=6.9\times10^{-4}$	$\frac{b}{q_{temp}}=0.1$ ml	6.9×10^{-5} ml	Own estimation Laboratory temp. 15–25 °C
5	MPE of the pH meter	q_{pH}	1	≈ 0[e]	–	–	–	–	Instrument specifications. Own judgement

$$u(b)=\sqrt{(0.004)^2+(6.9\times10^{-5})^2}=4.0\times10^{-3}\text{ ml}$$

[a] No Kjeldahl mixture or acid was used. The digestion was not performed

[b] Addition of water and NaOH; distillation time; leaks; incomplete transmission of NH_3 to the receiving vessel; temperature of the vapour: All components are assumed to be negligible

[c] A standard deviation of 0.004 ml was found experimentally

[d] Water density at 15 °C is 0.99913 g/ml, at 20 °C 0.99823 g/ml and at 25 °C 0.99707 g/ml. It is assumed that the temperature in the laboratory (and the temperature of the acid) on average is 20 °C, but in *worst case* may vary ±5 °C. The largest change in water density from 20 to 25 °C is 0.00116 g/ml, or 0.12%

[e] The pH-meter specifications stipulate a MPE of 0.02 pH units. The form of the titration curve indicates that an uncertainty of 0.02 pH corresponds to a negligible uncertainty in the volume of titrant. Thus, the value of this uncertainty component is not included in the uncertainty budget

Table B4 Estimation of the uncertainty of *c*

No.	Input quantity	Symbol (x_i)	Value of input estimate	Stated or evaluated uncertainty	Type (A/B) and type of distribution	Standard uncertainty $u(x_i)$	Sensitivity coefficient $\frac{\partial f}{\partial x_i}$	Contribution to the standard uncertainty of the output estimate $u(x_i)\frac{\partial f}{\partial x_i}$	Reference
1	Digestion[a]	q_{digest}	1	0.002	B rectangular	$\left(\frac{0.002}{\sqrt{3}}\right)=1.2\times10^{-3}$	$\frac{c}{q_{digest}}=4.5$ ml	5.2×10^{-3} ml	Own estimation
2	Distillation[b]	q_{distil}	–	–	–	–	–	–	Own estimation
3	Volume of titrant	v	4.5 ml	0.004 ml	A normal	0.004 ml[c]	$\frac{c}{v}=1$	4.0×10^{-3} ml	Instrument qualification report
4	Temperature of titrant	q_{temp}	1	0.0012	B rectangular	$\left(\frac{0.0012}{\sqrt{3}}\right)=6.9\times10^{-3}$	$\frac{c}{q_{distil}}=4.5$ ml	3.0×10^{-3} ml	Own estimation Laboratory temp. 15–25 °C
5	MPE of the pH meter	q_{pH}	1	≈ 0 ml[d]	–	–	–	–	Instrument specifications. Own judgment

$$u(c)=\sqrt{(5.2\times10^{-3})^2+(4\times10^{-3})^2+(3.0\times10^{-3})^2}=7.22\times10^{-3}\text{ ml}$$

[a] The uncertainty components: Amount of catalysator, the amount of Kjeldahl mixture and H_2SO_4, block temperature, digestion time and boiling are assumed not to contribute significantly to the uncertainty. Transfer of sample to digestion vessel is assumed to contribute with 0.1 mg or 0.2%

[b] Addition of water and NaOH; destillation time; leaks; incomplete transmission of NH_3 to the receiving vessel; temperature of the vapour: All components are assumed to be negligible

[c] A standard deviation of 0.004 ml was found experimentally

[d] The pH-meter specifications stipulate a MPE of 0.02 pH units. The form of the titration curve indicates that an uncertainty of 0.02 pH corresponds to a negligible uncertainty in the volume of titrant. Thus, the value of this uncertainty component is not included in the uncertainty budget

N: Number of quantities or uncertainty components
N_{HCl}: The normality of the hydrochloric acid
N_{total}: Content of nitrogen in the sample (average of 2 samples)
q_i: Correction factors
$u(x_i)$: Standard uncertainty of the input estimate x_i
$U(y)$: Expanded uncertainty of y
$u(y)$: The combined standard uncertainty of y
x_i: Input estimates
y: Output estimate (the result of a measurement)
$\frac{\partial f}{\partial x_i}$: Sensitivity coefficient for x_i

2. Equation for the standard uncertainty

In addition to Eq. (3) the following equations evaluating the standard uncertainty were applied: If the input estimates x_i in Eq. (2) is related to y only by multiplication's and divisions, Eq. (3) can be simplified to [2, 4]:

$$\left(\frac{u(y)}{y}\right)^2 = \left(\frac{u(x_1)}{x_1}\right)^2 + \left(\frac{u(x_2)}{x_2}\right)^2 + \ldots + \left(\frac{u(x_N)}{x_N}\right)^2 \quad (A.1)$$

with sensitivity coefficients given by:

$$\frac{\partial f}{\partial x_i} = \frac{y}{x_i} \quad (A.2)$$

If the input estimates x_i is related to y by additions and subtractions, Eq. (3) becomes:

$$u(y)^2 = u(x_1)^2 + u(x_2)^2 + \ldots + u(x_N)^2 \quad (A.3)$$

with all sensitivity coefficients equal to 1.

Appendix B

1. Estimation of the uncertainty of N_{HCl}

N_{HCl} is calculated from the amount of Tris base used, the molar mass of Tris, the volume of titrant, and correction factors for the water content of Tris, the accuracy of the pH meter, the temperature of the titrant, and the purity of Tris (Table B1):

$$N_{HCl} = f(x_i) = q_{contr} q_{pH} q_{temp} q_{pur} \left(\frac{m_2 - m_1}{M_{Tris} v}\right).$$

2. Estimation of the uncertainty of a

The input estimate a is obtained from weighing the sample (m_2) and sample cup alone (m_1), i.e. $a = f(x_i) = m_2 - m_1$. Thus, the standard uncertainty of a is given by the equation $u_a^2 = u_{m_1}^2 + u_{m_2}^2$ (Table B2).

3. Estimation of the uncertainty of b

In evaluating the uncertainty of b it was assumed that the volume of titrant used for titration of a blank depends on factors associated with digestion, distillation, pH measurement and the assessment of the volume of titrant used. The combination of uncertainty components was based on the following relation between the volume (b) and the various factors:

$$b = f(x_i) = q_{digest} q_{distil} q_{temp} q_{pH} v,$$

where v is the estimated volume of titrant read from the burette and all the correction factors (q_i) are equal to 1 (Table B3).

4. Estimation of the uncertainty of c

The uncertainty components and equations used are the same as for b (Appendix B3), i.e.

$$c = f(x_i) = q_{digest} q_{distil} q_{temp} q_{pH} v,$$

where v is the estimated volume of titrant read from the burette and the correction factors (q_i) are all equal to 1 (Table B4).

References

1. International Standard ISO/IEC 17025 (1998) General requirements for the competence of testing and calibration laboratories (Draft). International Organization for Standardization, Geneva
2. BIPM, IEC, IFCC, ISO, IUPAC, IUPAP, OIML (1993) Guide to the expression of uncertainty in measurement. International Organization for Standardization, Geneva
3. EURACHEM (1995) Quantifying uncertainty in analytical measurement, 1st edn. EURACHEM
4. Miller JC, Miller JN (1993) Statistics for analytical chemistry, 3rd edn. Ellis Harwood, New York
5. Kristiansen J, Christensen JM, Nielsen JL (1996) Mikrochim Acta 123:241–249
6. Hansen AM, Kristiansen J, Nielsen JL, Byrialsen K, Christensen JM (1990) Talanta 50:367–379
7. ISO (1993) International vocabulary of basic and general terms in metrology (VIM), 2nd edn. International Organization for Standardization, Geneva

John Fleming
Bernd Neidhart
Christoph Tausch
Wolfhard Wegscheider

Glossary of analytical terms*

John Fleming
LGC, Queens Road, Teddington,
Middlesex TW 11 0LY, UK

Bernd Neidhart, Christoph Tausch
Philipps-Universität Marburg,
Hans-Meerwein-Strasse,
D-35032 Marburg, Germany

Wolfhard Wegscheider
Montanuniversität Leoben,
Franz-Josef-Strasse 18,
A-8700 Leoben, Austria

Introduction

Analytical data play a vital role in our daily lives, with increasing influence on both economy and ecology. The harmonisation of the European market – including the Eastern European countries – and the opening of the international borders for trade and communication have led to serious problems with terminology in analytical chemistry. We can identify the three main reasons that have caused this situation. These can be classified as "linguistics", "semantics", and "acceptance".

Frequent translations of a term through a chain of languages, and the use of terms by non-native speakers, may lead to a misuse of terms followed by grave misunderstandings. In addition, the co-existence of different meanings of terms due to their independent definition by national and international bodies or authorities, together with recommendations given by international organisations like IUPAC, leads to problems of semantics and confusion resulting in reduced acceptance.

A strategy on terminology

During the last 5 years, the EURACHEM Education and Training Working Group (E&TWG) has analysed this situation and has developed a strategy which is expected to resolve the dilemma. The first, and most important, step in this concept is to provide a forum which initiates and enables international discussions among experts in the field. The catalyst for these discussions will be a dictionary-like "glossary of terms" which will be published as a series in this journal. Each term in the glossary is provided with a definition (taken from the highest international level, if possible ISO) followed by a scientific description of the meaning of the definition and one or more examples explaining its practical use. In addition, translations of the term into other European languages are given. This structure will facilitate translation of the glossary into other languages, and errors will be minimised if not excluded. The translation will be performed by the E&TWG members, who are experts in the field and native speakers of the respective language, and will finally be published in a suitable national journal.

Feedback will be sought at both national and international levels to enable a dynamic development of the glossary at the highest scientific and linguistic levels possible. This might also include the deletion of existing and the creation of new words, if, in the latter case, the scientific definition and meaning has no linguistic equivalent in a given language. Let us take as an example the term traceability, which by definition describes a way to achieve quality (accuracy, comparability) in chemical measurements. The equivalent in German would be Rückführbarkeit but the term Rückverfolgbarkeit is used as the respective DIN Standard, the linguistic meaning of which is "follow the way (track) back". Consequently, the term Rückverfolgbarkeit is part of providing assurance of quality and not of creating quality. Unfortunately, there is no English word for Rückverfolgbarkeit. There are two ways of solving this problem: one is to create a new English word and the other to introduce the German word into the English language.

We are willing to "grasp the nettle" and open the debate on this issue by proposing the term trackability to cover this concept.

Discussion forum

It is proposed that the EURACHEM E&TWG should be the catalyst which will promote a wider debate of the issues raised by this glossary of terms. All analytical scientists

* EURACHEM Education and Training Working Group

are urged to contribute to the debate and work towards a consensus on the usage of the key terms covered by the glossary. This debate can be pursued either by corresponding with the editor of this journal or by sending an email message to jwf@lgc.co.uk for consideration by the working group.

Repeatability

Wiederholpräzision (**D, A, CH**); Répétabilité (**F, B**); Repetibilidad (**E**); E.panalhcimo? thta (**GR**); Ripetibilitá (**I**); Herhaalbarheid (**NL**); Powtarzalnosc (**PL**); Toistettavuus (**SF**); Ismételhetõség (**H**); (**RUS**); Repetibilidade (**P**)

Definition

Precision under repeatability conditions.[1]

Description

Repeatability is the closeness of the agreement between the results of independent measurements of the same analyte carried out subject to all of the following conditions:
the same method of measurement, the same observer, the same measuring instrument, the same location, the same conditions of use, repetition over a short period of time.[2]

Independent measurements are made on distinct subsamples of a test material. If possible, at least 8 measurements should be performed.

Repeatability is a characteristic of a method not of a result.

Example

Successive measurements under the above conditions gave eight single results from which a standard deviation is calculated. The standard deviation multiplied by 2.8 gives the repeatability at 95% confidence level.

Suppose that an analyst uses a method for which the repeatability has been established as 2 mg/mL.

If, in a real case, the same analyst reported results of a measurement repeated over a short time interval as 50 and 56 mg/mL, there would be a question over the validity of these results as they are very unlikely to have differed by 6 mg/mL as a result of random variability.

[1] ISO 3534-1 (1993)

[2] International vocabulary of basic and general terms in metrology, 1993, (BIPM, IEC, IFCC, ISO, IUPAC, IUPAP, OIML); ISO central secretariat, 1 rue de Varambé, CH-1211 Geneva 20

Reproducibility

Vergleichpräzision (**D, A, CH**); Réproductibilité (**F, B**); Reproducibilidad (**E**); Anaparagvgýmóthta (**GR**); Riproducibilitá (**I**); Reproduceerbarheid (**NL**); Odtwarzalnosc (**PL**); Uusittavuus (**SF**); Reprodukálhatós 'ag (**H**); (**RUS**); Reprodutibilidade (**P**)

Definition

Precision under reproducibility conditions. [1]

Description

Reproducibility is the closeness of the agreement between the results of measurements of the same analyte in distinct subsamples of a test material, where the individual measurements are carried out changing conditions such as: observer, measuring instrument, location, conditions of use, time, but applying the same method.[2]

Example

In a laboratory intercomparison samples (e.g. a surface water) were sent to a number of laboratories for determination of e.g. nitrite. Each laboratory reports its results as single values. The standard deviation from all accepted individual results multiplied by 2.8 gives the reproducibility at 95% confidence level.

Suppose that the reproducibility of a method has been determined to be x. If two of the laboratories in a real case reported results for subsamples of the same sample which differed by ¡x there would be a question concerning the quality of performance.

Methods which have a large reproducibility may not be suitable for making valid comparisons in a given real situation. In this case either the method must be improved or another method with a smaller reproducibility must be applied.

[1] ISO 3534-1 (1993)

[2] International vocabulary of basic and general terms in metrology, 1993, (BIPM, IEC, IFCC, ISO, IUPAC, IUPAP, OIML); ISO central secretariat, 1 rue de Varambé, CH-1211 Geneva 20

Traceability

Rückführbarkeit (**D, A, CH**); Tracabilité (**F, B**); Trazabilidad (**E**); Ixnhla' thsh (**GR**); Riferibilitá (**I**); Herleidbarheid (**NL**); Rastreabilidade (**P**); Jaeljitettaevyys (**SF**); Visszavezethetõség (**H**); Zgodnosc (**PL**); (**RUS**)

Definition

The property of a result of measurement whereby it can be related to appropriate standards, generally international or national standards, through an unbroken chain of comparisons.[1]

Description

For each analytical measurement, it should be possible to relate the result of the measurement back to an appropriate national or international measurement standard through an unbroken chain of comparisons. For measurement of weight, this would be the kilogram standard in Paris, or for amount of substance it should be the SI unit, the mole. If calibrated by an accredited body, the balance is an instrument which can provide measures of weight which are traceable to national measurement standards. Instruments for chemical analysis must be calibrated by the use of certified reference materials, or other suitable reference materials.

Example

Determination of lead in water by atomic absorption spectrometry (AAS): The AAS instrument has be to calibrated using reference solutions made up by dissolving known amounts (balance) of a certified reference material (CRM) or a pure substance such as $Pb(NO_3)_2$ in a defined volume of pure water; in the latter case the pure substance has to be compared with a CRM. A calibration graph which covers the concentration range of the analyte in the sample should be prepared.

For more complicated analyses, which might involve extraction and other analytical procedures, the traceability of the result of a measurement can be established by subjecting a certified reference material – with similar composition to the unknown – to the same analytical procedures. If for example the measurement-standard used has not been compaired with a CRM of the same type the chain of comparison is broken.

[1] ISO 3534-1 (1993)

Trackability

Rückverfolgbarkeit (**D, A, CH**), Relacionabilidad (**E**), Sporbarhet (**NOR**)

Definition

The property of a result of a measurement whereby the result can be uniquely related to the sample.

Description

Each step of an analytical method has to be documented in a way that the result of a measurement can be linked unambiguously to the sample to which it refers.

Example

All samples must be uniquely labelled. All operations performed on a sample must be recorded in a notebook or computer system. Chromatograms, spectra and other instrumental outputs must be labelled with the sample identification.

Track: Result

⇩

Calculation

⇩

Determination

⇩

Separation

⇩

Preparation

⇩

Storage

⇩

Sampling

Uncertainty of measurement

Meßunsicherheit (**D, A, CH**), Incertitude de mesure (**F, B**), Intercidumbre de la medida (**E**), Ab baióthta th m΄trhsh (**GR**); Meetonzekerheid (**NL**); Incerteza da medida (**P**); Mérési byzonytalansá g (**H**); incertezza di misura (**I**); Mittauksen epaevarmuus (**SF**)

Definition

Parameter, associated with the result of a measurement, that characterizes the dispersion of the values that could reasonably be attributed to the measurand.[1]

Description

Uncertainty sets the limits within which a result is regarded accurate, i.e. precise and true. Uncertainty of measurement comprises, in general, many components. Some of these components may be evaluated from the statistical distribution of the results of series of measurements and can be characterized by experimental standard deviations. The other components, which can also be characerized by standard deviations, are evaluated from assumed probability distributions based on experience or other information.[2]

Example

Overall uncertainty can be estimated by identifying all factors which contribute to the uncertainty. Their contributions are estimated as standard deviations, either from repeated observations (for random components), or from other sources of information (for systematic components). The combined standard uncertainty is calculated by combining the variances of the uncertainty components, and is expressed as a standard deviation. The combined standard uncertainty is multiplied by a coverage factor of 2 to give a 95% level of confidence (approximately).

The Uncertainty Estimation Process

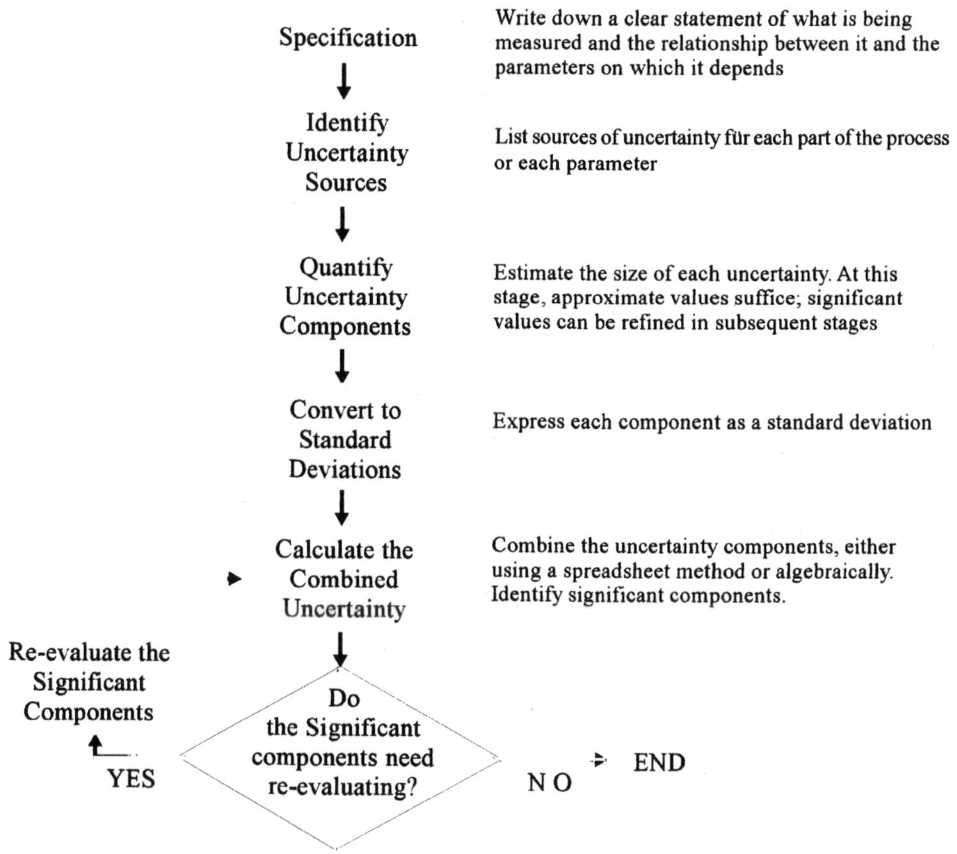

The uncertainty for the determination of e.g. atrazine in water consists of the calibration of several components of uncertainty, such as the uncertainty of the true content of the atrazine standard, uncertainty from dilution of this standard, uncertainty regarding the loss of atrazine in sampling and storage prior to analysis, as well as that associated with the preconcentration step after correction for recovery. The result would be expressed as:
1.02B0.13 mg/L

[1] International vocabulary of basic and general terms in metrology, 1993, (BIPM, IEC, IFCC, ISO, IUPAC, IUPAP, OIML); ISO central secretariat, 1 rue de Varambé, CH-1211 Geneva 20

[2] Quantifying uncertainty in analytical measurement, EURACHEM, Queens road, Teddington, Middlesex TW11 OLY UK

MIX
Papier aus verantwortungsvollen Quellen
Paper from responsible sources
FSC® C105338

If you have any concerns about our products,
you can contact us on
ProductSafety@springernature.com

In case Publisher is established outside the EU,
the EU authorized representative is:
Springer Nature Customer Service Center GmbH
Europaplatz 3, 69115 Heidelberg, Germany

Printed by Libri Plureos GmbH
in Hamburg, Germany